Apparent and Microscopic Contact Angles

APPARENT

AND

MICROSCOPIC

CONTACT ANGLES

Editors:

J. Drelich, J.S. Laskowski
and K.L. Mittal

CRC Press
Taylor & Francis Group
Boca Raton London New York

CRC Press is an imprint of the
Taylor & Francis Group, an **informa** business

First published 2000 by VSP Publishing

Published 2018 by CRC Press
Taylor & Francis Group
6000 Broken Sound Parkway NW, Suite 300
Boca Raton, FL 33487-2742

© 2000 by Taylor & Francis Group, LLC
CRC Press is an imprint of Taylor & Francis Group, an Informa business

First issued in paperback 2019

No claim to original U.S. Government works

ISBN 13: 978-0-367-44742-7 (pbk)
ISBN 13: 978-90-6764-321-4 (hbk)

Visit the Taylor & Francis Web site at
http://www.taylorandfrancis.com

and the CRC Press Web site at
http://www.crcpress.com

Contents

Apparent and Microscopic Contact Angles, pp. ix–x
J. Drelich, J. S. Laskowski and K. L. Mittal (Eds)
© VSP 2000.

Preface

This book chronicles the proceedings of the International Symposium on *Apparent and Microscopic Contact Angles* held in conjunction with the American Chemical Society Meeting in Boston, August 24–27, 1998.

The symposium was truly international and interdisciplinary and provided a forum for almost 100 chemists, metallurgists, physicists and biologists from 17 countries (Australia, Belarus, Belgium, Brazil, Canada, France, Germany, Israel, Italy, Japan, Poland, Russia, Singapore, Sweden, Turkey, the United Kingdom and the United States). The program included almost 70 oral and poster presentations.

The symposium in Boston provided an opportunity to discuss several controversial issues associated with interfacial phenomena that govern the behavior of the three-phase systems. Particularly welcomed were contributions on the molecular and nanoscopic effects in wetting phenomena. We are on the verge of understanding the arrangements of molecules at interfaces and their effect on microscopic properties of the three-phase systems, such as interfacial forces, interfacial tension, linear excess energy, etc. The interfacial transition regions between immiscible phases are the subject of current experimental and theoretical investigations, and the progress in imaging and understanding of these regions was reported in Boston. Significant improvements in examination of the line tension values were demonstrated, although it is still uncertain whether the linear excess energy is of any importance in practical situations.

It is well recognized in our scientific community that among the most significant topics are the ones concerning the determination of the surface free energy of solids through contact angle measurements. The models developed to determine the surface free energy of solids from contact angle measurements, as pointed out at the forum, still require improvements.

The methodologies and interpretations of the contact angle measurements at heterogeneous, rough and/or powdered solids are vital in the area of applied surface chemistry, and the number of papers presented at the symposium proves this point very well. Although the material presented was 'state-of-the-art', both modeling and experimentation with heterogeneous and rough surfaces are in an early stage development, and we hope that this symposium stimulated the progress in this area.

The lecture room in which sessions of the International Symposium on *Apparent and Microscopic Contact Angles* were held was overcrowded up to the very last presentation. This is another indication that advances in understanding of the wetting phenomena attract the attention of a very broad audience.

As for this book, it contains 28 papers which were earlier published in three issues of the *Journal of Adhesion Science and Technology* (JAST) as follows: Volume 13, No. 10 and No. 12, and Volume 14, No. 2. However, a large number of people indicated interest in acquiring these three issues separately; concomitantly, we decided to bring out this hard-bound book. It should be noted that in the book the papers have been rearranged in a more logical manner vis-à-vis the order in which they were published in the special issues of JAST.

The book is divided into four parts as follows: Part 1. Nanoscopic and Molecular Effects on Contact Angles; Part 2. Surface Forces and Surface Free Energy; Part 3. Wetting of Heterogeneous, Rough and Curved Surfaces; and Part 4. Dynamic Effects in Contact Angle Measurements. The topics covered include: contact line tension measurement; liquid drop surface topography; molecular mechanism of hydrophobic transitions; stereochemical and conformational aspects of polymer surfaces; determination of acid–base properties of metal oxides and polymers by contact angle measurements; van Oss–Good theory of acid–base surface free energies; AFM measurement of forces; thin liquid films; wettability of flat and curved surfaces; contact angle hysteresis; factors affecting contact angle measurements; dynamic wetting behavior; and effect of surfactants on wetting.

Now comes the pleasant task of thanking those who contributed in many and varied ways. First, we would like to thank the session chairpersons: Terry D. Blake, Ana M. Carmona-Ribeiro, Jaroslaw Drelich, Robert J. Good, Karina Grundke, Robert Hayes, Janusz S. Laskowski, Lieng-Huang Lee, Glen McHale, John Ralston and Claudio Della Volpe, for doing an excellent job in moderating the sessions. The authors should be thanked for their contribution, cooperation, interest and enthusiasm. The time and efforts of the reviewers in providing valuable comments is gratefully acknowledged. We would like to acknowledge the assistance of the American Chemical Society through the Petroleum Research Fund Program and Division of Colloid and Surface Chemistry; University of Utah through the College of Mines and Earth Sciences; Michigan Technological University through the Department of Metallurgical and Materials Engineering; and University of British Columbia through the Department of Mining and Mineral Process Engineering, for support of the symposium. Last, but not least, we are appreciative of the excellent job done by the VSP staff in producing this book.

We sincerely hope this book summarizing the cumulative wisdom of a legion of researchers and commenting on the contemporary research will serve as a fountainhead for new ideas in the wonderful (but complex) world of contact angles.

Jaroslaw Drelich
Janusz S. Laskowski
K. L. Mittal

Part 1

Nanoscopic and Molecular Effects on Contact Angles

Apparent and Microscopic Contact Angles, pp. 3–12
J. Drelich, J. S. Laskowski and K. L. Mittal (Eds)
© VSP 2000.

Measurement of contact line tension by analysis of the three-phase boundary with nanometer resolution

T. POMPE, A. FERY [1] and S. HERMINGHAUS [2,*]

[1] *MPI für Kolloide und Grenzflächen, D-14424 Potsdam, Germany*
[2] *Abteilung Angewandte Physik, Universität Ulm, Albert-Einstein-Allee 11, D-89081, Ulm, Germany*

Received in final form 11 February 1999

Abstract—Liquid structures on solid substrates have been imaged with a resolution in the nanometer range by scanning force microscopy in the tapping mode. Using substrates with an artificially patterned wettability, characteristic features in the three-phase contact line were induced, which allow the contact line tension to be determined. The values in the range of -1×10^{-10} N obtained for sessile droplets of hexaethylene glycol are consistent with theoretical predictions.

Key words: Contact line tension; scanning force microscopy; modified Young equation.

1. INTRODUCTION

The contact line tension, i.e. the excess free energy of a three-phase contact line, is probably the most controversial quantity in wetting science. This is mainly due to its poor accessibility. The theory predicts that typical values of the contact line tension, τ, should be in the range of a few tens of piconewtons [1–7]. The characteristic length scale, l, at which these forces become important can readily be obtained by comparison with typical values of interfacial tensions, γ, yielding $l := \tau/\gamma \approx 10^{-11}$ N/10^{-2} N m^{-1} = 1 nm. At much larger scales, interfacial tensions are expected to dominate all measurements of line tension effects, and particular care must be taken in experiments. The classical approach for tackling the problem is to investigate the dependence of the contact angle of small sessile droplets on their size. The smaller a droplet becomes, the larger the effect of the contact line tension on its shape. For a positive line tension, the droplet base is contracted, which gives rise to a larger value of the contact angle. This is expressed by the modified Young equation for a liquid droplet on a plane solid surface, which

---

*To whom correspondence should be addressed. E-mail: stephan.herminghaus@physik.uni-ulm.de

reads

$$\cos(\Theta) = \cos(\Theta_{\text{Young}}) - \frac{1}{\gamma}\frac{\tau}{R}, \tag{1}$$

where Θ is the actual contact angle that the droplet forms with the substrate and Θ_{Young} is the contact angle which would be derived from Young's equation (i.e. for a straight contact line or an infinitely large drop). R is the radius of the (circular) contact line and γ is the surface tension of the liquid. This nomenclature will be used throughout this paper. Assuming the theoretically predicted values of the contact line tension to be true, one is faced with the necessity of investigating extremely small droplets in order for the change in contact angle to be measurable. For instance, for a water droplet, a radius of contact line curvature of 20 nm on a substrate with Young's contact angle of 30° gives rise to a change in the contact angle (with respect to its value for a straight contact line) of only 1° when a contact line tension of 10^{-11} N is assumed. In marked contrast, experiments have been performed for a long time using optical techniques on droplets with typical sizes of at least several tens of micrometers. The results show an exceptionally large scatter in the magnitude of the contact line tensions obtained: values ranging from 10^{-5} to 10^{-12} N, both positive and negative, are reported in the literature [8–12].

From the diagram presented in Fig. 1, it is quite straightforward to imagine how such a large scatter in the experimental values of the contact line tension may come about. Since every surface is, as far as its surface energy is concerned, inhomogeneous at least at a very small scale, the contact line is expected to exhibit a corrugation as shown in the figure, which is not accessible with optical imaging techniques. As a consequence, the total free energy of the contact line which is measured using a coarse-graining experimental technique (such as optical microscopy) consists not only of the genuine contact line tension or its sum over the corrugations as proposed by Neumann and co-workers [13], but also of the excess free energy connected to the increased total surface near this line. Since the typical length scale at which the wettability of the substrate varies is likely to exceed the above-mentioned critical length l, the measured line tension may well exceed its

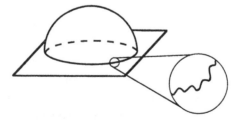

Figure 1. Schematic diagram of a macroscopic liquid droplet on a microscopically heterogeneous surface. Optical techniques would yield an apparent radius of curvature of the contact line from the base of the macroscopic droplet. The real radius of curvature of the contact line (from the local corrugations) is beyond their resolution limit.

genuine (unperturbed) value. This may at least explain large positive apparent line tensions.

It follows directly that high-resolution imaging of the three-phase contact line and the local liquid surface is crucial for an exact determination of the contact line tension. In order to achieve nanometer spatial resolution, scanning force microscopy (SFM) would be the method of choice. Although one could not expect to image liquid structures by SFM techniques with high resolution straightforwardly, the imaging of surface profiles, even of liquids with low viscosity, using tapping mode SFM and electrical force microscopy was recently reported by several groups [14–17]. A lateral resolution of 10 nm and a vertical resolution of about 1 nm were achieved [14, 17, 18].

2. EXPERIMENTAL

Investigations of the liquid structures were carried out on a Nanoscope IIIa scanning force microscope (Digital Instruments). Typical working conditions were cantilever oscillation amplitudes of 10–20 nm with very small damping values of 1–5%. As tips, commercial silicon cantilevers were used (point probe: Si-cantilever, non-contact; resonance frequency: 200–400 kHz; spring constant: 20–40 N/m; tip radius as specified by the manufacturer ≈ 10 nm; purchased from Nanoprobe). The liquid structures were prepared by depositing liquid droplets on a substrate which was provided with a pattern of stripewise wettability, as proposed by Neumann and co-workers [19]. In contrast to other experiments on such substrates [9] with a periodicity of the stripes of some micrometers, and optical investigations of the liquid structure, the periodicity of our wettability structure was well below 1 μm, in order to enhance the role of the contact line tension.

For patterning the substrate, microcontact printing (μCP) [20–23] was used. To avoid the effects of surface roughness of the solid substrate, perfluorinated alkylsilanes [(heptadecafluoro-1,1,2,2-tetrahydrodecyl)dimethylchlorosilane] were stamped from a hexane solution onto a silicon wafer. The roughness of the silicon wafers used, which had a native oxide layer on top, was in the range of 0.2–0.3 nm, as revealed by SFM. The μCP process leads to stripes of self-assembled monolayers on the silicon wafer, which are covalently bound and have a typical thickness of 0.5 nm. Hence, the topography of this structure and of the solid substrate in general can be neglected in comparison with the larger liquid structures that are investigated.

The periodicity of the stripes ranged from 400 to 1000 nm. The patterned substrates are characterized by hydrophobic stripes of self-assembled monolayers of perfluoro-alkylsilanes with a width of 50–400 nm and corresponding hydrophilic stripes of the silicon dioxide with a width of 300–700 nm. A typical image of a structure of this kind is presented in Fig. 2. Due to the preparation process during μCP, the hydrophilic regions are not as hydrophilic as the freshly prepared silicon dioxide surfaces. However, the contrast in wettability between the stripes is large

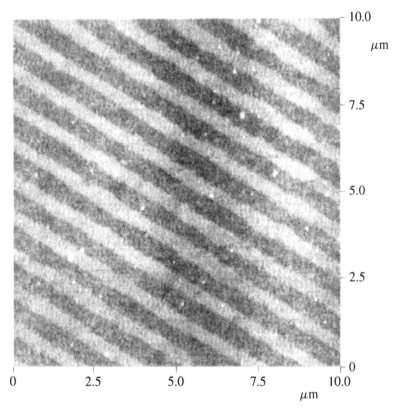

Figure 2. SFM topography image of a patterned silicon wafer. The stripes of self-assembled monolayers of perfluorinated monochloroalkylsilane (height: 0.5 nm) are prepared by μCP. A wettability contrast of hydrophilic (non-stamped) and hydrophobic (stamped) stripes is obtained (periodicity: 900 nm).

enough to generate pronounced periodic structures in the contact lines of the sessile droplets on these surfaces.

The next preparation step is the deposition of small liquid droplets on the wettability structure of the substrate. In order to achieve liquid structures in thermal equilibrium with the solid substrate, the droplets were created by means of an atomizer (Fisher Scientific) and were deposited on the substrate out of the aerosol phase. For the small droplets of interest here, this method minimizes the contact line pinning effects, as opposed to other methods [16]. It can be simply shown based on an energy consideration that the surface energy of the system is much larger than the kinetic energy of the droplet deposition process, leading to a surface energy-controlled equilibration process of the solid–liquid–vapor system. The surface energy of a liquid droplet in the aerosol phase can be written as $E_{\text{surf}} = 4\pi\gamma r^2$ (γ is the surface tension of the liquid and r is the radius of the droplet). By using the gravitational force of the droplet, $F_{\text{G}} = \rho g(4/3)\pi r^3$ (ρ is the density of the liquid and g is the gravitational constant) and its Stokes friction force, $F_{\text{F}} = 6\pi\eta r v$ (η is the viscosity of the air and v is the velocity of the droplet), one can calculate the ratio between the surface energy and the kinetic energy, $E_{kin} = (1/2)mv^2$ (m is the mass of the droplet), of the droplet. The ratio of E_{kin}/E_{surf} is proportional to r^{-5} and shows a negligible kinetic energy for small droplets. For the liquid used in this experiment (hexaethylene glycol, Aldrich), this results in a ratio of 10^{-6} when one assumes a droplet of 5 μm radius.

Hexaethylene glycol was used in our experiments because it fulfils several conditions. First, the low viscosity ensures a very fast transition to an equilibrium state on the surface. Second, due to its low vapor pressure, the droplets can be considered to be very close to the liquid–vapor co-existence. Furthermore, the imaging process requires a liquid with a not too small surface tension [γ(hexaethylene glycol) = 45 mN/m) and a contact angle on the substrate below 80° (because of the tip geometry) [14].

3. RESULTS AND DISCUSSION

After preparation of the liquid structures in the described manner, one could already see with an optical microscope that the structured wettability of the substrate leads to a deformation of the general shape of the liquid droplets. Figure 3 shows an

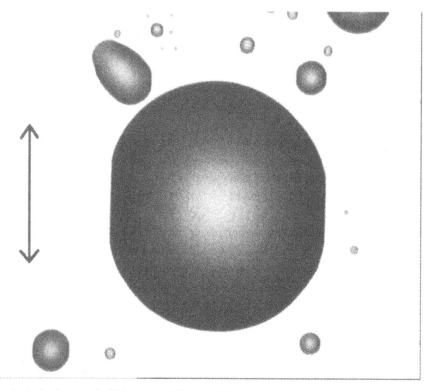

Figure 3. Optical micrograph (150 μm × 110 μm) of hexaethylene glycol droplets deposited on a patterned substrate. The underlying substrate with stripes in the direction of the arrow drawn (periodicity: 400 nm) induces elongated droplets even on the macroscopic scale.

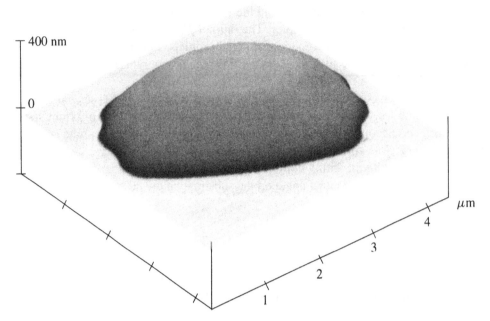

Figure 4. SFM topography image of a small droplet on a patterned substrate (periodicity: 800 nm). The elongation of the droplet in the direction of the stripes and the distinct corrugation of the contact line perpendicular to the stripes can be clearly seen.

optical micrograph of hexaethylene glycol droplets on a silicon wafer with a stripe periodicity of 400 nm. The droplets can be clearly seen to be elongated in the direction of the stripes and are facetted along them. Obviously, the artificial wetting heterogeneities in the range of 100 nm induced strong changes in the macroscopic wetting characteristics of the droplets. By imaging smaller droplets with the SFM, images of small droplets and the liquid profile at the three-phase boundary (contact line region) can be obtained with a nanometer resolution. A typical image of a small droplet is shown in Fig. 4, while in Fig. 5, the contact line region of a larger droplet is shown in detail. Images like Fig. 5 can be used to determine the local value of the contact angle and the local radius of curvature of the contact line. Images (3 μm $\times$ 3 μm) obtained with ≈ 10 nm lateral resolution were processed numerically in order to obtain the local radius of curvature of the contact line and the local contact angle. The radius of curvature of the contact line was obtained by a second-order polynomial fit in the vicinity of the point of interest along the contact line. The local contact angles were determined directly from the topographical data perpendicular to the local direction of the contact line (cf. Fig. 6).

In order to connect our data to equation (1), we have plotted the cosine of the contact angle Θ versus the local curvature of the contact line, $1/R$, in Fig. 7. As can be clearly seen, the data group into two distinct clusters, one at higher and the other at lower cosines of the contact angle. They correspond to the hydrophilic (non-stamped) and hydrophobic (stamped) regions, respectively.

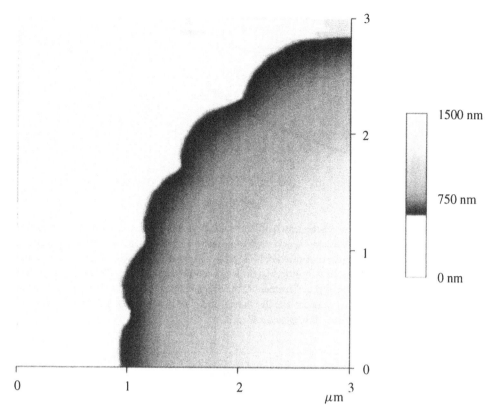

Figure 5. High-resolution ($\approx$ 10 nm per pixel) SFM topography image of the contact line region of a large hexaethylene glycol droplet. The corrugation of the contact line due to the stripewise wettability contrast of the substrate is used to determine the dependence of the local contact angle on the local curvature of the contact line. The direction of the stripes is from left to right in this image.

The occurrence of negative curvature on the hydrophilic stripes may be an artifact due to the observed larger scatter in the contact line position in this region. This deserves further investigation.

Quite obviously, the cosine of the contact angle varies linearly with the curvature of the contact line for both the hydrophilic and the hydrophobic domains. This dependence is expected from equation (1). It is now straightforward to determine the contact line tension by a linear fit, which is then found to be -6×10^{-11} N for the hydrophilic domains ($\Theta_{Young} = 18°$) and -3.5×10^{-10} N for the hydrophobic domains ($\Theta_{Young} = 34°$).

By comparing our results with that of other experiments [8, 9] one can see that the absolute values of the contact line tension are much lower. However, our values lie in the theoretically predicted range. One could conclude that experimental verification of the theoretical predictions is a problem of resolution of the imaging technique, by means of an exact determination of the local curvature of the contact line and the local contact angle.

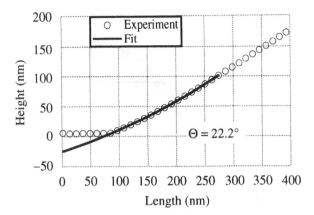

Figure 6. An example for the determination of the local contact angle. At the point of interest of the contact line, the profile of the liquid droplet is taken perpendicularly to the local direction of the contact line. The experimental profile is fitted by a second-order polynomial for the exact determination of the contact angle at the intersection with the substrate surface. The curvature of the liquid droplet profile is due to the global structure of the droplet and not to the van der Waals tails near the contact line, which decay on a much smaller lateral scale. The contact angle derived in this way can thus be considered the correct 'asymptotic' value.

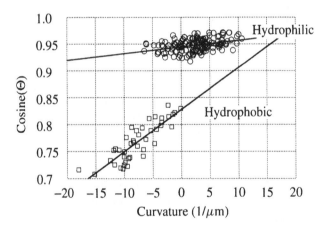

Figure 7. The experimental values of the cosine of the local contact angle are plotted versus the local curvature of the contact line. Two distinct clusters of data are seen, one at higher and the other at lower cosines of the contact angle, corresponding to the hydrophilic and hydrophobic regions, respectively. A linear dependence of the cosine of the contact angle on the curvature of the contact line is observed, as expected from the modified Young equation [equation (1)]. With a linear fit, a contact line tension of -6×10^{-11} N and -3.5×10^{-10} N for hydrophilic and hydrophobic surfaces, respectively, is determined.

In theoretical simulations of the contact line tension for a similar system by Dietrich and co-workers [2, 24] negative values were also obtained, which compare well with our results. Even the ratio between the contact line tension values of the hydrophobic and hydrophilic substrates was the same as in the simulations. The absolute values of the contact line tension in the simulations are a factor of 3–5 lower. This is within the range of choices for intercation parameters of van der Waals forces in the simulation. Furthermore, the dependence of the contact line tension on the Young contact angle is similar to theoretical predictions [25, 26], indicating increasing contact line tension with decreasing Young's contact angle.

As the substrates used are patterned in wettability, the observed local contact angle Θ is a function of both the local wettability of the substrate and the local curvature of the contact line. Ideally, Θ_{Young} has one distinct value on hydrophilic and another (higher) on the hydrophobic stripes. Due to the preparation process, this assumption could not be satisfied completely. The change in wettability from hydrophobic to hydrophilic is not discontinuous, but rather smeared out over a finite region with intermediate values of Θ_{Young}. Therefore, the data have been analyzed to exclude influences of such an effect. First, data points in a region of about 50 nm width were not taken into account. Indeed, the values of Θ that can be explained only by assuming an intermediate value of Θ_{Young} were found close (distance < 15 nm) to the idealized borderline between the hydrophilic and hydrophobic regions.

4. CONCLUSION

By using the SFM technique for imaging liquid structures, it was possible to trace the contact line, to determine its local curvature, and to measure the local contact angle with very high precision. These high-resolution data were used to extract, on the basis of the modified Young equation, a numerical value for the contact line tension. The values of -6×10^{-11} N and -3.5×10^{-10} N for hydrophilic and hydrophobic surfaces, respectively, are in the theoretically predicted range.

REFERENCES

1. A. Marmur, *J. Colloid Interface Sci.* **186**, 462–466 (1997).
2. T. Getta and S. Dietrich, *Phys. Rev. E* **57**, 655–671 (1998).
3. J. A. de Feijter and A. Vrij, *J. Electroanal. Chem.* **37**, 9–22 (1972).
4. W. D. Harkins, *J. Chem. Phys.* **5**, 135–140 (1937).
5. F. Bresme and N. Quirke, *Phys. Rev. Lett.* **80**, 3791–3794 (1998).
6. J. S. Rowlinson and B. Widom, *Molecular Theory of Capillarity*, p. 240. Oxford University Press, New York (1984).
7. N. V. Churaev, V. M. Starov and B. V. Derjaguin, *J. Colloid Interface Sci.* **89**, 16 (1982).
8. D. Li and A. W. Neumann, *Colloids Surfaces* **43**, 195–206 (1990).
9. J. Drelich, J. L. Wilbur, J. D. Miller and G. M. Whitesides, *Langmuir* **12**, 1913–1922 (1996).
10. J. Drelich, *Polish J. Chem.* **71**, 525–549 (1997).
11. R. Aveyard, J. H. Clint and D. Nees, *J. Chem. Soc., Faraday Trans.* **93**, 4409 (1997).
12. A. Amirfazli, D. Y. Kwok, J. Gaydos and A. W. Neumann, *J. Colloid Interface Sci.* **205**, 1 (1998).
13. D. Li, F. Y. H. Lin and A. W. Neumann, *J. Colloid Interface Sci.* **142**, 224–231 (1991).
14. S. Herminghaus, A. Fery and D. Reim, *Ultramicroscopy* **69**, 211–217 (1997).
15. S. S. Sheiko, G. Eckert, G. Ignateva, A. M. Muzafarov, J. Spickermann, H. J. Räder and M. Möller, *Makromol. Rapid Commun.* **17**, 283–297 (1996).
16. J. Hu, R. W. Carpick and M. Salmeron, *J. Vac. Sci. Technol. B* **14**, 1341–1343 (1996).

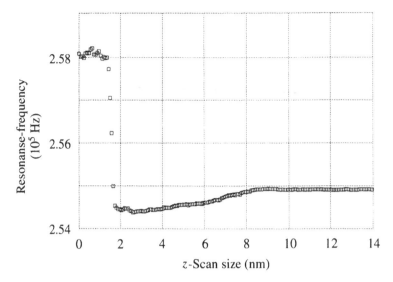

Figure 6. Dependence of the resonance frequency ω_{Res} of the cantilever on the distance between the tip and the liquid surface. ω_{Res} was calculated from the measured quantities as explained in the text. ω_{Res} first decreases with decreasing distance and then increases discontinuously (within the accuracy of the measurement) to a higher resonance frequency which shows no distance dependence in the range investigated. For distances in the latter regime, the amplitude of the tip is still measurable (≈ 2 nm), but is strongly reduced ($A_f = 12$ nm).

We have carried out measurements of the frequency shift with a tip hydrophobized by electron beam deposition of carbon [36, 37] on aqueous solutions of P_2O_5 [14]. The tip parameters were $Q_f = 166$, $k = 20$ N/m, and $\omega_{Res,f} = 254\,900$ Hz. The shift of the resonance frequency was calculated from the phase and the amplitude measured in DFS measurements, as described above. Figure 6 shows the resonance frequency as a function of the tip–liquid distance. Qualitatively, one finds a regime of shift towards lower resonance frequencies at large distances and a sharp transition towards a higher resonance frequency regime at smaller distances. The first regime is used for imaging and is the one described by the INFM model. At amplitudes typically used for imaging, one finds resonance frequency shifts in the range of 300 Hz.

From the simultanously measured dissipation, a capillary radius of 5 nm, resulting in an attractive force of 1.57×10^{-9} N, and a momentum transfer of 9.3×10^{-16} N s were calculated from the INFM model (model parameters were $S = 0.4$ and $\rho = 10$ nm). Such a momentum transfer can also be understood as an average force of 2.37×10^{-10} N acting on the liquid surface during the imaging process. The model predicts for this momentum transfer a frequency shift of 300 Hz, in perfect agreement with the experimental data. The regime of resonance frequency shift towards higher frequencies can be explained by the formation of a permanent neck, as reported in ref. [14]. In agreement with these measurements, the transition between the regimes is sharp. In the DFS measurments, a hysteresis of 12 nm was observed, if the permanent neck region was entered. In DFS measurements in the purely attractive regime, no hysteresis was observed.

4. PREDICTIONS OF THE MODEL

The INFM model can be used to make predictions on the feasibility of imaging for any liquid–tip combination. The tip properties (S and ρ) and the liquids properties (σ) are varied over a wide range. The value of D_{break} gives an estimation of the length scale of distortions. In Fig. 7, D_{break} is plotted for different S and ρ, for water ($\sigma = 72$ mN/m, $\tau = -1 \times 10^{-10}$ N). Except for small capillary radii ($R < 3$ nm), where the line tension τ has a significant influence, the free energy is in good approximation proportional to σ. Therefore, the surface tension of the liquid alone defines the energy scale and Fig. 7, except for small R, can be used for any liquid. The influence of the spreading parameter S is as expected: D_{break} grows monotonically with S. For $S > 1.6$ (not shown in the plot) D_{break} diverges; this means that the asymptotic behavior of the system becomes hydrophilic (a macroscopic capillary is energetically favorable). Figure 7 also shows that the radius of the tip ρ has a strong influence on the capillary formation. D_{break} grows monotonically with ρ. This fact should be kept in mind when imaging liquid structures. So, for example, coating tips with hydrophobic material reduces S, but increases ρ at the same time.

From these considerations, one can derive a parameter range where the imaging of liquids should be possible, as indicated in Fig. 7. The solid line corresponds to a dissipation of 20 eV per cycle, which we found to be sufficient for successful operation of the feedback loop. Beyond this line, imaging is difficult, if possible at all, due to the lack of dissipation energy. On the other hand, when the critical distance, D_{break}, becomes large, it becomes improbable for the liquid bridge to

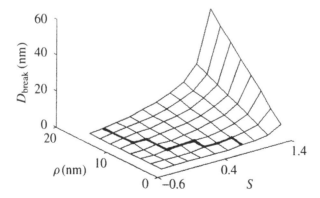

Figure 7. Distance D_{break} plotted as a function of the radius of the tip ρ and surface property S as calculated from the INFM model. For negative S and small radii, $D_{break} = 0$, which means that a capillary of any radius R and distance D is not stable. On increasing S or R, D_{break} becomes greater than 0 and capillaries are stable for $R \leqslant R_{max}$ and $D \leqslant D_{break}$. For $S \geqslant 1.6$, D_{break} diverges and any (even macroscopic) capillary is stable.

Apparent and Microscopic Contact Angles, pp. 27–45
J. Drelich, J. S. Laskowski and K. L. Mittal (Eds)
© VSP 2000.

Instability of the three-phase contact region and its effect on contact angle relaxation

JAROSLAW DRELICH *

Department of Metallurgical and Materials Engineering, Michigan Technological University, Houghton, MI 49931, USA

Received in final form 15 March 1999

Abstract—Contact angle relaxation was measured for captive air bubbles placed on solid surfaces of varying degrees of heterogeneity, roughness, and stability, in water. The experimental results indicate that both advancing and receding contact angles undergo slow relaxation in these water–air–solid systems, due to instabilities of the three-phase contact line region. It is shown that the advancing contact angle decreases and the receding contact angle increases for many systems over a period of a few hours. Also, examples of reverse progressions are reported. Additionally, in extreme cases, the contact angle oscillates down and up, over and over again, preventing the system from stabilization/equilibration. Four different mechanisms are proposed to explain the contact angle relaxation. These include (i) pinning of the three-phase contact line and its slow evolution; (ii) the formation of microdroplets on the solid surface and their coalescence with the base of the gas bubble, which causes dynamic behavior of the three-phase contact line; (iii) deformation of the solid surface and its effect on the apparent contact angle; and (iv) chemical instability of the solid.

Key words: Contact angle relaxation; deformable solid; heterogeneous surface; rough surface; line pinning.

1. INTRODUCTION

For any multi-phase system of practical and fundamental significance, i.e. with distinct interfaces, wetting is a common phenomenon that frequently needs to be considered. For example, when an inert liquid is placed on the surface of a rigid solid, it spreads spontaneously or partially to form a film or a lens. The extent of spreading is, in general, controlled by the competition between the interfacial energies in the system. For the solid–liquid–vapor (or other immiscible liquid) system with a finite affinity between phases, the system at equilibrium (in thermodynamic terms) can be described by Young's equation (1), or its modification

---

*E-mail: jwdrelic@mtu.edu

(2), as follows [1]:

$$\gamma_{SV} - \gamma_{SL} = \gamma_{LV} \cos \theta \qquad (1)$$

$$\gamma_{SV} - \gamma_{SL} = \gamma_{LV} \cos \theta + \gamma_{SLV} \kappa_{gs} \qquad (2)$$

$$\kappa_{gs} = \frac{\cos \alpha}{r},$$

where γ_{SV}, γ_{SL}, and γ_{LV} are the solid–vapor, solid–liquid, and liquid–vapor interfacial tensions, respectively, for phases that are saturated with each other; γ_{SLV} is the line tension (the excess of energy associated with the three-phase contact line); θ is the contact angle as measured from the tangent to the liquid–vapor interface and the plane of the solid surface at the point of contact between three phases (solid, liquid, and vapor); κ_{gs} is the geodesic curvature of the three-phase contact line; α is the angle between the solid surface and the plane containing the wetting perimeter; and r is the radius of curvature of the three-phase contact line ($\kappa_{gs} = 1/r$ for a drop on a flat solid surface, where r is the drop base radius). The line tension component of equation (2) becomes important when the radius of curvature for the three-phase contact line is small, a few micrometers or less [2, 3].

Although the line tension contribution to three-phase systems was recognized many years ago, it is just during the last several years that this parameter has received greater attention from researchers working with solid–liquid–fluid systems. The analysis of the line tension phenomenon ends up with the old and unresolved problem in surface chemistry: what are the properties and significance of the microscopic region in the vicinity of the contact line, and what are the consequences of the specificity of this region on the dynamics and thermodynamics of the wetting phenomena?

The microscopic and molecular aspects of the three-phase system can only be targeted after a careful selection of the system(s). As is usually recognized, both of the above equations, (1) and (2), apply to systems with solids that are homogeneous, isotropic, smooth, rigid, and composed of phases that are inert to each other. Also, all of the three phases in the system should remain in perfect mutual saturation and at an equilibrium state. In the 'real world', we often work with heterogeneous, rough, and deformable solids, which obviously differ from the perfect system. Reaching the equilibrium state in such systems is more difficult than is currently recognized, as indicated by the results demonstrated in the next part of this paper.

The problem of the instability of the three-phase systems during contact angle measurements, and the contact angle relaxation associated with this instability, was recently brought back to our attention by Hayes and Ralston [4]. They demonstrated the contact angle relaxation on poly(ethylene terephthalate) (PET) and poly(methyl methacrylate) (PMMA) polymer surfaces using the Wilhelmy plate technique. Based on the results of contact angle measurements, Hayes and Ralston proposed that an imbalance in interfacial energies existed in the region of the three-phase contact line over a period of several hours. Through the mechanisms of evaporation, condensation, and absorption, the system approaches the equilibrium state.

The discussion by Hayes and Ralston inspired the experiments in our laboratory, although the work presented in this paper goes beyond that described in ref. [4]. We support the significance of the water condensation process in the vicinity of the three-phase contact line. The example of the oscillatory changes in the contact angle over a period of several hours recorded in our laboratory is probably one of the most intriguing ones. We also demonstrate a slow evolution of the three-phase contact line, probably caused by internal factors such as thermal fluctuations, surface tension forces, and/or vibrational energy. The deformation of the soft material in the region of the contact line and the instability of the organic monolayer on a mineral surface are additional examples of mechanisms that further affect and complicate the kinetics of contact angle relaxation.

2. EXPERIMENTAL PROCEDURES

A monolayer of 1-dodecanethiol (98% purity; Aldrich Chemical Co.) was prepared on a 100–200 nm gold film, supported on a silicon wafer, by adsorption and self-assembling from a 1 mM dodecanethiol-in-ethanol solution. The gold surface was washed with ethanol and water, and cleaned in a plasma chemistry reactor (Plasmod model; Tegal Co.) with argon plasma for 50–60 min before its immersion into the dodecanethiol-in-ethanol solution. The monolayer of thiol on gold was washed with ethanol and then dried in a stream of argon before contact angle measurements were taken.

A monolayer of oleate on the surface of a fluorite crystal was prepared as follows: a fluorite crystal (Optovac Inc.) was immersed into a 10 μM aqueous solution of sodium oleate (99% purity from Sigma Chemical Co.) of pH 9.5 ± 0.1. The solution was stirred using a magnetic stirrer for about 2 h. Next, both the fluorite crystal and the adsorption solution were placed in the glass cell and the contact angle was measured as described in the next part of this section.

Fluoroaliphatic copolymer coating FC-722 (3M Company) was prepared on a freshly cleaved surface of mica by spreading 3–4 drops of a 2% FC-722 solution in organic solvent and then the solvent was evaporated under a hood. The thickness of the coating was 500–1000 nm as determined by a profilometer. Only fresh coatings, prepared a few hours before taking the contact angle measurements, were used in this study.

A powder of low-density polyethylene (PE) (Scientific Polymer Product, Inc.) was dissolved in analytical grade toluene at a temperature of 70–75°C (0.2–1 wt% concentration). The PE–toluene solution was spread on a glass slide and kept in a vacuum oven for 3–4 h at a temperature of 60–65°C. The glass with the PE coating was used in contact angle measurements after cooling it down to room temperature.

Coal specimens (a few centimeters in size) from a Mexican power plant (Micare) were carefully selected for contact angle measurements to avoid specimens with cracks. The proximate analysis for this coal showed the following composition: 35.3 wt% ash, 22.0 wt% volatile matter, and 42.7 wt% fixed carbon. The coal specimens were cut and polished with a series of silicon carbide abrasive papers, from #60 to #1200 grits, and 0.05 μm alumina powder, as described previously [5].

The above-described solid substrates were used in contact angle measurement experiments. Each solid sample was placed in a rectangular glass chamber on two stable supports. The glass chamber was filled with deionized water. An air bubble produced at the tip of a U-shaped needle using a microsyringe was made to touch the sample surface. The size of the bubble was enlarged or reduced in order to create advancing or receding conditions for the three-phase contact line. After regulating the size of the bubble, the needle was immediately detached from the bubble, and the container was covered with a polyethylene film to isolate the system from the laboratory environment. Contact angles were measured using an NRL Rame-Hart goniometer. The contact angles measured on one side of the bubble are reported, as the contact angle variation with time sometimes differed at the two bubble sides and an average contact angle value could not represent a real relaxation process. The time reported in this paper is that from the moment when the air bubble size was adjusted to the moment that the contact angle was measured.

All contact angle measurements were done in a temperature-controlled room, at 20–23°C.

3. RESULTS

The variety of samples used in this study included those with smooth and rough surfaces, homogeneous and heterogeneous surfaces, and stable and unstable surfaces. In this way, a broad view of the kinetics of contact angle relaxation could be obtained.

3.1. Self-assembled thiol monolayer on a gold film ('smooth' and 'homogeneous' surface)

Thiols are capable of forming relatively uniform and stable (over several hours) monolayers on a gold surface [6]. Because uniform and relatively smooth gold films can also be easily deposited on a smooth silicon wafer, the thiol monolayers serve as model systems in studies of wetting phenomena [7–10].

It is shown in Fig. 1 that the advancing (θ_A) and receding (θ_R) contact angles for water on a freshly prepared dodecanethiol monolayer were about 111° and 98°, respectively, when measured 15–20 s after regulating the size of the air bubble. Therefore, the contact angle hysteresis ($\Delta\theta = \theta_A - \theta_R$) for the system was 13°. We experienced a slight variation in the quality of the fabricated monolayers. A few degrees smaller and larger contact angle hysteresis values were often observed in our laboratory for these thiol-on-gold monolayers. For example, we noted a contact angle hysteresis as small as 5–7° for the monolayer made of dodecanethiol [9].

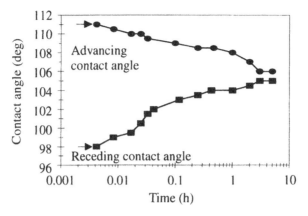

Figure 1. The relaxation of water contact angles at a monolayer of dodecanethiol prepared on a gold surface. A substrate with a dodecanethiol monolayer was immersed in water and then two air bubbles (3.5–4.5 mm in diameter) were attached to the hydrophobic monolayer. The size of one bubble was slightly enlarged by adding air to the bubble using a microsyringe and the receding contact angle was measured on one side of the bubble about 15 s after the bubble size changed. The size of the second bubble was slightly reduced by removing the air from the bubble using a microsyringe and the advancing contact angle was measured. Then, during the next 7 h, the contact angle measurements were repeated several times on the same sides of both bubbles. Markers indicate experimental data and the solid lines are shown as a guide to the eye.

We found that a lack of reproducibility of the contact angle hysteresis value for the air bubble–water–dodecanethiol monolayer system was mainly associated with a variation in the receding contact angle. The advancing contact angle from 110° to 112° (the error associated with contact angle measurement was about 1° in our experiments) was well reproduced (not shown). We expect that monolayer defects are responsible for the limited reproducibility of the receding contact angle. Although homogeneous monolayers of thiol can be prepared on a gold surface, they are not free of molecular defects. As demonstrated by Schönenberger *et al.* [11], dodecanethiol forms domains on the gold surface with linear and point defects. Additionally, nanoroughness of the gold film could contribute to an imperfection of the model sample surface. In this regard, slight imperfections of this system should be kept in mind in order to understand the results in Fig. 1.

As demonstrated in Fig. 1, both the advancing and the receding contact angles changed over a period of a few hours. The advancing contact angle decreased and the receding contact angle increased by 6–8° in about 3 h. No changes in the advancing or receding contact angles were observed after 3 h, and both contact angles stabilized at approximately the same value of 105–106°. Similar kinetics of the contact angle variation was observed in other experiments involving similar thiol monolayers. The average contact angle always stabilized at a value of 105–108° (not shown). It should be noted that similar observations on contact angle relaxation to an average value from advancing and receding contact angles were made by Decker and Garoff [12] in other systems with chemically heterogeneous solid surfaces. The relaxation process was also enhanced by Decker and Garoff by introducing external energy into the three-phase system through mechanical vibrations.

A hypothesis of destabilization of the monolayer and desorption of dodecanethiol, at least to any great extent, is rejected in the explanation of the results in Fig. 1. Also, the oxidation of the monolayer, as discussed in the literature [13], had rather a minor effect, if any, on the stability of the monolayer that was in contact with water and air during the time frame of the experiment. We found that practically the same advancing and receding water contact angles were measured at the same spot of the monolayer when the size of the bubble was again decreased and increased, respectively, after 3–7 h of study of the contact angle relaxation.

3.2. Carboxylate monolayer on a fluorite crystal (unstable system)

A relaxation of the contact angle for an air bubble attached to an oleate monolayer on a fluorite crystal is shown in Fig. 2. In this particular experiment, a solution of sodium oleate was used instead of deionized water. The pH of the solution was 9.5 at the beginning of the experiment and dropped to about 7.7 at the end of the experiment. This drift in pH value clearly indicates that the solution chemistry changed substantially during the experiment.

The low value of the contact angle hysteresis for this system at the beginning of the experiment, about 8°, indicates that the monomolecular structure of oleate adsorbed at the smooth surface of the fluorite crystal was relatively well organized and tightly packed. Similarly, as in the case of the dodecanethiol monolayer formed

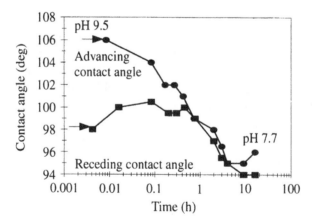

Figure 2. The relaxation of water contact angles at an oleate monolayer prepared on a fluorite crystal surface. A fluorite crystal coated with a monolayer of oleate was immersed in 10 μM sodium oleate solution (pH 9.5). Two air bubbles were placed on the monolayer surface. The size of one bubble was enlarged and the receding contact angle was measured on one side of the bubble at about 15 s after the bubble size changed. The size of the second bubble was reduced and the advancing contact angle was measured on one side of the bubble. Then, during the next 18 h, the contact angles were measured again on the same sides of both bubbles and contact angle relaxation was observed. At the end of the experiment, the pH value of the solution was measured again and was found to drop to 7.7. Markers indicate experimental data and the solid lines are shown as a guide to the eye.

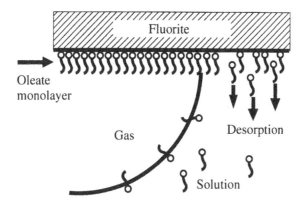

Figure 3. Illustration of oleate desorption from the fluorite surface. This mechanism is expected to occur in the system examined in this study, due to the instability of the pH of the solution.

on the surface of gold, small changes in the contact angles were observed over time. Specifically, the 'advancing' and 'receding' contact angles approached an average value of 100–101° during the first several minutes of the experiment. Next, both 'advancing' and 'receding' contact angles systematically decreased to a value of 94–95° over the period of 10–15 h. It was only at the end of the experiment that a small difference between the two contact angles was noted. This happened as the two curves in Fig. 2 were obtained based on the contact angle measurements for two air bubbles that were placed at different areas of the oleate-coated fluorite crystal. A small change in the surface characteristics of these two regions could also evolve with time.

It is well documented that carboxylates adsorb onto the surface of fluorite from alkaline solutions through chemisorption [14]. These monolayers are relatively stable in aqueous solutions [14]. Nevertheless, the substantial drop in the pH of the solution experienced during the experiment, from pH 9.5 to pH 7.7, cannot be ignored in the discussion of the experimental data of Fig. 2. As the pH decreased, this change in the environment probably forced very slow but systematic desorption of oleate molecules from the surface of fluorite (Fig. 3). This instability of the system probably provoked the difference between the contact angle vs. time results shown in Figs 1 and 2. It has already been well reported in the literature [15, 16] that chemical and morphological changes on the solid surface cause the contact angle to relax.

3.3. Polyethylene film (rough and homogeneous surface)

Polyethylene films prepared in this study had a relatively homogeneous but rough surface. We observed irregularities on the polyethylene film surface under an optical microscope using a magnification of × 500 and larger. However, we did not conduct detailed studies on the surface roughness parameters in this investigation.

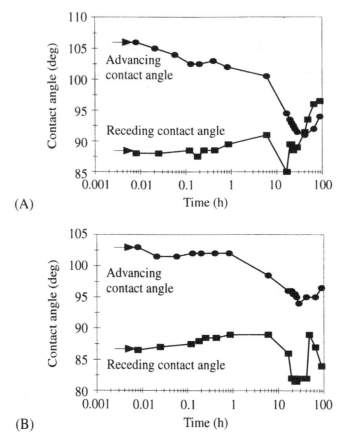

(A)

(B)

Figure 4. The relaxation of water contact angles on a polyethylene surface. Results of two independent experiments. Markers indicate experimental data and the solid lines are shown as a guide to the eye.

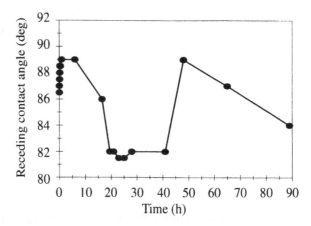

Figure 5. The relaxation of receding contact angles on a polyethylene surface. The results are the same as those in Fig. 4B but at a different scale. Markers indicate experimental data and the solid line is shown as a guide to the eye.

We measured a 16–18° contact angle hysteresis on the prepared polyethylene films. The advancing contact angle was from 103° to 106°, and the receding contact angle was from 87° to 88° (Figs 4 and 5).

The changes in contact angles observed in the air bubble–water–polyethylene system (Figs 4 and 5) were similar to those that we observed for the two previously discussed systems (Figs 1 and 2) during the first few hours of the experiment. The changes in contact angles were relatively systematic, although the kinetics was a little slower than observed for the previous systems. The advancing contact angle decreased and the receding contact angle increased by 3–5° in 5–6 h. The contact angle values did not reach any average value after 5–6 h of the experiment. Instead, further changes in both contact angles were observed. We found that these changes usually did not follow any smooth pattern. Quite often, the values of both contact angles jumped from one value to another (dropped and increased) by as much as a few degrees in a relatively short interval between the two measurements. This oscillation of the contact angle is quite clear from the results shown in Figs 4 and 5. Similar time-dependent contact angle values were measured in our laboratory for other polyethylene samples (not shown). We also found it interesting that during some of the experiments, the contact angles dropped below the values of the original receding contact angles (Figs 4 and 5). This happened only after several hours of experiments. The values for the 'receding' contact angles, reached after about 40 h, were larger than the 'advancing' contact angles (Fig. 4A), but they should not be misinterpreted as an error in the measurements. Different bubbles were attached to the polyethylene film on different regions of the same sample in each case. The surface roughness was not uniform over the entire sample, and this also caused a few degrees variation in contact angles when measurements were repeated at different spots on the surface.

3.4. Coal ('smooth' and heterogeneous surface)

The coal samples used in this study represented heterogeneous surfaces: primary hydrophobic organic maceral contaminated with hydrophilic inorganic inclusions having a size of a few micrometers or less. They were polished with a series of grit papers and 0.05 μm alumina powder as described in detail in our previous paper [5]. We found that such a polishing procedure usually left the coal surface with a microroughness at a level of several nanometers, with inorganic components rising slightly over the organic maceral, although a few scratches were difficult to avoid [5]. In this regard, the experiments with coal specimens should be understood as involving heterogeneous surfaces, and the possible effect of roughness on measured contact angles cannot be ignored.

The result of the examination of contact angles for the coal specimens revealed similar contact angle vs. time relationships to those observed for other systems (Figs 6 and 7). We observed a few degrees change in both contact angles after a few hours of experiments. Keeping the air bubble on the surface of coal over the long period of 80–100 h showed again interesting contact angle vs. time relationships.

J. Drelich

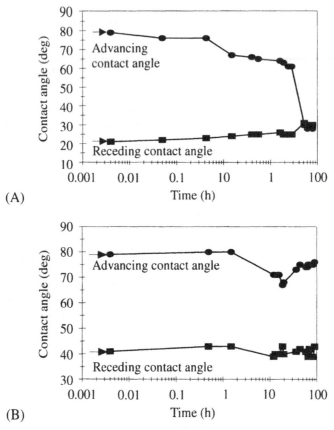

(A)

(B)

Figure 6. The relaxation of water contact angles on a coal specimen. Results of two independent experiments. Markers indicate experimental data and the solid lines are shown as a guide to the eye.

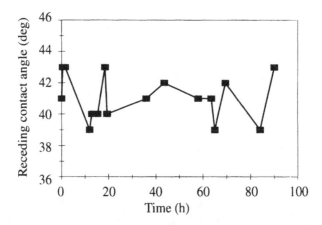

Figure 7. The relaxation of receding contact angles on a coal specimen. The results are the same as those in Fig. 6B but at a different scale. Markers indicate experimental data and the solid line is shown as a guide to the eye.

We found that in some experiments an average contact angle was approached (Fig. 6A). This average value was usually much closer to the receding contact angle than to the advancing contact angle. In other experiments, the contact angle vs. time curve became oscillatory in shape, with a 'no ending' value (Fig. 6B), such as was observed for the polyethylene film. Similar contact angle vs. time curves, such as in Figs 6 and 7, were also obtained for other coal specimens (not shown). We often observe in our laboratory that the wetting characteristics of a coal surface change during an extended period of exposure to water and air. This phenomenon, however, leads to a systematic decrease in the contact angle and no oscillation of the contact angle value.

3.5. Fluoroaliphatic copolymer coating (deformable surface)

The results of the contact angle relaxation are shown in Fig. 8. There are some similarities and differences in the contact angle vs. time relationships compared with the results discussed in the previous sections. As in any other system discussed

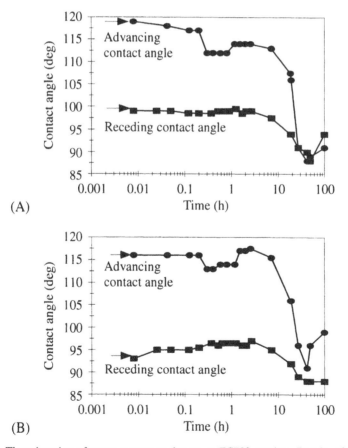

Figure 8. The relaxation of water contact angles on an FC722 coating. Results of two independent experiments. Markers indicate experimental data and the solid lines are shown as a guide to the eye.

so far in this paper, we noted slight changes in both contact angles, 2–3°, at the beginning of the experiments. Also, we observed significantly larger variations in both contact angles after several hours of experiments.

There were some differences in the contact angle vs. time relationships for this system compared with the previously discussed systems. First of all, already after several minutes, we reported a significant variation in the advancing contact angle (Fig. 8). Also, it was quite common that the receding contact angle decreased slightly at the beginning of the experiment instead of increasing (Fig. 8). If the 'final' contact angle value was approached by the relaxing advancing and receding contact angles, it was smaller than the receding contact angle (measured at the beginning of the experiment) (Fig. 8).

The fluoroaliphatic copolymer coating had a smooth surface. We did not observe any roughness features under an optical microscope. This coating, however, was not rigid. The attachment of the air bubble to the coating over a period of many hours caused significant topographic changes on the surface. In time, the coating became waved with serious local distortion of the surface, visible even with a low-magnification optical microscope. The vertical displacement of the coating was particularly clear in the region of the gas bubble base. We assume that the local deformations propagating on the coating surface had a significant impact on the results in Fig. 8, as discussed in the next part of this paper.

4. DISCUSSION

The contact angle relaxation on surfaces of practical significance, which demonstrate some defects such as heterogeneity, roughness, deformation, and instability, is quite obvious from the results presented in Figs 1, 2, and 4–8. As a variety of experimental data for solid surfaces of different characteristics show, the contact angle relaxation cannot usually be explained by one simple mechanism, but rather a combination of at least two mechanisms.

We believe that the mechanisms that are important in the analysis of the contact angle data in Figs 1, 2, and 4–8 are (i) pinning of the three-phase contact line during its receding and advancing, and its slow evolution to a more stable configuration; (ii) nucleation/condensation of liquid microdroplets in the vicinity of the three-phase contact line and their coalescence with the bubble base; (iii) deformation of the solid at the contact line; and (iv) chemical changes in the system. The following is a brief (and general) discussion of mechanisms (i)–(iii). It has already been reported that chemical and morphological changes on the solid surface cause contact angle relaxation [15, 16]. In this regard, mechanism (iv) is not further discussed.

4.1. Line pinning effect

It is now well known that the three-phase contact line becomes pinned at any surface that is heterogeneous and/or rough when the solid surface imperfections

Relaxation of contact line
on heterogeneous surface

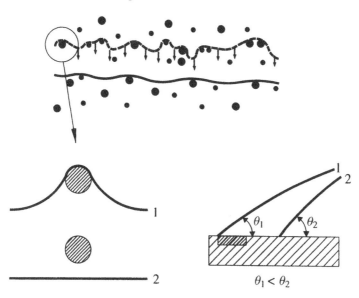

Figure 9. The nature of the three-phase contact line at a heterogeneous surface.

have linear dimensions much smaller than the length of the contact line. For example, Fig. 9 illustrates the pinning of the fragment of the three-phase contact line on a heterogeneity of the solid surface (a similar picture can be drawn for a rough surface where hills/asperities form mechanical barriers for a moving contact line). The solid surface in Fig. 9 is composed of two regions having different wetting characteristics: the guest material (representing heterogeneity), shown as a circle, is less hydrophobic than the primary (host) material. The distortion of the three-phase contact line occurs at such a heterogeneous surface when it passes through this region of the surface. This happens, for example, during measurement of the receding or advancing contact angle.

The pinning of the three-phase contact line affects the contact angle value as shown in Fig. 9. As might be expected for the example discussed, the contact angle is smaller for the pinned contact line (line 1) than the contact angle measured directly at the primary material (line 2). Whether a transition from position 1 to 2 takes place depends on the energy introduced to the system and the existing energetic state of the system. It was shown theoretically that if the size of heterogeneity [17] or roughness features [18] was larger than about 0.1 μm, the three-phase contact line had a tendency to be stopped by a trapping mechanism on a local surface imperfection(s) and could not move on without introducing additional external force. In other words, the system demonstrates so-called metastable states, and the contact angle hysteresis is a common phenomenon of such systems. When the size of the solid surface imperfection is less than about 0.1 μm, the solid surface is expected to be recognized by the probe liquid during contact angle measurements

J. Drelich

as a 'homogeneous' surface, having the same wetting properties in every direction. Although the limit of imperfection size that affects the contact line contortion is still a subject of debate, it is very likely that the internal energy of the system associated with thermal fluctuations, surface tension forces, and/or vibrational energy might significantly affect the shape of the three-phase contact line for the liquid at a heterogeneous/rough solid surface, as well as its relaxation. For example, we can imagine the heterogeneous solid surface composed of small imperfections having a size of much less than 0.1 μm (preferentially only a few nanometers) and randomly distributed. The liquid edge, whether receding or advancing, becomes contorted on such a surface, at least at the first moment of the liquid movement. Specifically, the advancing/receding contact angle relates to a temporary contact line position where the contact line gets trapped at a particular time. However, whenever the movement of the liquid edge is stopped, the three-phase contact line tries to smooth out. This is caused, for example, by surface tension forces tending to smooth the contorted surface of a bubble or drop base. Further, any thermal fluctuation and/or vibrational energy in the system might cause the change in the position and the shape of the contact line, at least in a particular local area (this effect can be enhanced by introducing additional energy to the system as well demonstrated in ref. [12]). Next, the locally trapped contact line can be pulled away from this position by neighboring parts of the contact line and so on, until a stable position (associated with an equilibrium state, or a metastable state that cannot be overcome by the internal energy of the system) is reached.

The three-phase contact line can be trapped at random places, where it remains trapped until it escapes, probably after a certain (but random) time, to the next trap position, and so on. This mechanism causes a slow movement of the contact line, through a stick–slip mechanism, and consequently the contact angle relaxes. If the solid surface is only composed of nanoscopic-sized imperfections, such as the monolayer of dodecanethiol on the gold film (Fig. 1), the relaxation takes place from both the receding and the advancing modes towards an average ('equilibrium') position. This equilibrium position is located in between the advancing contact angle and the receding contact angle.

At a surface having macroscopic imperfections, the three-phase contact line never relaxes to an equilibrium position and the contact angle hysteresis remains unchanged. The contact line usually falls into one of the metastable state positions and never escapes from it in the time frame of the measurement. This is because the energetic barrier is larger than any thermal fluctuations, surface tension forces, and/or vibrational energy existing in the system.

We believe that the kinetics of the contact line position evolution depends on the size of the solid surface imperfections, their density, distribution, and wetting characteristics. It appears from Figs 1 and 4–8 that the line position evolution is fully responsible for the contact angle relaxation during the first 0.5–2 h of the experiments. The kinetics of these changes appears to follow almost a linear relationship described by $\theta = n \log(t) + m$, where θ is the contact angle, t is the time, and n and m are the constants. We calculated the value of n from the slope of the θ vs. $\log(t)$ line in Figs 1 and 4–8 and found it to vary from ± 0.5 to ± 2.0. The negative values refer to advancing contact angles and the positive values to receding contact angles. The magnitude of n probably depends on energetic barriers faced by the evolving three-phase contact line and the energy available to this line. This suggestion is, however, very speculative at this moment, as we do not have a detailed picture of imperfections of the solid surfaces used in our experiments.

Finally, it should be recognized that although our discussion in this section was concentrated on a heterogeneous surface, the same concepts should, in general, apply to homogeneous but rough surfaces.

4.2. Evaporation–condensation–coalescence cycle

The relaxation of the contact line through a stick–slip mechanism over the surface imperfections cannot be used to explain the experimental results shown in Figs 4–8, particularly when the systems were aged over 7–15 h. The thermal fluctuations, surface tension forces, and/or vibrational energy in the systems examined were not enough to create the oscillatory character of the contact angle vs. time relationship. We found, after microscopic observations, that the change in the appearance of the contact angle vs. time relationship is associated with the nucleation/condensation of the microdroplets at the solid surface inside the air bubbles. The concept of the

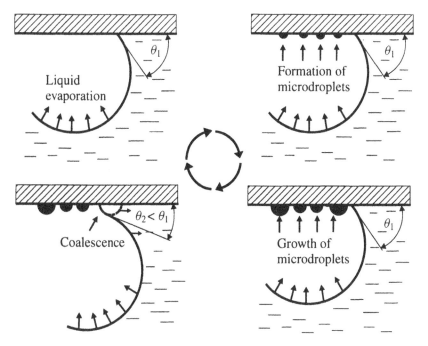

Figure 10. A hypothetical mechanism of liquid evaporation, condensation, and coalescence inside a gas bubble placed on a solid surface.

entire mechanism is illustrated in a simplified drawing in Fig. 10 and discussed as follows.

The air inside the bubble, attached to the sample surface, was saturated with water vapor, probably no later than a few minutes after bubble deposition. The water vapor adsorbed/condensed at active sites of the solid surface and water droplets were formed. These droplets slowly increased their size to microscopic dimensions. Microdroplets that had been formed in the vicinity of the bubble base coalesced with the neighboring liquid–gas interface. In the coalescence process, the microdroplets were sucked into the water bulk surrounding the gas bubble. After the coalescence process, the contact angle always relaxed to a smaller value. This is represented by several minima in the contact angle vs. time curves in Figs 4–8. Whether the three-phase contact line returned to the previous position, remained in the new location, or reached some intermediate state, depended on the energetic barrier associated with the new position. This is why we noted different contact angle values for various systems in our experiments. In an extreme case, the contact angle value changed from a value close to the advancing contact angle to the value of the receding contact angle and never returned to the original state. This is because the three-phase contact line became trapped in a new position, after the coalescence process, and could never escape from this position using internal energy. For example, we report such a case for the coal sample in Fig. 6A. In most of the other situations, we observed an oscillatory nature of the contact angle relaxation (Figs 5 and 7), where the three-phase contact line returned to an original or close-to-original position after the coalescence of the gas bubble base with a microdroplet(s).

4.3. Deformation of the soft material

The analysis of contact angle relaxation for soft materials is even more complex than discussed in the previous sections. The placement of a gas bubble or a liquid drop on the surface of the soft material causes the deformation of this surface at the contact line region [19, 20]. This is shown schematically in Fig 11. We observed a deformation of the fluoroaliphatic copolymer coating used in this study under an optical microscope. However, a detailed description of the coating deformation and its effect on contact angle variation cannot be drawn from our preliminary experiments. In this regard, the discussion below is speculative and very general.

As shown schematically in Fig 11, the deformation of the coating changes the topography of the solid surface in the vicinity of the contact line. The system at any moment in the aging process probably tries to establish the same internal contact angle, θ_M. If the solid surface is smooth, the internal contact angle is exactly the same as the apparent contact angle, θ_{A1}, measured through an optical instrument. After deformation, if the internal contact angle remains the same, the apparent contact angle changes, as shown in Fig 11: it decreases or increases depending on the location of the three-phase contact line on the deformation. The system will remain unstable as long as the deformation of the material and the position of the contact line establish a compromising (energetically favorable) configuration.

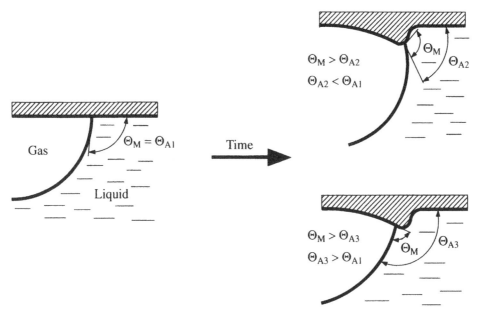

Figure 11. Simplified illustration of instability of the three-phase contact line region (probably a several micrometers-wide region, at most) on a soft solid. It is expected that two different mechanisms take place in this system. The first mechanism is associated with the deformation of a soft material. This deformation might also slowly 'flow' in time (deformation increases its diameter) under the pressure induced by the gas bubble. The second mechanism is the movement of the three-phase contact line at and around the deformation due to mechanical (and consequently energetic) changes in a microregion of the contact line. Θ_M and Θ_A represent microscopic (local) and apparent contact angles, respectively.

The experimental data in Fig. 8 show that the contact angle vs. time relationship is not smooth for a deformable material. Some jumps of the contact line from the 'top' of the vertical displacement to its 'base' are expected. After the jump, the process of material deformation and contact angle relaxation starts again, as long as a 'stable' configuration is reached. This is what probably took place on the surface of the fluoroaliphatic coating after a few minutes (Fig. 8). A detailed analysis of the contact angle data for the fluoroaliphatic coating is difficult as both the contact line pinning effects and the coalescence processes took place in this system as well.

5. CONCLUSIONS

The contact angle results presented in this paper indicate that both advancing and receding contact angles undergo changes over time in the imperfect water–air–solid systems — systems that differ from the system described by Young's equation on the macroscopic scale of the observations. In particular, the kinetics of contact angle relaxation is sensitive to solid surface imperfections such as heterogeneity, roughness, deformation, and instability. The region of the three-phase contact line

remains in a dynamic state for several hours or more. We propose that at least four different mechanisms, specified below, cause the contact angle relaxation.

The first mechanism is due to the pinning of the three-phase contact line. The pinned line is trapped temporarily, on the solid surface area having small imperfections such as nanoscopic-sized heterogeneities and roughness, until the contact line escapes and changes its shape. We expect that the escape of the three-phase contact line from a metastable position is caused by the internal energy (thermal fluctuations, surface tension forces, and/or vibrational energy) present in the system.

The second mechanism responsible for the contact angle relaxation is caused by the formation of microdroplets on the solid surface. Air inside the gas bubble becomes saturated with water vapor. Over time, water microdroplets are formed on the active sites of the solid surface. The microdroplets located in the contact line region coalesce with bulk liquid and trigger the dynamic behavior of contact angles. We observed in some experiments that the contact angles relaxed to values that were close to the receding contact angles. Interesting cases occurred when the coalescence process caused the contact angle vs. time curve to have an oscillatory character and sometimes, the stable state was 'never' reached.

The third mechanism proposed in this contribution to explain the contact angle relaxation data is associated with the deformation of the solid surface. The microscopic deformation of the soft material occurred in the contact line region and altered the topography of the surface of the fluoroaliphatic coating in the vicinity of the contact line. We believe that this microscopic deformation caused a variation in the contact angle.

Finally, the instability of the system is also believed to cause the contact angle relaxation. Specifically, for the oleate monolayer on the fluorite crystal studied in this contribution, we attribute the changes in contact angle to desorption of the molecules initiated by the drift of the pH value of the solution from 9.5 to 7.7.

REFERENCES

1. L. Boruvka and A. W. Neumann, *J. Chem. Phys.* **66**, 5464–5476 (1977).
2. J. Drelich, *Colloids Surfaces A: Physicochem. Eng. Aspects* **116**, 43–54 (1996).
3. J. Drelich, *Polish J. Chem.* **71**, 525–549 (1997).
4. R. A. Hayes and J. Ralston, *Colloids Surfaces A: Physicochem. Eng. Aspects* **80**, 137–146 (1993).
5. J. Drelich, J. S. Laskowski, M. Pawlik and S. Veeramasuneni, *J. Adhesion Sci. Technol.* **11**, 1399–1431 (1997).
6. A. Ulman, *Chem. Rev.* **96**, 1533–1554 (1996).
7. J. Drelich, J. D. Miller, A. Kumar and G. M. Whitesides, *Colloids Surfaces A: Physicochem. Eng. Aspects* **93**, 1–13 (1994).
8. J. Drelich, J. L. Wilbur, J. D. Miller and G. M. Whitesides, *Langmuir* **12**, 1913–1922 (1996).
9. J. Drelich, J. D. Miller and R. J. Good, *J. Colloid Interface Sci.* **179**, 37–50 (1996).
10. A. Ulman, *Thin Solid Films* **273**, 48–53 (1996).

11. C. Schönenberger, J. Jorritsma, J. A. M. Sondag-Huethorst and L. G. J. Fokkink, *J. Phys. Chem.* **99**, 3259–3271 (1995).
12. E. L. Decker and S. Garoff, *Langmuir* **12**, 2100–2110 (1996).
13. A. B. Horn, D. A. Russell, L. J. Shorthouse and T. R. E. Simpson, *J. Chem. Soc., Faraday Trans.* **92**, 4759–4762 (1996).
14. J. Drelich, A. A. Atia, M. R. Yalamanchili and J. D. Miller, *J. Colloid Interface Sci.* **178**, 720–732 (1996).
15. R. E. Johnson, Jr. and R. H. Dettre, in: *Surface and Colloid Science*, E. Matijevic (Ed.), Vol. 2, pp. 85–153. Wiley-Interscience, New York (1969).
16. R. E. Johnson, Jr. and R. H. Dettre, in: *Wettability*, J. C. Berg (Ed.), pp. 1–73. Marcel Dekker, New York (1993).
17. A. W. Neumann and R. J. Good, *J. Colloid Interface Sci.* **38**, 341–358 (1972).
18. J. D. Eick, R. J. Good and A. W. Neumann, *J. Colloid Interface Sci.* **53**, 235–248 (1975).
19. G. R. Lester, *J. Colloid Sci.* **16**, 315–326 (1961).
20. M. E. R. Shanahan, *J. Phys. D: Appl. Phys.* **20**, 945–950 (1987).

Apparent and Microscopic Contact Angles, pp. 47–93
J. Drelich, J. S. Laskowski and K. L. Mittal (Eds)
© VSP 2000.

Molecular mechanisms of hydrophobic transitions

V. V. YAMINSKY *

*Department of Applied Mathematics, Research School of Physical Sciences and Engineering,
Institute of Advanced Studies, Australian National University, Canberra A.C.T. 0200, Australia*

Received in final form 3 June 1999

Abstract—This review article provides new insights into the molecular mechanisms of hydrophobicity that emerge within the phenomenological framework of classical capillarity. The thermodynamic analysis of surface phenomena via the Gibbs adsorption equation that forms the basis of the experimental physical chemistry of liquid–fluid interfaces extends with its rigorous implications towards the wetting of solid surfaces. Observations on equilibrium contact angles, contact angle hysteresis, and the dynamics of wetting hitherto based on qualitative results that received fragmentary and eclectic interpretations are accounted for in terms of the values of adsorption pressures that act at the three-phase line. Hydrophobic surfaces and hydrophobic transitions induced by reversible solute adsorption qualify as two distinct limiting cases in this thermodynamic context. The equilibrium and kinetic aspects of the wetting of uniform and chemically heterogeneous surfaces by pure liquids and surfactant solutions is discussed with reference to reversible liquid–fluid interfaces. The importance of the solvent and solute adsorption at the solid–vapor interface for proper assessment of static and dynamic dewetting transitions is particularly addressed.

Key words: Wetting transitions; hydrophobic effect; contact angles; adsorption; surface pressure.

1. INTRODUCTION

The subdivision of surfaces into hydrophilic and hydrophobic is an important consideration of classical and modern colloid science. Clean surfaces of polar inorganic materials such as the basal plane of mica (and clays) or hydroxylated quartz (silica glass, sand) are completely wetted by water with a zero value of the contact angle. On nonpolar hydrocarbon and fluorocarbon substrates (paraffin wax, Teflon), water forms contact angles in excess of 90°. Other solids according to the chemical nature of their surfaces cover the entire spectrum between these extremes. For example, talc, graphite or molybdenum disulfide are at the hydrophobic end.

---

*On leave from the Advanced Technologies Center, Moscow, Russia. E-mail: vvy110@rsphysse.
anu.edu.au

Alumina, silver iodide, and many other water-insoluble metal oxides and salts are more hydrophilic, but still the contact angles can be larger than zero.

It is tempting to classify different materials according to the contact angle values they show with water. Such a compilation would be of immense value for diverse engineering applications, for example, cleaning, flotation, tertiary oil recovery, and biocompatibility. However, no extensive tables of contact angles exist in the comprehensive handbooks on the physical and chemical properties of materials.

Semi-empirical equations that enable one to define the dispersion and polar ('acid–base') components of the surface free energy of solids in its relations to contact angles have been introduced over the past several decades. Such 'equations of state' suggested by several authors help us to order and systemize a vast body of empirical data on the wetting properties of different materials. Still, they lack predictive powers and fundamental clarity of the issue.

The wetting of solids by liquids and solutions reveals complex patterns of contact angle hysteresis. The simplified theories of contact angles do not account for this fundamental effect. They cannot even be tested accurately because the associated uncertainty in the experimental contact angle values is too large. The fact that contact angles suffer such a great deal of ambiguity explains why they do not occur in tabulated form. And this, in turn, is a difficulty for a better theory that might account for the hysteresis effect.

To resolve this problem, one should be able to establish an unambiguous scale of hydrophobicity, in the manner of scales of viscosity, saturated vapor pressure, and other bulk phase properties. Such a scale indeed occurs also in the realm of interfacial phenomena. Chemical thermodynamics at the foundation of the molecular kinetic theory of matter does not restrict its exact results to the properties of bulk phases. Interfaces can be characterized quantitatively in exact thermodynamic and molecular terms. This is the realm of classical capillarity that deals with the surfaces of liquids. Surface tension as the basic parameter in this context can be measured accurately. Tables of exact values of the surface tension of pure liquids and solutions do exist. Such compilations supplement databases of bulk phase properties such as density, heat capacity, melting points, etc. They played a crucial role in the formulation of basic concepts of the physical chemistry of interfaces. Surface entropy and enthalpy follow from temperature derivatives of the surface tension, and concentration derivatives define the exact molecular compositions of liquid–fluid interfaces in terms of values of solute adsorption. Unless the task of a similar quantification of contact angles is complete, the ambitious goal of the foundation of an exact science of wetting transitions that would explain the contact angle phenomenon might be practically difficult to achieve.

The difficulties associated with the interpretation of the contact angles of liquids on solids do not occur when a liquid wets another liquid. Here contact angles are defined in terms of the three surface tensions applied at the three-phase line. While a similar definition follows from the Young equation, only for liquids can all three tensions be directly measured. Contact angles here are unique, can be measured accurately, and can be verified in terms of the surface tensions. This can be done for any particular liquid–liquid–fluid system of interest. Only when the substrate is a solid do difficulties begin to emerge.

One might recall in this connection that the results of force measurements (SFA, AFM) between mica and other kinds of solid surfaces suffer a similar ambiguity. Surface force data are almost as uncertain as the values of contact angles. One can argue in this connection that surface force measurement is a rather new technique. With its future developments quantitative results may become possible. But this argument does not apply to contact angle measurements. The latter might in fact have an even longer historical tradition than the measurement of the surface tension of liquids. In the latter case, the problem of accuracy had already been resolved by the beginning of this century.

The reason for this difficulty in the quantification of surface forces and contact angles might be due to the fact that the surfaces of solids are involved in both cases. However, the mere fact that a material is in the solid state is not an obstacle for thermodynamic quantification. Indeed, bulk phase properties of solids can be defined in precise terms of chemical thermodynamics. Of course, quantum chemistry still might have some difficulties explaining why sodium chloride is soluble in water and silver chloride is not. But this theoretical difficulty does not obstruct our ability to measure the associated parameters experimentally, and precise results can always be interpreted in exact thermodynamic terms. The solubility products for crystalline substances such as salts, bases and acids in water and other solvents are in fact often known with an even higher accuracy than the mutual solubility of liquids.

Apparently, the contact angle uncertainty cannot be attributed to some fundamental principles that might limit the applicability of chemical thermodynamics and chemical kinetics to solid states of matter. Still the phenomenon of contact angle hysteresis is clearly related to the fact that solid substrates are involved. To explain the situation that has analogies in the theories of fracture and tribology, we need to consider the basic concepts behind the thermodynamics of wetting in some greater detail. We would like to outline in this connection several considerations:

(1) We notice that contact angles and surface forces are routinely studied with insoluble solid materials, otherwise one simply might not be able to conduct measurements with pure water, but rather would have to deal with saturated solutions. Recondensation phenomena in the manner of Ostwald ripening that develop in these situations violate the mere principle of contact angle and surface force measurements.

(2) Also for insoluble substrates, the occurrence of contact angle hysteresis makes the measurement too uncertain. The question about the existence of equilibrium contact angles confused literature on the subject for almost 200 years. This important question is about contact angles that can be justified thermodynamically in conjunction with the Young equation.

(3) As if to complicate the issue further, some surfaces not only manifest static and dynamic contact angle hysteresis, but also change their hydrophobicity over time, for example by storage in air or under water. Furthermore, the fact that ubiquitous environmental contaminants can have dramatic effects on wetting makes the situation even more difficult.

The failure to quantify the surfaces of solids shows that there should be fundamental drawbacks in our approaches. And there is, indeed, a paradox: apparently, notions of chemical thermodynamics are difficult to apply here; still they should stay valid. To resolve the mystery of contact angle uncertainty, we need to analyse more carefully the basic concepts behind the thermodynamics of wetting.

2. THERMODYNAMIC ASPECT OF 'WETTABILITY'

The Young equation is fundamentally important in the theories of wetting. In its common form,

$$\gamma_{SV} - \gamma_{SL} = \gamma_{LV} \cos \theta, \tag{1}$$

it defines the contact angle (θ) of a liquid (L) on a solid (S) in a vapor (V) or some other fluid in terms of the corresponding interfacial tensions (γ).

While the mechanical interpretation of the tension vector balance is quite simple, one can readily notice a fundamental drawback of this straightforward derivation. Indeed, the vectors are balanced only in the plane of the solid, while the normal component of the surface tension of the liquid is not counteracted by the other two tensions.

This is in great contrast to the equilibrium of a liquid droplet of oil on water. For a three-fluid phase system (two immiscible liquids + gas or a third liquid), the sum of the three tension vectors is inevitably set to zero. The condition of equilibrium between the interfaces is fulfilled not just in the lateral plane, but also in the normal direction. By this mutual balance of forces that act along the surfaces on the line where they meet each other, the tension effect does not cause any stress in the surrounding bulk phases. The values of the chemical potential are maintained constant for all components throughout the entire system including bulk phases and interfaces. Tensor stress effects in the form of surface and line tensions associated with interfaces and the three-phase line over which the interfaces intersect balance each other. They do not violate the condition of chemical equilibrium. Both mechanical equilibrium and chemical equilibrium conditions are satisfied here.

Typically, no contact angle hysteresis occurs for a droplet of oil that floats on the surface of water. The contact angle values are uniquely defined in terms of known values of the three surface tensions that act at the line. The tensions can be conveniently measured in this case. The changes in contact angles induced by solutes follow from concentration dependences of the surface tensions which, in turn, can be related to the adsorption.

The equilibrium shape of the droplet follows from the free energy minimization procedure, $\Sigma \gamma_s = \min$. This condition is consistent with the condition for chemical equilibrium. By the mere notion of Pascal's law, there are no local stresses in the bulk liquid phases. Such a stress that might violate the chemical equilibrium does not occur here, even in the vicinity of the three-phase line.

Of course, at short lifetimes of the interfaces, the contact angles of surfactant solutions can change over time. This is because dynamic values of the surface tensions deviate from their static values; the ultimate balance is reached once the interfaces are in adsorption equilibrium with the bulk phases. Before this, the conditions for mechanical equilibrium and for chemical equilibrium remain separated. The mechanical principle of vector balance holds at any instant of time, even when the system has not yet reached the state of chemical equilibrium. The latter implies adsorption equilibrium between the surface and the solution.

In transient states when the values of adsorption continue to change over time, current values of the contact angles are related to the acting values of the dynamic surface tensions. The associated time dependences of contact angles follow from the corresponding time dependences of the surface tensions. The dynamic surface tensions can, in turn, be conveniently explained by kinetic factors that control the rates of the solute adsorption. The chemical equilibration is fast for pure (mutually equilibrated) liquids, but can be much slower when surfactant solutions are involved. In the latter situation, the rate of the tension/adsorption equilibration can be controlled by diffusion of the solute to the interfaces. This is a slow process at low bulk concentrations. At the time when adsorption equilibrium is reached, both chemical equilibrium and mechanical equilibrium conditions become satisfied.

The energy minimization procedure used to account for the droplet (or a meniscus) geometry substitutes for the mechanical condition of pressure balance; the Laplace equation that accounts for the capillary pressure can be conveniently derived by this principle of minimization of surface free energy. Whichever way the mechanical equilibrium of liquid–fluid interfaces is envisaged is a matter of mathematical convenience. The final result is unique, independent of the method used, and can be tested experimentally. All parameters that occur in the equations can be measured. The principle explains why in the simplest situation of the absence of external forces, liquid droplets are spherical, or why an oil droplet on the surface of water occurs in the form of a biconvex lens.

The same principle applies to the surfaces of solids. However, the condition for mechanical and chemical equilibrium becomes more involved in this situation. There is no universal geometrical solution like the one that demands a droplet to be a sphere. For each particular substance, the equilibrium crystal has its own geometry. It is determined by a compromise between the restrictions imposed by the long-range crystalline order and the requirement of the minimum total surface free energy.

The simplest situation is that of a crystal embedded in an isotropic space of a fluid with which it occurs in dynamic molecular exchange by dissolution (evaporation) and condensation. The situation becomes more involved when the crystal is floated at the liquid–fluid interface, or when a droplet of a liquid attaches to one of the crystal facets. These are precisely the situations when wetting of the solid by the liquid becomes important, but it cannot be dealt with by the Young equation!

And, indeed, the Young equation conflicts with this condition for the free energy balance in a very severe way. By assuming that the three-phase line resides on a plane, it requires the droplet to attain the shape of a monoconvex lens (the droplet geometry may be more complicated when the surface is non-planar). This shape of a spherical segment is what is indeed observed when a liquid droplet is placed on a flat surface. The Young equation just acknowledges this fact; it does not allow this geometry to be changed. Still, the geometry does not conform to the condition for the free energy minimum. The droplet is in mechanical equilibrium, but chemical equilibrium cannot be reached as long as the conditions of the Young equation apply.

This is an interesting situation. When a liquid–fluid interface adjoins the planar surface of a solid at an angle, a normal stress is inevitably induced within the solid body in the vicinity of the three-phase line. The normal component of the surface tension of the liquid is balanced not by the other two tensions (as would be the case for a droplet on a liquid substrate), but by the elastic response of the solid.

By this tensile stress, the chemical potential of the solid increases locally above the co-existence value. The material begins to dissolve from the three-phase line in order to release this stress. It then recondenses at other parts of the body. To restore chemical equilibrium, the entire geometry of the substrate has to alter. Only by a macroscopic rearrangement of the crystal into a new shape that would conform with the changed thermodynamic condition of the minimum surface energy might the mechanical condition of tension balance also be satisfied.

This is a different situation from that described by the Young equation. The condition of thermodynamic equilibrium requires the three tensions to act in three different planes, the same way as it occurs for the wetting of liquid substrates. But for the wetting of solids, this equilibrium geometry becomes more complex. The equilibrium shape of a crystal might depend on its own volume and the amount of the liquid attached to it. This configuration inevitably changes by going from one liquid to another, or from the pure solvent to a surfactant solution. This is simply because the values of the tensions here are changed. For amorphous solids such as silica glasses and other vitreous materials, this geometry would not be different from that of a liquid on another liquid substrate. Because of the long-range crystallographic order and associated tension anisotropy, different facets of one and the same crystal have different tensions, and the tensions change differently with a change of the environment. This imposes a new condition that should be further taken into account.

The situation becomes rather involved. The principle by which this reconstruction occurs is basically the same as that for a three-fluid system. But mathematical solutions that would define the ultimate equilibrium geometry become very complex once one of the phases is in the crystalline state. The simplest situation is for a crystalline cluster in a uniform medium (solution or vapor) with which it is maintained in chemical equilibrium. The Pascal–Laplace condition for the capillary equilibrium of fluid phases here substitutes for the well-known Wolf theorem. It requires the distances of the facets to the center of the crystal to be in proportion to the values of their surface free energies. Already here a search for the equilibrium crystal geometry (which in principle is uniquely defined) is complicated by a multiplicity of planes that might occur at the surface with different surface energies. The equilibrium shape of a monocrystal changes when it floats at a liquid–fluid interface. Formal mathematical solutions for the new geometry follow from equations of phenomenological capillarity. These, however, become even more complicated. Among the phenomena in which the solid phase becomes explicitly involved in chemical equilibria of capillary phases is the effect of Ostwald ripening, the simplest and well-known example.

We do not try to extend here the question of the ultimate configuration of the solid substrate after chemical equilibrium is reached. These questions are largely beyond the scope of our present review and we simply do not know the answers. Obviously the Young equation is of little help here. This is first of all for the mere reason that the three-phase line stays at an edge between two crystalline planes when the solid is in chemical equilibrium with the other two phases. And for this situation the contact angle is uncertain, the way that it is defined by the Young equation (cf. Section 4). We continue to base our analysis of contact angles on the Young approximation, in which chemical activity of the solid phase is effectively disregarded.

We have now reached an important point that is essential for our argument: the Young equation is based on a *quasi-thermodynamic* approximation. It is valid as long as the reconstruction effects in the manner of Ostwald ripening under the effect of the stress that occurs at the three-phase line are not allowed to develop. Indeed, the processes of recrystallization that might allow the solid to undergo an evolution towards a new geometry that would accord with the condition for the absolute free energy minimum are typically very slow for insoluble and involatile solid materials.

Under ordinary conditions of contact angle measurements, changes in the substrate geometry required by the condition of absolute free energy minimum do not occur on the experimental time scales, or at least they do not develop macroscopically. It might take much longer times before this inevitable chemical evolution becomes obvious. Even for soluble materials in equilibrium with the solution such effects might be difficult to observe.

In the quasi-thermodynamic approximation, the solid substrate is not allowed to change macroscopically over time. We then impose a stricter condition that does not allow molecules or atoms that constitute the solid to leave the body or join with it during the observation time. This is a quite common assumption that occurs implicitly or explicitly in connection with contact angle measurements. Then local deformations of the substrate induced by the unbalanced interfacial stress may remain purely reversible (elastic), unless fracture damage caused by the effect of the three-phase line stress occurs. It should be recalled in this connection that by the mere nature of the surface tension effect the interfacial stress is indeed very large. It stays on the ultimate spinodal value of the critical tensile stress above which the condensed phase becomes unstable and inevitably collapses. At the interface such a stress occurs within a layer effectively one molecule thick.

However, if the surface is not just physically (macroscopically or atomically) but mathematically uniform, not only is the system mechanically reversible, but there should also be no energy barriers in the way of the liquid front. By the principle of virtual work, there should be no change in the free energy by a displacement of the droplet along such an ideal substrate. The simple mechanical model behind the Young equation remains valid. The value of the contact angle is unique and there should be no contact angle hysteresis.

One can notice, however, that even for this ideal approximation the contact angle does not occur as a function-of-state of the system as would be the case for the wetting of liquid substrates. This is not just because ordinary methods of measurement of surface tension of liquids do not apply for solid surfaces. The surface free energy of a crystalline facet is a physically meaningful quantity. It can be measured under appropriate conditions. The relevant experimental methods can be based on the Wolf theorem that enables us to define relative values of γ_i once the equilibrium crystal configuration has been established, and on the Kelvin equation that would define the absolute values of the tensions via the dependence of solubility on the crystallite size.

This consideration, however, does not yet show that these are the relevant values that are to be used in conjunction with the Young equation. Moreover, tensions that occur in this equation strictly speaking cannot be uniquely defined in terms of the properties of the solid bulk! This is in contrast with the liquid–fluid situation, for which the interfacial tension occurs as a function of state of the surrounding bulk phases. And this is not just because of the scarcity of relevant experimental techniques, but rather is a matter of principle.

Indeed, as soon as the environment changes, the equilibrium surface structure might change in response, in order to conform to the new condition for the (surface) free energy minimum. This is a well-known effect in the realm of liquid–fluid interfaces. For example, at the n-octanol–vapor interface the amphiphile molecules remain with their hydrocarbon chains oriented towards air, but turn the other way round with their hydroxyls outwards to contact water. Like the liquid n-octanol, some solid substances such as stearic acid when crystallized in air are hydrophobic. However, if the crystals were precipitated under water, they would remain hydrophilic even after they were dried.

For a liquid, such a reorientation towards the new equilibrium occurs almost instantly. The transition occurs each time the environment changes. But for solids, the associated hydrophobic–hydrophilic transitions occur much more slowly. Even for a material that possesses a finite solubility in water, such as stearic acid, it might take hours of conditioning of the originally hydrophobic surface under water for hydrophilization to occur. The surface can be maintained in a nonequilibrium state over a period of time that extends far beyond the typical duration of contact angle measurements.

The situation is further involved. As we have already noted, the condition for chemical equilibrium would require not just a change in the surface structure, but a change in the macroscopic geometry of the entire crystal. Placement of a droplet or attachment of a crystal to the surface of a liquid triggers recrystallization processes. There should be a new configuration of the crystal which meets the condition for the tension balance. New facets of the crystalline material can be exposed to the environment in this recrystallization process. And this new crystallite geometry has to conform to the $\Sigma(\gamma_i \cdot s_i) = \min$ condition that changes by a change in the environment. Only for a particular case when all the tensions change in the same proportion and no preferable planes with a lower surface free energy emerge can the crystal shape be preserved in the different media.

Unlike liquid interfaces that reshape rapidly by a mechanism of overflow, for solids, changes in shape required by capillary effects can occur only by dissolution and subsequent recondensation. These processes are typically very slow. We have already commented on this. Once the duration of the contact angle measurement is short on the scale of rates and times of the recrystallization events, macroscopic changes to the solid body by contact with the liquid do not occur.

This is the essence of the quasi-thermodynamic approximation that applies to the Young equation. Once the component that constitutes the solid phase does not participate in the dynamic molecular exchange throughout the system, its chemical potential, even if it changes from point to point within the body, can be disregarded in thermodynamic equations that describe chemical equilibria.

Once the condition of chemical equilibrium extends to the solid phase as well, the solid–vapor and solid–liquid interfaces inevitably meet each other at an angle. Only in this new physical geometry does the three-phase line contact stress, which is the driving force for the process of recrystallization, vanish. However, the resulting equilibrium system cannot be treated on the basis of the Young equation. The latter assumes the two interfaces to occur in one plane.

One cannot resolve this thermodynamic paradox within the Young approximation. It can further be noted that contact angles can be measured at any crystallographic facets, including those that might not occur in a crystal of equilibrium geometry in a given environment. Depending on the environment, the surfaces that conform to the condition of thermodynamic equilibrium can be different. The Young equation remains self-consistent only as long as the associated nonequilibrium geometry, for which the solid–vapor and solid–liquid interfaces reside in one plane, is maintained. But given this, the thermodynamic meaning of the two tensions that occur in the equation with reference to the solid phase (γ_{SV} and γ_{SL}) remains quite uncertain.

This consideration has far-reaching consequences. Polycrystalline, 'rough', and microheterogeneous surfaces are equal in this quasi-thermodynamic context to single crystalline facets. All the (quasi) thermodynamic definitions continue to apply to such 'imperfect' substrates, as long as macroscopic averaging along the plain can be done. Of course, for practical reasons it often might be preferable to deal with smooth and uniform surfaces. By the use of such 'ideal' substrates one might hope to diminish the specific problems associated with surface fractality of real solid materials. This might help to reduce the uncertainties associated with contact angle hysteresis or definitions of 'actual' and 'apparent' areas of contact between the two phases. The latter consideration might be important when one would like to compare the result with that for a smoother surface. A similar consideration remains true, of course, about the lengths of the three-phase line. This perimeter is, in fact, a more relevant parameter in the wetting context than the area of the liquid–solid contact; the latter does not make any explicit manifestation in the magnitude of the contact angle.

This argument is very universal. It shows why so many attempts to deduce values of the surface free energy of solids from contact angle data have failed. They are erroneous in their essence. The entire concept of the surface energy of a solid that is not in chemical equilibrium with the environment was based on confusion. It is essential that a distinction should be made between these two fundamental limits. In one limit, absolute thermodynamic equilibrium, the chemical potential of the solid phase occurs explicitly in the free energy minimization procedure. For such chemically active solids that exchange components with the environment by dissolution and/or vaporization, the surface energy can be uniquely defined. In the quasi-thermodynamic approximation of the Young equation, this is not true.

The contradiction disappears once one recalls the well-known fact that while the values of surface tensions of a solid may be unknown and difficult to define, they occur in the Young equation in the form of their difference. By this mere arithmetical argument their absolute values can already be ignored. Whether physically meaningful or not, they are irrelevant in the wetting context. The difference can be defined as a single physical quantity. Often known as the wetting tension, this force, which drives the spreading of a liquid along the surface of a solid, also occurs in the literature under other names: adhesion tension, spreading coefficient, etc.

The wetting tension can be defined in alternative ways that avoid any references to absolute values of the surface energies of the solid. By the effect of the Dupre equation, the wetting tension is placed in correspondence with the values of the work of cohesion of the liquid and the work of its adhesion to the solid. While the Young equation and the Dupre equation follow from each other by an elementary rearrangement, this is not just a tautology. The work of adhesion can conveniently be defined in terms of surface forces that arise by separation of the surfaces in the normal direction.

An alternative definition is based on the Gibbs equation. Here the wetting tension is related not to the surface forces that act along the normal to the surfaces, but to the values of tangential surface pressures. The latter act in the lateral direction in the plane of the solid. Applied to the two sides of the three-phase line, they push it to spread or to recede. The restoring force is the surface tension of the liquid. The balance between the surface pressures and the surface tension gives the condition of mechanical equilibrium.

Of these two alternative definitions of the wetting tension, the first was advocated by Adam, and the second by Frumkin and Derjaguin. Both interrelated definitions show that the wetting tension is a physically meaningful quantity. It can be unambiguously defined in (quasi) thermodynamic terms. No reference to the absolute values of the surface energies of the solid is required for these definitions. These definitions follow in terms of parameters that can be measured experimentally and calculated theoretically.

The surface pressure consideration is in some respects more general and more practical than the one that relates the contact angle to surface forces. Indeed, the disjoining pressure is a macroscopic concept that limits the consideration to thick liquid films. Normal forces might still be difficult to measure. On the other hand, surface pressure in its relation to the wetting tension follows from the adsorption isotherms. These can also be directly measured in the limit of zero film thicknesses that would include rarefied monolayers.

This, in fact, is very important because the co-existence values of adsorption in equilibrium with the liquid are low in many practical cases. We will use this surface pressure argument in many following places of this review. The surface force interpretation is a somewhat more abstract consideration. It might be more convenient to extend it in a theoretical rather than in an experimental aspect. However, we notice that the surface force approximation follows as a limiting macroscopic case of the surface pressure consideration. This way, by use of the Gibbs adsorption isotherm, we derive Lifshitz and DLVO theories and gain new insights into the origin of the long-range hydrophobic attraction (cf. also ref. [1] and Section 7 of this review).

By considering this situation further, it becomes almost obvious that while contact angles depend on the *state* of the surface, the surface state does not necessarily occur in a unique relation to the state of the solid bulk. The bulk properties, as we have already noticed, are largely irrelevant. This shows why a definition of the surface energy, or tension, which in the context of liquid–fluid interfaces occurs as the most basic property, as well as for solids in chemical exchange with the environment, becomes quite meaningless and unnecessary in the quasi-thermodynamic context of the Young equation.

The state of the surface of a solid cannot be derived from its bulk composition once the surface molecules are not in chemical exchange with the solid bulk and the environment. Only for a given surface state can the equation-of-state consistent with the Young equation be defined. It then follows in terms of molecular composition or other thermodynamic properties of the surrounding media — the liquid, or a solution, and the vapor or another liquid. Whatever the assumed absolute values of γ might be, they change when the state of the surface changes, for example, by chemical modification, etching, polishing, etc. There can be numerous variations of such procedures. They do not affect the bulk properties of the solid material but change its surface energy by changing the surface structure and composition.

For example, in the presence of chemisorbed species the wetting indeed changes dramatically. A hydrophilic solid can be converted into a hydrophobic one by binding just one molecular layer of alkyl chains on top of it. These procedures such as a silane treatment of glass or quartz by grafting it with various organic end-groups and even polymers are well known. These almost trivial considerations are still often ignored in a temptation to use contact angle data as a panacea for surface energy estimates.

As we have already established, the absolute values of the tensions that occur in the Young equation are irrelevant in the contact angle context. One might not be able to consider them separately from each other. What is more important is that the meaningful value to be used in conjunction with the equation is the tension difference. This is the difference between the two values of the surface energy of a solid in two different environments. Or, what essentially is the same, the difference between the two adsorption pressures.

Still it might be quite practical to definite a suitable reference state. This can be conveniently taken for the surface in vacuum, or in air, whatever the absolute value of the surface free energy might be. With respect to this level, a quasi-thermodynamic scale of absolute surface energies can be defined. Of course, this level can be ascribed with an absolute value of γ. This can be done arbitrarily or with a more careful consideration for such a nonequilibrium state. The consideration for a physically meaningful absolute value of the surface free energy can be done either theoretically or by some experimental means (based, for example, on the heats of dissolution of a powder compared to the macrocrystal which is an alternative calorimetric way to measure the interfacial energy). But it is important to stress again that such a definition has to be done for each particular surface state and only in that situation does it apply. Without changing the state of the solid in the bulk we can change its surface; this situation never occurs for liquids.

This logical construction shows that from the conceptual point of view, such absolute values of γ (often referred to as critical surface tension of wetting, acid–base components, surface equations-of-state, etc.) are not required. All the parameters of the Young equation that are essential for the contact angle calculation are already defined in a more consistent way. This concerns any quasi-thermodynamic theory of wetting, i.e. a theory that takes its origin from the Young equation.

These 'absolute tensions' can always be conveniently substituted for physically meaningful quantities of the work of adhesion, or the equilibrium wetting tension, etc. The notion of wetting tension then extends far beyond the condition for chemical equilibrium or even mechanical equilibrium of contact angles. As the driving force for spreading, it determines the rates of droplet expansion in fast dynamic regimes for which inertial acceleration terms dominate.

In quasi-static regimes of steadier spreading, the process is controlled by mechanisms of molecular migration rather than by inertia or viscosity. By substituting the value of the wetting tension (τ) for the tension difference in the Young equation, the mechanical status of the equation becomes ultimately recovered. We now proceed with an outline of these well-known definitions in consideration for equilibrium and kinetic hydrophobic phenomena.

3. SURFACE PRESSURE, SURFACE FORCES AND ADHESION

The wetting tension can conveniently be defined by the Gibbs adsorption equation in terms of values of the surface (adsorption) pressure: $\pi = \int \Gamma \, d\mu$. Here Γ is the adsorption at the surface and $d\mu \approx kT \, d\ln C$ is the change in the chemical potential. A sum can be conveniently taken over all chemical species involved in the adsorption process. This summation is already important for ionic surfactants and other solutes that dissociate in water. The simplest case, however, that illustrates the general principles is just adsorption of neutral molecules of the vapor of the pure liquid whose contact angle is being studied. Like for adsorption from solution, for vapor adsorption on a liquid substrate the surface pressure equals the change in the surface tension: $\pi = \gamma^{\circ} - \gamma$; here γ° is the surface tension of the pure liquid (the substrate) taken at zero concentration (zero partial pressure) of the second component (the vapor of the other liquid).

As the partial vapor pressure increases, the adsorbed layer builds up. At saturation the film is in equilibrium with the bulk phase of the liquid. The latter can be treated as an infinitely thick layer of the liquid on top of the substrate. Such a film that contains two independent physical boundaries can still be considered as the substrate–vapor interface. The tension of a thick film is then the sum of the values of the surface tension of the substrate–liquid and liquid–vapor interfaces.

Whether the substrate is a liquid or a solid, the result remains the same. In particular, the Young equation takes the form

$$\gamma(\cos\theta - 1) = \tau(\mu) - \gamma = \pi_{\infty} - \pi(\Gamma) = E(\Gamma) + \Gamma\Delta\mu. \qquad (2)$$

The above set of the Young and Gibbs equations solved together is often referred to as the Frumkin–Derjaguin equation. These equations have simple and clear physical meanings. They are stated in terms of parameters that are experimentally accessible not only for liquid but also for solid substrates. These parameters are the adsorption pressure or the energy of interaction between the film surfaces, chemical potential, and adsorption. The wetting tension and the surface tension of the liquid are expressed in terms of these parameters. No reference to absolute values of the surface tension of the solid is made.

The second line of this equation tells us that the contact angle is finite (τ is smaller than γ, the first equality) when the surface pressure of the adsorbed vapor of the liquid exceeds the surface pressure of the thick liquid film (the second equality). The same result restated in terms of disjoining pressures (the third equality) shows that this occurs when the film boundaries attract each other to such an extent that the energy of interaction between the solid–liquid and liquid–vapor interfaces attains negative values over a range of film thicknesses. At saturation ($\Delta\mu = 0$), the second right-hand term vanishes. Here the interaction potential has a minimum. This can be the absolute minimum, or a local minimum as well (as occurs, for example, for a metastable symmetric film).

The molecular forces responsible for this attraction can be different. Repulsive interactions might occur between the surfaces over other ranges of film thicknesses. However, the gross energy of this attraction at the point when the film thickness reduces to that of the adsorbed layer in equilibrium with the vapor follows from measurement of the contact angle. The contact angle is a measure of the interaction energy over the range of film instability. Here the disjoining pressure cannot be measured in a simple way. The surface pressure result might be difficult to obtain via adsorption measurements. The contact angle measurement is nevertheless straightforward for the macroscopic droplets that occur at saturation.

The contact angle formed at macroscopic co-existence shows in its value just the depth of the potential well. By variation of μ, the energy vs. distance (adsorption) profile can be obtained. This can be conveniently done by monitoring the adsorption over the stability range. The latter extends from zero adsorption at zero concentration (vapor pressure) up to the point of supersaturation for which spinodal instability occurs. Over a range of the supersaturation before the spinodal condition is reached, the surface pressure/interaction energy result can be verified in two independent ways: by macroscopic contact angle and by adsorption measurements; both of these complementary measurements can be carried out simultaneously and should agree with each other.

The contact angle measurement still remains important. The value of the contact angle carries important information that might be difficult to obtain on the basis of adsorption measurements alone. As we have seen, neither contact angle nor adsorption measurements can be conveniently done over this range of very high supersaturations at which the film is labile. The gross integral of the energy variation over the spinodal interval nevertheless can be measured: this value is shown in the magnitude of the macroscopic contact angle.

This is a quite appealing theory. It shows that adsorption and contact angle measurements are just alternative ways of surface force measurements. The latter are currently carried out by the use of the more common SFA and AFM techniques. Like the measurement of pull-off forces does not require measurement of distances, the free energy of adhesion in a liquid film follows from the contact angle measurement alone and does not require measurements of the adsorption.

The fact that for a film of water on a hydrocarbon the energy of attraction between the water–air and water–oil interfaces is several orders of magnitude larger than 'predictions' of the Lifshitz theory is almost obvious here. And it is also obvious that the attraction is not very long range. For more details regarding the relation of contact angles to long- and short-range DLVO and non-DLVO forces in the realm of Gibbs adsorption equation we refer the reader to a more extended review [1]. Here we just note that of equation (2) the last equality makes use of a macroscopic approximation. It treats the material of the film as an incompressible condensed continuum (the film itself, of course, occurs as a 'discontinuum' between the two bulk phases). In this approximation, Γ/ρ can be interpreted as the effective thickness of the film of an average density ρ, which typically is assumed to be equal to the bulk density of the liquid (a convenient but not necessary assumption). Then $\rho\Delta\mu$ occurs as the force of interaction between the film boundaries ('disjoining pressure'). By the mere notion of the incompressibility approximation ($\rho = $ const.) of the Kelvin equation the measurement of the force reduces simply to the control of vapor pressure.

Like the Kelvin equation itself, the definition $dE = \Pi\,dh$ of the energy of interaction between the film boundaries follows in the form of the general definition of the interaction potential in the continuum approximation of the Gibbs adsorption equation. The description in terms of adsorption pressures via the Gibbs adsorption equation is more general. It does not suffer the macroscopic limitation that restricts the disjoining pressure result to condensed films. The surface pressure result is exact and quite universal. It applies not just to condensed monolayers and multilayers on the transition to thick liquid films, but also to submonolayers and rarefied gas-like adsorbed layers.

The definition of contact angles, surface pressures, and surface forces in terms of adsorption isotherms can be readily generalized for multicomponent media. In particular, the theory of the electrical double-layer interaction across liquid films of aqueous electrolytes and solutions of ionic surfactants follows in the limit of large film thicknesses in the electrochemical Lippmann approximation of the Gibbs adsorption equation in the form of the DLVO result [2]. Any 'non-DLVO' surface forces (hydrophobic, hydration, steric, undulation, etc.), whenever these might be observed in reversible systems satisfying the condition of adsorption equilibrium, can be interpreted thermodynamically in this simple way [1].

By consideration of these basic equations, many other results can be explained. One can notice, for example, that compared with contact angles at macroscopic co-existence, the contact angles are smaller for capillary condensates that form in the slits at undersaturation and are larger for microdroplets that nucleate at free interfaces at supersaturation. This is because π_{SV} increases with increasing chemical potential. The conclusion is valid as long as the surface tension of the condensed liquid phase at both interfaces with the substrate and the vapor shows a much weaker dependence on the chemical potential (hydrostatic pressure) than the surface pressure of the adsorbed vapor. While this common assumption behind the Kelvin equation for a liquid–vapor interface might be more difficult to verify, for a liquid–liquid interface the pressure dependence of the surface tension can be studied in a macroscopic experiment.

When the size of a droplet or a capillary condensate reduces·below the range of surface forces, one is no longer allowed to use the macroscopic value of the surface tension in the Young–Gibbs equation. The corresponding corrections often occur in the form of 'line tension' in application to contact angles and 'curvature dependences of surface tension' in conjunction with the Kelvin equation. These corrections are rather insignificant for molecular clusters and aggregates that do not contain a very large number of molecules. They are small even compared with the effects of changed vapor adsorption on contact angles of small droplets and capillary condensates. The latter are not large when the system is not very far away from the macroscopic co-existence, so that the contact angle can still be defined. And even for a single molecule the surface tension result holds by at least an order of magnitude. The correlations of the surface tension and the standard part of the chemical potential via the molecular volume are indeed quite accurate.

In fact, these correction problems do not arise when the interaction is dealt with by the Gibbs equation. The Gibbs equation does not suffer conceptual limitations that follow from macroscopic definition of such quantities as the 'disjoining pressure'. Once rigorous treatment of curved surfaces in the same way as it applies to planar interfaces is achieved, the result substitutes for line tension and curvature-dependent surface tension effects. Indeed, exact results based on the Gibbs equation are valid in the entire range of distances from infinitely thick films to zero values of adsorption. The Kelvin, Laplace, and Young equations to which such corrections apply follow as the limiting macroscopic forms of the Gibbs adsorption equation. While the latter holds universally, the former are approximations that break down beyond the limit of the macroscopic postulate. This explains why such corrections are not needed once the equation is used in its general form.

The line tension effects and $\gamma(r)$ dependences are usually insignificant for most practical purposes. Even much larger chemical potential effects of vapor adsorption that arise on going down or up in chemical potential away from the macroscopic co-existence might be difficult to observe. Indeed, for a monocomponent liquid a significant change in the adsorption pressure implies a significant change in the vapor pressure. And according to the Kelvin equation this occurs only when the sizes of droplets/condensates are very small. Later we will see that for solutions these Gibbs effects arise at macroscopic co-existence. By adding a new variable (concentration), the dependent parameter (adsorption) can be changed. This changes the surface pressures. The resulting large changes in the contact angle are easy to observe in ordinary experiments by the use of macroscopic droplets, Wilhelmy plates, capillary rise and other traditional techniques.

Typically, however, the largest uncertainty that dominates the contact angle effect is associated with contact angle hysteresis. On a very general level, the hysteresis can be understood as a *microfriction* phenomenon, by an extension of the most fundamental property of solids — the ability to withstand static stress. For an ideal (defect-free) solid, the shear yield stress is very large. It typically occurs on the order of the modulus of shear elasticity. This is essentially the same order of magnitude as the tensile yield stress or the Young modulus, as well as the lateral spinodal stress responsible for the surface tension of liquids. While the latter effect is concentrated in a layer one molecule thick, in a critically strained bulk material the stress occurs over the entire volume of the body. The ability of solids to withstand static stress is basically the reason why the principle of virtual work, which enables the surface tension of liquids to be measured without invoking any changes in the state of the bulk material, does not apply to solids. And the same consideration explains the phenomenon of contact angle hysteresis as well.

Indeed, the shear stress associated with the resistance to lateral displacement of two parts of the body for simple liquids takes the form of Newtonian viscosity. The viscous resistance vanishes in the limit of zero shear rates. That is why there is no hysteresis associated with measurement of the surface tension of liquids. The tangential force acts under such quasi-static conditions only in the plane of the surface. It equals the work of cohesion, 2γ [or adhesion, $2(\gamma + \tau)$, if two liquids are involved], when two new surfaces are formed in the shear separation process by which a body of unit square cross-sectional area becomes disconnected. Or the force equals zero when one of the adhering bodies just slides on the top of the other without passing over the edge. In the latter situation, for example, of a floated droplet or a crystal, the surface area does not change in the process. When the disconnection occurs in the lateral direction, the work of adhesion (or cohesion) released on a distance of unit length equals numerically the force per unit length (therefore the two ways of presentating the one and same unit of the surface tension, for example $mN/m = mJ/m^2$).

However, if the body (a solid or a liquid) is disconnected in the normal direction, the same amount of work is released not on a distance of a unit of length (say 1 cm), but effectively just on a characteristic length of about one atomic size. While the same amount of energy is released in both modes after the process of separation is complete, the maximum force required to achieve the separation in the normal mode is ten billion times larger than the cohesional or adhesional tension. In other words, the effective distance over which this 'disjoining pressure' acts is billions of times smaller, while the normal yield stress needed to perform the same amount of work is billions of times larger than the shear yield. And while the latter is constant, the former changes with distance.

During the normal separation, all interatomic bonds rupture simultaneously, whether the body is a solid or a liquid. This force can be estimated by taking an atomic size (a few Å) for a characteristic length in the Hamaker–Lifshitz formula. Alternatively, this can be done simply by multiplication of the bond strength by the number of the bonds across a unit cross-section. The latter way, which ignores long-range contributions from atoms and molecules outside the first coordination sphere, is even more accurate.

For solids, such a cooperative rearrangement of all intermolecular bonds across the plane of disconnection inevitably occurs not only by normal separation, but also by activating a lateral displacement. The amount of energy that has to be supplied externally in order to overcome the barrier in the way of tangential translation, by which all the atoms have to be moved simultaneously from their equilibrium positions in the crystalline lattice to the next such position, would correspond to a displacement of the body just by a single crystallographic unit (atomic or molecular period of the lattice, of the order of 1 nm) and is of the order of the work of cohesion. Here, however, this amount of work has to be performed not on the unit of length, but over a short distance of the order of molecular size. Because of this, the force (shear yield) is again ten billion times larger than the corresponding adhesional or cohesional tension.

Only for liquids does the static shear yield stress equal zero. No energy barriers are encountered in this case by a low velocity gradient shear. Energy barriers towards mutual displacement of molecules occur also in a liquid. They are imposed by the short-range coordination in the liquid structure. But they are overcome not cooperatively under the effect of external force, but spontaneously one by one, assisted by the heat motion of the molecules. For solids, on the other hand, the shear yield is as large as the normal yield. Only after this low-energy, but huge force, barrier is overcome do the atoms move to their next equilibrium positions. Solids, unlike liquids, display static resistance to shear. And it is this property of the solid state of matter that, when translated to the three-phase line, explains the phenomenon of contact angle hysteresis.

After the elementary translation by a unit atomic or molecular length is complete, all disconnected bonds are re-established, except for the two single lines of atoms that emerged to the two sides of the body in the direction of shear after the process is complete. This molecular distance that separates the two equilibrium positions is the length by which the two parts of the body are shifted. The energy of the new state of mechanical equilibrium differs from the energy of the previous state by the energy of the broken bonds on this newly formed surface area. And this area is small, about ten billion times smaller than the unit area. The change in the energy that corresponds to the adhesional or cohesional tension is ten billion times smaller than the energy barrier in the way of the process.

This explains why shear fracture of a solid is essentially irreversible. There is no way by which the reversible cohesional or adhesional shear component could be measured. A real solid cracks or deforms plastically before it begins to shear. The surface tension effect that clearly is too small simply cannot be noticed on the background of the bulk stress involved.

Interestingly, this consideration does not apply to contacts between two dissimilar solids. Theoretically, even when the mismatched surfaces of one and the same crystal are contacted, and even given that the contact occurs over the entire macroscopic area, the two parts of the body slide laterally with zero friction coefficient. What typically makes friction occur is just a layer of adsorbed molecules in between. Such layers of water and other environmental components are almost inevitably present on most solid surfaces. The adsorption can be further enhanced in the manner of capillary condensation effects when the surfaces are in contact with each other. It is these adsorbed molecules that help to link the mismatched surface groups and atoms by bonding them together. Then the frictional interaction dismissed by the theory re-appears; and it is almost as strong as the strength of a consolidated body.

In a friction pair of two solids, the yield stress that helps to overcome the 'dry' friction is typically several orders of magnitude lower than the fracture yield stress of the materials involved. This is because the intimate contact of rough solid surfaces occurs only at microscopic asperities. The area of such contacts is just a minor part of the overall surface area. The latter for rough surfaces is even larger than the unit area of the apparent macroscopic contact. The frictional resistance that increases with increasing normal load is, on the other hand, several orders of magnitude lower than the value expected for ideal flat-to-flat contact. But at the same time it typically remains several orders of magnitude higher than the adhesional tension. The complexity of the dry friction phenomenon follows from both the geometrical irregularity of the surfaces and the occurrence of irreversible fracture deformations under normal load and shear that interfere with adhesion effects at the microscopic areas of contact.

Only for surfaces of liquids the tangential stress is maintained at a constant value. Such surfaces are mechanically reversible and the condition of thermodynamical equilibrium is fulfilled. The values of the cohesional or adhesional tension are uniquely defined. The only resistance here occurs by viscosity and it can be neglected at low speeds.

These considerations help us to explain the phenomenon of contact angle hysteresis. They show that, in principle, contact angle hysteresis can occur also on atomically flat and uniform substrates. Only if the substrate were mathematically uniform might there be no contact angle hysteresis. And this condition is indeed effectively fulfilled for surfaces of low viscosity liquids. Here random instant positions of the molecules are effectively averaged already over a short period of time. On larger time scales for which dynamic surface tension effects can be neglected, displacements of the three-phase line proceed under equilibrium conditions.

However, for a droplet on a solid substrate (Fig. 1, bottom) there is a static resistance to shear. It occurs not over the entire solid–liquid interface, but only at the three-phase line. But here it has the same stress magnitude as for the fracture stress at ideal solid–solid contacts. Otherwise there is no surface stress at other areas of the solid–liquid contact. This might sound like a paradox. Indeed, even for a rough substrate the solid–liquid contact occurs over the entire area. It is even larger than the apparent macroscopic area (a reduction in the actual area of the solid–liquid contact can be associated with the formation of cavitational Kelvin vapor pockets in microvoids of hydrophobic surfaces, but usually this is a minor effect). Here the solid–liquid bonds are essentially the same as those that occur at the three-phase line. Still this area does not contribute to the friction effect that occurs only at the three-phase line.

This paradox is easily resolved once one realizes that the liquid–solid interaction is in fact not involved in the process of overflow of liquids above solid surfaces. A boundary condition of zero shear velocity typically occurs even for liquid–liquid

V. V. Yaminsky

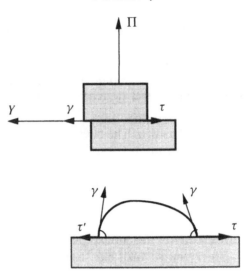

Figure 1. Static yield shear stress associated with dry friction (top) and wetting (bottom) phenomena is illustrated in this schematic drawing. In a plane across a solid body, the critical tangential yield stress is almost as large as the fracture stress of an ideal body separated in the normal direction. These are characteristically billions of times larger than the adhesion tension, in a ratio of the macroscopic body size (taken for the unit length) to a molecular dimension. In a friction pair between mismatched crystalline planes, like for liquid–liquid and solid–liquid contacts, this stress does not occur. However, it rises up to the ultimate high critical level in the presence of adsorbed layers that link the two solid surfaces. For a droplet on a solid substrate, a similar frictional resistance occurs only at the three-phase line. While the line is effectively one molecule wide, the associated wetting tension uncertainty is of the order of equilibrium values of this parameter.

contacts. Only for monocomponent liquids that do not develop surface tension gradients at the interface with the vapor does this condition not apply. Then the top surface layer can be moved at finite speeds. But even given that the strong binding condition does apply to solid–liquid interfaces, this does not prevent the upper layers of the liquid from flowing above the 'stagnant layer' of a gradient velocity. The movement of the liquid over the wetted areas occurs in the absence of static resistance. Interaction in the manner of dry friction occurs only at the three-phase line.

The wetting line where this static resistance to shear occurs is effectively one molecule thick. Its effective area is just a minor fraction of the wetted area, of the order of one ten billionths. This is a small value, even compared with typical values of static friction encountered in friction pairs of solid bodies; here the effective area of solid–solid contacts is much larger. However, static resistance towards lateral displacement experienced by the three-phase line stress is almost as large as the adhesional or the wetting tension.

The mode by which this line moves is by molecular jumps from one equilibrium position to another. The discrete positions are localized in space by the structure of the solid substrate. And like the situation in a solid body, these jumps occur cooperatively in order to avoid large local bends on the interface. Such bending is counteracted by the Laplace pressure effect. On the atomic level, an interface is almost as stiff as a solid. Such molecular scale bends can occur only at the expense of large energy losses associated with an increase of the area of the adjoining interface. For molecular radii of curvature of the bends, the restoring force is as large as the spinodal decompression in the plane of the interface or the yield stress in a critically expanded body.

For liquids far away from the critical point this surface tension stress that causes the interface to contract occurs in a layer effectively one molecule thick. Only as the interface becomes more diffuse at temperatures close to the critical point or on approaching the point of mutual miscibility of the two liquid phases does the tension that prevents the surface from curving become small. And under such conditions the effective width of the interface also becomes much larger than atomic or molecular sizes. The molecular and supermolecular granularity of the substrate becomes a less important consideration. The surface structure is not 'noticed' by the diffuse interface that is much larger in the effective width than the size of a single atom. In such situations, a hysteresis-free behavior reappears. Under the subcritical conditions, however, even the contact angle itself typically reduces to zero. This is simply because the surface tension becomes too low to counteract the wetting tension effect, and unless for the particular case when the wetting tension by coincidence is also set close to zero, complete spreading of one of the two phases occurs (θ equals 0 or 180°).

The details of associated molecular mechanisms responsible for the contact angle hysteresis in different situations are analyzed later in this review. But this general consideration already explains why the wetting tension uncertainty associated with contact angle hysteresis is typically quite large. Though the frictional resistance involved is much lower than for fracture and dry friction phenomena, it is essentially as large as the cohesion or adhesion. The latter are the parameters intended to be measured.

And indeed, this consideration shows that in the contact angle situation the value of the uncertainty occurs on the same order as the entire range of variation of the wetting tension. That this is actually the case is confirmed by the entire practice of contact angle measurements.

This is an interesting situation. The uncertainty is not too large for observation of thermodynamic contact angle effects to be totally prevented. Still, the error is comparable to the value of the parameter that is measured. This is a rather unpleasant situation for any kind of measurement. It diminishes the value of the result and complicates its interpretation. We will later see how the hysteresis effect, instead of being dismissed as an inevitable 'error', can in itself be explained and quantified by an extension of the thermodynamic argument. The hysteresis phenomenon becomes incorporated naturally in the generalized theory of wetting.

This static stress in the manner of dry friction that occurs at the three-phase line is quite real. By its effect, triboelectricity can be produced simply by retraction of a hydrophobized plate out of water or some other liquid that forms a high contact angle with the surface [3].

Following these preliminary remarks, we now proceed with a brief review of the common models that explain contact angle hysteresis in further detail.

4. MODELS OF THREE-PHASE STRESS

The traditional explanations for contact angle hysteresis in terms of surface roughness and chemical heterogeneity are well known. Both factors create energy barriers towards displacements of the three-phase line along the surface of the solid substrate. In the previous section we have shown that such barriers can arise even for surfaces that are smooth and uniform above the atomic level. They occur by the mere fact of molecular granularity of the surface of the substrate, and depend on its structure and the nature of the liquid. In Section 6, we show that this model extends with exact thermodynamic implications towards more complicated situations of wetting by surfactant and polymer solutions. In this and the following sections, we consider several results based on macroscopic models of geometrical and energy singularities in the form of line boundaries between mathematically uniform planes. These basic models explain macro-, micro-, and atomic scale effects.

The geometrical model of pinning of the three-phase line makes use of the fact that the contact angle is uncertain at the line where two planes intersect (Fig. 2, top). The simplest basic model shows the essence of the phenomenon. The value of the contact angle is not uniquely defined at the edge. Once the three-phase line reaches such a boundary, the contact angle changes from the value on one surface to the value on the other surface. The three-phase line stays 'pinned' at the edge while the liquid continues to overflow and the contact angle changes. Only after the boundary contact angle condition matches that for the other surface does the three-phase line begin to move again.

The range of contact angle uncertainty is obvious here. It equals the angle at which the planes intersect. The three-phase line sticks to the edge as long as the value of the contact angle remains within this uncertainty interval. The phenomenon is real. Just a single line asperity in the form of a scratch on the surface causes such an effect and can show up in the capillarographic scan in a Wilhelmy plate type set-up. The liquid front of the meniscus that advances steadily onto the surface suddenly stops at some point, stays there for a while while the contact angle increases, and then jumps ahead. This occurs even though the surface macroscopically might appear flat and uniform. And the role of such a line asperity can be played by a single array of atoms or a molecular step on a crystal surface. The mathematical singularity that occurs for the macroscopic model shows the essence of the molecular physical mechanism. An alternative view of this dry friction type resistance experienced by the line of one molecule thick is by a model of energy singularity that we consider next.

The geometrical model provides us with just an idea of how by the occurrence of microedges at polycrystalline and other kinds of rough surfaces hysteresis appears, even though it might not occur when the same surface is smooth. We do not go

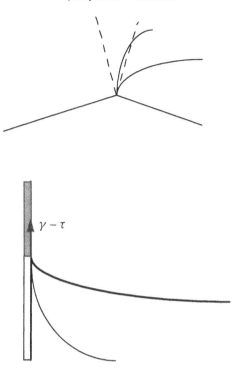

Figure 2. Geometrical (top) and 'chemical' (bottom) models explain the principle by which contact angle uncertainty occurs. The three-phase line becomes 'pinned' at an edge between two smooth planes where the contact angle is not uniquely defined. A boundary between two differently wetted areas acts as a similar stopper when the liquid advances onto the more hydrophobic area, or when it recedes from the hydrophilic area. In the reverse situations of advancing onto the hydrophilic area and receding from the hydrophobic area, the three-phase line becomes unstable; the contact angle transition occurs in a jump. These mechanisms that operate on microscales for rough and chemically heterogeneous surfaces provide an alternative insight into the molecular friction origin of contact angle hysteresis.

into detail here of the mathematical difficulties that occur in the application of our model. It is aimed to explain wetting implications of the geometrical fractality of real surfaces. General solutions have not yet been found and hardly can be purely analytical. But qualitative answers can be provided. Similar problems and solutions arise for the elementary model of chemical heterogeneity that we extend below in some further detail.

The model of energy discontinuity in a plane is particularly interesting in our hydrophobic context. Before we proceed to analyze this model, we point to just one more well-known effect associated with surface roughness. It is easy to notice that the wetting tension as the macroscopically averaged thermodynamic parameter changes in proportion to the 'roughness factor'. The latter should be understood as the ratio of the actual microscopic/molecular area of contact to the macroscopic (cross-sectional unit) area. This consideration can be done in a purely thermodynamic aspect without invoking explicitly the associated contact angle uncertainties.

The relevant thermodynamic quantities can, in principle, be measured via, for example, adsorption isotherms, and/or calorimetrically, for example, by comparing heats of wetting of the smooth and the rough substrates. As long as the same result follows using different liquids, the quantity is quite well defined even on the mathematical level, otherwise the physical definition still follows for each particular liquid. This consideration shows that there should be an equilibrium configuration of the droplet that conforms to the thermodynamic condition for the free energy minimum. In particular the equilibrium contact angle can be defined macroscopically on a scale larger than the characteristic length for which the roughness of the substrate can be averaged.

Once the equilibrium contact angle on the smooth surface is known, it follows also for the rough substrate once the value of the roughness factor is established. And this knowledge comes from independent thermodynamic data that are free of uncertainties associated with the contact angle hysteresis. The equilibrium contact angle is defined even though it might be difficult to measure it directly because of the hysteresis. The energy barriers that follow in the way of the three-phase line by the effects of surface roughness still might be too high to allow for spontaneous evolution of the system towards this equilibrium state. The associated relaxation times might simply be so long that in practice the equilibrium state is never reached. This purely thermodynamic consideration for the roughness factor explains the equilibrium contact angle, but this does not yet explicitly explain the hysteresis effect.

The other basic model that we are more interested in is a boundary between two uniform hydrophobic and hydrophilic areas (Fig. 2, bottom). When the liquid front reaches such a boundary from the side of the more hydrophilic area, it is stopped at this position. If the liquid is forced to advance, the contact angle increases above zero (or some other value that it has on the more hydrophilic part). This causes compression of the hydrophobic layer at which the three-phase line is stopped to a higher surface pressure. The magnitude of this additional compression is given by the value of $\gamma(1 - \cos\theta)$, or more generally by $\tau' - \tau''$ ($\tau = \gamma\cos\theta$) if the contact angle at the hydrophilic part (θ') is greater than zero. The limiting case $\tau = \gamma$ of zero contact angle is still very important as it applies to mica, hydroxylated silica, and many other hydrophilic surfaces. The value of the wetting tension here is about 70 mN/m, i.e. the surface tension of water.

The characteristic value of τ'' for the hydrophobic area of about -30 mN/m, according to the wetting tension of water on hydrocarbons, is also quite representative. And this is not just of solid paraffin for which it agrees with the experimental contact angle of $112°$. In the majority of practical situations, the hydrophobicity is due to hydrocarbon or fluorocarbon (in the latter case, the representative τ'' value may be even somewhat more negative) residues of molecules involved in the wetting process. These can be hydrocarbon tails of surfactants and lipids that induce dewetting transitions by deposition from solution or on a Langmuir trough, alkyl grafted surfaces, surfaces of nonpolar and amphiphile polymers, and lower-molecular-weight compounds that contain such nonpolar groups. We will use these reference values in our following examples.

Several consequences of this two-dimensional compression can be considered:

1. If the material of the hydrophobizing layer is insoluble in water but its translation mobility on the solid substrate is not arrested, the layer may slide forwards along the solid–vapor interface. It does so by being pushed away from the newly wetted areas by the advancing front of the liquid. In such a situation, the static contact angle may still remain at its original value θ', which may be zero. However, at finite speeds, effects of viscous or frictional resistance of the monolayer may occur. The monolayer becomes compressed to a higher surface pressure above the original value at which, for example, it has previously been deposited on a Langmuir trough. Then when the wetting meniscus reaches the monolayer boundary, the dynamic advancing angle increases above zero. The monolayer becomes compressed by the springing action of the meniscus.

As in statics, in this dynamic situation the magnitude of this compression is also described by our formula that accounts for the mechanical equilibrium. It is based on the Young equation, or simply on the Newton's third law: the change in the wetting tension equals the change in the surface pressure of the mono-layer. Alternatively, by the effect of this compression or simply by contacting water at zero value of the contact angle the monolayer can be spread back onto the surface of water from which it might earlier have been deposited by forced compression on the trough. Both mechanisms, which lead to a release of the supercompression, have their justification in the principle of Le Chatelier. This consideration repeated for other situations such as spontaneous and forced receding of water from the monolayer boundary explains the complicated and variable capillarographic patterns usually observed for Langmuir–Blodgett monolayers.

2. Monolayers of soluble surfactants also become compressed or decompressed depending on the direction of motion of the wetting front. One more degree of freedom arises here by molecular exchange with the bulk solution. For example, the advancing three-phase line raises the chemical potential within the monolayer ahead of it. This is because the surface pressure of the adsorbed layer is forced to increase above the equilibrium value for a stagnant three-phase line in co-existence with the solution. Also, here this compression is explained mechanically by the capillary springing effect of the interface when the contact angle changes.

The associated local increase of the chemical potential at the monolayer boundary triggers chemical re-equilibration. The molecules begin to desorb from the three-phase line and dissolve back into water, or even evaporate if the substance is volatile. Spontaneous contact angle relaxation occurs in the absence of external influences that force the liquid to move with respect to the substrate. The mechanism of the three-phase line re-adsorption continues to operate once the contact angle deviates from its equilibrium value. The relaxation process can be slow or fast depending on the direction and magnitude of the deviation, concentration, and molecular nature of the system. Ultimately it brings the system back to equilibrium. At equilibrium, the three-phase line comes to a new position on the surface while the surface pressure and the contact angle come back to their original values determined by the concentration of the solution.

However, if the forced movement of the liquid with respect to the surface is continued at a constant speed, a steady state is reached. While the monolayer becomes compressed above the equilibrium value for the stagnant three-phase line, the chemical potential increases and desorption proceeds faster. This mechanically induced shift in the adsorption equilibrium enables us to maintain the surface pressure at a constant value. This stationary dynamic value is higher than the equilibrium value at zero speed. The local three-phase line values of the chemical potential and the adsorption are also higher and do not change here over time. They remain constant while the three-phase line moves with respect to the surface at a constant speed. The contact angle also does not change. These constant values of the parameters which are interrelated via the equation-of-state depend in this situation both on the concentration of the solution and on the speed. In the steady state, the dynamic wetting tension stays at a stationary value that balances the dynamic surface pressure effect. We will consider this situation in more detail in Section 6.

3. Stable hydrophobic surfaces can be formed by chemical modification. If the covalent bonds that anchor the hydrophobic groups to the substrate do not hydrolyze by the presence of water or destabilize in some other way the three-phase line stops at the boundary which does not move with respect to the surface. When the Wilhelmy plate, for example, is lowered, the contact angle increases. When the value of the contact angle on the hydrophobic area is reached, the three-phase line goes off the boundary and begins to move on top of the hydrophobic monolayer. As on the hydrophilic area before the stop, the motion proceeds at the speed at which the plate is lowered, but this time at the higher value of the contact angle.

As for the geometrical model, the situation becomes very complicated in precise details for real (fractal) surfaces that may contain both hydrophobic and hydrophilic patches. It is this situation that is ordinarily dealt with by the theories of wetting. This is the approximation of irreversibility of adsorption. Basically such theories assume that whichever way the hydrophobic and hydrophilic domains are mixed together, the adsorption distribution pattern does not change when the three-phase line scans over the substrate. The original pattern is preserved after the surface is transferred from one environment to another, for example, from air into water. In our thermodynamic analysis of dewetting transitions, we begin with consideration of this fixed pattern approximation. This situation might be more or less relevant to experimental observations for other kinds of stable hydrophobic surfaces (Teflon, polystyrene, graphite, etc.).

5. HYDROPHOBIC AND HETEROGENEOUS SURFACES

Let us now consider the situation when the model surface is lowered into water with the hydrophobic end down. While the three-phase line moves over the hydrophobic area of the surface, the meniscus stays in its nonwetting configuration. At the moment when the boundary line between the hydrophobic and hydrophilic areas catches up with the three-phase line, the liquid suddenly jumps upwards. Water spreads spontaneously on the hydrophilic part of the surface. A wetting meniscus forms. The jump-up distance equals the difference between the hydrophilic and hydrophobic meniscus heights. In the receding situation (which occurs on the way up), the transformation of the meniscus over the same range of heights occurs, as we have described before. The instability on passing the boundary occurs also in this direction but at the accordingly higher position of the plate with respect to the liquid. This is the simplest and straightforward model that already possesses a hysteresis loop.

Real surfaces often represent complex patterns of an intricate fractal mosaic of hydrophilic and hydrophobic patches mixed together on the range of scales from microscopic to molecular. A simple semi-classical model of parallel strips already shows several basic features typical of real situations. By painting our cylinder (which we use here instead of a plate to avoid edge effects*) with alternating hydrophilic and hydrophobic bands our elementary model can readily be extended closer to the reality with preservation for the exact result.

For the 'zebra' cylinder, the exact sequence of wetting events is straightforward to devise. It follows theoretically in terms of the results of our basic model. Like for the latter situation, what occurs when such a cylinder is moved with respect to the water is completely defined in terms of the Young and Laplace equations that describe meniscus profiles at mathematically uniform cylindrical or some other surface. The exact expression for the capillary force is simply $2\pi r \tau$. The theory can be readily generalized from static towards dynamic situations. Acceleration and viscosity effects can be added in the form of associated equations of motion in order to describe the meniscus dynamics over the ranges of capillary instabilities. This is just a matter of finding the exact solution of fluid mechanics. While these might or might not exist, simple estimates in the manner of the Washborn equation for the speed of capillary rise are always possible. In fact, our consideration can be readily repeated for a capillary as well.

---

*By using a rod instead of a plate in the Wilhelmy set-up, one can avoid edge effects. The three-phase line curves on approaching the edges of the plate that inevitably is finite in width. The analytical formula that relates the height of the meniscus rise to the contact angle and the surface tension is more complicated for a cylinder than for an infinitely wide flat wall. But the edge corrections to the three-phase line profile cannot be resolved analytically, even in the simple situation when the surface is uniform. An associated problem that is avoided in the case of cylindrical geometry is edge meniscus instabilities. The jumps of the meniscus are very sensitive to the state of the edges that are very difficult to control. This results in a random distortion of capillarographic curves.

When the surface (a rod or a capillary) is lowered (the liquid advances) while the three-phase line stays pinned at the hydrophobic boundary, the contact angle first increases to the ultimate value that it has on the hydrophobic area. At this point, the liquid begins to move over the hydrophobic band in the hydrophobic configuration of the meniscus. The three-phase line scans over the surface at the speed with which the surface is moved. At the end of the hydrophobic path, when the three-phase line reaches the hydrophilic boundary, the liquid jumps ahead over this band. Provided the latter is narrower than the height of the wetting meniscus, the liquid passes over this hydrophilic area at a high inertia/viscosity controlled speed and stops at the next hydrophobic boundary.

However, if the hydrophobic band is narrow, the barrier hit by the accelerating liquid can be overcome by the inertia effect. The meniscus then jumps over the band to the next position of local free energy minimum. This is at the next boundary for which the condition for mechanical equilibrium is fulfilled. Or, on its way, the wetting line passes several such positions if the acceleration is high while the hydrophobic bands are narrow. For the position at which it stops, the condition of absolute free energy minimum might not yet have been reached. Then the mechanical equilibrium is established at an angle which is larger than the thermodynamically equilibrium contact angle.

The parameters needed to reproduce the entire sequence of dynamic wetting events are just the width of the strips and the contact angles on the uniform areas that are free of hysteresis. Similar trivial considerations to those for the advancing situation follow for the receding situation. Here the liquid moves at a constant (e.g. zero) contact angle over the hydrophilic band, then jumps over the hydrophobic band onto the next hydrophilic area, etc. The meniscus height/wetting tension oscillation continues in repeated cycles while the rod is moved at a constant speed.

In simple situations, when the hydrophobic bands are sufficiently wide, the meniscus is stopped at each next boundary. The arising hysteresis loop is easy to reconstruct in exact detail. The well-known analytical formulae based on the Young and Laplace equations that describe meniscus profiles on mathematically uniform planar [4] or cylindrical [5] surfaces can be used. The same principle on which we have based our elementary model of a single boundary extends towards this more complex situation.

The alternating band model already features more closely the wetting patterns observed for real substrates. Both the advancing and the receding branches of the loop exhibit oscillations, a quite common observation for many hydrophobic surfaces. These indeed follow with a regular periodicity once a rod is used in place of a plate. Just by changing the width of the bands the amplitude and the period can be changed and by placing them along the rod (or a capillary) in a more or less random manner quite sophisticated capillarographic patterns such as white noise or macroscopic tension gradients can be simulated.

One more common experimental observation is explained by our model: a tendency of the loop to narrow when the surface tension of the liquid decreases. An improvement of wetting is commonly achieved by adding a surfactant to water, the effect very important in detergency. And this makes the advancing and the receding traces of the loop converge. This, of course, is first of all a trivial consequence of the fact that the absolute magnitude of the wetting tension is limited by the value of the surface tension. There is, however, a more general Brownian consideration that allows the hysteresis to vanish by a reduction in the surface tension.

We notice first of all in this connection that hitherto the extension of our model was based on a phenomenological argument. We restricted ourselves to purely mechanical results supported thermodynamically, but without considering more explicitly the molecular implications that might follow from notions of chemical thermodynamics and statistical mechanics. Similar limitations of this mechanistic approach are obvious. Firstly, for our model the maximum receding and the minimum advancing tensions are always equal to the values on the corresponding hydrophilic and hydrophobic reference surfaces. Secondly, the current value of the contact angle, whatever it is, does not change with time if the surface is not moved. These features, however, are not necessarily observed in experiments with real surfaces.

These obvious limitations are inevitable in the view of oversimplifications that have been done. These were necessary to obtain simple and exact solutions. Beyond this point, mainly qualitative explanations can be reached. Real substrates, as we noted, represent a far more complex mosaic of alternating hydrophobic and hydrophilic patches. Nonpolar and polar molecular groups are mixed together in the plane of the surface, either in the form of single molecular groups or as clusters segregated in blocks. These hydrophobic islands or hydrophilic lakes can be distributed over the surface randomly or regularly, in the manner of a chessboard. Monodisperse or variable in shapes and sizes, on different ranges of scales from macroscopic to molecular, they can be joined in sophisticated fractal patterns. In each particular case, the structure of the surface depends on the method of preparation (chemisorption of alkylsilanes onto silica and reaction of gold with alkylthiols, oxidized polystyrene, block graft copolymers, hydrophobic planes, and hydrophilic edges of graphite, etc.). In such situations, a straightforward description might not be easy to achieve. The fact that real surfaces might indeed be even more complicated is considered in the last section of this review.

For the moment, we stick to a purely thermodynamic view of the surface. An approach very similar to the one that we used to define the wetting tension and other adhesion-related parameters of rough surfaces can be used here also. To simplify the result and make it more practical when relevant calorimetric data may not be available, we will make use of a simple additive approximation. In the manner of the 'roughness factor', it just adds more functionality to our model. We assume that $\tau = \varphi\tau' + (1 - \varphi)\tau''$; here φ is the effective hydrophobic area; $\tau' = \gamma(\theta = 0)$ is the particular example that we use here as an illustration. By this additive approximation we do not limit our model. This primitive formula can always be substituted by a more precise thermodynamic result. But this formula is not so poor. This basically is an extension of the principle of independent surface action of Langmuir. It holds surprisingly well in many practical situations related to wetting and adhesion. The formula tells us that the work of adhesion is a linear function of the extent of surface substitution, or the adsorption. This is justified as long as the adsorption is understood here in its irreversible chemical context, as chemisorption; otherwise, when the adsorption is reversible, the formula has to be substituted by the result based on the Gibbs equation (see Section 6). The formula is easier to justify by a formal argument in the macroscopic limit of larger patches. It still accounts very well for contact angles on heterogeneous substrates even when hydrophobic and hydrophilic groups are mixed on the molecular level.

Like the case of ordinary reversible Gibbs adsorption, the extent of substitution (φ) occurs in our formula as a macroscopically averaged thermodynamic quantity. The properties described by this equation are indeed not very sensitive to the precise way by which the chemisorbed material is distributed over the surface. As long as this distribution does not change over time, we do not make a large conceptual mistake. In itself, the immobilization in the absence of desorption or re-adsorption is a very important assumption. Dramatic consequences of its violation that occur in the form of hydrophobic forces and polywater are considered in the last section of this review. The other limiting case of totally reversible adsorption that leads to hydrophobic transitions on hydrophilic surfaces is extended in the next section.

In the model of parallel strips, the barriers that separate local equilibria are generally too high to be overcome spontaneously by the effect of the Brownian fluctuations of the three-phase line. Indeed, the length of the hydrophobic bands is the macroscopic perimeter of the rod. However, already for this oversimplified model, when the band width approaches zero, the barriers become penetrable. In a more realistic approximation, if the barrier is a hydrophobic spot of area s, its height is $s(\tau' - \tau'')$. The barrier height further decreases by $\tau = \gamma \cos \theta$ by increase of the acting value of the contact angle above zero that can be achieved through external compression.

A simple statistical result then follows by the Arrhenius equation (cf., for example, the molecular kinetic theory of wetting by Blake [6]). In the presence of such hydrophobic barriers, the rate of spreading reduces compared with its limiting value on a hysteresis-free uniform substrate which does not possess such energy barriers. The reduction depends on their density number and the Boltzmann factor $\exp(-\Delta\tau s / kT)$. One need not go into much detail of this derivation to make a rough estimate. For water on a hydrophilic surface, $\tau' \approx \gamma \approx 70$ mN/m. The value of τ'' is about -30 mN/m, according to the wetting tension of water on oil or grease (liquid or solid hydrocarbon). Then, assuming that $s = 1$ nm^2, which would be of the order of the area covered by a single hydrocarbon group, the barrier height when pure water approaches the hydrophobic spot at zero contact angle is about $30\,kT$. It reduces to zero when the advancing contact angle is larger than $112°$. The value of the contact angle decreases with time during the relaxation process, which slows down until the equilibrium (macroscopic average) value of the contact angle is reached. The process slows down because the height of the barrier increases when the contact angle decreases. Already at 90° the value of τ reduces to zero, i.e. by 30 mN/m compared with the value at the beginning of the relaxation process. Being at zero in the initial state, the barrier becomes of the order of $10\,kT$ when $\theta = 90°$.

The rate of the advancing angle relaxation reduces to negligibly small values long before the equilibrium angle is reached. For example, for a surface with 50% of substitution ($\varphi = 0.5$) the equilibrium contact angle is about 50°. It is never reached in such a situation. A similar consideration applies to the backward relaxation from a receding value of the contact angle that is smaller than the equilibrium one. This explains why for water on hydrophobic substrates that in most cases incorporate polar groups within the hydrocarbon matrix, contact angle hysteresis occurs as almost static. The initial rate of contact angle relaxation after forced movement of the liquid is stopped is fast, but the equilibrium is not reached even when observation times are extended to hours. Relaxation times are still orders of magnitude longer.

However, just a two-fold reduction in the surface tension, to a value of 35 mN/m, changes the situation quite dramatically. The value of τ'' changes to about 25 mN/m; this is by taking into account that the work of adhesion does not change significantly by the change in surface tension. The ultimate height of the barrier reduces by an order of magnitude, to a few kT units. The value of the exponential factor then rises significantly above zero, irrespective of the acting value of the wetting tension. The process of molecular exchange at the three-phase line proceeds here at high rates. Relaxation to the equilibrium value of the contact angle occurs over much shorter periods of time. These are of the order of minutes according to our estimates, in agreement with experimental observation [7], not hours, days or even ages as for pure water.

Such a rapid acceleration of molecular exchange is a general consequence of the fact that the activation energy occurs in the Arrhenius exponent. According to this kinetic equation, the relaxation rate is very sensitive to the parameter that controls the barrier height. Indeed, relaxation to the equilibrium contact angle proceeds almost instantaneously at still lower values of the surface tension of the liquid. Here, when the barrier reduces to values of the order of $1\,kT$ or less, the frictional retardation of the three-phase line almost disappears. The liquid begins to spread at an ultimate inertia controlled speed of the order of $10-100$ cm/s, as it would do on a mathematically uniform substrate.

This rate of fast spreading of low viscosity liquids such as water and ethanol is mainly controlled by inertia when the tensions are out of balance. No energy barriers occur when the contact angle equals zero. Under this condition of complete wetting, the solid surface is under a thick film of the liquid that forms at co-existence. This explains the fast contact angle equilibration on approaching the condition of wetting transition. The effect of relaxation acceleration by a reduction in the surface tension is readily explained by this theory. The effect of Marangoni retardation of spreading of surfactant solutions on hydrophilic surfaces is considered in the next section.

6. HYDROPHOBIC TRANSITIONS ON HYDROPHILIC SUBSTRATES

In the previous section we have considered the wetting of stable hydrophobic surfaces. Surfaces grafted with alkylsilanes apparently fall in the third category of surfaces. The chemisorbed layers are stable on contact with water and organic solvents. This is in contrast to monolayers of oil-soluble lipids deposited on surfaces of silica or mica by LB techniques. Unless laterally polymerized and/or chemically grafted to the surface, such monolayers are destabilized by water and easily washed away by liquids like ethanol. The same is even more true about the surfaces rendered hydrophobic by water-soluble cationic surfactants.

There are, however, similarities in wetting behavior between different hydrophobic surfaces, apart from just high contact angles that they form with water. For example, large contact angle hysteresis is typically observed for most of these systems. Even the fine structure of capillarographic diagrams is often quite similar for different categories of surfaces.

Silica and mica surfaces rendered hydrophobic by water-soluble cationic surfactants apparently have these features in common with other hydrophobic surfaces. As an example, we will consider CTAB (cetyltrimethylammonium bromide), a surfactant used in many studies of hydrophobic effects by contact angles and surface forces. Its very dilute solutions (concentrations orders of magnitude lower than the cmc, about 10^{-3} M) dewet hydrophilic surfaces of silica or mica that become negatively charged in pure water. The dewetting patterns at first sight might indeed not be very different from those observed for hydrophobic surfaces coated with LB monolayers, grafted silanes or even surfaces of bulk nonpolar materials such as Teflon or graphite.

For example, contact angles can be larger than 90°, even though clean surfaces of freshly cleaved mica or hydroxylated silica are completely wetted by water. On the surface retracted from CTAB solution, pure water also forms a large advancing contact angle. The hysteresis loop is wide; the difference between the advancing and the receding contact angles reaches tens of degrees. The relaxation times of the order of minutes around the cmc increase to hours or even longer at lower concentrations. High contact angles, wide hysteresis loop, and long relaxation times on going from cmc solutions back to pure water are what is observed for silane-treated and other hydrophobic surfaces as well.

In particular, the rewetting of glass occurs when the CTAB concentration increases to the cmc. This is a common detergency effect. Hydrophilic transitions with increasing surfactant concentration to the cmc occur for different surfactants and different kinds of hydrophobic surfaces. Adsorption from solution at the interfaces both with air and with the hydrophobic solid induces the hydrophilization. The effect is readily explained molecularly, in terms of preferential orientation, and thermodynamically by the Gibbs and Young equations, in terms of surface pressures or a reduction in the corresponding surface tensions. The molecular and thermodynamic explanations of the hydrophilic effect do not contradict each other in this case.

From this point of view, the only difference between surfactant-induced hydrophilic transitions on silane-treated glasses or other kinds of hydrophobic surfaces and the CTAB effect is that in the latter case the hydrophobicity is due to CTAB itself, the same substance that causes hydrophilization at higher concentrations. But once one assumes that CTAB strongly binds to the surface and forms a hydrophobic monolayer, the similarity is explained. However, this explanation, which often occurs in the literature, contains some internal contradictions.

A more careful consideration shows that molecular mechanisms by which these similarities between CTAB and silane grafting effects can be explained are different. To account for the hydrophobic transitions induced by cationic surfactants in an ordinary way, one has to postulate first of all that like silanes, CTAB indeed forms hydrophobic monolayers and binds irreversibly to surfaces of silica and mica. The latter condition apparently can be relaxed, but still the binding has to be strong enough. The monolayer, once adsorbed, should not be washed away by pure water. The associated condition would demand CTAB monolayers not to desorb or re-adsorb to a significant extent in the course of contact angle measurements.

Even trace amounts of CTAB induce the hydrophobic transition. Contact angles and adhesion forces begin to increase at concentrations that are hundreds and even thousands times lower than the cmc. This observation might be taken as a strong argument in support of the 'theory' of strong binding. The lower the concentration at which the monolayer forms, the higher the energy of adsorption. Each decimal order down in concentration contributes about $2.3\ kT$ to the standard adsorption free energy. Interpreted in this way, the free energy of adsorption at the solid–liquid interface should be at least $10\ kT$ higher than at the liquid–air interface. For the latter, the monolayer forms only on approaching the cmc. Such a high energy of adsorption should justify the irreversibility postulate. Desorption times should be quite long in such a situation. The 'theory' apparently explains why the contact angle for pure water on the surface retracted from CTAB solution is high, and many other features of the phenomenon, including the mere fact of hydrophobicity.

However, the interpretation of molecular mechanisms behind the hydrophobic transitions based on irreversible monolayer postulate is inconsistent with experimental data on CTAB adsorption. A self-consistent explanation of the CTAB effect and similar hydrophobic transitions induced by other cationic surfactants on negatively charged hydrophilic surfaces was proposed by Ter-Minassian-Saraga in a series of papers published more than 30 years ago. The contact angle and adsorption results were considered with their thermodynamic implications in the context of the Young and Gibbs equations. To start arguing our thesis, we point to a more recent relevant observation. There is evidence that CTAB adsorption from solution is in fact entirely reversible. In particular, desorption on contact with pure water proceeds almost instantly [8].

And there is another general observation that has been largely ignored in the literature. As we might have already noticed, the hysteresis patterns, however complicated they might appear at first sight, depend on the concentration of the

CTAB in the solution. It is quite clear from the literature data that contact angles as well as pull-off forces measured between such surfaces in CTAB solutions increase with increasing concentration. The presumption of a strong irreversible adsorption does not explain the concentration effect. It would rather imply that once such a monolayer is formed, the contact angle should not change further with concentration, at least as long as concentrations remain orders of magnitude lower than the cmc. Indeed, at such low surfactant concentrations, adsorption at interfaces of water with air, oil or hydrophobic solids is negligibly small. The surface tension of the solution is almost identical to that of pure water. The theory does not explain why the contact angle increases with concentration and does not decrease if it changes at all.

There are other experimental observations that might be difficult to explain by this theory. Among theses the experimental fact that CTAB simply does not adsorb at the solid–liquid interface at such low concentrations might be the most unpleasant.

But this then poses a question: if the hydrophobic monolayer does not form at the solid–liquid interface, then why does the surface not wet? Why does the hydrophobic transition take place? Should not the contact angle be small or remain zero as it is for the clean substrate? This is what the 'ordinary' theory would predict.

To avoid such questions to which the theory does not provide clear answers, one might take them as irrelevant. This apparently is the way this theory works. But even if by ignoring experimental facts one were to continue to assume that CTAB does adsorb at the solid–liquid interface, would this save the situation? Can this help us to explain the hydrophobic transition? Apparently not, at least as long as we are not ready to sacrifice another experimental fact: at much higher concentration, at which adsorption indeed proceeds at the solid–liquid interface, it is entirely reversible [8].

And reversible adsorption, by whichever molecular mechanism it occurs, inevitably contributes to a reduction in the interfacial tension. This indisputable argument occurs on a purely phenomenological level. It follows from the mere notion of the Gibbs adsorption equation. By adsorbing at the solid–liquid interface, CTAB should reduce the corresponding interfacial tension, Then, according to the Young equation, it should contribute to a reduction, not to an increase of the contact angle. And, similarly, it should decrease, not increase, pull-off forces. Both conclusions contradict the experimental observations on contact angles and surface forces.

There is one more paradox: unlike the situation for surfaces considered in the previous section, which are hydrophobic from the start, the surfaces of freshly cleaved mica or hydroxylated silica are hydrophilic. The receding and the advancing contact angles equal zero. By the mere notion of this fact should not the wetting process be entirely reversible? The fact that this is not the case is evidenced by the large hysteresis loop. But then, once adsorption is reversible and the desorption proceeds fast how do we explain this contact angle hysteresis? Should not the wetting process be reversible for the solution the way it is for pure water when the contact angle is zero? What makes the contact angle relaxation be so slow? Why does contact angle hysteresis occur under static conditions? Why is hydrophobicity maintained in contact with pure water that desorbs CTAB almost instantly?

There are apparently too many questions. Some answers that would shed some light on this issue can be found in the old paper by Ter-Minassian Saraga published, we note here again, more than three decades ago [9]. Relevant considerations were made later by some other authors [10–13]. In spite of this effort, the confusion in the literature continues to persist.

To start unfolding our argument along these lines, we first note that although the relaxation is slow, contact angle equilibration occurs over experimentally accessible times. These times are of the order of several hours at low and moderate CTAB concentrations and much faster closer to the cmc [13]. At shorter times, before the equilibrium is reached, the droplet might take an irregular shape. However, on the elapse of some time, which depends on the type and concentration of the surfactant, the three-phase line becomes perfectly smooth. Here the value of the contact angle is unique; it depends only on the concentration. The meniscus that spans a rod, a capillary, or a droplet attains its mathematically perfect shape in the geometry demanded by the Young and Laplace equations.

Just a quick look at the condensate that forms by temperature instability on the wall of a stoppered flask in which CTAB solution is stored readily convinces one of this effect. All droplets appear here in the form of spherical segments that have a regular circular base. This is not the case once one shakes the flask; then the smeared droplets left on the walls are irregular in shape. These imperfections correspond to a large variation of the contact angle along the wetting perimeter. But after a period of subsequent evolution, the droplets attain their ideal equilibrium shapes back again.

The value of the contact angle and the time needed to reach the equilibrium depend on the concentration of the solution. The fact that the equilibration occurs over experimentally accessible times practically means that the wetting tension as a function of the concentration can be measured. This measurement can be done as accurately as the measurement of the surface tension of water, or of a soap solution, at their interfaces with air. The latter measurement is an experimental routine and can be done to quite a high precision. Then, simply by differentiating the Young equation in basically the same manner that traditionally applies to surface tension isotherms, one obtains

$$d\tau/d\mu = \Gamma_{SV} - \Gamma_{SL}. \tag{3}$$

This form immediately follows from the Gibbs equation by taking further into consideration the Young equation. The latter equation postulates that the wetting tension, $\tau = \gamma \cos\theta$, equals $\gamma_{SV} - \gamma_{SL}$. This result understood (quasi) thermodynamically for the equilibrium situation then extends, as we will later show, towards kinetic contact angle phenomena.

What is important at this stage is that the difference between the values of adsorption at the solid–vapor and solid–liquid interfaces can be determined by contact angle measurements. This is done in essentially the same way by which

V. V. Yaminsky

the absolute values of adsorption at air–water and water–oil interfaces follow from surface tension measurements. The latter principle of adsorption analysis of surface tension isotherms is well known. This is the basic principle of the surface chemistry of surfactant solutions. An exact science of wetting phenomena, or capillarography, the way we extend it here, follows in essentially the same way.

For CTAB solutions that dewet silica or mica, this adsorption difference is positive and large starting from low concentrations. While adsorption at the solid–liquid interface remains negligible in this range, a condensed monolayer with an area of about 1 nm^2 per molecule adsorbs at the solid–vapor interface.

Once one acknowledges the importance of this adsorption outside the solution, the paradoxes associated with the dewetting transitions become consistently resolved. Thermodynamic analysis of the adsorption of surfactants and other chemical species at solid–vapor interfaces deserves as much attention in the context of physical chemistry of surfaces as the adsorption at water–air and water–oil interfaces.

The main driving force for the adsorption of CTAB and similar surfactants on surfaces of silica and mica in air is the free energy gain associated with the transfer of hydrophobic tails of the surfactant molecules out of water. This hydrophobic contribution to the free energy of adsorption is supplemented by a favorable ion exchange interaction that follows when alkylammonium ions replace surface protons. While the hydrophobic contribution is approximately the same for monolayers adsorbed at the solid–air and water–air interfaces, the electrostatic contributions are quite different. For adsorption at the water–air interface the electro-osmotic self-energy is essentially unfavorable. It results from a repulsive interaction associated with the formation of the electrical double layer in the process of adsorption that requires bringing similarly charged ions closer together. For ion exchange, this unfavorable osmotic contribution does not occur. This is because the plane of charge is already present in the initial state for the surface in pure water.

This, then, explains why the layer at the solid–vapor interface begins to adsorb starting from much smaller concentrations in the solution than the adsorbed layer at the water–air interface. An adsorbed layer at the solid–liquid interface forms at further higher concentrations on approaching the cmc. Like the bulk micelle formation, the process of this bilayer adsorption in the form of 'two-dimensional micelles' is essentially cooperative in the manner of a subcritical phase transition. It is driven by the tendency of the hydrocarbon moieties of the surfactant molecules to hide out of water. Unlike bulk micelles, the bilayer, like the monolayer at the water–vapor interface, has effectively zero curvature. The curvature effect explains why the monolayer condensation at the interface of the solution with air proceeds at somewhat lower concentrations than the cmc, when micelles begin to form in the bulk solution. While micelles are completely surrounded by water with electrical double layers extending into the bulk solution, for the bilayer the double layer is present only on the outer part of it. For the inner part of the bilayer where the ion exchange interaction occur, the electrostatic contribution is rather favorable.

The curvature effects also make a more favorable contribution to the free energy of adsorption compared with the free energy of micelle formation.

This explains why the bilayer, like the monolayer at the water–air interface, begins to form at concentrations lower than the cmc. It should be noted, however, that unlike the processes of bilayer adsorption and micelle formation where the hydrophobic association is cooperative, at the water–air interface the ability of hydrophobic tails to escape from water does not depend on the presence of other molecules of the surfactant. The free energy of transfer into a rarefied gas-like layer can here be even higher than for adsorption within the condensed monolayer. This is because the charged groups of the surfactant molecules that repel each other are further away. The unfavorable double-layer energy is lower at lower concentrations. At the water–air interface, the adsorbed layer gradually builds up in the quasi-Henry regime starting from much lower concentrations than for the cooperative bilayer adsorption at the solid–liquid interface.

The bilayer adsorption is responsible for the rewetting transition that occurs near the cmc. Once this adsorption becomes larger than the monolayer adsorption at the solid–vapor interface, the adsorption difference changes sign and becomes negative. It is important to stress that adsorption at the solid–liquid interface always proceeds in a bilayer form. Hydrophobic monolayers here never occur. They adsorb only onto the other side of the three-phase line, i.e. onto the surface out of the solution in contact with air. And this adsorption proceeds starting from very low concentrations, for which the adsorption at the other two interfaces is negligible.

Molecular mechanisms that follow from adsorption analysis of wetting tension isotherms can be extended further in detail. This is essential to complete the extended picture of surfactant monolayers at water–air and water–oil interfaces. In this way, physical mechanisms behind the dewetting effects of surfactant monolayers become consistently explained. Consider, for example, the integral form of the Young–Gibbs equation:

$$\tau^{\circ} - \tau = \pi_{\text{SV}} - \pi_{\text{SL}}. \tag{4}$$

Restated in this way, the equation shows explicitly that the three-phase line essentially acts in the role of the barrier on a Langmuir trough. Being pushed to the two sides of it by the two opposing surface pressures, it moves in the process of contact angle relaxation. The change in the values of surface pressure by the surfactant re-adsorption shows up in the corresponding changes in contact angles. By this gauge effect of the three-phase line that acts as a molecular selective pump the liquid–vapor interface readjusts itself by changing the contact angle. This self-regulation that proceeds in a feedback mode occurs each time that the surface pressure difference in the plane of the solid changes. This balance between the wetting tension and the surface pressure effects is the essence of the mechanism that continuously maintains mechanical equilibrium at the three-phase line.

And operation of this physical mechanism extends beyond the condition of chemical equilibrium. The principle continues to apply even though chemical equilibrium throughout the system might not yet have been reached. One might recall in this connection that the translational mobility of CTAB molecules on the dry solid substrate is very restricted. It occurs by solid-state mechanisms of interfacial diffusion that are very slow. The monolayer that self-assembles near the three-phase line does not spread to more distant areas of the solid–vapor interface. In fact, even the three-phase line is not in equilibrium with the bulk solution as long as the process of contact angle relaxation continues. In this process, the contact angle changes in such a way that the restoring force of the surface tension response compensates for the change in local value of the surface pressures that act on the three-phase line in the plane of the solid.

At low CTAB concentrations, the π_{SV} term dominates. The wetting tension decreases with increase of concentration. This decrease equals the increase in the surface pressure. The contact angle increases with increasing concentration in this range, while the change in the surface tension at the liquid–air interface remains small.

Over a wide range of small surfactant concentrations in the bulk solution, Γ_{SV} remains approximately constant. It corresponds to an area per molecule of the order of 1 nm^2, typical of a condensed monolayer. This shows that the adsorbed layer at the solid–air interface becomes saturated starting from very low concentrations. At such concentrations the contact angle already becomes finite. The wetting tension then decreases with increasing concentration at a constant value of adsorption. The contact angle changes even though the adsorption does not. This is a universal chemical potential effect. Indeed, according to the Gibbs equation, π changes logarithmically with concentration provided that Γ remains constant.

While the thermodynamic argument is based on consideration of the equilibrium situation, it is quite general. It explains plainly and in exact thermodynamic terms dynamic wetting phenomena, contact angle hysteresis, and kinetic aspects of contact angle relaxation. Consider, for example, what happens when a clean surface comes in contact for the first time with the solution. The liquid jumps up onto the plate when it touches the surface of the liquid. The solution at the instant of the jump acts essentially as pure water and forms a wetting meniscus. From here the surfactant begins to adsorb. This adsorption proceeds above the three-phase line. By this effect the surface pressure of the adsorbed layer increases. It pushes the liquid down, away from the area already wetted by the solution. This adsorption-driven recession of the three-phase line makes the contact angle increase above zero. The changed angle at which the surface tension is applied generates surface pressure in the opposite direction, $\pi_{SV} = \gamma(1 - \cos\theta)$. This elastic response of the liquid meniscus compresses the monolayer further and further as the contact angle increases (Fig. 3).

This is an interesting situation. By this compression, which increases during the spontaneous process of descending of the meniscus, the chemical potential of the surfactant that adsorbs at the three-phase line is not constant; it changes in this process. Initially it is much lower than the co-existence value determined by the

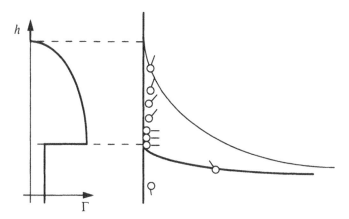

Figure 3. Contact angle hysteresis on a uniform surface, for which no hysteresis may occur in contact with a pure liquid, can be induced by equilibrium solute adsorption. A solute that preferentially adsorbs at the solid–vapor interface induces surface pressure that acts on the three-phase line. This makes the contact angle increase. The arising restoring force from the side of the meniscus, in turn, compresses the monolayer to a higher value of the surface pressure. The rates of contact angle relaxation are related to the rates of the three-phase line re-adsorption controlled by interfacial and bulk solute diffusion and rates of monolayer re-dissolution. The associated gradient adsorption deposition patterns follow in exact relation to acting values of the dynamic contact angles via the equation-of-state of the monolayer. The three-phase line that essentially performs in the role of the barrier on a Langmuir trough and the gauge response of the liquid–vapor interface compress or expand the monolayer by changing the contact angle.

bulk concentration of the solution. The adsorbed amount is low while the rate of the contact angle relaxation is quite high; the rate is limited only by the rate of the supply of the surfactant to the three-phase line.

The relation between the acting values of the adsorption density and the chemical potential at the three-phase line is given by the equation of state, $\pi[\Gamma(\mu)]$. This surface co-existence diagram establishes the rules that relate the three-phase line magnitude of the adsorption to the value of the contact angle. At each given instant of time, nonequilibrium local values of the adsorption correspond to a point of the equation-of-state that in itself describes equilibrium properties of the self-assembled monolayer. Once the corresponding $\Delta\pi(C)$ isotherm is known experimentally, by measurement of the wetting tension at equilibrium, the gradient adsorption profile left behind the retreating meniscus follows from the change of the contact angle in the process of wetting relaxation.

This is a general principle. It explains why the contact angle relaxation is almost as slow in this situation as it is for pure water on chemically grafted hydrophobic silica surfaces, even though molecular mechanisms behind the relaxation phenomena are different for the two cases. The latter mechanism has already been explained in the previous section of this review in terms of activational water re-adsorption from the three-phase line. It proceeds under the condition for which the original adsorption distribution pattern of the irreversibly grafted alkylsilane species does not change in the wetting process. In the present situation, it is the three-phase line re-adsorption of the surfactant itself that causes the dewetting.

Exact theoretical solutions that explain the observed rates of the relaxation can be based on diffusion laws that enable us to derive the associated kinetics equations. The condition for dynamic molecular equilibrium follows from the mass action law and associated rates of adsorption and desorption. This consideration shows, in particular, that the adsorption density within the self-deposited layer of the surfactant changes with height. It increases as the liquid recedes.

At some point the layer that self-assembles behind the three-phase line undergoes a condensation transition. Here the saturation level of adsorption is reached. If this occurs before the contact angle reaches equilibrium, and this depends on the equilibrium value of the surface pressure, which, in turn, depends on the concentration of the solution, the surface pressure continues to increase. This leads to a further increase of the contact angle. Over this range, the self-deposition occurs at a constant value of the monolayer adsorption, but there is a gradient of surface pressure across the monolayer.

The initial rate of the receding relaxation is controlled by the rate of interfacial and bulk diffusion of the surfactant through the liquid wedge to the three-phase line. The rate is low at low concentrations for which the hydrophobic transition occurs. This explains the long relaxation times observed in this system.

The receding stops once the local value of the chemical potential of the surfactant within the monolayer adjacent to the three-phase line becomes equal to the chemical potential in the bulk solution, and at the water–air interface. The dynamic equilibrium between adsorption and desorption maintains the wetting tension at a constant value. This equilibrium value does not change with time unless the concentration is changed or forced movement of the solution over the substrate is resumed. If, for example, after the adsorption/contact angle equilibrium is reached the liquid is forced to advance into the solution, the contact angle increases above the equilibrium value. This corresponds to a supercompression of the monolayer by the advancing wetting front. The chemical potential within the monolayer increases above the co-existence value. The layer begins to dissolve from the three-phase line more rapidly. A stationary level of the advancing wetting tension may be reached when by increase of the rate of the desorption/dissolution the surface pressure stabilizes at an increased constant value that is higher at higher speeds. At further higher speeds, the maximum value of the advancing wetting tension may be limited by the collapse pressure of the monolayer.

If the forced advancement is stopped, relaxation to the equilibrium contact angle occurs. The rate at which the contact angle decreases reflects the rate of the desorption of the surfactant from the three-phase line.

Like the rate of bulk dissolution from the surface of a three-dimensional crystal depends on solubility and changes from substance to substance, the rate of two-dimensional monolayer dissolution from the line boundary of the monolayer depends on its material properties. While the initial rate of adsorption behind the receding three-phase line is diffusion-controlled, the kinetics of advancing re-dissolution for a given material depends on the state of compression of the monolayer near the three-phase line. One should bear in mind that these changes in surface pressure are very large, of the order of hundreds of atmospheres in the bulk pressure equivalent.

For solid-like monolayers of CTAB adsorbed at a solid–vapor interface, the molecular mobility is apparently low. It is restricted by the electrostatic interaction of the ionic head groups of the surfactant molecules with the hydrophilic surface and lateral cohesion between the long-chain hydrocarbon tails. The relaxation times during which the liquid advances while the contact angle decreases back to the equilibrium value is of the order of hours in the CTAB case.

The relaxation rates are very different, for example, for the symmetrical tetra-butylammonium ion (TBA$^+$) compared to CTA$^+$. This is in spite of the fact that not only the molecular weight, but also the chemical composition of the two tetra-alkylammonium salts are similar. Like CTA$^+$, TBA$^+$ ions adsorb in the form of condensed monolayers at the mica–vapor interface [14]. However, characteristic times of advancing relaxation are reduced to milliseconds in the case of TBAB, compared with hours for CTAB.

The electrostatic attraction to the substrate is weaker for the much larger radius TBA$^+$ ion compared with the trimethylammonium head-group of CTAB. The lateral repulsion of the charged spheres is not supplemented by van der Waals cohesion of the long electroneutral hydrocarbon chains. This lateral cohesion is an important factor in consolidating CTAB monolayers. The surface mobility of the large quasi-spherical ions of TBA$^+$ should be much higher compared with the long-chain CTAB with a much smaller ionic group. The advancing relaxation that is slow for CTAB proceeds fast in the case of TBAB.

This explains why solutions of tetrabutylammonium bromide (TBAB) are free of contact angle hysteresis. Droplets of TBAB solutions slide without friction on the surface of mica, like droplets of oil that float on the surface of water. Interestingly, spontaneous spreading of the adsorbed layer onto more distant areas of the surface outside the area on which the monolayer was originally self-assembled is quite slow also in this case. The molecular exchange is strongly enhanced at the three-phase line by contact with water, for TBAB solubility is much higher than that for CTAB, and the rate of the three-phase line dissolution is also much higher for TBAB than for CTAB monolayers. However, molecular rearrangement within the consolidated monolayers of the two nonvolatile substances, as in three-dimensional crystalline lattices, is effectively frozen. Re-adsorption on the dry areas of the surface by mechanisms of two-dimensional vaporization is effectively prevented. Also, the rate of receding relaxation is similar at similar concentrations for both substances. This diffusion-controlled process is slow at low concentrations for both cases.

We chose these examples just to illustrate the principle of the thermodynamic analysis of wetting phenomena. This shows how by considering the equilibrium wetting and adsorption effects, complicated dynamic wetting patterns can be explained. For example, when the first immersion in CTAB solution proceeds at a finite speed, the three-phase line moves in a stick–slip mode. This effect is in fact, much more pronounced than the noisy structure of the advancing curves that are noticed with chemically grafted surfaces. The amplitude of this contact angle oscillation depends on the relation between the rate of adsorption and the speed of immersion. It increases with decreasing immersion speed and then decreases again.

The model of hydrophobic bands on the cylinder surface that underpins the three-phase line has been considered in the previous section. This was done in connection with the wetting of heterogeneous surfaces, for which the surface distribution pattern of grafted hydrophobic chains was fixed and did not change in the wetting process. It is, in fact, more relevant to the situation that we consider now. Such cylindrical bands indeed form spontaneously above the three-phase line by the CTAB re-adsorption. In our situation, the effective band width and the resistance that it creates to the immersion process depend on the speed and concentration and change over time.

At high speeds the hydrophobic band ahead of the three-phase line might not have enough time to develop. The fast increase of the contact angle increases the surface pressure of the adsorption precursor that just begins to form. The associated increase of the chemical potential increases the rate of monolayer dissolution. The band is still very narrow and the boundary value of the three-phase line adsorption is not high when a critical level of surface pressure is reached. The monolayer collapses, washes away or the meniscus jumps over it to the adjacent hydrophilic area that has not yet been touched by the precursor adsorption. The maximum amplitude of these jumps is observed at intermediate speeds. Here the monolayer has more time to develop and can be compressed well above the equilibrium value of the surface pressure determined by the concentration of the solution. At still lower speeds, the critical level of the dynamic supercompression decreases towards the equilibrium value of π. The advancing wetting tension increases with decreasing speed. The supercompression does not develop under quasi-static conditions. In the limit of zero speed, the monolayer precursor forms faster than the three-phase line is displaced. It follows the displacement of the advancing three-phase line and no wetting tension oscillations occur. There is no contact angle hysteresis on this time scale.

Once the surface is under the solution, no surfactant is adsorbed. However, it becomes deposited on the surface during the retraction. In the limit of low retraction speeds, the monolayer densities of the layers deposited by retraction from solution are close to the equilibrium values of the adsorption. The entire area of the macroscopic surface becomes covered by the condensed CTAB monolayer. This deposition from solution occurs in the manner of LB deposition even though there are not many adsorbed molecules at the water–air interface. Still the deposition occurs at a high constant value of the surface pressure controlled by the concentration of the solution. When the plate is retracted significantly faster than the rate of the re-adsorption, the monolayer is decompressed below the equilibrium value and the receding contact angle becomes lower. The deposition density related to the acting magnitude of the contact angle by the equation-of-state also decreases. When the dynamic contact angle reduces to zero, a Marangoni film ('postcursor') is pulled out of the liquid. Destabilized by the attractive heteroelectrostatic DLVO effect, it can be subsequently split into microdroplets.

These examples demonstrate that the role of compression of a monolayer on a Langmuir trough prior to deposition on the solid surface is to increase the chemical potential of the monolayer material. Our consideration shows that the compression to a high value of surface pressure at the liquid–vapor interface is not a necessary prerequisite for high density LB deposition. It can occur at low values of π_{LV} provided that the co-existence π_{SV} values are high. For insoluble lipids on a LB trough, the same thermodynamic mechanism of spontaneous self-assembly operates as for CTAB and other water-soluble surfactants and simpler solutes [15]. LB deposition is just the limiting case of zero solubility. In line with other effects of two-dimensional self-compression at the three-phase line, it should not be envisaged just as a forced mechanical transfer. The controlling thermodynamic parameter, the coefficient of distribution of the surfactant between the two co-existing interfaces, determines the acting magnitude of the 'zipper' angle at which the deposition occurs.

It is important to note again that for dilute solutions, for which the equilibrium adsorption at the solid–liquid interface remains negligible, CTAB layers deposited by retraction are rapidly washed away on re-immersion back into the solution. The fact that this deposition on retraction really occurs is already clear by comparison of the results for the first and second immersion. These are indeed entirely different. The first immersion of the clean surface into the solution occurs as in the oscillatory mode. The second immersion does not show these wetting oscillations. There are no more bands in front of the meniscus. The entire macroscopic area behind the three-phase line is now uniformly covered with the hydrophobic monolayer. There is no hydrophilic space left ahead onto which the solution could spread at a low value of the contact angle to form a wetting meniscus.

We note that any change of the acting value of the wetting tension that can occur in a nonequilibrium situation is explained in terms of two local values of the surface pressures applied at the three-phase line. When the line moves, the monolayer on one side of it becomes compressed, and on the other decompressed. The hysteresis loop narrows with decreasing speed by the mechanism that we have already explained. Equation (4), which accounts for equilibrium contact angles in terms of the acting values of surface pressure, extends to nonequilibrium situations as well.

For surfactant solutions that wet hydrophilic surfaces, as would be the case for many nonionic surfactants in particular, dewetting transitions can still occur under dynamic conditions [16]. At quasi-static conditions, provided that the value of the contact angle is zero, the monolayer becomes spontaneously transferred onto the solid at the value of the surface pressure at the water–air interface. However, when the speed of immersion is finite, it becomes compressed by the advancing wetting front above the co-existence value of the adsorption pressure. The monolayer rheology and the rates of dissolution would then account for the actual magnitude of this compression. Then, according to the basic equation, the dynamic contact angle becomes finite and can be quite high. This is a classical example of the Marangoni effect, a tension gradient dynamic phenomenon that, among other well-known manifestations, explains not just the damping of waves on water by monolayers, but also contact angle hysteresis and dynamic spreading of surfactant solutions. The same principle accounts for many other details of the dynamic wetting patterns.

7. CONCLUSIONS

We have presented a brief review of wetting phenomena that occur for 'true', e.g. chemically modified hydrophobic surfaces, and of hydrophobic transitions induced by reversible solute adsorption. The first of these two cases correspond to irreversibility, the second to reversibility of the adsorption of the chemical species responsible for high contact angles. A broad variety of static and dynamic dewetting effects occur for each of the two situations. Between these two extremes, there are inevitably intermediate situations and this explains the even greater complexity of dynamic wetting effects. For example, surfaces covered by deposited LB monolayers of insoluble lipids are quite hydrophobic. Still, such coatings are typically destabilized by the presence of water. This surface mobility shows up in complicated capillarographic patterns that can be consistently explained in terms of three-phase line re-adsorption. The associated capillary condensation effects are responsible for the long-range 'hydrophobic attraction' between such surfaces in water.

Unlike the case for CTAB monolayers that dissolve when the surface is transferred under water, dissolution is hindered for LB-deposited lipid monolayers. According to the low solubility of these materials, desorption under water proceeds very slowly. But it is enhanced in the vicinity of the three-phase line. Here the monolayer is under the stress of the surface tension effect. This increase of the local value of the chemical potential makes the monolayer thermodynamically even less stable. And this chemically activated lipid material that cannot be easily desorbed from the solid–liquid interface can be readily transferred onto the water–air interface from the areas surrounding the three-phase line. The surface of water is initially free of the lipid and the high chemical potential gradient drives the transfer. This mechanism explains the hydrophilic kinks on the capillarographic curves at the positions of the previous stagnation of the three-phase line.

As we have already noted in Section 3, the material of the solid at the three-phase line is stressed and its chemical potential is increased. Line damage of the surface is inevitably incurred. This is a very general effect. Although the time scales on which this damage becomes apparent can be very different for different materials, this process of gradual restructuring of the solid always proceeds in the course of contact angle measurements. This evolution away from the quasi-thermodynamic condition of the Young equation becomes more pronounced as the three-phase line stays longer at a fixed position. This kind of effect is progressively more difficult to deal with within the quasi-thermodynamic approximation, and this inevitably sets the limit for the validity of the Young equation.

Our analysis explains why a high contact angle in itself is not a prerequisite for monolayer stability. Only detailed capillarographic scans interpreted in terms of dynamic surface pressure effects manifested in the dynamic wetting patterns can provide firm knowledge pertinent to the physical properties of surfactant monolayers on solid surfaces and their stability in different environments. This information cannot be substituted for the morphological characterization of the monolayers by routine means of fluorescence or neutron reflectometry or AFM scanography. Many of these complicated wetting phenomena in their immense diversity can be consistently understood in the generalized context of the Gibbs adsorption equation. The principle, not restricted to water or aqueous solutions, extends to any other kinds of wetting transitions with pure liquids, surfactants, and polymers.

There might be other complications to formulate wetting as an exact science that we have not touched on this review. We have already mentioned, with references to well-known examples, that hydrophobicity can depend on the environment to which the surface was exposed to prior to the measurement. The issue that is rarely addressed in the literature since the times of the polywater boom is that the surface of silica itself changes its hydrophobicity simply by storage in air or under water [8, 16]. The phenomenon is more complex than just surface hydroxylation–dehydroxylation. Surface hydrolysis that proceeds further towards depolycondensation of the top layers of silica is a well-known phenomenon.

This kind of evolution is not very surprising in view of the complicated chemistry of silica surfaces. Any solid surfaces that we are dealing with in contact angle and surface force measurements are in a nonequilibrium state. All such surfaces are unstable and one way or another undergo decay and find their way towards a more favorable situation. Only when the evolution is sufficiently long do we have enough time to perform our measurements. This problem does not occur for surfaces of liquids that from the start of the surface tension measurement already persist in a state of thermodynamic equilibrium.

The 'soft' gel-like layers of polysilicic acids on the surface of silica were once erroneously taken for 'polywater'. The development of the diffuse surface structure that occurs under water or even in contact with water vapor can be further enhanced by branching these polymers with alkyl chains. The resulting rubber-like material of the alkylpolysiloxane layer routinely forms by ordinary surface grafting reactions of silica with alkylchlorosilanes. The associated effects of steric repulsion between such hydrophilic and hydrophobic surfaces in good solvents and bridging attraction in poor solvents and in air are well known by studies of hydration and hydrophobic surface forces. In the structure of silane-grafted surfaces, not restricted to two dimensions the way it is usually envisaged, hydrophobic and hydrophilic groups mix together or segregate, and this can depend on the environment. Such layer preamble to air, water, and organic solvents is hydrolysable by the mere nature of the siloxane bond that is very prone to rearrange. The layer undergoes a wide range of irreversible and reversible structural transformations by changes of the environment and experiences large elongations under the effects of the three-phase line and contact stress. One might need to consider similar chemical consequences for other surfaces as long as the condition of chemical equilibrium does not apply. The result comes out with far-reaching consequences for the physical theories of surface forces. The latter tend to consider an interface a kind of unchangeable mathematical plane.

Whether bare or hydrophobically substituted, the polywater properties of 'live' polymers of silicic acid and its alkyl-branched derivatives can indeed be quite intriguing. Branched and cross-linked condensation structures of the soft hydrophilic and hydrophobic materials occur in dynamic chemical exchange for which the equilibrium can eagerly be shifted by reversible surfactant and electrolyte adsorption, a co-solvent, pH, temperature, and other factors. In both reversible and history-dependent aspects, such surfaces display 'strange' long-range forces. They occur in the form of hydrophobic attraction [17] that occurs also in air; and also in the form of hydration repulsion [18] which in extreme situations extends to equally large distances. These capillary springing effects are related to the three-phase line instabilities caused by the normal tension stress in the contact angle measurements [3, 7].

One always has to bear in mind these kinds of effects when dealing with real substrates. The observations are linked to the general problem of the capillarity of solid states of matter. Here the condition for absolute thermodynamic equilibrium is extended towards the material of the solid substrate. This principle, however, is inconsistent with the Young equation. In order to deal with soluble and/or volatile substances, one has to sacrifice the quasi-thermodynamic approach and use the Kelvin and Gibbs equations instead. For solids with a high vapor pressure or solubility, equilibration times can be reduced so that relevant experimental observations can be done. We briefly commented on solid-state capillarity in this review.

The entire arena of wetting phenomena appears more involved than common theoretical concepts or routine interpretations of contact angles. New ideas in this conceptually rich area of static and dynamic wetting transitions have just begun to emerge. Solid state capillarity is a more general physical discipline that rests on classical principles of chemical thermodynamics. These have already provided exact results over the past two centuries in the realm of liquid–fluid interfaces. The Young approximation is just a limiting case. Our analysis shows its importance for the quantitative analysis of wetting phenomena and points to limits of its applicability. The (quasi) thermodynamic analysis, powerful and exact, provides us with new insights that still cannot be based entirely on computer simulation and surface analysis techniques. In this review, we have tried to extend the (quasi) thermodynamic view of wetting phenomena.

REFERENCES

1. V. V. Yaminsky and B. W. Ninham, *Adv. Colloid Interface Sci.* (in press).
2. D. H. Everett and C. J. Radke, in: *Adsorption at Interfaces*, K. L. Mittal (Ed.), ACS Symposium Series No. 8, p. 1. American Chemical Society, Washington, DC (1975).
3. V. V. Yaminsky and M. B. Johnston, *Langmuir* **11**, 4153–4158 (1995).
4. A. W. Adamson, *Physical Chemistry of Surfaces,* 5th edn. Wiley, New York (1990).
5. D. F. James, *J. Fluid Mech.* **63**, 657–664 (1974).
6. T. D. Blake, in: *Wettability*, J. C. Berg (Ed.), p. 251. Marcel Dekker, New York (1993).
7. V. V. Yaminsky, P. M. Claesson and J. C. Eriksson, *J. Colloid Interface Sci.* **161**, 91–100 (1993).
8. K. Eskilsson and V. V. Yaminsky, *Langmuir* **14**, 2444–2450 (1998).
9. L. Ter-Minassian-Saraga, in: *Contact Angle, Wettability and Adhesion*, Adv. Chem. Ser. No. 43, pp. 232–249. American Chemical Society, Washington, DC (1964).
10. E. A. Vogler, D. A. Martin, D. B. Montgomery, J. Graper and H. W. Sugg, *Langmuir* **9**, 497–507 (1993).
11. B. Frank and S. Garoff, *Langmuir* **11**, 4333–4340 (1995).
12. S. Gerdes and F. Tiberg, *Langmuir* (in press); S. Gerdes, Ph.D. Thesis, Institute for Surface Chemistry (Stockholm) and Lund University (1998).
13. V. V. Yaminsky and K. B. Yaminskaya, *Langmuir* **11**, 936–941 (1995).
14. V. V. Yaminsky, B. W. Ninham and M. Karaman, *Langmuir* **13**, 5979–5990 (1997).
15. V. V. Yaminsky, T. Nylander and B. W. Ninham, *Langmuir* **13**, 1746–1757 (1997).
16. V. V. Yaminsky, K. Thuresson and B. W. Ninham, *Langmuir* **15**, 3683 (1999).
17. V. V. Yaminsky, *Colloids Surfaces* **129–130**, 415–424 (1997).
18. V. V. Yaminsky, B. W. Ninham and R. M. Pashley, *Langmuir* **14**, 3223–3235 (1998).

Apparent and Microscopic Contact Angles, pp. 95–109
J. Drelich, J. S. Laskowski and K. L. Mittal (Eds)
© VSP 2000.

Wetting of SiO₂ surfaces by phospholipid dispersions

LUIZ CARLOS SALAY [1] and ANA MARIA CARMONA-RIBEIRO [2,*]

[1] *Laboratório de Sistemas Integráveis, Departamento de Engenharia Eletrônica, Escola Politécnica,*
Universidade de São Paulo, CP 8174, 01065-970 São Paulo SP, Brazil
[2] *Departamento de Bioquímica, Instituto de Química, Universidade de São Paulo, CP 26077,*
05599-970 São Paulo SP, Brazil

Received in final form 12 February 1999

Abstract—Wetting of $SiO_2/Si/SiO_2$ slides by phospholipid dispersions has been determined over a range of phospholipid concentrations and times of interaction between vesicles and the solid surface. Both advancing and receding dynamic contact angles (θ) increased and the contact angle hysteresis decreased as a function of the phospholipid concentration (or interaction time), attaining a maximum (a minimum). Maximization of both the advancing and receding contact angles and minimization of the contact angle hysteresis are associated with an increase in the chemical homogeneity of the surface and one-bilayer deposition. Determination of contact angles is a powerful, quick, and simple technique to establish experimental conditions for bilayer deposition on solid surfaces in general. This result may be of importance for further advancements in rapidly developing research areas such as the design of biosensors, immunodiagnosis, and the development of biocompatible materials.

Key words: Interactions; phospholipids; SiO_2 surfaces; surface tension; dynamic contact angles; bilayer adsorption vs. vesicle adhesion.

1. INTRODUCTION

The interaction of water-soluble surfactants with solid polymeric or mineral surfaces has been intensively and extensively studied [1–3]. However, the potential of bilayer-forming amphiphiles as interface agents able to modify solid surfaces remains hitherto poorly explored. Bilayer-forming amphiphiles assemble on solid particles or planar surfaces either as a bilayer, a monolayer, or adhered vesicles, depending on the nature of the amphiphile and surface and on the interactions driving amphiphile deposition [4–16]. Bilayer deposition of dioctadecyldimethylammonium bromide (DODAB) and sodium dihexadecylphosphate (DHP) from vesicles onto oppositely charged polystyrene microspheres has been reported [4]. There is

---

*To whom correspondence should be addressed. E-mail: mcribeir@quim.iq.usp.br

electrostatically-driven vesicle adhesion to the latex that is followed by bilayer deposition [5]. Thereafter, as the amphiphile concentration increases, vesicles adhere to the bilayer-covered latex with or without disruption, depending on the nature of the synthetic amphiphile [5]. In general, the interaction between bilayer vesicles and solid surfaces in the form of particulates [6–15] or planar surfaces [16, 17] still lacks characterization from the point of view of exact physical parameters. Recently, the physical adsorption of bilayer-forming amphiphiles on hydrophilic silica, in particular, was shown to depend on previous centrifugation of the vesicle sample, the pH, the buffer, the temperature, and the physical state of the bilayer [14]. In fact, establishing suitable experimental conditions for the occurrence of bilayer deposition as well as an efficient and quick methodology to ascertain whether bilayer deposition has indeed taken place on a solid surface is still a problem for those interested in producing supported bilayers. Many applications of supported bilayers in biotechnology and biophysics are presently hampered by the poor reproducibility of various bilayer deposition methods. The process of bilayer deposition on solid substrates needs to be described in terms of exact physical parameters such as the contact angle [10] or surface roughness [16].

For some synthetic amphiphile vesicles [18], it is possible to distinguish mere vesicle adhesion from bilayer deposition on a solid surface simply by surface tension and contact angle determinations as a function of the amphiphile concentration and time of interaction between the amphiphile dispersion and the surface [10, 19]. In this case, contact angle maximization and contact angle hysteresis minimization can be used as criteria for bilayer deposition [10, 19]. We recently determined the influence of time and DODAB concentration on assembling one bilayer or adhered vesicles on planar SiO_2 surfaces [19]. Synthetic bilayer deposition was achieved at a maximum of advancing or receding contact angle and a minimum of contact angle hysteresis both for polymeric [10] and for mineral surfaces [19]. In this work, we generalize the criteria previously established for the synthetic vesicles to two phospholipids: phosphatidylcholine (PC) and dipalmitoylphosphatidylcholine (DPPC). In addition, phospholipid concentration and time effects on the phospholipid assembly at a SiO_2 surface are presented.

2. MATERIALS AND METHODS

Egg PC (type XIII-E) and DPPC of the highest purity available were from Sigma. Phospholipid concentrations were determined by inorganic phosphorus (P_i) analysis [20]. All other reagents were of analytical grade and were used without further purification. Water was Milli-Q quality.

Small unilamellar phospholipid vesicles were prepared by ethanolic injection [21, 22] in water or in 10 mM Tris buffer at pH 7.4. The injection was followed by 4 h dialysis (20 ml against 2 l, 2 ×) to eliminate ethanol and centrifugation (14 000 rpm for 1 h) to precipitate any multilamellar vesicles eventually present.

Typical vesicle mean diameter was 62 and 74 nm for PC and DPPC, respectively, as measured from photon correlation spectroscopy [6].

Monocrystalline silicon slides ($<$ 100 $>$ crystal orientation, p type, 10–20 Ω cm resistivity) of 0.4 mm mean thickness were cut with a diamond tip in order to obtain rectangular planar surfaces (25 $\times$ 15 mm). Before thermal oxidation, cleaning of the silicon surface was performed in accordance with standard procedures for silicon uses in microelectronics [23]. Rapid thermal oxidation [24] was carried out under dried oxygen (less than 5 ppm water) at 1150°C for 500 s using a Heatpulse 410T heater (A. G. Associates, Sunny Vale, CA, USA). A thickness of about 30 nm for the SiO$_2$ layer was determined using a Rudolph Ellipsometer (Autoel IV Nir-3/4D/SSI). Two SiO$_2$/Si/SiO$_2$ slides of the same dimensions were glued at their rough faces with the smooth silicon oxide surface to the outside. Routinely, this slide was cleaned with deionized water/ethanol/deionized water/isopropyl alcohol/deionized water in that order. Thereafter, the smooth SiO$_2$ surfaces were considered clean with respect to the occurrence of deposited organic films or particles and were attached to the dynometer balance.

Advancing and receding dynamic contact angles were determined using the extended Wilhelmy plate or wetting-balance method [25–27]. The Wilhelmy plate technique was introduced as a method of measuring the surface tension between air and a liquid [25]. This technique has been extended to measurements of dynamic contact angles [26]. The validity of this method was previously established by comparing its results with contact angle determinations using the sessile drop method on a number of different surfaces [26]. Force–depth profiles for the slide under immersion or emersion in water or in dioctadecyldimethylammonium (DODA) salt dispersions were obtained using a Dynometer (Byk-Labotron, Germany). The instrument consists of a recording balance, to which the SiO$_2$ slide is attached, and an elevator holding the liquid sample. The liquid surface is slowly moved along the SiO$_2$ slide and the forces exerted on the slide are continuously recorded as a function of the position. The surface tension at any location of the three-phase boundary causes a downward force, whereas buoyancy causes an upward force (Fig. 1). The maximum pressure acting on the transducer yields an output of + 10 V, whereas maximal tension yields − 10 V. The phospholipid dispersion was placed in a beaker on the platform. The SiO$_2$ slide holder was then adjusted to align the bottom edge of the slide with the liquid surface. Immersion and withdrawal of the SiO$_2$ slide were accomplished by raising and lowering the platform. In the present work, platform velocities available from the Dynometer were 1.5 and 15 mm/min. No velocity effect on wettability was observed for these two extreme speeds either in the presence or in the absence of phospholipid. The smaller velocity was adopted for all measurements, namely 1.5 mm/min. The receding or advancing contact angle θ was calculated from the equation [26]

$$\cos \theta = (1/p\gamma)\{F + \rho g Ah\}, \tag{1}$$

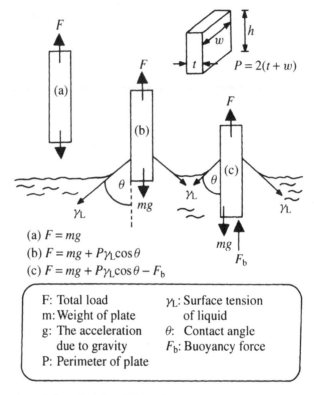

(a) $F = mg$

(b) $F = mg + P\gamma_L\cos\theta$

(c) $F = mg + P\gamma_L\cos\theta - F_b$

F: Total load	γ_L: Surface tension
m: Weight of plate	of liquid
g: The acceleration	θ: Contact angle
due to gravity	F_b: Buoyancy force
P: Perimeter of plate	

Figure 1. Illustration of the principle of dynamic contact angle measurements by the modified Wilhelmy method.

where p is the perimeter of the slide (in cm), γ is the surface tension (in dynes/cm), F is the force on the slide (in dynes), ρ is the density of water (in g/ml), $g = 980.2$ dynes/g), A is the cross-sectional area of the slide, and h is the depth of immersion (in cm).

Equation (1) predicts a linear relationship between F and h, if θ is constant. At a given constant and slow speed, for a smooth and homogeneous surface with a constant perimeter p, the contact angle remains constant along movement [26]. A rate dependence of contact angles determined by this method has been related to surface heterogeneity [26]. Because the SiO_2 surfaces in this work are rather smooth, no rate dependence is to be expected for the contact angle measurements. The force on the plate was obtained from the apparent mass weighed by the balance multiplied by the acceleration due to gravity. For determining θ, the dependence of the apparent mass on the immersed depth was obtained first for immersion and then for emersion of the slide. The linear portion of the mass–depth profile was extrapolated to yield the apparent mass at $h = 0$. From the intercept, the slide perimeter p (in cm), and the measured surface tension, the contact angle was obtained. Figure 1 shows a self-explanatory scheme where the principle of dynamic contact angle measurements by the modified Wilhelmy method is illustrated. One should note that as the velocity is constant and equal to 1.5 mm/min, the depth of immersion of the slide is directly proportional to time, so that measuring time after starting immersion (or emersion) is the same as measuring depth of immersion (h).

The equilibrium surface tension at 25°C was measured as a function of the PC or DPPC concentration from zero up to 1 mM phospholipid, using a Du Nouy ring and the Dynometer itself set to the surface tension mode. For surface tension measurements, 20 ml of phospholipid dispersion was added to a polypropylene beaker (4 cm diameter) so that equilibration of the phospholipid with the air–water interface was always over the same total air–water interfacial area of 12.56 cm^2. Surface tension was measured as a function of the phospholipid concentration either in pure water or in 10 mM Tris buffer at pH 7.4, or as a function of time after adding the phospholipid dispersion to the beaker.

3. RESULTS

3.1. Effect of the equilibration time and the phospholipid concentration on the surface tension at the air–water interface

Typical phospholipid dispersions composed of small unilamellar vesicles in 10 mM Tris buffer at pH 7.4 displayed equilibrium surface tensions close to that of pure water up to 1.0 mM phospholipid (Figs 2b and 2d). In water, there was a slight decrease of surface tension as a function of the phospholipid concentration that was

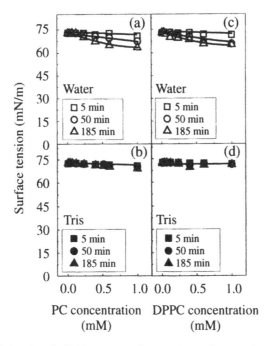

Figure 2. Effect of the phospholipid concentration on the surface tension γ (in mN/m) at the air–water interface for three different equilibration times of vesicle dispersion in the beaker. PC (a, b) or DPPC vesicles (c, d) were prepared in water (a, c) or in 10 mM Tris at pH 7.4 (b, d).

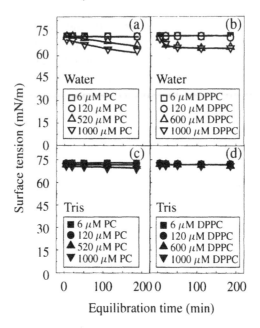

Figure 3. Effect of the equilibration time on the surface tension for four different phospholipid concentrations in the beaker equilibrating with the air–water interface. PC (a, c) or DPPC vesicles (b, d) were prepared in water (a, b) or in 10 mM Tris at pH 7.4 (c, d).

more pronounced at the highest equilibration times (Figs 2a and 2c). This is shown at three different equilibration times for equilibration of vesicles with the air–water interface (Fig. 2). In Tris as a buffer, at room temperature, the surface tension was not affected by the equilibration time as shown at three different equilibration times (Figs 2b and 2d). In water, the surface tension decreased slightly as a function of the phospholipid concentration at three different equilibration times (Figs 2a and 2c). The effect of the equilibration time on the surface tension is better seen in Fig. 3. Basically, in buffer, there is no equilibration time effect on the surface tension, whereas, in water, there is a small effect for both the phospholipids tested (Fig. 3). A slight but significant decrease in surface tension requires large equilibration times and phospholipid concentrations. The surface tension data in this subsection are required for calculation of the dynamic contact angles in the next sections.

3.2. Effect of the phospholipid concentration on the contact angles and contact angle hysteresis on the SiO_2 surface

The effect of the phosphatidylcholine (PC) concentration on the contact angles is shown in Fig. 4. At low PC concentrations, the presence of PC as vesicles in dispersion causes a significant decrease of the contact angle in comparison with that measured for the bare SiO_2 surface in water and in the absence of phospholipid (Fig. 4a). However, at phospholipid concentrations larger than 0.2 mM, the contact angle increases, attaining a value that is larger than the angle obtained for the bare surface (Fig. 4a).

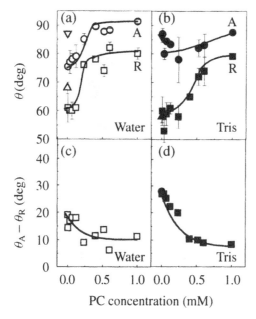

Figure 4. Effect of the PC concentration and medium on the advancing (A) or receding (R) dynamic contact angles (a, b) and on contact angle hysteresis, $\theta_A - \theta_R$ (c, d), for SiO₂ surfaces in the presence of PC vesicles. In a and c, the medium is pure water. In b and d, the medium is 10 mM Tris buffer, pH 7.4. The single different symbol in each figure represents the contact angle measured for the bare SiO₂ surface, in the absence of vesicles. Measurements were taken at 2 h interaction time.

Tris used as a buffer presented the interesting property of reducing the contact angle of the solid surface. This can be seen from the comparison between angles for the bare surface in 10 mM Tris, pH 7.4 (solid triangles in Fig. 4b) and angles for the bare surface in water (open triangles in Fig. 4a). Similarly to the measurements in water, in Tris, the contact angle also increased as a function of the PC concentration, becoming steeper at the PC concentrations, above 0.3 mM and attaining a maximum value at 0.5 mM PC (Fig. 4b).

The difference between the advancing and the receding contact angles is the contact angle hysteresis. Hysteresis attained a minimum as a function of the PC concentration both in water (Fig. 4c) and in Tris (Fig. 4d). The maximum values for both contact angles both in water and in Tris were attained at the same PC concentration where the contact angle hysteresis attained a minimum (Figs 4c and 4d). In water, this minimum was first observed at *ca.* 0.2–0.3 mM PC, whereas in Tris, it was attained at about 0.5 mM PC. There is a close relationship between hysteresis minimization and advancing or receding contact angle maximization.

The effect of dipalmitoylphosphatidylcholine (DPPC) vesicles on the wettability of the SiO₂ surface is shown in Fig. 5. In this case, both the advancing and receding contact angles increase steeply as a function of the DPPC concentration from the start at very low phospholipid concentrations (Figs 5a and 5b). Also, hysteresis minimization (shown in Figs 5c and 5d) is achieved at very low DPPC concentrations

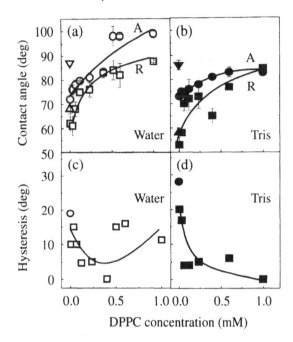

Figure 5. Effect of the DPPC concentration and medium on the advancing (A) or receding (R) dynamic contact angle (a, b) and on contact angle hysteresis, $\theta_A - \theta_R$ (c, d), for SiO$_2$ surfaces in the presence of DPPC vesicles. In a and c, the medium is pure water. In b and d, the medium is 10 mM Tris buffer, pH 7.4. The single different symbol in each figure represents the contact angle measured for the bare SiO$_2$ surface, in the absence of vesicles. Measurements were taken at 2 h interaction time.

(*ca.* 0.1 mM DPPC in water and *ca.* 0.05 mM DPPC in Tris). In water, hysteresis does not seem to remain low for the higher DPPC concentrations, possibly due to further vesicle adhesion to the bilayer-covered surface (see Discussion), but in Tris the phospholipid assembly on the surface seems to be able to guarantee a very high surface homogeneity, thereby nullifying the contact angle hysteresis (Figs 5c and 5d). One should note that the measurements above were done at a fixed time of interaction of 2 h.

3.3. Effect of the time of interaction between the vesicle and surface on the contact angles at a fixed phospholipid concentration

At a PC concentration of 0.52 mM, the contact angle increased, attaining a maximum as a function of the interaction time (Figs 6a and 6b). Hysteresis minimization was coincident with contact angle maximization and was observed from 120 and 80 min of interaction in water and in Tris, respectively (Figs 6c and 6d).

At a DPPC concentration of 0.235 mM, both contact angles increased, attaining a maximum as a function of the interaction time (Figs 7a and 7b). Hysteresis minimization was coincident with the advancing and receding contact angle maxi-

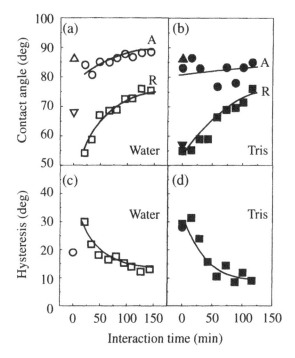

Figure 6. Effect of the time of interaction between vesicles (0.52 mM PC) and the SiO₂ surface on the advancing or receding contact angle (a, b) and contact angle hysteresis (c, d). Single symbols in each figure represent the contact angle measured for the surface in the absence of vesicles.

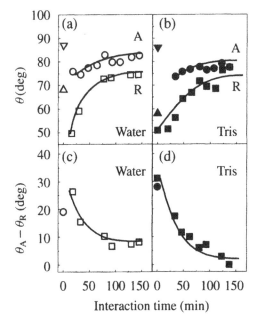

Figure 7. Effect of the time of interaction between vesicles (0.235 mM DPPC) and the SiO₂ surface on the advancing or receding contact angle (a, b) and contact angle hysteresis (c, d). Single symbols in each figure represent the contact angle measured for the surface in the absence of vesicles.

mization and was observed from 75 and 120 min of interaction time in water and in Tris, respectively (Figs 7c and 7d).

4. DISCUSSION

Table 1 summarizes the minimization of contact angle hysteresis, in degrees, depicted from the profiles of advancing and receding contact angles as a function of the phospholipid concentration at 2 h of interaction time, or as a function of the interaction time at 0.235 mM DPPC or at 0.520 mM PC. Contact angles previously obtained [19] for the same SiO_2 surface in the presence of 0.235 mM DODAB, a cationic bilayer-forming amphiphile, at 2 h interaction time are also included in Table 1 for comparison.

Table 2 summarizes the experimental conditions required for minimization of hysteresis and maximization of the advancing or receding contact angle. The two variables to be controlled are the time of interaction between the vesicles and the surface, and the phospholipid concentration. If contact angle hysteresis does not drop to values close to zero under a given experimental condition, the assembly on the surface cannot be a single bilayer. For example, DODAB vesicles in water,

Table 1.

Hysteresis minima (in degrees) obtained from equilibrium advancing and receding contact angle (θ) dependences on the phospholipid concentration (C) at an interaction time of 2 h or from advancing and receding contact angle dependences on the interaction time (t) at a fixed C (0.235 mM for DODAB or DPPC; 0.520 mM for PC). Minima values were taken from the curves θ vs. C and θ vs. t. Data for DODAB were taken from ref. [19]

Lipid	Hysteresis minimum ($\theta_A - \theta_R$), degrees			
	θ vs. C		θ vs. t	
	Water	Tris	Water	Tris
DODAB	20	3	20	5
PC	12	8	12	8
DPPC	2	2	7	2

Table 2.

Minimal lipid concentration (C_{min}) at $t = 2$ h and minimal time of interaction (t_{min}) at 0.235 mM DPPC or DODAB, or at 0.520 mM PC for hysteresis minimization, and maximization of advancing and receding contact angles

Lipid	C_{min} (mM)		t_{min} (h)	
	Water	Tris	Water	Tris
DODAB	—	0.20	—	1.7
PC	0.40	0.40	1.7	1.0
DPPC	0.40	0.25	1.7	1.7

which do not yield hysteresis minimization to values close to zero, hysteresis remaining high and equal to 20° (Table 1), do not deposit as one bilayer on the flat SiO_2 surface. However, in 10 mM Tris, the increased ionic strength might have induced fusion of vesicles [28, 29] which had previously adhered to the surface due to the electrostatic attraction between the cationic vesicles and the negatively charged SiO_2 surface. Consequently, the surface acquires the large chemical homogeneity of the DODAB bilayer surface and hysteresis becomes as small as 2–5° (Table 1) [19]. This is in very good agreement with adsorption isotherms for DODAB on silica both in water and in Tris [14]. In water, the surface charge on silica is low and this causes destabilization of the DODAB bilayer deposition which is not observed at the higher charge densities that occur in 10 mM Tris [14]. Therefore, electrostatically-driven adsorption such as that for DODAB on SiO_2 is similar both for particulates and for flat surfaces.

Phospholipid adsorption on spherical and flat SiO_2 surfaces can be compared because PC or DPPC adsorption isotherms on silica are also available from the literature for a range of experimental conditions [14]. From Table 1, the best amphiphiles to produce bilayer coverage under the experimental conditions specified are DODAB in Tris and DPPC in both water and Tris, although a not so large but still considerable reduction in hysteresis was obtained for PC in water or Tris. Deposition of neutral phospholipid bilayers on silica requires special experimental conditions such as the use of Tris as buffer and vesicles in the liquid-crystalline state instead of vesicles in the rigid gel state [14]. The affinity of the bare silica surface for neutral phospholipids is low and the usual amount of −OH groups is not sufficient to drive deposition of a phospholipid bilayer from a vesicle via hydrogen bridges. Because DPPC has a lower area per molecule at the air–water interface (*ca.* 0.5 nm²) than PC (*ca.* 0.7 nm²), DPPC has a tendency to form larger aggregates than does PC, thereby tending to deposit as bilayers that are less curved. This could explain the larger affinity that DPPC has presented for the flat surface in comparison with that exhibited by PC (Table 2). In fact, we have already made a similar observation on comparing the adsorption of dihexadecylphosphate (DHP) and DODAC on oppositely charged latex [4]. Because DHP has a smaller area per molecule at the air–water interface, it tends to deposit more easily onto surfaces that are almost flat, as are large polystyrene microspheres (850 nm mean diameter), whereas DODAC deposits with higher affinity on surfaces that are more curved, as are small latex beads [4]. DPPC has a much lower affinity for silica particles (50 nm diameter) [14] than it does for a flat surface where bilayer deposition at room temperature can be inferred from the contact angle criterion (Tables 1 and 2). For the flat surface, at 0.52 mM PC and 2 h of interaction time, PC bilayer deposition was barely achieved, whereas high surface homogeneity had already been attained at 0.235 mM DPPC/2 h interaction (cf. the hysteresis minimum achieved for both phospholipids presented in Table 1). Certainly, from the molecular geometry for the PC and the DPPC molecules, the affinity of a PC or DPPC bilayer-vesicle for a surface also should depend on the radius of curvature of the solid surface in question, since the molecular association in the bilayer has to conform to the curvature of the supporting solid surface. Our results from adsorption measurements on silica in ref. [14] (representing a very curved solid surface) and on the SiO_2 slide (planar surface) in this and previous work [19] strongly indicate the preference of DPPC bilayers for flat surfaces and the preference of PC bilayers for highly curved surfaces at room temperature.

The interpretation of the contact angle data is straightforward. Maximization of the advancing or receding contact angle and minimization of hysteresis are consistent with bilayer deposition since withdrawal of the surface implicitly causes the outer monolayer of the deposited bilayer to remain in the water phase, whereas the inner monolayer remains attached to the solid surface. This causes the increased surface hydrophobicity observed as a large contact angle at the plateau of the angle vs. concentration or angle vs. interaction time curves. Bilayer deposition can eventually be followed by further vesicle adhesion as the phospholipid concentration increases further. This would be due to the van der Waals attraction between the deposited bilayer and free bilayer vesicles in the dispersion. In this case, the advancing and receding contact angles and contact angle hysteresis should drop. This seems to be occurring for DPPC in water (Fig. 5c) and for DODAB in Tris [19] since hysteresis as a function of the amphiphile concentration presents an inverted bell-shaped profile with a minimum, whereas the advancing or receding contact angle presents a bell-shaped profile with a maximum.

Over the range of low phospholipid concentrations, the contact angles always decreased in comparison with the angles measured for a bare surface. This suggests that vesicle adhesion to the solid surface is the first interaction step. If vesicle rupture and fusion of vesicles at the surface do not take place, the contact angles should remain low and hysteresis should remain high because surface roughness for a surface with adhered vesicles (which collapse upon receding into air but still carry water) is expected to be high. This interpretation is confirmed from the dynamics of the interaction between phospholipid vesicles and the solid surface (Figs 6 and 7). PC vesicles in water at 0.52 mM PC take a long time to cover the surface with a bilayer since hysteresis decreased only to 12° after more than 2 h of interaction with the surface (Fig. 6c). Certainly, hysteresis minimization is more rapid for DPPC (Figs 7c and 7d) which has a higher propensity for deposition over a flatter surface compared with PC.

The effect of the interaction time on the advancing and receding contact angles (Figs 6 and 7) shows that time affects the angle by producing an equilibrium contact angle. Further increase of the interaction time does not affect this constant angle (Figs 6a, 6b, 7a, and 7b). The error bars in Figs 4 and 5 correspond to the mean standard deviations for measurements taken after equilibrium was attained from 1.5 h of interaction time.

It is interesting how dynamic aspects of the vesicle–surface interaction can be further elucidated from the point of view of the dynamic contact angle (Fig. 6). For PC vesicles in water or in Tris, the advancing and receding equilibrium contact angles attained after an interaction time of 80–100 min almost do not change in comparison with those obtained for the bare surface (Figs 6a and 6b). Hysteresis at equilibrium is also similar to that obtained for the bare surface, being slightly smaller than the 20° obtained for the bare surface (Fig. 6c). At equilibrium with the surface, PC vesicles in water at a relatively high concentration (0.52 mM PC) do not seem to have a significant effect on the wettability of the surface. Indeed, the affinity of PC for the SiO₂ slide indeed seems to be low as depicted from the sigmoidal dependences for the PC concentration effect on wettability (Figs 4a and 4b).

For DPPC, in water, hysteresis was still large at receding and advancing contact angle maximization (Figs 5c and 7c). However, at DPPC bilayer coverage in Tris (at maximal receding or advancing angle in Figs 5b and 7b), hysteresis not only drops considerably, but also approaches values very close to zero. At maximal angle, the advancing and receding angles differ by about 2°, largely within the limits of the experimental error of the technique (Table 1 and Figs 5d and 7d).

For poly(styrene/methacrylate) copolymer rods interacting with DODAB, the equilibrium contact angle increased as a function of the amphiphile concentration, attaining a plateau value which depended on the charge density of the polymer surface [10]. At the plateau, there was no hysteresis for DODAB dispersions interacting with the most charged poly(styrene/methacrylate) rods [10]. The same was observed for DODA chloride and DODA acetate [10]. Thus, two completely different materials such as the copolymer [10] and the mineral oxide [19] yielded a minimum in contact angle hysteresis and a maximum in equilibrium contact angle itself upon coverage with a DODAB bilayer.

For PC and DPPC, which are neutral phospholipids, no effect of the surface charge is to be expected but an effect of Tris as an adsorbed buffer on the surface could generate stabilization of bilayer adsorption via the formation of hydrogen bridges between −OH in the adsorbed buffer and −P=O in the phospholipid [14].

Contact angle hysteresis has also been related to mechanical and/or chemical surface heterogeneities. If surface roughness were the major factor accounting for hysteresis, coverage with one bilayer would not be expected to effectively reduce surface roughness and thereby bilayer coverage would have no effect on the hysteresis measured. On the other hand, chemical heterogeneity would indeed be eliminated by covering the surface with a chemically homogeneous DODAB bilayer. Thus, chemical heterogeneity on the SiO₂ surface is the most probable cause for contact angle hysteresis observed for the bare surface. This conclusion agrees fairly well with detailed work in the literature regarding polishing effects on the bare SiO₂ surface [30]. A remarkable decrease in contact angle hysteresis (from 50° to 10°) was obtained by chemomechanical polishing, a procedure known to eliminate chemical heterogeneity on the native surface. Considering the limits of the experimental error, an even more remarkable effect is described here: the reduction of contact angle hysteresis practically to zero by means of bilayer coverage.

5. CONCLUSIONS

Phospholipid vesicles cause changes in the wettability behavior of planar SiO_2 surfaces due to vesicle adhesion or bilayer deposition which yield smaller or larger contact angles, respectively, in comparison with the angles on the bare surface. The affinity of the bilayer for the surface depends on the surface curvature radius and on the geometrical parameter $v/(al)$ of the lipid. DPPC prefers to deposit as a bilayer on flat surfaces, whereas PC prefers to deposit as a bilayer on more curved surfaces. This can be understood from their different mean areas per molecule at the air–water interface (0.5 nm^2 for DPPC and 0.7 nm^2 for PC) which yield different geometrical parameters for the self-assembly. The tendency of DPPC to deposit as a flat bilayer is larger than that exihibited by PC at room temperature.

Phospholipid vesicles either adhere or open up upon contact with the surface. Bilayer deposition typically corresponds to the absence of contact angle hysteresis and maximization of equilibrium contact angles. The present results for planar SiO_2 surfaces have been compared with previously reported adsorption isotherms for PC or DPPC on silica in water or in 10 mM Tris buffer at pH 7.4 [14], where DPPC has a much lower affinity for the silica particle than PC. A major factor controlling stabilization of the deposited phospholipid bilayer is the co-operative formation of several hydrogen bridges between $-OH$ groups on the surface and $-P=O$ on the phospholipid.

Acknowledgements

We thank FAPESP and CNPq for financial support. L.C.S. thanks CAPES for a fellowship and Mr. Zambom and Mr. Mansano for their help. The use of facilities at the Departamento de Engenharia Eletrônica-LSI-Escola Politécnica da USP is gratefully acknowledged.

REFERENCES

1. J. F. Goodman and T. Walker, in: *Colloid Science*, D. H. Everett (Ed.), Vol. 3, pp. 230–252. The Chemical Society, London (1978).
2. T. D. Blake, in: *Surfactants*, Th. F. Tadros (Ed.), pp. 221–275. Academic Press, London (1984).
3. D. Attwood and A. T. Florence, *Surfactant Systems — Their Chemistry, Pharmacy and Biology*. Chapman & Hall, London (1983).
4. A. M. Carmona-Ribeiro and B. R. Midmore, *Langmuir* **8**, 801–806 (1992).
5. L. R. Tsuruta, M. M. Lessa and A. M. Carmona-Ribeiro, *Langmuir* **11**, 2938–2943 (1995).
6. A. M. Carmona-Ribeiro and T. M. Herrington, *J. Colloid Interface Sci.* **156**, 19–24 (1993).
7. L. R. Tsuruta, M. M. Lessa and A. M. Carmona-Ribeiro, *J. Colloid Interface Sci.* **175**, 470–475 (1995).
8. L. R. Tsuruta and A. M. Carmona-Ribeiro, *J. Phys. Chem.* **100**, 7130–7134 (1996).
9. S. M. Sicchierolli and A. M. Carmona-Ribeiro, *Colloids Surfaces B — Biointerfaces* **5**, 57–61 (1995).
10. M. Lessa and A. M. Carmona-Ribeiro, *J. Colloid Interface Sci.* **182**, 166–171 (1996).
11. M. Sicchierolli and A. M. Carmona-Ribeiro, *J. Phys. Chem.* **100**, 16771–16775 (1996).

12. M. Bayerl and M. Bloom, *Biophys. J.* **58**, 357–362 (1990).
13. C. Naumann, T. Brumm and T. M. Bayerl, *Biophys. J.* **63**, 1314–1318 (1992).
14. R. Rapuano and A. M. Carmona-Ribeiro, *J. Colloid Interface Sci.* **193**, 104–111 (1997).
15. M. Linseisen, M. Hetzer, T. Brumm and T. M. Bayerl, *Biophys. J.* **72**, 1659–1674 (1997).
16. J. Raedler, H. Strey and E. Sackmann, *Langmuir* **11**, 4539–4546 (1995).
17. P. Heyn, M. Egger and H. E. Gaub, *J. Phys. Chem.* **94**, 5073–5077 (1990).
18. A. M. Carmona-Ribeiro, *Chem. Soc. Rev.* **21**, 207–212 (1992).
19. L. C. Salay and A. M. Carmona-Ribeiro, *J. Phys. Chem. B* **102**, 4011–4015 (1998).
20. G. Houser, S. Fleischer and A. Yamamoto, *Lipids* **5**, 594–597 (1970).
21. S. Batzri and E. D. Korn, *Biochim. Biophys. Acta* **298**, 1015–1018 (1973).
22. J. H. Kremer, M. W. J. v. d. Esker, C. Pathmamanoharan and P. H. Wiersema, *Biochemistry* **16**, 3932–3936 (1977).
23. S. G. S. Filho, C. M. Hasenack, L. C. Salay and P. W. Mertens, *J. Electrochem. Soc.* **142**, 902–906 (1995).
24. Y. L. Chiou, C. H. Sow and G. Li, *Appl. Phys. Lett.* **57**, 881–883 (1990).
25. L. Wilhelmy, *Ann. Phys.* **119**, 177–180 (1863).
26. R. E. Johnson, Jr., R. H. Dettre and D. A. Brandreth, *J. Colloid Interface Sci.* **62**, 205–210 (1977).
27. A. Mennella and N. R. Morrow, *J. Colloid Interface Sci.* **172**, 48–54 (1995).
28. A. M. Carmona-Ribeiro and H. Chaimovich, *Biophys. J.* **50**, 621–628 (1986).
29. A. M. Carmona-Ribeiro, *J. Phys. Chem.* **97**, 11 843–11 849 (1993).
30. R. Thomas, F. B. Kaufman, J. T. Kirleis and R. A. Belsky, *J. Electrochem. Soc.* **143**, 643–647 (1996).

Apparent and Microscopic Contact Angles, pp. 111–128
J. Drelich, J. S. Laskowski and K. L. Mittal (Eds)
© VSP 2000.

Wettability and microstructure of polymer surfaces: stereochemical and conformational aspects

OLEG N. TRETINNIKOV

B. I. Stepanov Institute of Physics, National Academy of Sciences of Belarus, Minsk 220072, Belarus and Institute for Frontier Medical Sciences, Kyoto University, Sakyo-ku, Kyoto 606-8507, Japan

Received in final form 3 February 1999

Abstract—Stereoregular poly(methyl methacrylates) (PMMAs) solvent-cast in the form of films against a glass substrate were employed as model systems for a systematic study of the relationship between the molecular structure (as characterized by the stereochemistry and conformations of macromolecules), the functional-group composition, and the wettability of polymer surfaces. The water wettability of a syndiotactic surface was found to be highly sensitive to the polarity of the adjacent phase in the film-casting process, whereas the wettability of an isotactic surface was invariant to the polarity of the contacting medium. The tacticity-dependent wetting behavior arises from the difference in the extent of functional-group surface segregation or, in other words, from the different surface activity of the different tactic versions of the polymer. This difference, in turn, is associated with fundamental distinctions in the conformational structures energetically allowed for the isotactic and syndiotactic configurations of the polymer chain; the syndiotactic macromolecule is capable of adopting an amphiphilic surface conformation, whereas the energetically allowed conformational structures of the isotactic macromolecule do not possess amphiphilic character. In view of these findings, the isotactic surfaces can be regarded as 'ideal' model surfaces for research on the fundamentals of wetting phenomena. In addition, there is evidence for failure of the basic assumption of the Cassie approach, i.e. the assumption of *macroscopic* chemical heterogeneity, to describe adequately the wetting behavior of isotactic PMMA surfaces.

Key words: Contact angle; wettability; molecular structure; polymer surfaces.

1. INTRODUCTION

Wetting is governed by molecular interactions in the outermost surface layer of a few angstroms. Consequently, the forces dictating the wetting behavior of organic substances do not originate from the organic molecule as a whole, but rather from the outermost surface groups. Furthermore, for interfacial-energy minimization reasons, the molecules tend to arrange themselves in the surface layer in such a way that only their low-energy or high-energy portions come into contact with the surrounding phase. As a result, the wettability of an organic material is not related to its overall chemical structure, but depends on the chemical nature of energetically favored functional groups and on the extent to which these are exposed at the material surface. Langmuir [1–3] first developed this basic concept in his studies of the pure liquids of small molecules. More recently, it became apparent that the above considerations are equally applicable to long chain molecules (i.e. organic polymers) [4–6]. It is now widely accepted that the extent of the selective surface exposure of functional groups in polymers, unlike that in their low-molecular-weight analogues, must be largely controlled by the conformational characteristics of the molecular chains; however, very little work has been done to explore this statement [7–11]. Furthermore, since most of the fundamental research on the molecular mechanisms of wetting has been carried out with the surfaces of organic polymers, progress towards a molecular-based description of wetting phenomena has been hampered by the lack of microstructural understanding of the surfaces. In this paper, we report on the results of systematic experimental studies of the correlation between the molecular structure (as characterized by the stereochemistry and conformations of macromolecules), the extent of the selective surface exposure of functional groups, and the wettability of polymer surfaces.

1.1. Notes on terminology

Depending on the author, the surface structural alteration of a *homopolymer* arising from interfacial-energy minimization reasons has been termed surface reorientation, surface rearrangement, surface restructuring, or surface functional-group accumulation. We would like to note that these terms do not reflect properly the thermodynamic origin of the phenomenon or its molecular nature. To state this more clearly, let us recall that the surface process in which the lowering of interfacial free energy acts as the thermodynamic driving force is well known in the field of multicomponent molecular systems and there it is termed *surface segregation*. The component of a multicomponent system, which provides an effective reduction in the surface or interfacial free energy, segregates at the surface or interface, resulting in an effective minimization of the overall free energy. On the microscopic level, a material made by the same molecules can be regarded as a multicomponent one, provided the constituent molecule is composed of different types of functional groups. Obviously, such a system is more complicated than a simple multicomponent; however, the aforementioned principle still applies: the surface segregation may occur via surface enrichment in a certain functional group of the *multifunctional* molecule. One can see immediately that the phenomenon under consideration can be referred to as a particular case of surface segregation, that is, the *surface segregation of functional groups*. This, in turn, implies that basic terms used in the treatment of surface segregation in multicomponent systems, such as sign and extent of segregation, surface activity, pure component surface tension, etc., can be equally applied to the surfaces of multifunctional homopolymers. In view of these considerations, we use the term 'surface segregation of functional groups' or 'functional-group surface segregation' throughout this paper.

1.2. Experimental approaches

It should be obvious that functional-group surface segregation can only be detected by a technique which senses chemical groups not more than 0.5–1.0 nm deep into polymer. Among a number of surface characterization techniques that are available at this time, only contact angle and static secondary ion mass spectrometry (SSIMS) possess the required level of surface sensitivity. The applicability of SIMS analysis, however, is markedly limited by the qualitative nature of derived information, and much fundamental work is needed to improve the technique [12]. X-ray photoelectron spectroscopy (XPS) and high-resolution electron energy loss spectroscopy (HREELS) have also been used for probing the chemical functionality of the surfaces. However, since the sampling depth of these spectroscopies (> 2.5 nm) is significantly higher than the thickness characterizing the layer of surface-exposed functional groups, the tractability of surface analysis is hampered.

One of the most frequently used tools for the surface characterization of polymers is attenuated total reflection Fourier-transform infrared spectroscopy (ATR-FTIR). The sampling depth of an ATR-FTIR experiment is about $10^2 - 10^3$ nm. Since this thickness is 2–3 orders of magnitude higher than that examined by other surface techniques (contact angle, SIMS, XPS, etc.), the ATR-FTIR method has not been considered a sensitive surface characterization technique in a real sense. However, in polymer systems, a variety of surface phenomena extend up to a few hundred nanometers into the bulk. In these cases, ATR-FTIR spectroscopy can provide important information about the near surface region of the material. Relevant examples are the ATR-FTIR studies of surface segregation in polymer blends [13] and copolymers [14], surface crystallization [15], polymer/polymer interdiffusion [16], and surface-induced changes in the molecular structure [17]. Functional-group surface compositions are not accessible in this method. However, it offers a unique opportunity to establish trends in the surface-driven conformational changes and hence to correlate the extent of the selective surface exposure of functional groups (inferred by the use of a complimentary technique) with the conformational characteristics of polymer chains. There has been growing interest in applying external reflection infrared spectroscopies (IRS) to ultrathin polymer films (considered models of bulk polymer surfaces) in order to elucidate the surface orientation of functional groups and the surface conformation and orientation of polymer chains [11, 18, 19]. It should be noted, however, that the characteristics of ultrathin films are not directly relevant to those of the surface layers of bulk polymers.

In view of these circumstances, the measurement of contact angle appears to be the most surface-sensitive technique at our disposal [20]. Not surprisingly, the vast majority of studies on the surface segregation phenomenon in multifunctional polymers have employed contact angle measurements as a principal experimental tool. In 1975, Holly and Refojo [6] first ascribed a large contact angle hysteresis (i.e. the difference between advancing and receding contact angles) observed with poly(2-hydroxyethyl methacrylate) (PHEMA) hydrogels to the water-induced change in the proportions of hydrophobic and hydrophilic side-groups exposed at the hydrogel upper surface. Since then, numerous authors have intensively used the measurement of contact angle hysteresis as a probe for the environmentally induced change in the *sign* of functional-group surface segregation [21, 22]. However, this traditional approach cannot be used to quantify the *extent* of segregation, because general quantitative models of hysteresis have not been developed so far [23].

1.3. The Cassie and Israelachvili–Gee equations

In order to establish systematic, *quantitative* correlations between the polymer microstructure and the extent of functional-group surface segregation, one has to convert contact angle data into fractional surface exposure of functional groups. In principle, this can be done by using semi-empirical composition–contact angle relationships, such as the Cassie equation [24]:

$$\cos \theta = \sum_{j=1}^{N} f_j \cos \theta_j, \tag{1}$$

or the Israelachvili–Gee equation [25]:

$$(1 + \cos \theta)^2 = \sum_{j=1}^{N} f_j (1 + \cos \theta_j)^2, \tag{2}$$

where θ is the *equilibrium* contact angle on a heterogeneous surface composed of N components, f_j is the fractional exposure or coverage of component j, and θ_j is the equilibrium contact angle for the pure j surface. Despite the simple forms of equations (1) and (2), there is a fundamental difficulty associated with these approaches: the equilibrium contact angles, θ and θ_j, cannot be experimentally measured because of the nonideal effects of three-phase contact observed in all cases on heterogeneous surfaces [23]. This difficulty is usually circumvented by applying the above equations only to the quasi-equilibrium advancing contact angle or to the 'classical' static contact angle. In this way, *semiquantitative* information on the structural and chemical characteristics of complex, multifunctional surfaces can be obtained. Nonetheless, a principal question that remains to be answered is: Which of the two approaches should be used in each particular case? The main difference between the Cassie equation and the Israelachvili–Gee equation is that the former is based on the assumption of *macroscopic* compositional heterogeneity, whereas the latter has been derived for the case of atomic- or molecular-level *microscopic* heterogeneity. Since the chemical heterogeneity scale of organic surfaces is still to be decided, some authors have used both approaches in the quantitative treatment of their contact angle data. Although the Cassie equation consistently predicts a larger surface fraction of polar groups (and, accordingly, a lower fraction of nonpolar groups) than the Israelachvili–Gee equation, the overall trends in the surface coverages were found to be quite similar for the two approaches [26, 27]. An example of where the quantitative discrepancies between the two equations cannot be neglected will be presented here.

1.4. Stereoregular polymer surfaces as new models for surface structure–wettability studies

The microstructure–wettability correlations of polymer surfaces have been intensively studied over the last 40 years. An enormous amount of information has been accumulated about the effects of the chemistry and architecture of polymer repeat units, as well as of the constitution and length of side-chains, on the wettability of polymer surfaces [28, 29]. However, until very recently, data relating the chemical composition and the resultant wettability of the polymer surfaces with the conformational structure of polymer chains were practically lacking.

Our efforts have been aimed at bridging the above gap in microscopic structure and wetting properties by providing a systematic study of polymer surfaces on the molecular level. As a first step towards this objective, we have developed a new strategy that makes it possible to obtain a clear picture of the effects of molecular conformations on the extent of functional-group surface segregation and the resultant chemical composition of a polymer surface [30, 31]. The approach rests on the comparative study of stereochemical versions (i.e. iso- and syndiotactic forms) of the same polymer. The two tactic forms of a given polymer are built with the same monomer unit and, thus, are chemically identical. Therefore, any difference in their surface behavior must be due solely to the difference in conformational characteristics. Furthermore, samples of stereoregular polymers are prepared in the form of films cast on highly polar glass substrates, and the measurement of contact angle is used as a probe of the extent of the surface segregation. The higher the segregating capability (surface activity) of a macromolecule, the larger the difference in the contact angles on the air-side surface and the glass-side surface of the film. Finally, the ATR-FTIR technique is employed in order to obtain information on subtle details of the molecular conformations of stereoregular polymer chains in the vicinity of the surface.

2. EXPERIMENTAL

Poly(methyl methacrylate) (PMMA) was selected as a model system in this study. The characteristics of the stereoregular PMMA samples used are listed in Table 1. The characterization and synthesis/sources of these samples have been previously reported [30]. The polymers were purified by precipitation prior to use. Each sample of the polymers was first dissolved in benzene at a concentration of 2% (w/v). The films studied were prepared in a dust-free chamber by casting the solutions on chromic acid-cleaned glass substrates (Pyrex glass Petri dishes). The

Table 1.
Characteristics of the PMMA samples

Polymer	$M_n \times 10^{-3}$	M_w/M_n	Tacticity (%)		
			mm	*mr*	*rr*
s-PMMA-1	234	2.30	3	32	65
s-PMMA-2	62	3.37	4	21	75
s-PMMA-3	880	1.03	1	6	93
i-PMMA-1	160	2.05	74	15	11
i-PMMA-2	47	4.24	91	6	3

residual solvent was removed from the films formed at room temperature by drying films at 70°C for about 20 h in a clean oven. They were then separated from the substrate by using doubly distilled water and stored over desiccant under reduced pressure until required.

For the study of the blend films consisting of syndio- and isotactic PMMA macromolecules, the samples s-PMMA-2 and i-PMMA-2 were used as the syndiotactic and isotactic polymers, respectively. Each sample of the two tactic forms of PMMA was first dissolved in benzene at a concentration of 2% (w/v). On mixing corresponding volumes of these solutions under continuous stirring, solutions of i-PMMA/s-PMMA mixtures with weight ratios of the components of 25/75, 50/50, and 75/25 were prepared. The blend films were prepared in the same way as described above for the homopolymer films. The homopolymer and blend films were in the thickness range 9–12 μm, uniform to within 1 μm over the whole film area (ca. 20 cm^2).

Static contact angles of water on PMMA films were measured by the sessile drop method using a telescopic goniometer at 25°C and at about 65% relative humidity. These were determined 20–40 s after deposition of the drop. The volume of the water drops used was always 3 μl. In all cases, no appreciable changes in contact angle were observed in 2 min, suggesting that underwater surface restructuring did not play a role in the time frame of the measurements described here. All reported values are the average of at least eight measurements taken at different locations of the film surface and have a typical error in the mean of ± 1°.

The IR spectra were recorded on a Nicolet 7199 FTIR spectrometer equipped with a mercury–cadmium–telluride (MCT) detector at a resolution of 2 cm^{-1}. A Wilks model 50 ATR attachment and a Ge internal reflection element (52.5 × 20 × 2 mm^3) with an angle of incidence of 60° were used for ATR-FTIR measurements. Under these conditions, the penetration of the incident IR beam into the film surface is approximately 1/20 the wavelength (or about 0.6 μm at 750 cm^{-1}). The number of scans was 32 and 1000 for the transmission FTIR and ATR-FTIR measurements, respectively.

3. RESULTS AND DISCUSSION

3.1. Homopolymer films

3.1.1. Contact angle measurements.   Static contact angles were always used in this study. A static contact angle is defined here as the contact angle measured shortly after a drop of test liquid has stopped self-spreading along the solid surface and has attained its equilibrium shape. In the experiments reported here, water droplets attained their equilibrium shape in less than 20 s after being introduced on the PMMA surfaces. After that, the drop shape did not change within about 2 min, indicating that underwater surface alterations (e.g. restructuring, re-orientation, swelling) did not take place in the time frame of the measurements described here. One might have expected the advancing and receding contact angles to be more appropriate for the surface composition study, compared with the static contact angle. However, since the advancing angle is representative of the hydrophobic portion of a surface and the receding angle is representative of the hydrophilic portion, this approach would result in two sets of values of the surface fractions of hydrophobic and hydrophilic components for the same surface (see Section 1.3). The static contact angles have intermediate values between the advancing and receding contact angles and therefore better represent the actual surface composition.* The advancing and receding contact angles must be measured when a polymer interacts strongly with a wetting liquid and consequently the static contact angle is not accessible.

In Fig. 1, the water contact angles on the homopolymer PMMA films are plotted versus the percentage of syndiotactic triads, rr. For simplicity, we will use θ_{air} and θ_{glass} for the contact angles on the surfaces contacted with air and glass substrate, respectively, during the film formation. As can be seen, the syndiotactic PMMAs showed substantially greater contact angles at the air-side surfaces than at the glass-side surfaces. This contrasting low wettability/high wettability character on opposite surfaces of the syndiotactic films increased rapidly with triad syndiotacticity; the difference between θ_{air} and θ_{glass}, $\Delta\theta$, changed from $7°$ for $rr = 65\%$ to $24.5°$ for $rr = 93\%$. In stark contrast to the syndiotactic PMMAs, the value of $\Delta\theta$ observed with the isotactic surfaces was only $1.5°$, i.e. just within experimental error. Furthermore, the values of θ_{air} and θ_{glass} observed with the isotactic films were intermediate between those of the air-side surface and the glass-side surface of the syndiotactic polymers; that is, θ_{air} was lowered by $13°$ with respect to that of highly syndiotactic PMMA ($rr = 93\%$) and θ_{glass} was raised by $10°$ relative to the corresponding value of the highly syndiotactic polymer.

---

*A similar consideration applies to contact angle measurements aimed at establishing the surface free energy components of a polymer surface [32].

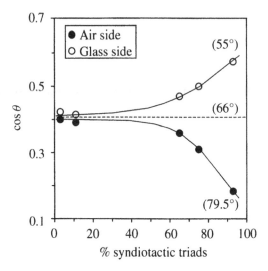

Figure 1. Water contact angles on PMMA films as a function of the percentage of syndiotactic triads in the polymer chain.

3.1.2. Surface chemical composition – wettability relationship.   All the polymer films were optically smooth and showed no differences in surface morphology examined on one and then on the other side of the films by a light microscope. The s-PMMA-1, i-PMMA-1, and i-PMMA-2 films appeared totally amorphous in wide-angle X-ray diffraction (WAXD) measurements and showed no signs of crystalline order when examined by transmission FTIR and ATR-FTIR. The bulk crystallinity of the s-PMMA-2 and s-PMMA-3 films estimated from WAXD measurements was 9 and 31%, respectively.   According to the ATR-FTIR and transmission FTIR results, the degree of molecular order at the film surfaces was practically the same as that in the film interior for the s-PMMA-2 film and slightly higher than the bulk value for the s-PMMA-3 film. In both cases, no appreciable differences in crystallinity between the air-side surface and the glass-side surface were detected.[†] Thus, the film surfaces were free from any effects of roughness or trans-crystallinity that might be responsible for the contact angle changes. It is also unlikely that these changes are due to surface contamination, because all five polymer samples were subjected to identical experiments. Therefore, the only factor that appears to be responsible for the difference in wettability is the difference in the functional-group surface composition. In PMMA, there are three types of functionalities present: the methylene group (CH_2) forming the backbone, and the α-methyl (α-CH_3) and ester ($COOCH_3$) groups existing as substitutes on the principal chain (Fig. 2). Thus, the PMMA repeat unit can be regarded as an amphiphilic molecule, with the hydrophobic part consisting of the methylene and the methyl group, and the

---

[†]The surface and bulk crystallinity results for the stereoregular PMMA films have been published elsewhere [33].

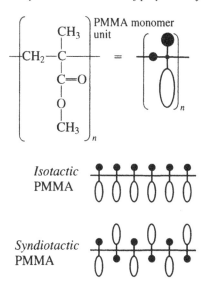

Figure 2. Schematic representation of the PMMA repeat unit and the stereochemical placement of pendant side-groups in the isotactic and syndiotactic macromolecules.

hydrophilic portion formed by the ester group. Accordingly, during the surface formation by casting the polymer solution, the repeat units tend to orient themselves in the surface region in such a way that their hydrophobic portions form the air-side surface of the polymer and the hydrophilic parts form the glass-side surface. As the solvent evaporates, these surface compositions are frozen in the polymer film formed. The contrasting hydrophobic–hydrophilic character observed on opposite surfaces of the syndiotactic films indicates that the above mechanism of functional-group surface segregation did actually take place in the syndiotactic polymer. On the other hand, the fact that the θ_{air} and θ_{glass} values for the isotactic films are very close to each other and represent an average for the θ_{air} and θ_{glass} values of the syndiotactic films indicates a nonselective, random distribution of the functional groups at the isotactic surfaces. Schematic models of the stereoregular PMMA surfaces illustrating the above chemical composition–wettability relationships are shown in Fig. 3.

3.1.3. Surface chemical composition–molecular structure relationship. It is apparent from the above analysis of contact angle data that the surface chemical composition of PMMA is tacticity-dependent. That is, the syndiotactic surfaces are covered preferably with hydrophobic or hydrophilic groups (depending on the polarity of the adjacent phase in the film-casting process), whereas the isotactic surfaces consist of comparable fractions of these groups. In order to understand this phenomenon, one has to consider the microstructural features of stereoregular polymers. Primarily, structural dissimilarity of the tactic versions of a given polymer arises from the difference in the stereochemical placement of pendant side-groups along the polymer backbone. This difference can be seen most clearly

O. N. Tretinnikov

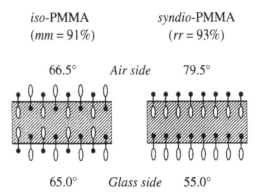

iso-PMMA   syndio-PMMA
($mm = 91\%$)   ($rr = 93\%$)

66.5°   Air side   79.5°

65.0°   Glass side   55.0°

Figure 3. Schematic representation of the relationship between the functional-group composition and the wettability (water contact angle) of stereoregular PMMA surfaces.

in a planar zigzag conformation of the polymer chain, as is shown for iso- and syndiotactic PMMA in Fig. 2. In the isotactic chain, the α-methyl groups lie on one side of the polymer backbone and the ester groups are situated on the other side of it. On the basis of this highly amphiphilic structure, one would have expected the isotactic polymer to construct a hydrophobic CH_3 surface at the polymer–air interface and a hydrophilic $COOCH_3$ surface at the polymer–glass interface. On the contrary, in the syndiotactic chain, the α-CH_3 groups are sandwiched between the $COOCH_3$ groups along the polymer backbone and, accordingly, a random mixture of the hydrophilic and hydrophobic groups could be expected for the syndiotactic surface. Thus, the stereochemical analysis suggests a tacticity–surface composition relationship which is just opposite to what is observed experimentally (Fig. 3). Obviously, more subtle structural factors must be taken into account in order to rationalize the experimental findings. In particular, it should be realized that the above stereochemical consideration assumes a planar zigzag conformation of the polymer chain, which may not necessarily be a characteristic of a given polymer. Specifically, isotactic polymers do not take the planar form; they typically adopt helical conformations instead [34].

According to experimental and theoretical studies, the most probable conformations of isotactic PMMA are (10/1) and (5/1) helices [35, 36]. The molecular models of these helical forms viewed along the helix axis are shown in Fig. 4. It can be seen that in the two models the polar and nonpolar side-groups radiate from the helix, forming an array of spatially mixed α-CH_3 and $COOCH_3$ groups. Therefore, the macromolecule may expose both the polar and the nonpolar groups but not one or the other — hence, the observed non-surface-active behavior of isotactic PMMA.

For syndiotactic PMMA, the theoretical and experimental results indicate that the macromolecule may have a planar zigzag and a glide-plane structure formed by the nearly *trans–trans* and *trans–gauche* backbone conformations, respectively [36–38] (Fig. 5). It has already been shown above that the planar zigzag form of syndiotactic PMMA is non-surface-active. In contrast, the glide-plane structure has clear amphiphilic character: the α-methyl and ester groups lie on different sides

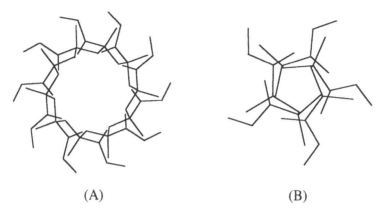

Figure 4. Minimum-energy conformational structures of isotactic PMMA: 10/1 (A) and 5/1 helix (B) viewed along the helix axis.

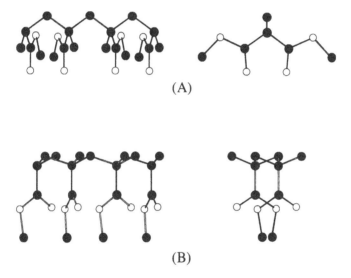

Figure 5. Planar zigzag (A) and glide-plane (B) structures of syndiotactic PMMA: left: side views; right: cross-sectional views. Open and filled circles represent oxygen and carbon atoms, respectively.

of the plane passing through the chain axis. Thus, the model analysis reveals that the glide-plane conformation is responsible for the observed functional-group segregation at the syndiotactic surfaces.

3.1.4. The 'ideal' PMMA surface. For studies aimed at understanding the fundamentals of wetting, it is highly desirable to use model surfaces that exhibit a nonselective, random distribution of the functional groups. The chemical composition of such an 'ideal' surface would resemble the chemical composition in the bulk and, consequently, could be easily derived from the chemistry of the monomer unit. In this regard, syndiotactic PMMA is not an appropriate model system. Its wetting behavior does not originate from the polymer chain as a whole, but rather from the energetically favored functional group. Accordingly, any attempt to interpret the wetting results in terms of the overall polymer chemistry would be misleading. At this point, it is appropriate to note that practically all commercial polymers that have been used as model surfaces in research on wetting are predominantly syndiotactic. On the contrary, the isotactic PMMA surface can be regarded as an ideal model surface. Its chemical composition is identical to that of the polymer bulk and is invariant to the polarity of the contacting medium. Obviously, the water contact angle on the ideal PMMA surface (θ_{id}) must be in the range of θ_{air} and θ_{glass} values of the isotactic PMMA films, i.e. $\theta_{id} = 66 \pm 1°$. Furthermore, the nonselective (isotropic) surface distribution of functional groups on the ideal PMMA surface means that the surface fraction of CH_2 and CH_3 groups (f_{CH_2, CH_3}) equals that of the ester group (f_{COOCH_3}), i.e. $f_{CH_2, CH_3} = f_{COOCH_3} = 0.5$. Besides, from the known values of $\theta = 94°$ and $110°$ on pure CH_2 and CH_3 surfaces, respectively [28, 39], we estimate the static contact angle for water on the pure CH_2, CH_3 surface (θ_{CH_2, CH_3}) to be $101°$. The last parameter to be determined in order to correlate the water wettability of the ideal surface of PMMA with its subunit properties is the contact angle on the pure $COOCH_3$ surface (θ_{COOCH_3}). The literature suggests a θ_{COOCH_3} value in the range 30–40° [40]. It is of interest to compare this experimental θ_{COOCH_3} value with that predicted by the Cassie equation [equation (1)] and the Israelachvili–Gee equation [equation (2)]. Using the known values of $\theta_{id} = 66°$, $\theta_{CH_2, CH_3} = 101°$, and $f_{CH_2, CH_3} = f_{COOCH_3} = 0.5$ for the ideal PMMA surface, we find $\theta_{COOCH_3} = 0°$ and 35° from equations (1) and (2), respectively. The θ_{COOCH_3} prediction based on the Israelachvili–Gee equation is in excellent agreement with the experimental values, whereas the θ_{COOCH_3} value predicted by the Cassie equation is in strong disagreement with experiment. Moreover, the Cassie prediction ($\theta_{COOCH_3} = 0°$) appears highly unrealistic, since hydrophobic interactions of terminal CH_3 groups with water molecules must result in a θ_{COOCH_3} value markedly higher than 0°. In other words, the experimentally derived contact angle on the ideal (isotactic) PMMA surface can be achieved in the Cassie model with an unrealistic value of θ_{COOCH_3}. This result likely indicates failure of the basic assumption of the Cassie approach, i.e. the assumption of *macroscopic* chemical heterogeneity, to describe adequately the wetting behavior of the PMMA surface.

3.1.5. Quantitative analysis of the surface functional group composition. The above analyses of the wettability of ideal (isotactic) PMMA surface in terms of the individual characteristics of constituent functional groups using the Cassie and Israelachvili–Gee equations revealed good predictability by the latter equation but unrealistic wetting prediction based on the former. Therefore, only the Israelachvili–Gee equation was used to calculate the fractional coverage of functional groups on the surfaces of PMMA films. In this particular case, equation (2) can be written in the following form relating the fractional coverage and contact

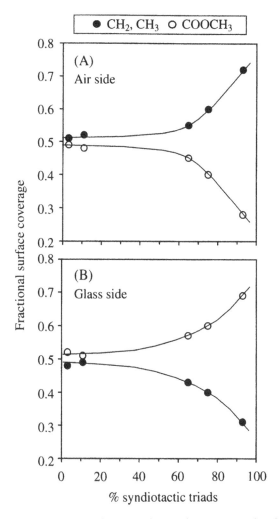

Figure 6. Fractional surface coverage of polar and nonpolar groups at the air-side (A) and glass-side (B) surfaces of PMMA films as a function of the percentage of syndiotactic triads in the polymer chain.

angle on a given PMMA surface:

$$f_{\text{COOCH}_3} = \frac{(1+\cos\theta)^2 - (1+\cos\theta_{\text{CH}_2,\,\text{CH}_3})^2}{(1+\cos\theta_{\text{COOCH}_3})^2 - (1+\cos\theta_{\text{CH}_2,\,\text{CH}_3})^2} \tag{3}$$

$$f_{\text{CH}_2,\,\text{CH}_3} = 1 - f_{\text{COOCH}_3}, \tag{4}$$

where $\theta_{\text{CH}_2,\,\text{CH}_3} = 101°$ and $\theta_{\text{COOCH}_3} = 35°$. The results of these calculations are presented in Fig. 6. The values of $f_{\text{CH}_2,\,\text{CH}_3}$ and f_{COOCH_3} show that the surfaces of isotactic films consist of comparable fractions of nonpolar and polar groups, both at the air-facing side and at the glass-facing side of the films. On the contrary, the syndiotactic surfaces are preferentially covered with nonpolar groups at the air side and with polar ones at the glass side of the films. The surface enrichment

O. N. Tretinnikov

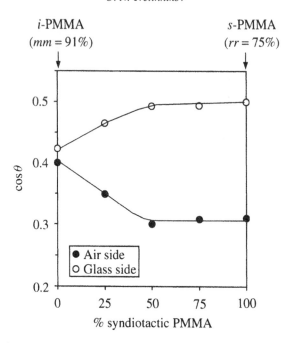

Figure 7. Cosine of the water contact angles on i-PMMA/s-PMMA blend films as a function of the bulk concentration of s-PMMA.

increased with triad syndiotacticity and reached the highest observed value for the highly syndiotactic ($rr = 93\%$) PMMA film, with 72% of the surface area consisting of CH_2, CH_3 groups on the air-side surface and 69% covered with $COOCH_3$ groups on the glass-side surface of the film. It should be noted that if syndiotactic PMMA macromolecules in the outermost surface layer were to adopt only the surface-active, *trans–gauche* conformation, the polymer chains by a simple rotation would orient themselves at the interface so that only ester groups or only α-methyl and methylene groups would be exposed (Fig. 3). Since the quantitative analysis reveals that this complete functional-group segregation does not actually take place, we conclude that there is a fraction of non-surface-active, *trans–trans* conformational sequences in the polymer chain which lower its surface segregating capability. In fact, in the bulk, syndiotactic PMMA shows a very strong preference for the *trans–trans* conformation; the fraction of chain units occurring in the *trans–gauche* conformation is only 0.2 [36]. Obviously, the surface concentration of *trans–gauche* conformers must substantially exceed this bulk value in order to account for the observed high selectivity in the surface exposure of functional groups. Indeed, the increased content of *trans–gauche* conformers in the surface regions of syndiotactic PMMA films has been observed in our previous work using attenuated total reflection infrared spectroscopy [41, 42]. However, this surface-driven increase in the content of *trans–gauche* conformers is likely to be counteracted by the bulk thermodynamics. Hence, a fraction of the *trans–trans* conformers is maintained in the polymer chains.

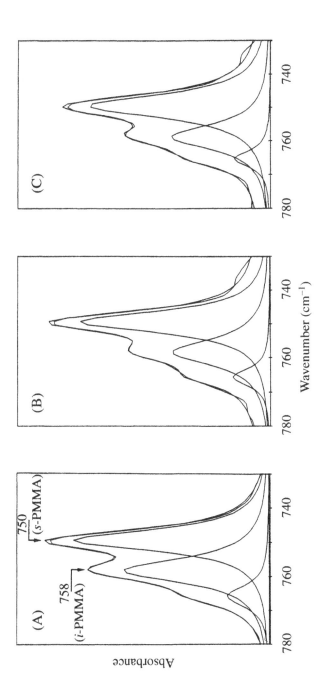

Figure 8. FTIR transmission (A) and ATR-FTIR (B, C) spectra in the 700–800 cm^{-1} region of the 50/50 i-PMMA/s-PMMA blend film. The ATR-FTIR spectra were obtained with the air-facing side (B) and glass-facing side (C) of the film.

3.2. PMMA blend films

One of the most striking findings of this study is the capability of the contact angle measurement with a water droplet to distinguish the macromolecules of syndiotactic PMMA from those of isotactic PMMA. This makes it possible to apply water drop contact angle measurements to probe the surface composition of blend films composed of the two tactic versions of PMMA (i.e. i-PMMA/s-PMMA blend films).

In Fig. 7, the water contact angles on the homopolymer and blend films are plotted versus the concentration of s-PMMA. As can be seen, the i-PMMA/s-PMMA blend films composed of 75% and 50% s-PMMA showed almost the same contact angle values as those of the pure s-PMMA film. This indicates that both the air-side and the glass-side surfaces of 75% and 50% s-PMMA blend films are composed almost totally of the syndiotactic component. For the 25% s-PMMA blend, the observed contact angles are intermediate between those of the corresponding syndio- and isotactic surfaces, suggesting that the surfaces of the blend are composed of a mixture of s-PMMA and i-PMMA macromolecules.

This finding of the contact angle study is supported further by the results of ATR-FTIR spectroscopy. As an example, Fig. 8 shows the ATR-FTIR and transmission spectra between 700 and 800 cm^{-1} for the 50/50 i-PMMA/s-PMMA blend film. The bands at 750 and 758 cm^{-1} are purely due to the syndiotactic and isotactic components, respectively [31]. It can be seen that in comparison with the transmission spectrum, the relative intensity of the syndiotactic peak to the isotactic peak is increased in the ATR spectra for both the air-side and the glass-side surfaces of the film. This finding suggests that the blend film contains a surface excess of s-PMMA at both surfaces.

We believe that the phenomenon of surface enrichment by s-PMMA in the solvent-cast s-PMMA/i-PMMA blends is a direct consequence of the high surface activity of s-PMMA and the non-surface-active behavior of i-PMMA described in the preceding section. The surface segregation of the blends occurs during casting as preferential s-PMMA adsorption at the polymer–air and polymer–glass interfaces. The driving force for segregation is the lowering of the *interfacial* free energy that occurs when nonpolar CH_2 and α-CH_3 groups of the surface-active polymer concentrate at the polymer–air interface and polar $COOCH_3$ groups concentrate at the polymer–glass interface. The surface-active syndiotactic macromolecules thus segregate at both the air-side and the glass-side surfaces, expelling their non-surface-active isotactic counterparts.

4. CONCLUSIONS

Solvent-cast films of stereoregular PMMAs were employed as model systems for a systematic study of the relationship between the microstructure and the wettability of polymer surfaces. A number of important conclusions can be drawn from this study.

(1) The wetting behavior of stereoregular PMMAs is tacticity-dependent. The water wettability of the syndiotactic surface is highly sensitive to the polarity of the adjacent phase in the film-casting process, whereas the wettability of the isotactic surface is invariant to the polarity of the contacting medium.

(2) The tacticity-dependent wetting behavior arises from the difference in surface activity of the tactic versions of the polymer. Syndiotactic PMMA is highly surface-active, due to its capability of adopting an amphiphilic surface conformation. Isotactic PMMA is non-surface-active, because its energetically allowed conformational structures do not possess amphiphilic character.

(3) The phenomenon of the tacticity-dependent surface activity of polymer chains causes another very interesting phenomenon: blends of two tactic versions of the same polymer segregate at the surface, favoring the syndiotactic form.

(4) The isotactic PMMA surfaces can be regarded as 'ideal' models for research on the fundamentals of wetting phenomena. They exhibit a nonselective, random distribution of functional groups and thus resemble the chemical composition in the bulk of the polymer.

(5) There is evidence for failure of the basic assumption of the Cassie approach, i.e. the assumption of *macroscopic* chemical heterogeneity, to describe adequately the wetting behavior of the isotactic PMMA surfaces.

Finally, we note that the non-surface-active behavior of isotactic PMMA is due to the *helical* structure of the main chain, whereas the high surface activity of syndiotactic PMMA macromolecule arises from its *planar* conformational structure. Since, in general, the helical and the planar conformations are intrinsic to isotactic and syndiotactic polymers, respectively [34], syndiotactic polymers should be inherently more surface-active than their isotactic counterparts.

REFERENCES

1. I. Langmuir, *J. Am. Chem. Soc.* **38**, 2221 (1916).
2. I. Langmuir, *J. Am. Chem. Soc.* **39**, 1848 (1917).
3. I. Langmuir, *Trans. Faraday. Soc.* **15** (Part 3), 62 (1920).
4. V. A. Pchelin and I. I. Korotkina, *Zh. Fiz. Khim.* **12**, 50 (1938).
5. J. P. Reardon and W. A. Zisman, *Macromolecules* **7**, 920 (1974).
6. F. J. Holly and M. F. Refojo, *J. Biomed. Mater. Res.* **9**, 315 (1975).
7. R. E. Baier and W. A. Zisman, *Macromolecules* **3**, 70 (1970).
8. R. G. Zhbankov, O. N. Tretinnikov and G. K. Tretinnikova, *Vysokomol. Soedin.* **B26**, 104 (1984).
9. O. N. Tretinnikov and R. G. Zhbankov, in: *Polymer–Solid Interfaces*, J. J. Pireaux, P. Bertrand and J. L. Bredas (Eds), pp. 361–370. IOP Publishing, Bristol (1992).
10. C. Vergelati, A. Perwuelz, L. Vovelle, M. A. Romero and Y. Holl, *Polymer* **35**, 262 (1994).
11. Y. Grohens, M. Brogly, C. Labbe and J. Schultz, *Polymer* **38**, 5913 (1997).
12. H. Vanden Eynde, L. T. Weng and P. Bertrand, *Surface Interface Anal.* **25**, 41 (1997).
13. J. M. G. Cowie, B. G. Devlin and I. J. McEwen, *Polymer* **34**, 501 (1993).
14. C. S. Paik Sung, C. B. Hu, E. W. Merrill and E. W. Salzman, *J. Biomed. Mater. Res.* **12**, 791 (1978).

15. J. B. Huang, J. W. Hong and M. W. Urban, *Polymer* **33**, 5173 (1992).
16. E. Jabbari and N. A. Peppas, *Macromolecules* **26**, 2175 (1993).
17. A. E. Tshmel, V. I. Vettegren and V. M. Zolotarev, *J. Macromol. Sci., Phys.* **B21**, 243 (1982).
18. R. H. G. Brinkhuis and A. J. Schouten, *Langmuir* **8**, 2247 (1992).
19. Y. Grohens, M. Brogly, C. Labbe and J. Schultz, *Eur. Polym. J.* **33**, 691 (1997).
20. S. R. Holmes-Farley, R. H. Reamey, R. Nuzzo, T. J. McCarthy and G. M. Whitesides, *Langmuir* **3**, 799 (1987).
21. J. D. Andrade (Ed.), *Polymer Surface Dynamics*. Plenum Press, New York (1988).
22. O. N. Tretinnikov and Y. Ikada, *Langmuir* **10**, 1606 (1994) and refs cited therein.
23. J. Drelich, *Pol. J. Chem.* **71**, 525 (1997).
24. A. B. D. Cassie, *Discuss. Faraday Soc.* **75**, 5041 (1952).
25. J. N. Israelachvili and M. L. Gee, *Langmuir* **5**, 288 (1989).
26. R. C. Chatelier, X. Xie, T. R. Gengenbach and H. J. Griesser, *Langmuir* **11**, 2576 (1995).
27. S. V. Atre, B. Liedberg and D. L. Allara, *Langmuir* **11**, 3882 (1995).
28. W. A. Zisman, in: *Contact Angle, Wettability, and Adhesion*, Adv. Chem. Ser. No. 43, p. 1. American Chemical Society, Washington, DC (1964).
29. S. Wu, *Polymer Interface and Adhesion*. Marcel Dekker, New York (1982).
30. O. N. Tretinnikov, *Langmuir* **13**, 2988 (1997).
31. O. N. Tretinnikov and K. Ohta, *Langmuir* **14**, 915–920 (1998).
32. M. A. Adao, B. Saramago and A. C. Fernandes, *Langmuir* **14**, 4198 (1998).
33. O. N. Tretinnikov, K. Nakao, R. Iwamoto and K. Ohta, *Macromol. Chem. Phys.* **197**, 753 (1996).
34. O. Vogl and G. D. Jaucox, *Polymer* **28**, 2179 (1987).
35. H. Kusanagi, Y. Chatani and H. Tadokoro, *Polymer* **35**, 2028 (1994).
36. M. Vacaletto and P. J. Flory, *Macromolecules* **19**, 405 (1986).
37. S. Havriliak and N. Roman, *Polymer* **7**, 387 (1966).
38. A. M. Liquori, G. Anzuino, V. M. Coiro, M. D'Alagni, P. De Santis and M. Savino, *Nature* **206**, 358 (1958).
39. E. G. Shafrin and W. A. Zisman, in: *Contact Angle, Wettability, and Adhesion*, Adv. Chem. Ser. No. 43, p. 145. American Chemical Society, Washington, DC (1964).
40. A. H. Hogt, D. E. Gregonis, J. D. Andrade, S. W. Kim, J. Dankert and J. Feijen, *J. Colloid Interface Sci.* **106**, 289 (1985).
41. O. N. Tretinnikov and R. G. Zhbankov, *J. Mater. Sci. Lett.* **10**, 1032 (1991).
42. O. N. Tretinnikov, K. Nakao and K. Ohta, *Polym. Prepr. Jpn.* **43**, 1570 (1994).

Apparent and Microscopic Contact Angles, pp. 129–145
J. Drelich, J. S. Laskowski and K. L. Mittal (Eds)
© VSP 2000.

Contact angles on quartz induced by adsorption of heteropolar hydrocarbons

X. XIE and N. R. MORROW *

Department of Chemical and Petroleum Engineering, University of Wyoming, P.O. Box 3295, University Station, Laramie, WY 82071, USA

Received in final form 10 February 1999

Abstract—The wetting behavior induced by the adsorption of crude oil components onto mineral substrates initially covered with brine is of special importance to petroleum engineering. The wettability of reservoir rock is a controlling factor in the efficiency of oil recovery from the swept zone of a waterflood. The effect of the adsorption of polar components from crude oil on the wetting properties of quartz plates was investigated by the dynamic Wilhelmy plate technique. Force–distance relationships were measured for treated quartz plates passing through oleic–aqueous interfaces. Water receding and advancing contact angles under dynamic conditions were obtained. Changes in the wetting of the quartz plate from a completely water-wet state were induced by adsorption from crude oil or solutions of its components. Different wetting states were developed by varying the oil composition, aqueous phase pH, temperature, time of adsorption, the solvents used to remove excess crude oil, and the probe oil used in contact angle measurements. Some important features of wetting, contact angle hysteresis and slippage of the three-phase contact line, are presented. Organic films adsorbed from the crude oil were examined by atomic force microscopy.

Key words: Wilhelmy plate; contact angle; wettability; hysteresis; pinning and slippage; three-phase contact line; aging time and temperature.

NOTATION

D distance, cm
D_p theoretical pinning distance, cm
D_s slippage distance, cm
F total force measured for Wilhelmy plate, dynes
F_B buoyancy force, dynes
p perimeter, cm
T_a aging temperature, °C
t_a aging time, h or days

---

*To whom correspondence should be addressed. E-mail: morrownr@asuwlink.uwyo.edu

Greek letters

θ contact angle, degrees
θ_A advancing contact angle, degrees
θ_R receding contact angle, degrees
σ_{o-w} oil–water interfacial tension, dynes/cm

1. INTRODUCTION

Wetting states induced by crude oil at mineral surfaces initially covered with brine are of special importance to oil recovery by waterflooding. This paper mainly concerns the wetting changes that occur when a crude oil–brine interface is brought into very close proximity to the interface formed by a quartz surface covered with brine. The properties of the two interfaces can have a dominant control on the subsequent changes in wetting caused by the adsorption of polar components from the crude oil onto the quartz surface [1].

When quartz, which is amphoteric, contacts an aqueous solution, H^+ will react with the negative oxygen sites on the quartz surface, and OH^- with the positive silicon sites [2]. The dissociation of H^+ from quartz surface silicic acid will generate a negative surface charge. The H^+ concentration in the aqueous solution will determine the extent of this dissociation, i.e. the pH value determines the quartz surface charge. An electric field will be associated with the charged surface. In the presence of this electric field, the counter-ions, ions of opposite charge to that of the surface, will approach the surface and form an electrostatic double layer. Similarly, for the crude oil–brine interface, as noted by Buckley *et al.* [3], many crude oils have both acidic and basic ionizable sites. When crude oil contacts brine, the oil–brine interface will also be charged. When the brine–solid and crude oil–brine interfaces are sufficiently close, the interaction of their electric charges controls the presence and thickness of the water film between them. This, in turn, controls adsorption onto the mineral surface of polar organic material from the crude oil. The effect of adsorption on wetting is commonly studied through contact angle measurements.

The wettability of reservoir rock surfaces can be altered by contact with crude oil across water films. Three mechanisms for the adsorption of crude oil components onto solid surfaces initially covered by brine have been identified [4]. The first depends mainly on the solvency and composition of the crude oil. Heavy polar components, commonly identified as asphaltenes, can precipitate as aggregates at the rock surface and change the wettability to strongly oil-wet. The second mechanism arises because heteropolar components at the crude oil–brine interface can behave as acids or bases and become charged. Ionization of acids and bases can produce net surface charges at the solid–brine and brine–oil interfaces that can result in unstable water films followed by adsorption from the crude oil. The third, ion binding between sites of similar electric charge, depends on the presence of multivalent cations in the brine. Factors which govern these three, sometimes competing, mechanisms and the extent and rate of wettability change have been discussed in detail by Buckley and Liu [5].

In this study, the dynamic Wilhelmy plate technique was employed to measure contact angles and related surface properties of quartz after treatment with crude oil or solutions of its components.

2. THE WILHELMY PLATE METHOD AND CONTACT ANGLE MEASUREMENT

The Wilhelmy plate method [6] was first used to measure surface tensions between air and liquids. The plate hangs vertically from a microbalance. This technique was extended to obtain dynamic contact angles, i.e. contact angles for a moving plate, and to monitor the behavior of the oil/brine/solid three-phase line of contact by recording the change in force after the plate meets the liquid–liquid interface. The oil–brine interface was defined as the reference height (zero). The microbalance measures a combination of interfacial and buoyancy forces. The general equation for the force on the Wilhelmy plate is [7]

$$F = p\sigma_{\text{o-w}} \cos \theta + F_{\text{B}}(D),$$ (1)

where F is the force on the plate (in dynes), p is the perimeter of the plate (in cm), $\sigma_{\text{o-w}}$ is the oil–brine interfacial tension (in dynes/cm, measured with the Du Noüy ring method in this study), θ is the oil/brine/solid contact angle (in degrees), and $F_{\text{B}}(D)$ is the buoyancy force (in dynes), which is a function of the plate position, D (in cm).

During the experiment, the force acting on the plate and the distance moved were recorded automatically. Because of the high density of data points (1000 data points for each cycle), all the results are presented as continuous curves in this study. Advancing and receding contact angles at each point were determined from equation (1).

Observation of straight-line force–distance relationships (constant slope) implies that the contact angle is independent of the plate height and time at which it is measured and also that the interfacial tension is constant. The force F on the plate is then affected only by the change in buoyancy force determined by the distance of plate movement D, and the force–distance relationship is referred to as the buoyancy line. If the buoyancy line is not straight, for example, the contact angle varies with the plate height or time, a point-by-point method can be used to obtain contact angles, because equation (1) still holds at each point [8].

The advantage of the dynamic Wilhelmy plate method is that it gives a direct quantitative determination of wetting preference and qualitative visual indication of wetting tendencies [9]. Results obtained by this technique have been published in relation to various investigations [10–13]. The application of the method to crude oil/brine liquid pairs with special emphasis on mineral surfaces initially covered by brine has been described in detail recently [7, 8, 14, 15]. The results presented in this paper complement these studies.

3. EXPERIMENTAL

3.1. Materials

3.1.1. Solid surface. Quartz plates ($\sim 2 \times 0.1 \times 4$ cm) with a quartz rod (diameter ~ 0.2 cm) fused onto one end were used as the solid substrate [14]. Before use, the plates were cleaned first with a detergent and thoroughly rinsed with tap water and then distilled water. They were next immersed in a solution of 30% H_2O_2 + 20% NH_4OH (9 : 1), agitated in an ultrasonic bath for 30 min, soaked overnight, and washed thoroughly with distilled water. Clean plates were soaked in the test brine for about 1 week under ambient conditions to reach ionic equilibrium.

3.1.2. Brines. The brine used in most of this work was a synthetic Prudhoe Bay reservoir water (PB) which has a density of 1.012 g/cm^3, a pH of 6.3, and an ionic strength of 0.375. This brine contained NaCl (0.3641 M), KCl (0.0013 M), $CaCl_2$ (0.0027 M), and $MgCl_2$ (0.001 M); total dissolved solids were 21 798 ppm. In addition, aqueous solutions of 0.01 M Na^+, buffered to give fixed values of pH, were tested. The buffers are listed in Table 1. 100 ppm of the biocide NaN_3 was added to all brines to prevent bacterial growth.

3.1.3. Crude oils. Two asphaltic crude oils from Prudhoe Bay, designated Alaska 93 (A93) and Alaska 95 (A95), were used in this study. Asphaltene solids

Table 1.

Properties of oils under ambient conditions

Oil	Density (g/cm^3)	Viscosity (cP)	C_7-asphaltene content (wt%)	IFT with PB (dynes/cm)	IFT with 0.01 M Na^+ buffer solutions (dynes/cm)	
A93	0.8950	39.25	3.98	31.7	24.9	pH 4: ($NaCH_2COOH$ + CH_3COOH buffer)
					30.6	pH 6: (Na_2HPO_4 + NaH_2PO_4 buffer)
					26.8	pH 10: (Na_2CO_3 + $NaHCO_3$ buffer)
A93 maltenes	0.8940	—	0	—	—	
A95	0.9077	44.89	6.55	31.6	26.1	pH 4: (same solution as above)
n-Decane	0.7269	0.92	—	52.0	—	
S220	0.7877	3.98	—	48.7	—	

were prepared from A93 crude oil by the ASTM method (ASTM D2007-80, 1980) with n-heptane as the precipitant. The deasphalted crude oil A93 (A93 maltenes) and solutions of the separated asphaltenes in toluene were also tested. The selected physical properties of the oils are shown in Table 1. Light components were removed from the crude oils by evaporation under vacuum (1–2 h) to minimize density changes during the course of an experiment.

3.1.4. Probe oils. Besides crude oil, normal decane (n-decane) and Soltrol 220 (S220) mineral oil (supplied by Phillips Petroleum Company), referred to as probe oils, were also used in the contact angle measurements. Decane and S220 were purified by flow through silica gel and alumina columns before use. The relevant physical properties of the oils are listed in Table 1.

3.1.5. Solvents. Toluene, n-heptane, tetrahydrofuran (THF), and benzene were used to wash excess crude oil from the solid surface prior to measurement of contact angles. All solvents were purified either by passing through silica gel and alumina columns or by distillation.

3.2. Procedures

3.2.1. Initially clean plates. Quartz plates were soaked in the test brine for about 1 week and were immersed in the brine before pouring the crude oil onto the brine.

3.2.2. Plates initially treated with crude oil. The clean quartz plates were soaked in brine for about 1 week, then drained but not dried. The plates were submerged in crude oil for an aging time, t_a, at an aging temperature, T_a. After aging, excess crude oil was removed from the plate by rinsing with a volatile solvent which was, in turn, removed by air-drying. The plate was then immersed in the test brine and the probe oil was poured onto the brine. The force–distance relationships were measured under ambient conditions. Full details of the experimental procedures are available elsewhere [15].

4. RESULTS AND DISCUSSION

4.1. Solid substrate

Figure 1 shows perfect water-wetness as given by clean quartz, purified decane, and PB brine. The water receding curve was retraced by the advancing curve and the calculated contact angle was 0°. There was no hysteresis in this case. A mirror-like sheen on the quartz plate indicated the presence of a brine film. Under comparable conditions, the results obtained for the wettability alteration of fused quartz by adsorption from crude oil are in satisfactory agreement with contact angle measurements for drops on soft glass (borosilicate) microscope slides [3]. Also,

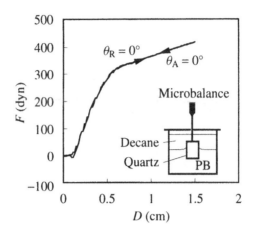

Figure 1. Example of the force–distance diagram for perfect wetting by water.

Liu and Buckley [16] tested two types of mica (muscovite and biotite) as the solid substrate. The trends in water film stability were similar to those for glass; the contact angles were somewhat lower for the two types of mica than for glass.

4.2. Contact angle hysteresis and slippage

Contact angle hysteresis can arise because of surface roughness, a heterogeneous distribution of adsorbed impurities on the solid surface, and adsorption/desorption or other changes during interface movement. When the contact angle switches from advancing to receding, or from receding to advancing, the liquid/liquid/solid three-phase contact line is usually assumed to be pinned until the contact angle transition is complete. If the established contact angle is independent of the plate height, definitive advancing and receding contact angles are obtained. A comparison of the theory given by the pinning condition with crude oil/brine/solid data obtained by the dynamic Wilhelmy plate method shows that for some wetting conditions, the contact line moves along the surface during the transition. This phenomenon is referred to as slippage [7, 8, 14]. In dynamic contact angle measurements, slippage results in a deviation from the buoyancy line and this can seriously affect the interpretation of the experimental data.

Both pinning and slippage are associated with contact angle hysteresis. The calculation of the theoretical pinning distance D_p and the relationship between pinning and slippage have been described in detail [7, 8]. An example of force–distance data that includes slippage is shown in Fig. 2; the predicted curve for no slippage and the pinning distance, D_p, and the slippage distance, D_s, are indicated. Slippage can be eliminated by allowing an equilibration period, usually about 20–60 min, before reversing the direction of motion of the plate in changing from receding to advancing conditions. The ultimate receding and advancing contact angles are unchanged by the equilibration period, but the three-phase line

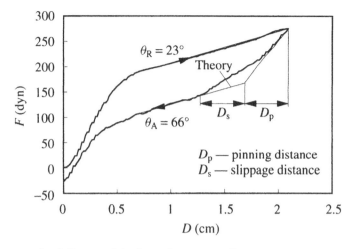

Figure 2. An example of slippage of the three-phase contact line.

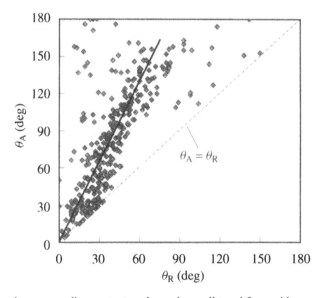

Figure 3. Advancing vs. receding contact angles and overall trend for a wide range of examples of wetting conditions induced by adsorption from crude oils.

of contact now remains pinned while the plate traverses the theoretical pinning distance D_p. Examples of slippage and its elimination have been reported [7].

Wetting conditions, defined by advancing and receding contact angles, can be determined from force–distance relationships that give well-defined buoyancy lines. The plot of advancing versus receding contact angles presented in Fig. 3 illustrates the wide variety of wetting conditions that can be induced by adsorption from crude oils [15]. Variables included the crude oil and modifications to its composition, the composition and pH of the brine phase, the aging temperature and time, the solvent used to remove excess crude oil from the plate, the choice of probe oil, the temperature at which the contact angles were measured, and the speed of plate movement. Most of the results show substantial hysteresis with θ_A, on average, approximately equal to twice θ_R for receding angles up to 70°. However, there are many specific circumstances where the hysteresis behavior is well removed from this average (e.g. very high advancing angles and very low receding angles or very low hysteresis). Examples of how these variables can affect wetting are presented in this paper.

4.3. Brine composition and pH — initially clean plates

Adhesion tests for numerous crude oils show a remarkably consistent trend for the effect of pH on wetting by crude oil [17, 18]. Transition from non-adhesion (water-wetness) to adhesion (oil-wetness) is observed with a reduction in pH below neutral. A comparison of the force–distance relationships for 0.01 M Na^+ buffer solutions with pH 4, 6, 10, and A93 crude oil for three separate initially clean plates is shown in Fig. 4. At pH 10, the contact angle was 6° and no hysteresis was measured. At pH 6, hysteresis of 31° was observed ($\theta_A = 52°$ and $\theta_R = 21°$). At pH 4, very high contact angle hysteresis, $\theta_A = 157°$ and $\theta_R = 13°$, was observed. During the course of measuring the force–distance curve for advancing conditions, the direction of plate movement was cycled four times between points A and B (see Fig. 4). This part of the curve was reversible; thus, the three-phase contact line was pinned during the contact angle transition.

Another example of very high contact angle hysteresis is shown in Fig. 5 for an initially clean plate and A95 crude oil. After raising the plate to its highest position in the oil (point A) for the first time, an equilibration time of 50 min was allowed. The three-phase contact line remained pinned until an advancing contact angle of 180° was achieved (curve 1). This result was consistent with the observation that the plate retained a thin film of oil during the water advancing

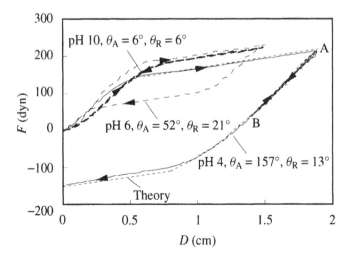

Figure 4. Effect of the brine pH on A93/brine/quartz. Clean plate initially in brine.

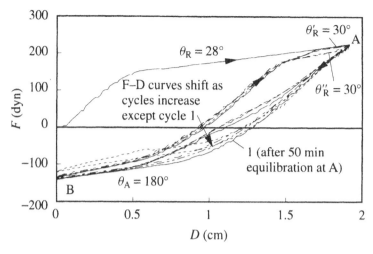

Figure 5. Multiple cycles for A95/pH 4 brine/quartz. Clean plate initially in brine.

process. The plate was then raised and lowered 12 times between points B and A without any equilibration periods. The secondary receding contact angle, θ'_R, was transient from 180° down to 30°. In subsequent cycles, the force–distance curve for water receding conditions was reproduced fairly closely. The water advancing curves showed slippage, which tended to decrease with each cycle, as indicated in Fig. 5. The nature and cause of large contact angle hysteresis are under further study.

The effect of pH and salinity on changes in surface properties induced by crude oil has been investigated in detail [1, 4, 6, 15, 19, 20]. At low salinity, high pH increases the negative surface charge densities on both the solid–brine and the brine–crude oil interfaces, and a relatively thick and stable film of brine remains between the solid surface and the oil phase. This film prevents the water-insoluble asphaltene components in crude oil from reaching the solid surface [1]. When the brine pH decreases, the water film between the crude oil and the solid surface is reduced to where the polar components in crude oil may contact the solid surface. At low pH, the forces between the brine–solid surface and crude oil–brine interfaces become attractive and high advancing contact angles are observed as a result of adsorption.

4.4. Plates pretreated with crude oil

4.4.1. Wash solvents. Study of the wetting changes after pretreatment of plates with crude oil is a convenient way to investigate many features of the adsorption process [19]. After a quartz surface was treated by aging in crude oil, excess oil was removed by rinsing the plate with a solvent. The solvent was chosen with the objective of removing the bulk crude oil with minimal change in the state of surface adsorption. The results in Fig. 6 show that different wash solvents achieved different wetting properties. Plots of force vs. distance were obtained with decane as the probe oil and PB brine. The plates were treated identically except that they

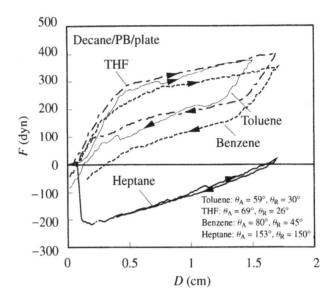

Figure 6. Effect of wash liquids on the contact angles. Plates were aged in A93 crude oil; $t_a = 20$ h; $T_a = 88\,°C$.

were rinsed with different solvents: benzene, toluene, tetrahydrofuran (THF), and heptane. There are significant differences in the results for toluene, THF, and benzene but the contact angles are all below 90°, whereas the plate washed with heptane exhibited a high contact angle, 153°, and there was essentially no contact angle hysteresis. The high contact angles measured after the solvent wash with heptane are ascribed to the precipitation of asphaltenes resulting from the change in crude oil composition upon addition of alkane. The mechanism of wettability change is described as surface precipitation [5]. As the solvency of the wash liquid for adsorbed asphaltenes increases, an opposite effect is expected whereby an increasing fraction of the adsorbed polar organic material is removed from the surface.

Buckley [4] has shown that the onset of precipitation of asphaltenes is related to the solvency of the crude oil and that the refractive index of the crude oil provides a useful indication of the intermolecular forces that determine solvency. Removal of excess crude oil without changing the state of adsorption at the surface is, therefore, more likely to be achieved by using a solvent that has a refractive index close to that of the crude oil. For example, A93 and A95 crude oils have refractive indices that are close to that of toluene (1.51). So toluene was adopted as the wash solvent for all of the subsequent results in this paper.

4.4.2. Probe oils. For most of the contact angle measurements on treated plates, the probe oil was decane. The rationale for use of a probe oil, rather than the crude oil, to examine the changes in wetting resulting from adsorption from crude oil has been discussed in detail previously [4]. After excess crude oil has been removed by

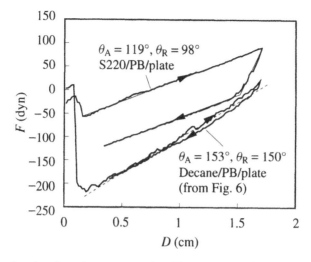

Figure 7. Effect of probe oil on the contact angles. Plates were aged in A93 crude oil; $t_a = 20$ h; $T_a = 88\,^\circ$C; then washed with n-heptane.

the solvent wash, alteration of the wettability by surface precipitation is no longer a factor and the adsorbed asphaltenes will be stable in the presence of decane.

In laboratory experiments on oil recovery for porous rocks, relatively large volumes of oil are needed. Refinery oils given by distillation temperature cuts are often used and are referred to as refined oils. Various core preparation procedures have been adopted with the objective of retaining or restoring crude oil components adsorbed at rock surfaces in order to reproduce the wetting conditions of a reservoir. In some cases, refined oil is then used as the oleic phase; it is assumed that wettability is essentially determined by the adsorbed polar compounds. The use of a probe oil to study the effect of adsorption on the wetting properties of smooth mineral surfaces is somewhat analogous in approach to that for rocks, and the question of the effect of the probe oil on the measured wetting behavior is of interest. One of the curves in Fig. 7 was measured for a plate washed with heptane but with S220 as the probe oil. The use of the refined oil as a probe gave a generally smoother force–distance curve, lower contact angles, and moderate hysteresis, whereas decane indicated a strongly oil-wet surface with very little contact angle hysteresis. Thus, in oil recovery tests on preserved or restored cores, the oleic phase chosen for displacement tests may have a significant effect on the wetting behavior and oil recovery.

4.4.3. Aging time and temperature.   Both the aging temperature and the aging time have been shown to be key factors in the wettability changes induced by crude oil. When contact angles on plates aged at elevated temperature are measured at elevated temperature, they are somewhat lower than those measured under ambient conditions [21]. The role of temperature in the adsorption mechanism that results

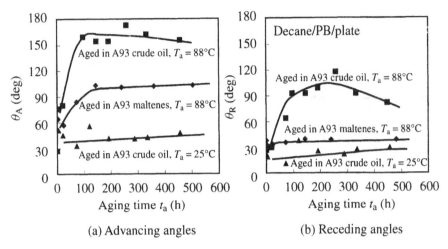

Figure 8. Effect of aging temperature and oil components on wetting.

in changes in wettability is probably related to kinetic effects but is not well understood.

A question of practical importance in laboratory tests on oil recovery relates to slow adsorption processes and the length of time needed to establish equilibrium between the crude oil and the rock surface that corresponds to that attained in an oil reservoir over geological time. Displacement tests indicate that aging times in the range of 14–40 days are needed [22].

Plots of the contact angle vs. the aging time derived from force–distance relationships, for aging at 25°C and 88°C with A93 crude oil are shown in Fig. 8. For plates aged at 25°C, the contact angles remained low (all less than 60°), even after long aging times. For high-temperature aging, advancing contact angles of more than 150° were achieved within 100 h of aging. A possible contributing factor, especially in the wetting behavior of smooth surfaces, is the effect of temperature on the thickness of any water film retained at the solid surface. For a water/vapor/quartz system, the water film thickness on the quartz surface decreases as the temperature increases [23]. At around 70°C, only a few layers of water molecules remain adsorbed at the surface. If the thickness of the water film between the solid surface and crude oil decreases, the polar components from crude oil are more likely to penetrate the water film and contact the solid surface. It has also been suggested [24] that an increase in temperature activates crude oil surfactants, and changes the thermal motion of the components, so that a change in wettability towards oil-wetness is promoted.

4.4.4. Thickness and uniformity of adsorbed layers. The thickness and distribution of adsorbed asphaltenes over a small area can be examined by atomic force microscopy (AFM) [3, 25]. However, this application is not yet routine and problems with tip damage are frequent. The softness of the adsorbed layer sometimes results in artifact as evidenced, for example, by smeared images and shadowing.

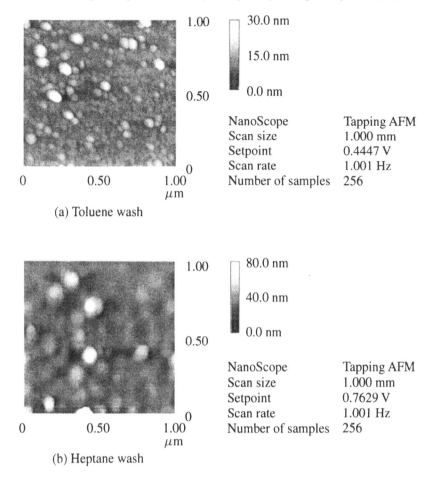

(a) Toluene wash

(b) Heptane wash

Figure 9. AFM pictures of adsorbed asphaltenes after removal of excess crude oil by (a) toluene and (b) heptane. Quartz aged in A93 crude oil; $T_a = 80\,^\circ C$; $t_a = 7$ days.

AFM images show the smoothness of the solid substrate to be 4 nm for soft glass and to be of molecular smoothness for mica [3, 25]. The images shown in Fig. 9 are for quartz surfaces after adsorption from A93 crude oil (the quartz was first equilibrated in PB brine). The only difference in the treatment conditions was that one plate was washed with toluene whereas the other was washed with heptane. If it is assumed that the lowest features of the image correspond to the quartz surface, the thickness of the adsorbed layer after the toluene wash ranged up to 30 nm, with much of the coverage being close to the average thickness of 15 nm. For the plate washed with heptane, there is evidence of surface precipitation. The adsorbed layer is much thicker (up to 80 nm) and much more heterogeneous. These images were representative of several 1 μm^2 areas of observation.

The contact angles corresponding to the plate washed with toluene were $\theta_A \sim 160^\circ$ and $\theta_R \sim 100^\circ$. For a plate aged with A93 crude oil but aged for only 20 h and then washed with heptane, θ_A was 153° and θ_R was 150° (high contact angles and low hysteresis are also expected for the 7-day aging time used in the AFM work). The average values of thickness determined by AFM (15 and 40 nm) and the apparently high surface coverage shown by the AFM images imply that the solid substrate plays a greatly diminished role in the wetting behavior. Contact angles and contact angle hysteresis should, therefore, be dependent on the chemical nature of the adsorbed aggregates (in particular, their intrinsic wetting properties) and the surface topology and pore structure developed by the adsorbed aggregates. The smooth appearance of the sample washed with toluene indicates that differences in chemistry of the adsorbed asphaltenes make a major contribution to the observed high contact angle hysteresis. Adsorption by surface precipitation produces adsorbed aggregates which are inherently more hydrophobic than those adsorbed by other mechanisms. The volume fraction of adsorbed asphaltene aggregates within the adsorbed layer is referred to by Dubey and Waxman [26] in the modeling of asphaltene adsorption as the packing factor. The hydrophobicity and pore shapes within the adsorbed asphaltenes will determine whether brine is retained in the adsorbed layer or is displaced by the oleic phase.

4.4.5. Crude oil composition. Asphaltenes in A93 crude oil were precipitated by addition of heptane. After separating the precipitate by filtration, the heptane was evaporated off to give A93 maltenes (deasphalted crude oil). Advancing contact angles measured for plates aged in maltenes at 88°C and different aging times, included in Fig. 8a, were closely reproduced for aging times higher than about 100 h. They were consistently lower than for aging in A93 crude oil at this temperature, but were still in the range of 80–100°. Thus, wettability alteration, although reduced, is still substantial after asphaltenes had been removed from the crude oil. However, the receding angles shown in Fig. 8b were all around 40°, much lower than for aging in A93 crude oil at the same temperature.

The adsorption of the n-heptane A93 asphaltenes was also examined. Plates initially soaked in PB brine were aged in solutions of 250, 400, and 700 mg/l asphaltene in toluene at 25°C for 10 days. The force–distance relationships are shown in Fig. 10. At the concentrations of 400 and 700 mg/l, the advancing contact angles were close (152° and 143°, respectively). The contact angles are about the same as those observed for aging in A93 crude oil (see Fig. 8), for which the asphaltene content was about 50 times higher than in the solutions in toluene.

For treatment with 250 mg/l solutions, the advancing contact angles were significantly lower than for the high concentration solutions, but still slightly higher than for the A93 maltenes under equivalent aging conditions (see Fig. 8). A feature of the results shown for solutions of asphaltenes in toluene is that the force–distance curves for receding conditions fluctuate markedly about the buoyancy line whereas the advancing angles are relatively smooth. Fluctuations were occasionally observed for particular combinations of brine and crude oil but were of distinctly lower frequency [7, 15]. An unexpected feature of asphaltene adsorption from solution was that the adsorbed layer sometimes desorbed from

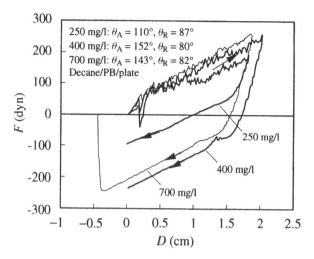

Figure 10. Force–distance diagrams of decane/PB/treated plate systems with fluctuation in receding curves. Plates were aged in A93 asphaltene solutions; $T_a = 25\,°C$; $t_a = 10$ days.

the solid surface and moved to the oil–water interface within a few seconds of contact by water to leave a strongly water-wet surface [3]. Such complications limit the usefulness of wettability studies with solutions of separate components such as asphaltenes. The solvency of a crude oil is determined by all components in combination and is a critical factor in its adsorption behavior. Solvency can be varied by the addition of known compounds or by the removal of light ends [4, 27]. Matching the solvency of an oil under reservoir conditions by addition of alkanes to crude oil may give an oil which mimics reservoir wetting properties at ambient pressure. This would avoid the problems and expense of working with recombined fluids at high pressure.

5. CONCLUSIONS

The changes in wettability induced by adsorption from crude oil can be investigated by the dynamic Wilhelmy plate method. A distinct advantage of the plate method is that dynamic receding and advancing contact angles are defined precisely and are constant with respect to the plate height if force–distance measurements match the buoyancy line and the three-phase contact line on the plate is fixed during transitions between angles. Slippage of the three-phase line is sometimes evident, but can be avoided by allowing an equilibration time before reversing the direction of plate motion. Fluctuations in the force–distance relationships for receding conditions were observed when wettability was altered by the adsorption of asphaltenes from solution. The main limitation of the method is that the solid substrate must be available in the form of a plate. In this respect, the method is less versatile than direct measurements of contact angles.

A wide variety of wetting conditions, as characterized by advancing and receding contact angles and related phenomena, can be induced by adsorption from crude oil. These depend on the crude oil, the brine, the solid substrate, and the conditions of preparation and measurement. Various mechanisms are involved in the adsorption process [5]. For initially clean plates suspended in brine and drawn into crude oil, acid–base interactions cause almost instantaneous change to oil-wetness at moderately low pH, but very little or no immediate change at moderately high pH.

For pretreated plates (initially coated with brine before aging in crude oil), the solvent used to remove the excess crude oil can result in a change to a strongly oil-wet condition by surface precipitation. The choice of the refined oil used as a probe in contact angle measurements can also have a dominant influence on the measured wetting properties of adsorbed organic films. Buckley [4] recommends that the refractive index of the wash solvent match that of the crude oil, and that once the excess crude oil is removed, the use of a probe oil of low solvency (decane) is least likely to disturb the adsorbed layer of polar organic material.

Atomic force microscopy indicated that the adsorbed layers of asphaltenes had an average thickness of about 15 nm after removal of excess crude oil by washing with toluene. Asphaltene precipitation, resulting from removal of excess crude oil by washing with heptane, induced a strongly oil-wet surface with little contact angle hysteresis. AFM measurements showed an average film thickness of 40 nm after the heptane wash. These results are qualitatively consistent with previously reported values [3, 25].

Acknowledgements

Support for this work was provided by ARCO, British Petroleum, Chevron, Elf-Acquitane, Exxon, Marathon, NorskHydro, Phillips, Shell, Statoil, EORI of the University of Wyoming, and Oak Ridge National Laboratory and Idaho National Engineering and Environmental Laboratory through the US Department of Energy Partnership Program.

REFERENCES

1. J. S. Buckley, K. Takamura and N. R. Morrow, *Soc. Pet. Eng., Formation Evaluation* 332–340 (Aug. 1989).
2. T. W. Healy and L. R. White, *Adv. Colloid Interface Sci.* 9, 303–345 (1978).
3. J. S. Buckley, Y. Liu, X. Xie and N. R. Morrow, *Soc. Pet. Eng. J.* 2, 107–119 (June 1997).
4. J. S. Buckley, Ph.D. thesis, Heriott-Watt University, UK (Sept. 1996).
5. J. S. Buckley and Y. Liu, *J. Pet. Sci. Eng.* 20, 155–160 (1998).
6. L. Wilhelmy, *Ann. Phys.* 119, 177–217 (1863).
7. X. Xie and N. R. Morrow, *Colloids Surfaces A* 132, 97–108 (1998).
8. A. Mennella and N. R. Morrow, *J. Colloid Interface Sci.* 172, 48–55 (1995).
9. M. A. Andersen, D. C. Thomas and D. C. Teeters, *Log Analyst*, 372–381 (Sept.–Oct. 1989).
10. H. M. Princen, *Aust. J. Chem.* 23, 1789–1799 (1970).

11. G. Giannotta, M. Morra, E. Occhiello and F. Garbassi, *Polym. Composites* **14**, No. 3, 224–228 (1993).
12. D. C. Teeters, J. F. Wilson, M. A. Andersen and D. C. Thomas, *J. Colloid Interface Sci.* **126**, 641–644 (1988).
13. J. M. Fleureau, *Soc. Pet. Eng., Formation Evaluation* 132–138 (June 1992).
14. A. Mennella, N. R. Morrow and X. Xie, *J. Pet. Sci. Eng.* **13**, 179–192 (1995).
15. X. Xie, Ph.D. thesis, University of Wyoming, Wyoming (1996).
16. L. Liu and J. S. Buckley, *Proceedings of the 5th International Symposium on Evaluation of Reservoir Wettability and its Effect on Oil Recovery*, Trondheim, Norway, pp. 25–35 (22–24 June 1998).
17. J. S. Buckley and N. R. Morrow, SPE 20263, *SPE/DOE Seventh Symposium on Enhanced Oil Recovery*, Tulsa, Oklahoma, pp. 871–877 (22–25 April 1990).
18. J. S. Buckley and N. R. Morrow, in: *Physical Chemistry of Colloids and Interfaces in Oil Production*, H. Toulhoat and J. Lecourtier (Eds), pp. 39–45. Editions Technip, Paris (1991).
19. Y. Liu and J. S. Buckley, in: *Proceedings of the 3rd International Symposium on Evaluation of Reservoir Wettability and its Effect on Oil Recovery*, N. R. Morrow (Ed.), pp. 27–32. University of Wyoming, Laramie, WY (1996).
20. J. Israelachvili and C. Drummond, *Proceedings of the 5th International Symposium on Evaluation of Reservoir Wettability and its Effect on Oil Recovery*, Trondheim, Norway, pp. 1–12 (22–24 June 1998).
21. X. Xie, N. R. Morrow and J. S. Buckley, *Proceedings of the 1997 International Symposium of the Society of Core Analysts*, Calgary, Paper SCA-9712, Sept. 7–10 (1997).
22. L. E. Cuiec, in: *Interfacial Phenomena in Petroleum Recovery*, N. R. Morrow (Ed.), pp. 319–375. Marcel Dekker, New York (1990).
23. B. V. Derjaguin, *Theory of Stability of Colloids and Thin Films*. Consultants Bureau, New York (1989).
24. W. G. Anderson, *J. Pet. Technol.* 1125–1144 (Oct. 1986).
25. S. Y. Yang and G. J. Hirasaki, *Proceedings of the 5th International Symposium on Evaluation of Reservoir Wettability and its Effect on Oil Recovery*, Trondheim, Norway, pp. 13–24 (22–24 June 1998).
26. S. T. Dubey and M. H. Waxman, *Soc. Pet. Eng. Reservoir Eng.* 389–395 (Aug. 1991).
27. G. Q. Tang and N. R. Morrow, *Soc. Pet. Eng., Reservoir Eng.* 269–276 (Nov. 1997).

Part 2

Surface Forces and Surface Free Energy

Apparent and Microscopic Contact Angles, pp. 149–170
J. Drelich, J. S. Laskowski and K. L. Mittal (Eds)
© VSP 2000.

Determination of the acid–base properties of metal oxide films and of polymers by contact angle measurements

E. McCAFFERTY [*,†] and J. P. WIGHTMAN

Department of Chemistry, Center for Adhesive and Sealant Science, Virginia Polytechnic Institute and State University, Blacksburg, VA 24061, USA

Received in final form 10 March 1999

Abstract—The surface isoelectric point for native air-formed oxide films on various metals has been determined by measurement of contact angles at the hexadecane/aqueous solution interface as a function of the pH of the aqueous phase. Application of Young's equation, the Gibbs equation, and surface equilibria conditions for hydroxylated oxide films leads to a mathematical expression which shows that the contact angle goes though a maximum at the isoelectric point of the oxide. The experimentally determined values for the oxide films on aluminum, chromium, and tantalum are within one to three pH units of the reported isoelectric points for the corresponding bulk oxide powders. The surface isoelectric point of an oxide-covered Ta-implanted Al surface lies between that of Al_2O_3 and Ta_2O_5. The acid–base properties of various polymers, including a commercially available pressure-sensitive adhesive, were determined by measuring the contributions γ_S^+ and γ_S^- to the solid surface free energy using the contact angle approach of van Oss and Good. Adhesion measurements for a pressure-sensitive adhesive having a positive surface charge show that the peel strength is greatest when the metal substrate has a surface oxide film of basic character.

Key words: Contact angles; oxide films; polymers; acid–base; isoelectric point; van Oss and Good method; peel strength.

1. INTRODUCTION

This work is directed towards understanding the adhesion of polymers onto various metal surfaces in terms of the acid–base properties of the adhering polymer and of the oxide-covered metal. The acid/base nature of the oxide film on various metal surfaces and the acid/base nature of selected polymers were determined by contact angle methods. The surface isoelectric point for native air-formed oxide

---

*To whom correspondence should be addressed.

†Visiting Scientist. Permanent address: Naval Research Laboratory, Code 6134, Washington, DC 20375-5343, USA. E-mail: mccafferty@anvil.nrl.navy.mil

films on various metals was determined by measurement of contact angles at the hexadecane/aqueous solution interface as a function of the pH of the aqueous phase [1]. The acid–base properties of the various polymers, including a pressure-sensitive adhesive (Scotch 610 Magic Tape®), were determined using the method of van Oss and Good [2–5]. Adhesion measurements for the pressure-sensitive adhesive on various metal substrates are interpreted in terms of the acid/base nature of the adhering polymer and of the surface oxide films on the metal substrates.

1.1. Oxide films on metals

It is well known that the surface oxide film on a metal substrate terminates in an outermost layer of hydroxyl groups [6–8]. These hydroxyl groups can be acidic or basic in nature, depending on the identity of the metal cation present in the oxide film. To date, there have been numerous determinations of the isoelectric point of bulk oxides, but only a limited number of studies have reported the isoelectric point of oxide films on metals [9–13].

In aqueous solutions, surface hydroxyl groups may remain undissociated, in which case the pH of the aqueous solution is the same as the isoelectric point of the oxide. If the pH is less than the isoelectric point, the surface will acquire a positive charge:

$$-MOH_{surf} + H^+_{(aq)} \rightleftharpoons -MOH^+_{2\,surf}. \tag{1}$$

If the pH is greater than the isoelectric point, the surface will acquire a negative charge:

$$-MOH_{surf} + OH^- \rightleftharpoons -MO^-_{surf} + H_2O$$

or

$$-MOH_{surf} \rightleftharpoons -MO^-_{surf} + H^+_{(aq)}. \tag{2}$$

There have been several studies on the effect of pH on the wetting of solid surfaces. Whitesides and co-workers [14] used the variation in contact angle with pH to follow the ionization of carboxylic acid surface groups in organic films. Chau and Porter [9] recently used a similar approach to determine the isoelectric point of silver oxide films on evaporated silver, and Gribanova et al. [15] conducted similar experiments on quartz and glass. In a related study, Billett et al. [16] measured the contact angle of an air bubble against thin films of silver iodide in an electrolyte solution as a function of pAg to determine the isoelectric point of the silver iodide film. Vujicic and Lovrecek [11] measured the contact angle of a paraffin oil drop against an oxide-covered aluminum surface in an electrolyte solution as a function of pH to determine the isoelectric point of the aluminum oxide film.

Chau and Porter [9] have derived an expression showing that the variation in contact angle with pH goes though a maximum at the isoelectric point of the surface. However, these authors considered only the case of a basic surface, such as silver oxide, in which the hydroxyl ion is the charge-determining ion. A more general

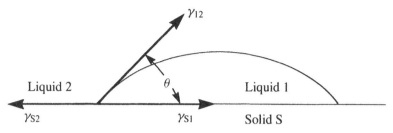

Figure 1. Schematic diagram for the two-liquid–solid system.

treatment applicable to any oxide film has been published recently [1] and is given only in brief here.

For a solid surface S in contact with liquids 1 and 2 as shown in Fig. 1, Young's equation gives

$$\pi + \gamma_{S2} = \gamma_{S1} + \gamma_{12} \cos\theta, \tag{3}$$

where π is the film pressure for liquid 2 on the solid surface S. If liquid 2 is a hydrocarbon and liquid 1 is an aqueous phase of varying pH, then differentiation with respect to pH gives

$$0 = \frac{d\gamma_{S1}}{dpH} + \gamma_{12}\frac{d\cos\theta}{dpH} + \cos\theta\frac{d\gamma_{12}}{dpH}. \tag{4}$$

It has been shown experimentally that there is only a slight variation with pH in the interfacial tension of hexadecane vs. aqueous solutions of various pH values [1]. As an approximation, then, we can take $d\gamma_{12}/dpH \cong 0$. Then equation (4) becomes

$$0 = \frac{d\gamma_{S1}}{dpH} + \gamma_{12}\frac{d\cos\theta}{dpH}. \tag{5}$$

The interface between the oxide-covered metal surface and the aqueous phase consists of the following surface species: adsorbed protons MOH_2^+, dissociated hydroxyl groups MO^-, undissociated hydroxyls MOH, and adsorbed water molecules. The Gibbs equation for the interface between the solid and liquid 1 is

$$-d\gamma_{S1} = \Gamma_{H^+}^{S1}d\mu_{H^+}^{S1} + \Gamma_{O^-}^{S1}d\mu_{O^-}^{S1} + \Gamma_{OH}^{S1}d\mu_{OH}^{S1} + \Gamma_{H_2O}^{S1}d\mu_{H_2O}^{S1}, \tag{6}$$

where $\Gamma_{H^+}^{S1}$, $\Gamma_{O^-}^{S1}$, Γ_{OH}^{S1}, and $\Gamma_{H_2O}^{S1}$ refer to the surface excesses of adsorbed protons MOH_2^+, dissociated hydroxyl groups MO^-, undissociated hydroxyls MOH, and adsorbed water molecules, respectively, at the solid/liquid 1 interface, and μ_i are electrochemical potentials for charged species or chemical potentials for uncharged ones.

When the surface reactions (1) and (2) are at equilibrium,

$$\Delta G_1 = \mu_{OH}^{S1} + \mu_{H^+}^{aq} - \mu_{H^+}^{S1} = -RT \ln K_{a1} \tag{7}$$

and

$$\Delta G_2 = \mu_{O^-}^{S1} + \mu_{H^+}^{aq} - \mu_{OH}^{S1} = -RT \ln K_{a2}. \tag{8}$$

The surface charge density, σ, for the oxide/solution interface is defined by

$$\sigma = zF\left(\Gamma^{S1}_{H+} - \Gamma^{S1}_{O-}\right),$$ (9)

where F is the Faraday constant. Combination of equations (5)–(9) along with $\mu^{S1}_{H+} = (\mu^{S1}_{H+})^0 + RT \ln[H^+] + zF\psi$, where ψ is the surface potential and the ionic charge $z = 1$ (but is carried along for completeness) gives

$$\frac{d\cos\theta}{dpH} = \frac{1}{\gamma_{12}}\left[\frac{\sigma}{zF}\left(-2.303RT + zF\frac{d\psi}{dpH}\right) + \Gamma^{S1}_{O-}zF\frac{d\psi}{dpH}\right].$$ (10)

At the isoelectric point, the surface charge σ is zero and the surface concentration of dissociated hydroxyl groups Γ^{S1}_{O-} is also zero, so that equation (10) gives

$$\left(\frac{d\cos\theta}{dpH}\right)_{\sigma=0} = 0.$$ (11)

At the isoelectric point, the cosine of the contact angle goes though a minimum and the contact angle through a maximum.

1.2. Polymer surfaces

The acid–base properties of polymer surfaces were determined by the method of van Oss and Good [2–5]. As is well known, this approach combines the Young–Dupre equation,

$$-\Delta G^a_{SL} = W_{SL} = \gamma_L(1 + \cos\theta)$$ (12)

(where ΔG^a and W_{SL} are the free energy and work of adhesion, respectively, in creating a solid/liquid interface) with the Fowkes assumption that the surface free energy of a phase γ_i consists of the sum of Lifshitz–van der Waals γ^{LW}_i and acid–base γ^{AB}_i contributions:

$$\gamma_i = \gamma^{LW}_i + \gamma^{AB}_i,$$ (13)

which leads to

$$\Delta G^a_{SL} = \Delta G^{LW}_{SL} + \Delta G^{AB}_{SL}.$$ (14)

The Good–Girifalco combining rule expression for the Lifshitz–van der Waals forces across an interface is

$$\Delta G^{LW}_{SL} = -2\left(\gamma^{LW}_S \gamma^{LW}_L\right)^{1/2}.$$ (15)

For the acid–base forces across an interface, van Oss and Good introduced the concept of the acid and base components of the surface free energy, γ^+_i and γ^-_i respectively, and the combining rule:

$$\Delta G^{AB}_{SL} = -2\left[\left(\gamma^+_S \gamma^-_L\right)^{1/2} + \left(\gamma^-_S \gamma^+_L\right)^{1/2}\right].$$ (16)

Table 1.
Surface tension parameters (in mJ/m^2) for various wetting liquids from van Oss and Good [4, 5]

Liquid	γ_L	γ_L^{LW}	γ_L^-	γ_L^+
Methylene iodide	50.8	50.8	0	0
Water	72.8	21.8	25.5	25.5
Formamide	58	39	39.6	2.28
Ethyene glycol	48	29	47.0	1.92
Glycerol	64	34	57.4	3.92

Equations (12) and (14)–(16) give the result

$$\left(\gamma_S^{LW}\gamma_L^{LW}\right)^{1/2} + \left(\gamma_S^+\gamma_L^-\right)^{1/2} + \left(\gamma_S^-\gamma_L^+\right)^{1/2} = \frac{1}{2}(1 + \cos\theta)\gamma_L. \qquad (17)$$

Equation (17) is an equation in three unknowns: γ_S^{LW}, γ_S^+, and γ_S^-. The first of these is determined by using an apolar liquid, such as methylene iodide, for which both γ_S^+ and γ_S^- are zero. Then equation (17) reduces to

$$\gamma_S^{LW} = \gamma_L\frac{(1 + \cos\theta)^2}{4}, \qquad (18)$$

which allows evaluation of the parameter γ_S^{LW} from the surface tension γ_L of methylene iodide (see Table 1). After this step, equation (17) contains two unknowns, γ_S^+ and γ_S^-, which are determined by using two different polar liquids and solving the resulting two equations simultaneously. The polar liquids used here were water, formamide, ethylene glycol, and glycerol. The surface tension parameters for these liquids are taken from van Oss and Good [4, 5] and are also given in Table 1.
   Equation (17) can be written in matrix form as

$$\begin{bmatrix} \left(\gamma_{L1}^{LW}\right)^{1/2} & \left(\gamma_{L1}^-\right)^{1/2} & \left(\gamma_{L1}^+\right)^{1/2} \\ \left(\gamma_{L2}^{LW}\right)^{1/2} & \left(\gamma_{L2}^-\right)^{1/2} & \left(\gamma_{L2}^+\right)^{1/2} \\ \left(\gamma_{L3}^{LW}\right)^{1/2} & \left(\gamma_{L3}^-\right)^{1/2} & \left(\gamma_{L3}^+\right)^{1/2} \end{bmatrix} \begin{bmatrix} \left(\gamma_S^{LW}\right)^{1/2} \\ \left(\gamma_S^+\right)^{1/2} \\ \left(\gamma_S^-\right)^{1/2} \end{bmatrix} = \frac{1}{2}\begin{bmatrix} \gamma_{L1}(1 + \cos\theta_1) \\ \gamma_{L2}(1 + \cos\theta_2) \\ \gamma_{L3}(1 + \cos\theta_3) \end{bmatrix}, \qquad (19)$$

where the numerical subscripts refer to the wetting liquid. Equation (19) can be written as

$$Ax = b, \qquad (20)$$

where A is the matrix containing the various surface energy parameters.
   As is the recommended practice [5, 17], one of the polar wetting liquids used in solving the two equations was water and was taken pairwise in turn with formamide, ethylene glycol, or glycerol. Hollander [17] has given an algebraic argument to show that this combination of wetting liquids is the appropriate one to employ. Following Della Volpe and Siboni [18], the proper choice of wetting liquids can be better shown by using the concept of the condition number of the matrix A [19].

Table 2.
Condition numbers for the matrix A containing the surface tension parameters γ_L^{LW}, γ_L^-, and γ_L^+ calculated for surface tension parameters from van Oss and Good [3, 4]. Condition numbers are based on the infinity-norm

First liquid	Second liquid	Third liquid	Condition number
Methylene iodide	Water	Formamide	6.8
		Ethylene glycol	6.7
		Glycerol	7.3
	Formamide	Ethylene glycol	123.9
		Glycerol	231.9
	Ethylene glycol	Glycerol	72.8

A matrix A is ill conditioned if relatively small changes in the entries of A can cause relatively large changes in the solutions to $Ax = b$. The smaller the condition number, the more well conditioned is the matrix. The condition number of a matrix is defined as the product $\|A\|\|A^{-1}\|$, where $\|\cdot\|$ refers to the matrix norm. Several different matrix norms can be used, but the one used here was the infinity-norm, in which the matrix norm is the maximum sum of the absolute values of the row entries, i.e. $\|A\|_\infty = \max_i \sum_j |a_{ij}|$. The condition numbers calculated using this norm for the various combinations of wetting liquids are given in Table 2. It can be clearly seen that for the surface tension parameters given by van Oss and Good, the appropriate combinations of wetting liquids are the ones utilized here.

It should be noted that there have been recent criticisms of the method of van Oss and Good [18, 20], and these will be taken into account later in this paper.

2. EXPERIMENTAL

2.1. Materials

Aluminum foil of 0.13 mm thickness and 99.9995% purity, zirconium foil of 0.25 mm thickness and 99.94% purity, and tantalum foil of 0.25 mm thickness and 99.95% purity were obtained from Johnson Matthey. Chromium was used in the form of chromium-plated steel (Ferrotype plates) with a mirror-like surface finish and was obtained from Apollo Metals. The Ta-implanted aluminum sample was prepared as described earlier [21]. In brief, a 1.9 cm diameter aluminum rod (99.999% pure) was polished to a 3 μm diamond finish and implanted with Ta ions to a dose of 4×10^{16} ions/cm^2 at 100 keV. Surface analysis by X-ray photoelectron spectroscopy (XPS) showed that the surface concentration of Ta^{+5} in the oxide film was approximately 5 atomic %.

Sheets of unplasticized poly(vinyl chloride) (0.2 mm thick) and of poly(methyl methacrylate) (1 mm thick) were obtained from Goodfellow. These two polymers were used because they are prototypical acidic and basic polymers, respectively. Poly(vinyl acetate) powder was obtained from Polysciences, Inc., and a poly(vinyl acetate) film of approximately 1 mm thickness was cast onto a clean glass slide from a solution of 5 g of poly(vinyl acetate) in 30 ml of acetone. A commercially available pressure-sensitive adhesive (Scotch 610 Magic Tape®) was also used.

2.2. Surface treatment

Before measurement of the contact angles, all metal surfaces were cleaned by argon plasma treatment, as previous work had shown that this procedure was effective in reducing the thickness of the overlayer of carbon contamination [1, 22]. Argon plasma treatment was done using a March Instruments Plasmod unit. The chamber containing the sample was evacuated to approximately 10^{-2} Torr, then filled with argon at a pressure of 1 Torr, and the sample was treated at 50 W for 10 min.

Before measurement of the contact angles on the various polymers, the polymer surfaces were wiped gently with methanol using a tissue, except for the pressure-sensitive adhesive which was used as received.

2.3. Contact angles

Contact angles were measured with a Rame-Hart 100-00 115 NRL contact angle goniometer equipped with a video monitor.

Two approaches were used in measuring the contact angles on the various substrates. In measuring the contact angles on various metals as a function of the pH of the wetting liquid, the two-liquid method was employed. The initial liquid was a droplet of an aqueous solution of a given pH and the surrounding liquid was hexadecane. Following argon plasma treatment, a metal sample was transferred within 15 s through air into hexadecane for contact angle measurements. A 5 μl drop of the aqueous phase was placed onto the surface of the immersed metal sample using a syringe needle held slightly above the metal surface. Contact angle measurements were made on each side of the drop and the sample was rotated 90° to observe the drop from both major axes. Data were recorded for symmetrical drops only. Six to ten contact angle measurements were made for each drop and were averaged. In the case of the tantalum samples, prior to argon plasma treatment, the specimens were first chemically etched in a 25 : 10 : 10 mixture (by volume) of sulfuric, nitric, and hydrofluoric acids for 5 s and then washed in a stream of ultrapure water prepared in a Millipore water system.

Contact angles were also measured for the chromium-plated steel (Ferrotype) at the hexadecane/aqueous solution interface by the Wilhelmy balance technique [23–25] using a Cahn electrobalance. The Ferrotype samples were approximately 1.3 cm in perimeter by 2.0 cm long and were given the same argon plasma treatment as that for the samples used in the goniometry measurements. The metal sample was attached to the arm of the electrobalance and then lowered at a constant rate of 5 μm/s first at the air/hexadecane interface and then at the hexadecane/aqueous interface. This rate was provided by a linear translation stage which moved the two-liquid reservoir relative to the metal sample.

All the aqueous solutions used in this study were prepared from reagent-grade chemicals or stock solutions using water from a Millipore water system containing a reverse osmosis pretreatment unit and a UV final treatment. All glassware was acid-cleaned in a mixture of nitric and sulfuric acids and washed with copious amounts of the Millipore ultrapure water (18 MΩ cm resistivity).

In the second approach, which utilized the van Oss and Good method, the wetting liquids were methylene iodide, water, formamide, ethylene glycol, and glycerol. Water was prepared in the Millipore system, and the other liquids in this series were reagent–grade chemicals. For each wetting liquid on a given polymer, 3–5 drops were measured. Contact angle measurements were made on each side of the drop and the data were averaged.

2.4. Adhesion measurements

The adhesion of the pressure-sensitive adhesive (Scotch 610 Magic Tape®) on various metal substrates was determined by measuring the peel force. Peel forces were measured using an Instrumentors Inc. slip/peel tester at peel rates of 0.2–13.6 cm/min. The peel force was measured for a 1.9 cm wide strip of the pressure-sensitive adhesive on aluminum, chromium, tantalum, and zirconium for various contact times before the peel measurement. In each case, the metal sample was treated in an argon plasma prior to application of the pressure-sensitive adhesive, which was rolled in place using a 5-lb (2.27 kg) cylindrical roller.

2.5. Atomic force microscopy (AFM)

The surface roughness of the aluminum, chromium, and tantalum substrates used in this study was determined from AFM measurements. AFM images were obtained in air using a Digital Instruments Dimension 3000 instrument with a Nanoscope IIIa controller operating in the tapping mode with a silicon tip. The surface roughness was determined in two ways: (i) from a cross-sectional analysis which compares the total distance along the trace between two points on the surface with the horizontal distance between the same two points and (ii) from the ratio of the 'three-dimensional' area of the imaged sample to the planar (projected) area.

3. RESULTS AND DISCUSSION

3.1. Acid–base properties of oxide-covered metals

Figure 2 shows the contact angles for argon plasma-treated chromium (chromium-plated steel) at the hexadecane/aqueous solution interface as a function of the pH of the aqueous phase. These contact angles were measured with a goniometer, as

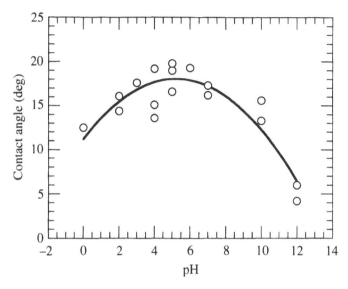

Figure 2. Contact angles as measured by the goniometry technique for chromium-plated steel (Ferrotype) at the hexadecane/aqueous solution interface, as a function of the pH of the aqueous phase.

described earlier. The standard deviation in the contact angle for a set of measurements on oxide-covered metals at the hexadecane/aqueous solution interface was typically $\leqslant 2.5°$. For example, in a set of ten measurements on oxide-covered chromium where the pH of the aqueous phase was 5, the contact angle varied from 17° to 23°, the average value was 19.7°, and the standard deviation was $\pm 2.0°$. The contact angle in Fig. 2 clearly exhibits a maximum and by fitting a second-order polynomial to the data, it can be seen that the maximum occurs at a pH of 5.2.

Figure 3 shows a typical force–distance trace from a Wilhelmy balance experiment for an argon plasma-treated chromium sample cut from the same sheet as in the samples used in the goniometry work above. Both advancing and receding directions are shown in Fig. 3. The initial jump in force when the sample enters the hexadecane phase is used to calculate the perimeter P of the sample. Use of interfacial tensions for hexadecane/aqueous solutions determined separately [1] and the advancing or receding jumps in force at the hexadecane/aqueous solution interface gives the advancing or receding contact angle, respectively, using the relation $F = P\gamma \cos \theta$.

The cause of the fluctuations in force in Fig. 3 for the advancing direction when the sample is immersed in the aqueous phase is not clear. It has been suggested, however, that small variations in the surface energy of the solid lead to local modifications in the contact angle so as to perturb the triple line and thus the overall form of the meniscus and the resulting force [26].

The receding contact angles as a function of the pH of the aqueous phase for chromium-plated steel (Ferrotype) are shown in Fig. 4. Receding contact angles were used because they proved to be more reproducible than advancing contact

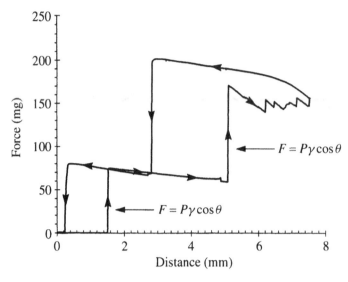

Figure 3. Force–distance plot for chromium-plated steel (Ferrotype) in the hexadecane/aqueous solution (two-liquid system).

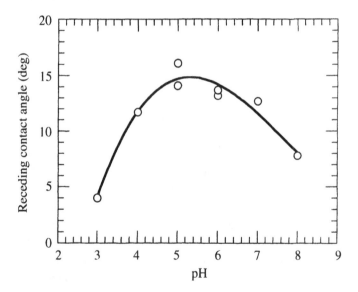

Figure 4. Receding contact angles as measured by the Wilhelmy balance technique for chromium-plated steel (Ferrotype) at the hexadecane/aqueous solution interface, as a function of the pH of the aqueous phase.

angles. The maximum in Fig. 4 occurs at pH 5.3, as can be seen by fitting a third-order polynomial through the data. (A better fit resulted than with a second-order polynomial.) Thus, the results for the goniometry and Wilhelmy balance measurements taken together show that the isoelectric point for the oxide-covered chromium surface occurs at a pH of 5.2–5.3. This value is quite close to the value of 6.35 for bulk α-Cr_2O_3 powder determined by potentiometric titration [27].

Figure 5 shows contact angles measured by the goniometry technique for argon plasma-treated aluminum at the hexadecane/aqueous solution interface as a function of the pH of the aqueous phase. The contact angle for aluminum in Fig. 5 goes through a maximum at a pH of 9.5. This result is in good agreement with the values of 8.7–9.2 reported by a number of investigators [10, 28–30] for the isoelectric point of bulk aluminum oxide (powders). Vittoz *et al.* [13] recently reported a value of approximately 7.0 for aluminum oxide (sapphire) as determined by contact angle measurements.

The few previous studies in the literature for the isoelectric point of oxide-covered aluminum report variable results. Campanella *et al.* [10] observed a value of 9.8 for anodized aluminum as determined by potentiometric titration. However, Vujicic and Lovrecek [11] reported an isoelectric point of 6.8 for oxide-covered aluminum based on contact angle measurements and electro-osmosis experiments. Using streaming potential measurements on oxidized aluminum powders in 10^{-3} M sodium chloride, Morfopoulos and Parreira [12] observed isoelectric points between 7 and 9 depending on the method of oxidation.

Figure 5 also shows contact angles measured by the goniometry technique for tantalum at the hexadecane/aqueous solution interface as a function of the pH of the aqueous phase. The tantalum samples were first chemically polished and argon plasma-treated, as described earlier. The contact angle for tantalum does not go through a well-defined maximum, but instead increases with decreasing pH down to a pH of -0.7. Data could not be obtained at lower pH values as the tantalum subtrate appeared to react with the acid droplet, as evidenced by discoloration of the droplet. Thus, the isoelectric point can be taken to be -0.7, so the oxide-covered tantalum surface is very acidic. Reported values for the isoelectric point of bulk tantalum oxide range from 2.7 to 3.0 [31].

Thus, contact angle measurements as a function of the pH have proved useful in determining the surface isoelectric point of oxide films on flat metal substrates. The results in Table 3 show that the isoelectric points of oxide films are similar to

Table 3.
Summary of the experimental isoelectric points for oxide-covered metals

System	Present work	Literature	Bulk oxides
Cr/Cr_2O_3	5.2–5.3	—	6.35 [27]
Al/Al_2O_3	9.5	9.8 [10] (titration)	8.7–9.2 [28–30]
		6.3 [11] (contact angles, electro-osmosis)	7.0 [13] sapphire
Ta-implanted Al	5.0	—	—
Ta/Ta_2O_5	*ca.* -0.7	—	2.7–3.0 [31]

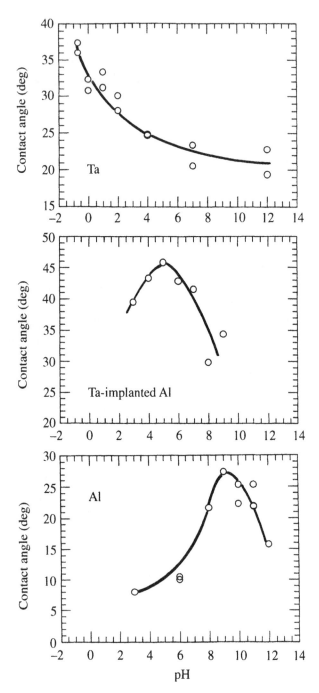

Figure 5. Contact angles as measured by the goniometry technique for tantalum, Ta-implanted aluminum, and aluminum at the hexadecane/aqueous solution interface, as a function of the pH of the aqueous phase.

Table 4.

Surface free energy components (in mJ/m^2) for oxide-covered chromium and aluminum as determined by the method of van Oss and Good

Metal	Water plus wetting liquid:	θ (degrees)	γ_S^{LW}	γ_S^+	γ_S^-
Chromium	Methylene iodide	26.0	45.8	—	—
	Water	7.1		—	—
	Formamide	5.5		0.70	54.8
	Ethylene glycol	12.1		0.0014	66.3
	Glycerol	12.8		1.89	45.3
				0.86	55.5
Aluminum	Methylene iodide	23.5	46.7	—	—
	Water	6.5		—	—
	Formamide	5.3		0.40	55.0
	Ethylene glycol	18.5		0.0047	68.4
	Glycerol	22.3		1.08	49.2
				0.50	57.5

(although not always the same as) the isoelectric points for the corresponding bulk oxides.

Figure 5 also shows contact angles measured by the goniometry method for aluminum which was ion-implanted with tantalum. With Ta implantation, the isoelectric point of the host aluminum oxide film decreased from 9.5 to 5.0. Athough the surface concentration of Ta^{5+} in the oxide film is only 5 at.%, the resulting isoelectric point has decreased by approximately 4.5 pH units. To our knowledge, this result provides the first clear evidence that the process of ion implantation changes the isoelectric point of a host oxide film.

As shown in Fig. 5, the isoelectric point of the oxide-covered Ta-implanted Al surface lies between that of Al$_2$O$_3$ and Ta$_2$O$_5$. This same trend has been observed for a number of mixed oxide powders, including Al$_2$O$_3$–TiO$_2$, Al$_2$O$_3$–MgO, Al$_2$O$_3$–SiO$_2$, RuO$_2$–IrO$_2$, Fe$_2$O$_3$–SiO$_2$, and Zn$_2$SiO$_4$ [29, 32–37].

The acid/base nature of the oxide film on chromium and aluminum was also determined by the method of van Oss and Good. It is realized that there may be inherent errors in applying this method to high energy surfaces in that the contact angles for the various wetting liquids are all low (except for methylene iodide), but it is interesting to examine the results. Table 4 lists the contact angle data and the resulting values of γ_S^{LW}, γ_S^+, and γ_S^-. For each metal, the value of γ_S^{LW} was calculated from equation (18) using the contact angle data for methylene iodide. The quantities γ_S^+ and γ_S^- were calculated from the data for water taken pairwise in turn with the data for formamide, ethylene glycol, and glycerol. For both chromium and aluminum, the γ_S^+ values obtained using ethylene glycol are substantially lower than those obtained using formamide or glycerol. However, for both chromium and aluminum, the average value for γ_S^+ is consistent with that of an acidic surface [3–5]; at the same time the value of γ_S^- is consistent with that for a basic surface [3–5]. Thus, the result of the van Oss and Good treatment reflects the fact that the oxide-covered metal surface is amphoteric in nature.

3.2. Acid–base properties of polymer surfaces

The method of van Oss and Good was used to determine the acid/base nature of the following polymers: poly(vinyl chloride), poly(methyl methacrylate), poly(vinyl acetate), and a pressure-sensitive adhesive (Scotch 610 Magic Tape®). Table 5 lists the contact angle data and the resulting values of γ_S^{LW}, γ_S^+, and γ_S^- for the various polymers.

For each polymer, the value of γ_S^{LW} was calculated from equation (18) using the contact angle data for methylene iodide. The quantities γ_S^+ and γ_S^- were calculated from the data for water taken pairwise in turn with the data for formamide, ethylene glycol, and glycerol. For any given polymer, there is some variation in the values of γ_S^+ and γ_S^-. Average values of γ_S^+ and γ_S^- are listed in Table 6 along with a comparison of literature values. As noted by Good and van Oss [4, 5], none of the solid surfaces displays high values of γ_S^+, but the acidic polymer surfaces have higher values of γ_S^+ than do the basic polymer surfaces. The basic surfaces clearly have higher values of γ_S^- than do the acidic surfaces.

A comparison of the γ_S^+ and γ_S^- values for the pressure-sensitive adhesive (Scotch 610 Magic Tape®) with those for the acidic and basic polymers shows that the pressure-sensitive adhesive has an acidic surface.

As mentioned earlier, the method of van Oss and Good has been criticized recently [18, 20]. These criticisms include the following: (i) the results depend on the choice of wetting liquids; (ii) the surfaces analyzed all have values of γ_S^- greater than those of γ_S^+; and (iii) inconsistent results appear in the literature. Della Volpe and Siboni [18] have recently suggested that account should be taken of the fact that water is about 6.5 times a stronger Lewis acid than a Lewis base. These authors then constructed a set of equations stemming from equations (13) and (17) and used the experimental data to solve the system of simultaneous equations for the various surface tension components of the wetting liquids γ_L^{LW}, γ_L^+, and γ_L^- in addition to the surface free energy of the various solids γ_S^{LW}, γ_S^+, and γ_S^-. They collected data for 14 different polymers, including poly(vinyl chloride) and poly(methyl methacrylate), and for ten different wetting liquids, including the ones used in this study. Their values of the surface tension parameters for the wetting liquids used in this study are given in Table 7.

Table 8 shows the calculated condition numbers for the matrix A defined earlier. It can be seen that the best conditioned data sets for the Della Volpe and Siboni parameters are for methylene iodide plus the following pairs of polar liquids: water plus formamide, water plus ethylene glycol, formamide plus glycerol, and ethylene glycol plus glycerol.

Table 5.
Surface free energy components (in mJ/m^2) for various polymers as determined by the method of van Oss and Good

Polymer	Water plus wetting liquid:	θ (degrees)	γ_S^{LW}	γ_S^+	γ_S^-
Poly(vinyl chloride)					
	Methylene iodide	38.8 ± 2.6	40.2	—	—
	Water	87.8 ± 4.2		—	—
	Formamide	69.2 ± 2.9		0.33	4.8
	Ethylene glycol	64.7 ± 2.6		0.15	4.1
	Glycerol	84.2 ± 0.9		0.78	6.3
				0.42	5.1
Poly(methyl methacrylate)					
	Methylene iodide	23.9 ± 0.8	46.5	—	—
	Water	64.3 ± 1.8		—	—
	Formamide	47.0 ± 1.8		0.00040	16.1
	Ethylene glycol	47.1 ± 0.8		0.13	19.3
	Glycerol	64.3 ± 2.2		0.11	19.0
				0.080	18.1
Poly(vinyl acetate)					
	Methylene iodide	33.8 ± 3.6	42.6	—	—
	Water	60.6 ± 2.9		—	—
	Formamide	46.2 ± 3.8		0.064	19.9
	Ethylene glycol	47.1 ± 1.9		0.059	24.6
	Glycerol	61.6 ± 2.9		0.00082	22.5
				0.041	22.3
Pressure-sensitive adhesive					
	Methylene iodide	90.2 ± 3.9	12.6	—	—
	Water	97.7 ± 5.3		—	—
	Formamide	97.6 ± 4.8		0.096	10.7
	Ethylene glycol	82.3 ± 5.7		0.53	5.0
	Glycerol	108.0 ± 6.2		0.63	14.1
				0.42	9.9

Table 9 lists the resulting values of γ_S^+ and γ_S^- for poly(vinyl chloride) and poly(methyl methacrylate) calculated by applying the Della Volpe and Siboni parameters in Table 7 to the present contact angle data. It can be seen that there is considerable inconsistency within the set of values of γ_S^+ and γ_S^- for both poly(vinyl chloride) and poly(methyl methacrylate). Based on the limited comparison made

Table 6.
Comparison of the surface free energy components γ_S^+ and γ_S^- for various polymers

Polymer	Type of surface	γ_S^+ (mJ/m^2)			γ_S^- (mJ/m^2)		
		This study	Literature values	Source	This study	Literature values	Source
PVC or CPVC	Acidic	0.42	0.24	Lloyd *et al.* [38]	5.1	3.1	Lloyd *et al.* [38]
PVF	Acidic	—	0.04	Good and Van Oss [4]		3.5	Good and van Oss [4]
			0.19	Lloyd *et al.* [38]	—	4.5	Lloyd *et al.* [38]
PSA	Acidic	0.42	—		9.9	—	
PMMA	Basic	0.08, 0.10^a	0	Lloyd *et al.* [38]	18.1, 13.2^a	12.2	Lloyd *et al.* [38]
			0.02	van Oss and Good [3]		9.5–22.4	Good and van Oss [4]
PVA	Basic	0.04	—		22.3	—	

^{a} Duplicate measurements.

PVC = poly(vinyl chloride); CPVC = chlorinated poly(vinyl chloride); PVF = poly(vinyl fluoride); PSA = pressure-sensitive adhesive (Scotch 610 Magic Tape®); PMMA = poly(methyl methacrylate); PVA = poly(vinyl acetate).

Table 7.
Surface tension parameters (in mJ/m^2) for various wetting liquids from Della Volpe and Siboni [18]

Liquid	γ_L	γ_L^{LW}	γ_L^-	γ_L^+
Methylene iodide	50.8	50.8	0	0
Water	72.8	21.8	10.0	65.0
Formamide	58.0	35.6	65.7	1.95
Ethylene glycol	48.0	31.4	42.5	1.58
Glycerol	64.0	34.4	12.9	16.9

Table 8.
Condition numbers for the matrix A containing the surface tension parameters γ_L^{LW}, γ_L^-, and γ_L^+ calculated from Della Volpe and Siboni [18]. Condition numbers are based on the infinity-norm

First liquid	Second liquid	Third liquid	Condition number
Methylene iodide	Water	Formamide	4.0
		Ethylene glycol	4.8
		Glycerol	16.1
	Formamide	Ethylene glycol	221.5
		Glycerol	8.4
	Ethylene glycol	Glycerol	7.7

here, it appears that no advantage is gained in using the Della Volpe and Siboni set of liquid surface tension parameters.

In attempting to determine the acid–base properties of polymers or oxide-covered metals, several different independent experimental techniques should be used rather than relying on any one given method. With metals, inverse gas chromatography [39], X-ray photoelectron spectroscopy [40, 41], and streaming potential measurements [42] are useful techniques. With polymers, streaming potential measurements [43] have proven useful.

3.3. Adhesion measurements

Figure 6 shows the peel force measured at 27°C for the pressure-sensitive adhesive (Scotch 610 Magic Tape®) on chromium and aluminum surfaces as a function of the peel rate. The measured peel force increases with the peel rate due to the higher dissipation of energy due to viscoelastic processes at the higher peel rates [44]. Thus, to estimate the interfacial adhesion for the various metal substrates, peel forces for the other metals were measured at one of the slower peel rates used in this study, i.e. 1.2 cm/min.

Table 9.
Surface free energy components γ_S^+ and γ_S^- for poly(vinyl chloride) and poly(methyl methacrylate) from the method of van Oss and Good using the surface tension parameters of liquids from Della Volpe and Siboni [18]

First liquid	Second liquid	Third liquid	Poly(vinyl chloride)	
			γ_S^+	γ_S^-
Methylene iodide	Water	Formamide	3.6	1.0
		Ethylene glycol	0.18	1.4
	Formamide	Glycerol	0.095	0.56
	Ethylene glycol	Glycerol	0.016	0.13
First liquid	Second liquid	Third liquid	Poly(methyl methacrylate)	
			γ_S^+	γ_S^-
Methylene iodide	Water	Formamide	0.36	5.2
		Ethylene glycol	0.024	6.7
	Formamide	Glycerol	0.78	0.43
	Ethylene glycol	Glycerol	0.0066	1.8

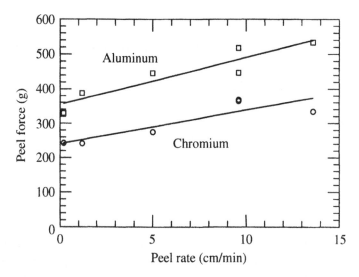

Figure 6. Measured peel force vs. peel rate for a pressure-sensitive adhesive on aluminum and chromium.

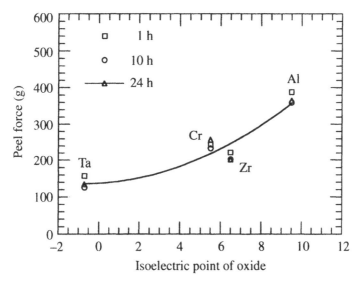

Figure 7. Measured peel force for various contact times of a pressure-sensitive adhesive on various oxide-covered metals vs. the isoelectric point of the oxide film, at a peel rate of 1.2 cm/min.

Figure 7 shows the data for a peel rate of 1.2 cm/min plotted as a function of the surface isoelectric point of the oxide film. Isoelectric points for oxide-covered tantalum, chromium, and aluminum were determined from the present work. The isoelectric point for oxide-covered zirconium is taken from the literature value for zirconium oxide [45]. The peel force for the positively charged pressure-sensitive adhesive clearly is the highest for the negatively charged (basic) oxide film.

Figure 7 also shows that the peel force was approximately constant with contact time for all the metal substrates studied. The mode of failure in all cases appeared to be interfacial in that visual examination showed that the pressure-sensitive adhesive peeled cleanly from all the substrates studied. However, a more careful analysis of the metal substrate using XPS after peeling would be required to substantiate this claim.

Finally, it is necessary to show that differences in the measured peel force for the various metal substrates cannot be attributed to differences in surface roughness. Roughness factors were determined from AFM, as mentioned earlier, for the aluminum, chromium, and tantalum oxide-covered surfaces. A typical AFM image is given in Fig. 8, which shows an AFM image for oxide-covered chromium. For each substrate, the roughness factor was determined from the AFM image in two ways, as described earlier, i.e. from a cross-sectional analysis and from the 'three-dimensional' surface area of the imaged sample. The first method gave roughness factors of 1.02, 1.00, and 1.06 for aluminum, chromium, and tantalum, respectively. The second method gave values of 1.02, 1.00, and 1.11. Thus, the roughness factors were close to 1.0 for each of the three metal substrates, so that differences in roughness factors can discounted as a possible complicating factor in interpreting the peel force measurements.

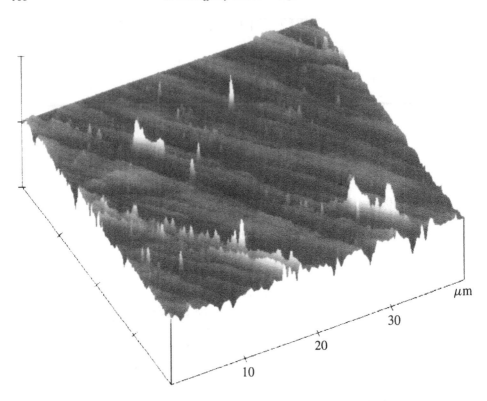

Figure 8. AFM image of a chromium surface. Vertical scale is 200 nm per division.

4. CONCLUSIONS

(1) The surface isoelectric points for native air-formed oxide films on various metals have been determined by measurement of contact angles at the hexadecane/ aqueous solution interface as a function of the pH of the aqueous phase.

(2) The experimentally determined values for the oxide films on chromium, aluminum, and tantalum are within one to three pH units of the reported isoelectric points for the corresponding bulk oxide powders.

(3) The surface isoelectric point of the oxide-covered Ta-implanted Al surface lies between that of Al_2O_3 and Ta_2O_5.

(4) The acid–base properties of several polymers, including a pressure-sensitive adhesive (Scotch 610 Magic Tape®), have been determined from contact angle measurements using the method of van Oss and Good.

(5) Adhesion measurements for a pressure-sensitive adhesive (Scotch 610 Magic Tape®) having a positive surface charge show that the peel strength is highest when the metal substrate has a surface oxide film of basic character.

Acknowledgements

E. McCafferty is very much appreciative of helpful discussions on the surface properties of polymers with Dr. Nursel Dilsiz, formerly a Research Associate in the Department of Chemistry at the Virginia Polytechnic Institute and State University and currently a member of the technical staff of Roketsan in Ankara, Turkey. We are grateful to Mr. Steve McCartney for obtaining the AFM images. E. McCafferty is especially appreciative of the opportunity provided him by the Naval Research Laboratory to spend a sabbatical year at the Virginia Polytechnic Institute and State University.

REFERENCES

1. E. McCafferty and J. P. Wightman, *J. Colloid Interface Sci.* **194**, 344–355 (1997).
2. C. J. van Oss, M. Chaudhury and R. J. Good, *Chem. Rev.* **88**, 927–941 (1988).
3. C. J. van Oss and R. J. Good, *J. Macromol. Sci. — Chem. A* **26**, 1183–1203 (1989).
4. R. J. Good and C. J. van Oss, in: *Modern Approaches to Wettability*, M. E. Schrader and G. I. Loeb (Eds), pp. 1–27. Plenum Press, New York (1992).
5. R. J. Good, *J. Adhesion Sci. Technol.* **6**, 1269–1302 (1992).
6. A. C. Zettlemoyer, F. J. Micale and Y. K. Lui, *Ber. Bunsenges. Phys. Chem.* **71**, 286–291 (1967).
7. E. McCafferty and A. C. Zettlemoyer, *Discuss. Faraday Soc.* **52**, 239–254 (1971).
8. J. C. Bolger, in: *Adhesion Aspects of Polymeric Coatings*, K. L. Mittal (Ed.), pp. 3–18. Plenum Press, New York (1983).
9. L.-K. Chau and M. D. Porter, *J. Colloid Interface Sci.* **145**, 283–286 (1995).
10. L. Campanella, F. Croce and P. Mazzoni, *Oberflaeche-Surface* **19**, 224–226 (1978).
11. V. Vujicic and B. Lovrecek, *Surface Technol.* **17**, 29–35 (1982).
12. V. C. P. Morfopoulos and H. C. Parreira, *Corrosion Sci.* **7**, 241–254 (1967).
13. C. Vittoz, P. E. Dubois, J. C. Joud and M. Mantel, *Proceedings of the 20th Annual 'Anniversary' Meeting of the Adhesion Society*, pp. 545–547. The Adhesion Society, Blacksburg, VA (1997).
14. S. R. Holmes-Farley, R. H. Reamy, T. J. McCarthy, J. Deutch and G. M. Whitesides, *Langmuir* **1**, 725–740 (1985).
15. E. V. Gribanova, L. I. Molchanova, K. B. Mazitova, G. N. Rezakova and N. A. Dmitrieva, *Colloid J. U.S.S.R.* **45**, 273–277 (1983).
16. D. F. Billett, D. B. Hough and R. H. Ottewill, *J. Electroanal. Chem.* **74**, 107–120 (1976).
17. A. Hollander, *J. Colloid Interface Sci.* **169**, 493–496 (1995).
18. C. Della Volpe and S. Siboni, *J. Colloid Interface Sci.* **195**, 121–136 (1997).
19. S. J. Leon, *Linear Algebra with Applications*, pp. 350–352. Macmillan, New York (1990).
20. M. Morra, *J. Colloid Interface Sci.* **182**, 312–314 (1996).
21. P. M. Natishan, E. McCafferty and G. K. Hubler, *Nuclear Instrum. Methods* **B59/60**, 841–844 (1991).
22. M. Mantel and J. P. Wightman, *Surface Interface Anal.* **21**, 595–605 (1994).
23. J. Schultz, L. Lavielle and C. Martin, *J. Adhesion* **23**, 45–60 (1987).
24. A. W. Neumann and R. J. Good, in: *Surface and Colloid Science*, R. J. Good and R. R. Stromberg (Eds), Vol. 11, pp. 31–91. Plenum Press, New York (1979).
25. B. Miller, L. S. Penn and S. Hedvat, *Colloids Surfaces* **6**, 49–61 (1983).
26. M. T. E. Shanahan, *Surface Interface Anal.* **17**, 489–495 (1991).
27. D. E. Yates and T. E. Healy, *J. Colloid Sci.* **52**, 222–228 (1975).
28. J. A. Yopps and D. W. Fuerstenau, *J. Colloid Sci.* **19**, 61–71 (1964).

29. P. Roy and D. W. Fuerstenau, *Surface Sci.* **30**, 487–490 (1972).
30. M. Tschapek, C. Wasowski and R. M. Torres Sanchez, *J. Electroanal. Chem.* **74**, 167–176 (1976).
31. S. Ardizzone and S. Trasatti, *Adv. Colloid Interface Sci.* **64**, 173–251 (1996).
32. S. Subramanian, J. A. Schwarz and Z. Hejase, *J. Catal.* **117**, 512–518 (1989).
33. J. A. Schwarz, C. T. Driscoll and A. K. Bhanot, *J. Colloid Interface Sci.* **97**, 55–61 (1984).
34. M. Tschapek, L. Tcheichvili and C. Wasowski, *Clay Miner.* **10**, 219–229 (1974).
35. R. Vigano, J. Taraszewska, A. Daghetti and S. Trasatti, *J. Electroanal. Chem.* **182**, 203–209 (1985).
36. G. A. Parks, in: *Equilibrium Concepts in Natural Water Systems*, W. Stumm (Ed.), pp. 121–160. American Chemical Society, Washington, DC (1967).
37. R. Sprycha, *Colloids Surfaces* **5**, 147–157 (1982).
38. T. B. Lloyd, K. Ferreti and J. Lagow, *J. Polym. Sci.* **58**, 291–296 (1995).
39. J. H. Burness and J. G. Dillard, *Langmuir* **10**, 1894–1897 (1994).
40. G. W. Simmons and B. C. Beard, *J. Phys. Chem.* **91**, 1143–1148 (1987).
41. M. Delamar, *J. Electron. Spectrosc. Relat. Phenom.* **53**, c11–c14 (1990).
42. D. Fairhurst and V. Ribitsch, in: *Particle Size Distribution II*, T. Provder (Ed.), ACS Symp. Ser. No. 472, pp. 337–353. ACS, Washington, DC (1991).
43. H.-J. Jacobasch and J. Schurz, *Prog. Colloid Polym. Sci.* **77**, 40–48 (1988).
44. D. W. Aubrey and S. Ginosatis, *J. Adhesion* **12**, 189–198 (1981).
45. M. A. Butler and D. S. Ginley, *J. Electrochem. Soc.* **125**, 228–232 (1973).

Apparent and Microscopic Contact Angles, pp. 171–208
J. Drelich, J. S. Laskowski and K. L. Mittal (Eds)
© VSP 2000.

Acid–base surface free energies of solids and the definition of scales in the Good–van Oss–Chaudhury theory

C. DELLA VOLPE* and S. SIBONI

Department of Materials Engineering, University of Trento, Via Mesiano 77, 35080 Trento, Italy

Received in final form 30 June 1999

Abstract—The overwhelming basicity of all analysed surfaces strongly dependent on the choice of liquid triplet used for contact angle measurements and the negative values sometimes obtained for the square roots of the acid–base parameters can be summarized as the main problems arising from the application of the Good–van Oss–Chaudhury (GvOC) theory to the calculation of Lewis acid–base properties of polymer surfaces from contact angle data. This paper tries to account for these problems, namely: (1) the Lewis base, or electron donor component, is much greater than the Lewis acid or electron-acceptor component because of the reference values for water chosen in the original GvOC theory. A direct comparison of the acidic component with the basic one of the same materials has no meaning. A new reference scale for water which is able to overcome this problem is suggested. For the calculation of acid–base components, a best-fit approach is proposed which does not require any starting information about the liquids or polymers and can yield estimates of the acid–base parameters for both the liquids and the polymers involved; (2) the strong dependence of the value of the acid–base components on the three liquids employed is due to ill-conditioning of the related set of equations, an intrinsic and purely mathematical feature which cannot be completely cured by any realistic improvement in experimental accuracy. To reduce or eliminate the effect, one only needs a proper set of liquids, representative of all kinds of different solvents; (3) the negative coefficients appear as a simple consequence of measurement uncertainty, combined with the possible ill-conditioning of the equation set. We cannot exclude, however, that in some cases they could have a different origin.

Key words: Contact angle; acid–base surface free energy; Good–van Oss–Chaudhury theory.

1. INTRODUCTION

The characterization and quantitative description of forces at interfaces constitute one of the most important problems in materials surface and interfacial science [1, 2]. Its solution would make possible the analytical prediction and explanation of the materials behaviour at interfaces by the quantification of interfacial interactions

---

*To whom correspondence should be addressed. E-mail: devol@devolmac.ing.unitn.it

and, as an immediate consequence, the technological capability to design material or surface structures for a specific purpose on the basis of a precisely defined surface structure/properties relationship.

It is well known that both permanent and temporary dipoles give rise to three kinds of interactions, usually named van der Waals forces: dipole/dipole (Keesom), dipole/induced dipole (Debye), and dispersion (London) forces. According to relatively recent achievements [3–7], only dispersion forces are reputed to be relevant in condensed matter, because of the rather large number of nearest neighbours which yield conflicting local fields and strongly reduce the mean dipole interactions. This means that, except for the case of dilute gases, 'polar' interactions involving permanent dipoles can typically be neglected, although it has long been assumed that non-dispersive forces between condensed phases should be ascribed to 'polar' interactions, and the role of the 'polar' component has been frequently invoked to account for several different interfacial phenomena. Nevertheless, it is also known that the water wettability of apolar polymers, such as polyolefins, greatly increases if chemical functions endowed with polar bonds are introduced by appropriate surface treatements. Since polar contributions cannot be responsible for any significant enhancement of interfacial actions, such a behaviour requires a quite different explanation of the forces involved. The currently accepted answer is acid–base interactions and, specifically, the particular sub-set of Lewis acid–base interactions known as hydrogen bonding [6, 7]. The fundamental role of acid–base interactions in the interfacial behaviour of condensed phases is now largely recognized and it has led to a large amount of work on this subject, particularly in the last 10 years or so [3–6, 8].

In the present paper, we discuss the interpretation of data obtained from the measurements of acid–base properties of polymer surfaces, with particular attention on wetting measurements. Indeed, although the correct measurement and interpretation of acid–base properties are obviously necessary in order to translate the acid–base approach to interfacial interactions into a useful tool for materials and surface design, everyday practice and a survey of the related literature show that the situation is not yet satisfactory [9–13]. This seems particularly true for approaches based solely on wetting measurements, which are the most commonly used methods to characterize the surface properties of polymers [13].

Our goal is to encourage discussion and to single out some common basis and starting point to help remove some of the obstacles that still hinder the full and coherent application of the acid–base approach to interfacial interactions.

The paper is organized as follows: after a brief introduction of some definitions and basic concepts related to acid–base properties, the main problem is stated, i.e. the inconsistency among acid–base properties determined by different methods and the unsatisfactory state of the results obtained from wettability measurements. Finally, some discussion on the nature of these problems and some proposals for possible solutions are presented.

2. SOME BASIC CONCEPTS

The most fundamental concept in the theory of interfacial acid–base interactions is certainly the definition itself. Among the different definitions of acids and bases, the Lewis theory is the most satisfactory for applications to polymers, and it is to this theory that all experimental approaches naturally refer in order to calculate the acid–base components of polymer surfaces. Note, however, that the original definition by Lewis [14] must be adapted to a more modern point of view, since the share of an electron pair provided by the base to the acid is no longer necessary [6]: any substance capable of furnishing electron density must be considered a base, while an acid is any substance available to accept electron density [15, 16]. According to the preceding definition, the sites that can act as electron acceptors are acidic: metal atoms of organometallic compounds, electrophilic carbons (i.e. carbon atoms covalently linked to a more electronegative element, such as oxygen or fluorine), hydrogen atoms in hydroxyl or carboxyl groups. In contrast, Lewis bases are electron donors: atoms containing lone-pair electrons (such as oxygen), or aromatic rings, where the π electrons act as a basic site. This broader definition best describes the kind of acid–base interactions of interest in polymer surface and interfacial science. And it is within this same notion that the terms 'electron donor' and 'electron acceptor' fit, and are frequently used in the literature as synonyms for 'Lewis base' and 'Lewis acid', respectively.

It is worthy of note that many compounds contain both acidic and basic sites and are, therefore, self-associated substances: water provides an important example, because of the balanced basicity of the oxygens and the acidity of hydrogens.

The water example suggests further comments on the term 'polar': it is obviously correct to define water as a 'polar' compound, owing to the polarity of the $-OH$ bond, but the associated dipole, as measured, for instance, by the relative dipole moment, does not contribute to the intermolecular interactions in a condensed phase [3]. The correct interpretation of 'polar' (i.e. non-dispersive) interactions of water molecules comes from the co-existence of electron-rich (i.e. basic) and electron-poor (acidic) sites. They will interact with neighbouring molecules in a Lewis acid–base way, and this particular acid–base interaction constitutes hydrogen bonding. The particular features of this bond have been well known since 1960 [17]: existence of preferred bond lengths and angles; bond strength dependent on the acidic and basic strength of the hydrogen donor and acceptor, respectively; and complete absence of correlations of the bond strength with the dipole moment. Although the bonds involved in hydrogen bonding are polar and hydrogen-bonded liquids are often referred to as 'polar', it cannot be assumed that their properties stem from dipoles [3–5]. In a recent series of papers, Besseling and co-workers [18–22] described a statistical model based on the orientation-dependent features of water molecules resulting from the presence of electron donor/acceptor sites which correctly accounts for the bulk and interface properties of water.

The most striking characteristic of acid–base interactions is the complementary nature of the interacting sites involved: a molecule endowed with an acidic site will exhibit its acidic behaviour only if a basic counterpart is available either in a molecule of the same kind (in the case of self-associated compounds) or in a different one, and the same considerations obviously apply to basic sites. Such a complementarity cannot be explained by the simple presence of molecular dipoles, because of the complete symmetry expected in the dipole interaction of two different molecules i and j, where the separate contributions of i and j appear formally similar to that of the interaction between two i (or j) molecules.

3. THE CALCULATION OF ACID–BASE PROPERTIES FROM WETTING MEASUREMENTS

3.1. General introduction

The calculation of acid–base properties by wetting measurements involves estimating the fundamental acid–base properties of solid surfaces by their ability to interact with liquids, as manifested through wetting phenomena. The basic idea of this approach [3–7] consists in the assumption that the surface free energy splits into components describing, respectively, the contribution γ^{LW} due to electrodynamic interactions (dominated by dispersion forces) and the acid–base contribution γ^{AB}:

$$\gamma^{tot} = \gamma^{LW} + \gamma^{AB}, \tag{1}$$

where the superscript LW stands for Lifshitz–van der Waals. Equation (1) basically states that 'dispersive' and acid–base interaction are independent of each other, and is formally similar to the dispersive/polar approach widely used before the role of acid–base properties was recognized.

The complementarity of acid–base interactions explicitly appears in the expression introduced by Good, van Oss and Chaudhury to describe the acid–base components the surface free energy of solids or liquids of [7, 23–26]:

$$\gamma_{L,S} = \gamma_{L,S}^{LW} + 2\left(\gamma_{L,S}^{+}\gamma_{L,S}^{-}\right)^{1/2}, \tag{2}$$

which, combined with the Young equation, leads to the following relationship for the work of adhesion between a liquid and a solid:

$$W_{adh} = \gamma_L(1 + \cos\theta) = 2\left(\gamma_L^{LW}\gamma_S^{LW}\right)^{1/2} + 2\left(\gamma_L^{+}\gamma_S^{-}\right)^{1/2} + 2\left(\gamma_L^{-}\gamma_S^{+}\right)^{1/2}. \tag{3}$$

Here γ^{LW} is the previously defined Lifshitz–van der Waals contribution, whereas γ^{+} and γ^{-} are electron-acceptor (Lewis acid) and electron-donor (Lewis base) parameters. The subscripts L and S refer to solid and liquid, respectively. From the practical point of view, equation (3), which will be referred to as the Good–van Oss–Chaudhury (GvOC) equation throughout the paper, accounts for both the acidic and the basic behaviour of liquids in wetting phenomena and allows one to calculate the Lifshitz–van der Waals, the electron-donor, and the electron-acceptor parameters of a solid by contact angle measurements using (at least) three liquids of known surface free energy components.

The GvOC equation is, in principle, the tool which allows us to measure the acid–base properties of polymer surfaces, to account for the results of interfacial interactions, and to design a given surface modification treatment for a given application, but its practical use shows that it is still very far from this goal.

One of the major problems encountered is the systematic overestimate of the basic component of the surface free energy by the GvOC approach, as appears by a comparison with other techniques for the measurement of acid–base strength, such as ζ-potential or inverse gas chromatography (IGC) [27–30]. This has been found, for instance, by Jacobasch *et al.* [13], who compared the acid–base properties of polypropylene (PP) and flame-treated PP deduced from contact angle and ζ-potential measurements. Similar results have been obtained by Tate *et al.* [12] by employing IGC to measure the acid–base properties of nylon 66: while ICG proves the existence of high-energy acidic sites on the polymer surface, the GvOC treatment of the wetting data on the same surfaces leads to the odd conclusion that the nylon surface is overwhelmingly basic. This kind of conclusion is typical for polymers and seems rather unsatisfactory, although several possible explanations have been suggested by Good, van Oss and Chaudhury (in particular, that the occurrence of monopolar basic polymer surfaces could be a general law [24, 25]).

Moreover, in spite of the relatively simple mathematical form (the set of GvOC equations is linear in the square roots of the polymer acid–base and dispersive parameters), the results from the GvOC method depend strongly on the choice of the three liquids used for the contact angle measurements and in some cases, these roots assume negative values, which obviously cannot be justified [11].

This state of the art can be tackled in different ways. One can simply ignore these difficulties and keep applying the GvOC approach in the firm belief that its results can be at least useful for direct comparisons, in spite of the intrinsic shortcomings.

Such a point of view, however, can only represent a temporary working hypothesis and in our opinion it simply represents a new abuse perpetrated against a correct use of the contact angle technique and theory. Originally introduced to describe the behaviour of an ideal liquid on an ideal surface [31], the Young equation has been applied indiscriminately to any kind of contact angle measured on all kinds of surfaces [32], so it is not astonishing to meet, in the (otherwise cautious) scientific literature, very critical expressions such as 'the comedy of errors' [33] or 'disposable theories' [34] to define the unsatisfactory status of interfacial thermodynamics and contact angle theory. Even if the GvOC equation can quantitatively account for a large number of complex interfacial interactions in aqueous media [26], no clear answer is available to some fundamental questions, such as which is the appropriate choice of the test liquids; how to interpret and manage the strong dependence of the surface free energy components on the liquid set employed for measurements; what experimental error and uncertainty is associated with the calculated components; which experimental angle (advancing, receding or whatever) should be more suitable for calculations and why?

Several authors have stressed various shortcomings in the current applications of the existing theories and claimed the necessity of a serious reflection [9, 35], and some others have been prompted to attempt solutions borrowed from the related fields [11, 36–38].

The following paragraphs are devoted to an in-depth analysis of the quantitative aspects of the GvOC theory, in order to avoid or rationalize some of the present problems in the characterization of the acid–base behaviour of polymer surfaces by wetting measurements. In the following treatments, the quantity commonly named the spreading pressure, π_e, defined as the difference between the surface free energy of a solid in equilibrium with vacuum and the surface free energy of the same solid in equilibrium with its (or a different) vapour, will not be considered. It represents the effect of the adsorption of a vapour on the solid surface, a process which decreases the surface free energy.

There is no doubt that this quantity must be considered in a general formulation of the Young equation and that a certain degree of adsorption can be present in many cases, influencing the values of contact angles. This fact is not in contrast with the general ideas of acid–base theory; on the contrary, the acid–base theory can easily rationalize that also in cases in which the total surface free energy of a solid is lower than that of the liquid, adsorption can take place, due to the favourable effect of only *one* of its components. However, the most common conclusion is that its value is very low when the contact angles are greater than zero. Moreover, as can be found in fundamental textbooks (e.g. p. 401 of ref. [1]), there is no general agreement on the values that it assumes in experimental situations similar to those used for contact angle measurements (i.e. flat surfaces).

For the above reasons, in the original GvOC theory (and *consequently* in this paper), this quantity is not explicitly considered. It is worthy of note that its presence, however, would not change all the considerations that we are currently developing on the subject; in fact, if the values of the spreading pressure were exactly known, they would simply be inserted, through their correct mathematical formulation, in the *known terms* of all the equations that we are treating.

A different situation would arise if one had the aim of developing a wider theory, including explicitly the spreading pressure in terms of acid–base quantities, pertaining to the liquids and solids considered.

Up to now, in the acid–base theory no author has used explicitly the experimental values of the spreading pressure (as part of the known terms of the equations used) or has tried to develop an acid–base formulation of the term π_e (thus introducing new variables in the theory).

As a general rule and as a mathematical limit for this second case, it is impossible to introduce in the present formulation of the acid–base theory a new *ad-hoc* term for each solid–liquid pair, because this will introduce an out-of-control number of variables; one should use the already existing variables, introducing, so, a function $\pi_e = \pi_e(\gamma^+, \gamma^-, \gamma^{LW})$, whose formulation is beyond the scope of the present paper.

We agree that the existence of spreading pressure can be invoked as a reason for the problems of the acid–base theory, but we are expecting more solid experimental measurements. This paper is devoted to the analysis of problems of acid–base theories which are well documented and are *independent* of each other. When acceptable values of the spreading pressure are measured, they will be introduced, without problems, in the present proposed *correct* mathematical formulation of the acid–base theory.

3.2. Calculation of acid–base components according to the Good–van Oss– Chaudhury theory

From a mathematical point of view, the GvOC theory is formally analogous to the theories proposed by Drago [39], Taft and co-workers [40], Abraham [41], and others, where a suitable thermodynamic potential F_{th}, related to the acid–base properties of two materials X and Y, is written as a sum of pairwise products of n dissimilar parameters (some of them possibly constant), i.e.

$$F_{th} = \sum_{i,\, j} X_i Y_j. \tag{4}$$

In equation (4), the typical form of a linear free energy relationship (LFER) is immediately recognized, although F_{th} may or may not be a free energy term.

Douillard [34] has recently suggested that the GvOC relations represent only a rough approximation and that they would be theoretically more acceptable if written in terms of surface enthalpy instead of free energy. Such a point of view is certainly interesting and may be partially correct, keeping in mind that the well-known Drago enthalpy approach works better than other models, but nevertheless the GvOC theory seems rather accurate in comparison with other LFERs applied to non-enthalpy quantities. Moreover, as we will discuss in a subsequent section, similar relationships can be introduced for both the enthalpy of adhesion and the work of adhesion (a free energy). The enthalpy of adhesion is experimentally accessible by wetting measurements at different temperatures (at least three, and estimating the enthalpy for the central one with the usual thermodynamic approaches). This makes the considerations presented here valid anyway, independently of Douillard's criticism.

The computation of free energy components of a solid by the GvOC method is made possible by the measurement of the contact angles of L ($L \geqslant 3$) suitable liquids on the test solid. All kinds of liquids should be employed in order to avoid systematic errors [11], including purely dispersive, prevalently or monopolar acidic, and prevalently or monopolar basic compounds. Denoted with γ_s^{LW}, γ_s^+, γ_s^-, the dispersive, acidic and basic components, respectively, of the solid and with $\gamma_{l,i}^{LW}$, $\gamma_{l,i}^+$, $\gamma_{l,i}^-$, those of the i-th liquid, $i = 1, 2, \ldots, L$, a slight adaptation of the subscripts in equation (3) allows one to write the work of adhesion W_i^{adh} in the following form:

$$W_i^{adh} = \gamma_{1,i}^{Tot}(1 + \cos\theta_i) = 2\left(\sqrt{\gamma_{1,i}^{LW}\gamma_s^{LW}} + \sqrt{\gamma_{1,i}^+\gamma_s^-} + \sqrt{\gamma_{1,i}^-\gamma_s^+}\right), \qquad (3a)$$

where $\gamma_{1,i}^{Tot}$ is the surface free energy of the i-th liquid and θ_i is the contact angle of the same liquid on the solid. The surface free energy of the same liquid then becomes

$$\gamma_{1,i}^{Tot} = \gamma_{1,i}^{LW} + 2\sqrt{\gamma_{1,i}^+\gamma_{1,i}^-} \qquad (2a)$$

and an analogous expression holds for the solid. Provided that the acid–base components of the liquids are known with sufficient accuracy, the overdetermined set of linear equations (3a) can be solved in terms of the variables $\sqrt{\gamma_s^{LW}}$, $\sqrt{\gamma_s^+}$, $\sqrt{\gamma_s^-}$ by the standard methods [37, 42, 43] and it is also possible to estimate the standard deviations of the results. This has been done for the flamed surfaces of polypropylene–ethylene-propylene-rubber blends (PP–EPR) flamed surfaces by using a set of seven liquids [37, 42] with the free energy components proposed by Good and van Oss.

Of course, a smaller set of liquids may be used as well, with the only condition that their number L is at least three. The case $L = 3$ is the most frequently used in the literature but it is critically affected by the choice of test liquids, since the square roots of the polymer surface free energy components stem from the solution of a linear set of three equations in three variables and serious problems of ill-conditioning may arise. This is a rather typical occurrence if no attention is paid to the choice of the liquid triplet. Denoting with A the 3×3 matrix whose rows are the coefficients $\sqrt{\gamma^{LW}}$, $\sqrt{\gamma^+}$, $\sqrt{\gamma^-}$ of each liquid, the set of equations (3a) in the case $L = 3$ takes the form

$$\begin{pmatrix} \gamma_{1,1}^{Tot}(1 + \cos\theta_1)/2 \\ \gamma_{1,2}^{Tot}(1 + \cos\theta_2)/2 \\ \gamma_{1,3}^{Tot}(1 + \cos\theta_3)/2 \end{pmatrix} = A \begin{pmatrix} \sqrt{\gamma_s^{LW}} \\ \sqrt{\gamma_s^-} \\ \sqrt{\gamma_s^+} \end{pmatrix} \qquad (3a')$$

and an estimate to the conditioning of the set is given by the condition number $C_n = ||A||_1 \cdot ||A^{-1}||_1$, where $||A||_1 = \max_j \sum_{i=1}^3 |A_{i,j}|$: a large C_n means a strong sensitivity of solutions to round-off or data errors (the relative error in the solution is bounded by multiplying by C_n the relative error in the wetting data, expressed in terms of the usual ℓ^1 vector norm*).

---

*We recall the definition of ℓ^1, ℓ^2 and ℓ^∞ vector norms. If x is a real vector of components $x_1, x_2, \ldots, x_n$, the above norms are defined as follows:

$$||x||_1 = \sum_{i=1}^n |x_i|, \qquad ||x||_2 = \left[\sum_{i=1}^n |x_i|^2\right]^{1/2}, \qquad ||x||_\infty = \max_{i=1,\ldots,n} |x_i|.$$

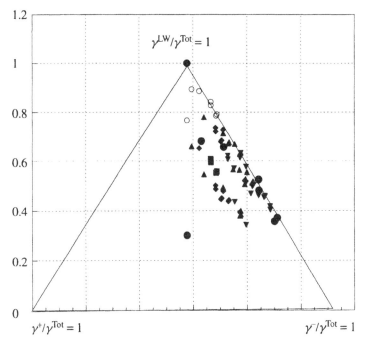

Figure 1. Condition numbers for some liquid triplets, according to the acid–base components by Good and van Oss. Each liquid is identified by a black circle, the distances from the lower, right and left sides corresponding to the values of the suitably normalized dispersive, acidic and basic components, respectively (division by the sum of the components provides the required normalization). The location of a triplet on the diagram (different symbols) is given by the centre of mass of the triangle having the liquid circles as vertices. In particular, singular triplets including two or more purely dispersive liquids are denoted by a white circle. The condition numbers C_n of the other triplets are calculated with respect to the matrix norm $\| \ \|_1$ and represented according to the following symbol code: squares ($5 < C_n < 10$), rhombus ($10 < C_n < 20$), triangles ($20 < C_n < 50$), and reversed triangles ($50 < C_n < 500$).

To better illustrate this point, the condition number has been calculated for all the possible triplets of ten commonly used liquids: water, glycerol, formamide, diiodomethane, ethylene glycol, α-bromonaphthalene, dimethyl sulphoxide, bromoform, pyrrole and hexadecane. Most of the acid–base components used for the calculations are those proposed by Good and van Oss, when available. Each liquid of the set can be represented by means of its free energy components in a triangular diagram, as shown in Fig. 1. A black circle stands for a given liquid; the distances from the lower, right, and left sides correspond to the values of the suitably normalized dispersive, acidic and basic components, respectively (division by the sum of the components provides the required normalization). In a similar way, each triplet is imagined as a triangle having as vertices the black circle symbols of its liquids; the symbol representing each triplet is then located at the centre of mass of the triangle (different symbols as explained in the caption). The general predominance of the basic components is evident in this kind of graph, since most

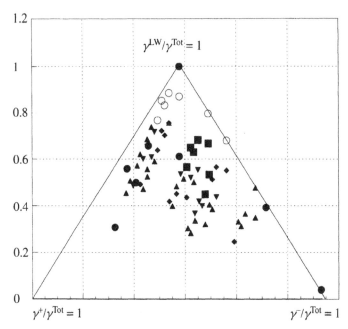

Figure 2. The same as in Fig. 1, but using DVS values of the acid–base components for liquids. One can easily note the more uniform distribution of black and of other symbols in the whole diagram, which corresponds to a more equilibrated distribution of the acid–base components, whose values are, in general, more acidic. The condition numbers C_n of the other triplets are calculated with respect to the matrix norm $\|\ \|_1$ and represented according to the following symbol code: squares ($5 < C_n < 10$), rhombus ($10 < C_n < 20$), triangles ($20 < C_n < 50$), and reversed triangles ($50 < C_n < 500$).

points accumulate near the right side. It is noticeable that for the surface free energy components estimated by Della Volpe and Siboni (DVS), whose approach will be described below, the acidic contributions become significant and both the liquid and the triplet circles scatter much better over the whole triangular diagram (see Fig. 2). In any case, different symbols denote different intervals of the C_n values as specified in the appropriate captions and, as a rule, the smallest values of the condition number are generally related to 'proper' triplets, where purely dispersive and 'polar' liquids are present as well. As an example, for the water–glycerol–α-bromonaphthalene triplet the condition number is about 5.98, whereas $C_n = 7.54$ in the case of water–diiodomethane–formamide. In contrast, bromoform–dimethyl sulphoxide–glycerol leads to the unacceptably large value $C_n = 28.6$, clearly due to the improper choice of the triplet, which includes no dispersive liquid [11]. The opposite case of triplets with at least two purely dispersive liquids is even more striking: no solution exists because two or more equations are linearly dependent and the matrix A turns out to be singular. An interesting example of the critical role played by ill-conditioning in the usual application of the GvOC theory is provided in ref. [44] about the characterization of the acid–base properties of fluorocarbon FC-721. Here the strong variability of the acid–base components calculated

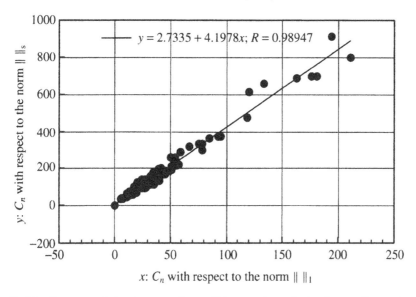

Figure 3. The linear correlation between the condition number calculated according to the matrix norm $\| \ \|_1$ and that obtained from the $\| \ \|_s$ norm. Liquid triplets with Good and van Oss values of the acid–base components.

by means of different liquid triplets, as reproduced in Table 14 on p. 328 of ref. [44], is interpreted as a problem of the theory, but it must be noted that all the employed triplets exhibit relatively large condition numbers and cannot be considered proper. This is likely due to the absence of purely dispersive liquids in such triplets.

The value of C_n depends on the choice of the matrix norm which, strictly speaking, should be that induced by the vector norm which expresses the global error in the wetting data. The matrix norms induced by the well-known ℓ^∞ or ℓ^2 vector norms do not significantly alter the condition number in the cases considered here. This is true even for the matrix norm defined by $\|A\|_s = \sum_{i,j=1}^{3} |A_{i,j}|$, which is very simple to compute but is not induced by any vector norm. As a remark, a very good linear correlation found among the C_n values obtained from different matrix norms is shown in Fig. 3 for the Good–van Oss parameters; a similar result is found for the parameters obtained by DVS as shown in Fig. 4.

Problems of ill-conditioning may also arise for larger sets of test liquids, when the GvOC equations provide an overdetermined linear set whose solution must be obtained by a best-fit procedure. Nevertheless, for proper sets the effect is expected to be comparatively small.

Since the choice of possible test liquids is limited by chemical reasons (to avoid absorption, swelling or chemical reactions, for instance), once a set of suitable liquids has been selected, the GvOC method should be applied by using the liquid subset with the smallest condition number.

From the above discussion it is clear that an appropriate choice of the liquids employed and an accurate determination of the surface free energy parameters, along with the total surface free energy, of such liquids are absolutely critical in

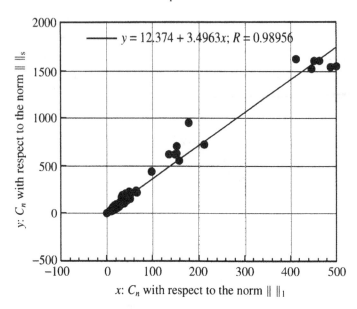

Figure 4. The same as in Fig. 3, but using DVS values of the acid–base components for liquids.

order for the GvOC approach to work well and to provide good estimates of polymer acid–base components. However, the determination of the acid–base components of liquids is not trivial and it requires rather arbitrary assumptions regarding the nature of the surfaces involved, or complex experimental procedures difficult to be reproduced and handled, with a significant effect on the estimate of the errors. This is the case of the values measured by Good and van Oss, achieved by employing alternatively some solid surfaces 'reasonably' monopolar or liquids encased in gels [26]. That is why a more general approach may be useful, as described below.

3.3. The DVS approach to the calculation of acid–base components by the Good–van Oss–Chaudhury theory. Comparison with the Drago model

In general, for the overdetermined set (3a) no exact solution is expected; this is even truer if the experimental errors and the approximate nature of the acid–base theory are taken into account. We can then choose a set of S solids, each denoted with the index $j = 1, 2, \ldots, S$, and introduce a larger set of equations including the relations below for every liquid i or solid j:

$$\gamma_{1,i}^{\text{Tot}} = \gamma_{1,i}^{\text{LW}} + 2\sqrt{\gamma_{1,i}^{+}\gamma_{1,i}^{-}}, \qquad \gamma_{s,j}^{\text{Tot}} = \gamma_{s,j}^{\text{LW}} + 2\sqrt{\gamma_{s,j}^{+}\gamma_{s,j}^{-}} \qquad (2b)$$

together with the work of adhesion equalities for each liquid–solid pair:

$$W_{i,j}^{\text{adh}} = \gamma_{1,i}^{\text{Tot}}(1 + \cos\theta_{i,j}) = 2\left(\sqrt{\gamma_{1,i}^{\text{LW}}\gamma_{s,j}^{\text{LW}}} + \sqrt{\gamma_{1,i}^{+}\gamma_{s,j}^{-}} + \sqrt{\gamma_{1,i}^{-}\gamma_{s,j}^{+}}\right), \qquad (3b)$$

where θ_{ij} is the contact angle of the i-th liquid on the j-th solid and a further index j distinguishes the acid–base components of different solids. Whenever the number of liquids and solids is large enough, we are led to an overdetermined set of $L + S + LS$ non-linear equations in the $3L + 4S$ variables:

$$\sqrt{\gamma_{1,i}^{LW}}, \quad \sqrt{\gamma_{1,i}^{+}}, \quad \sqrt{\gamma_{1,i}^{-}}, \qquad\qquad i = 1, 2, \ldots, L$$
$$\sqrt{\gamma_{s,j}^{Tot}}, \quad \sqrt{\gamma_{s,j}^{LW}}, \quad \sqrt{\gamma_{s,j}^{+}}, \quad \sqrt{\gamma_{s,j}^{-}}, \qquad j = 1, 2, \ldots, S \tag{5}$$

and if no further equation can be invoked (e.g. a condition of null acid–base components for non-polar solids or liquids), a purely formal optimization problem consisting in equations (2b) and (3b) must be solved with respect to the variables (5). Notice that the above set (5) does not include the total surface free energies $\gamma_{1,i}^{Tot}$ of the liquids, directly measurable quantities, so it seems more reasonable to treat them as given parameters. In contrast, no commonly accepted experimental technique is available for solids, so their total surface free energy must be taken as an unknown.

It is worthy of note that a similar non-linear approach has been developed also in the case of the two-component theory by Erbil and Meriç [45]. Some of their results will be commented on later.

Although physically reasonable, the existence of a best-fit solution to the optimization problem is far from being trivial, owing to the relatively large number of variables and the mathematical form of the GvOC equations, which makes inapplicable the most general results of optimization theory [46]. The question of the existence of a solution will not be faced in the present paper, but it will simply be assumed that the optimization problem admits such a best-fit solution, as suggested by physical considerations and also shown by numerical simulations.

If, in addition, the solution is unique (and the acid–base model is correct), we can be confident that the physical meaning given to the components γ^{LW}, γ^{+}, γ^{-} is essentially correct and useful for a direct description of surface interactions. But if the uniqueness of the best-fit solution fails, some attention must be paid to the calculation of the components, because of the indeterminacy involved in the best-fit procedure: a multiplicity of scales can be defined, according to the particular best-fit solution selected. This entails a serious problem about the interpretation of acid–base components, which can be solved essentially in two alternative ways.

(i) On the one hand, we can reject any physico-chemical interpretation of γ^{LW}, γ^{+}, γ^{-} calculated according to any particular scale, and state consistently that the acid–base components are significant only when inserted into equations (3b) or (2b) to compute the work of adhesion or surface free energies. This choice has been originally made by GvOC and implies the impossibility of comparing acidic and basic components of the same material and of gauging their respective contributions to the total interaction. The only possible comparison is that between the acidic or basic component of a certain material with the corresponding one of a different material [11]; actually, no definition of 'strength' was proposed by GvOC for acids and bases. Unfortunately, this fact has not been sufficiently emphasized and the current use of the GvOC theory in the literature is often incorrect; many authors compare directly the magnitude of the acidic and basic components of the same material, but they use the original scale and coefficients of GvOC theory, with the obvious consequence of obtaining very low acidic components.

(ii) On the other hand, if the extent of indeterminacy of the best-fit solution is large enough, we can hope to harness it so as to define an appropriate scale of acid–base components suitable to compare with other and more direct scales of acid–base strength.

In the case of the GvOC model, it can be shown that the solution of the best-fit problem actually implies the existence of an infinite number of solutions, a feature of the mathematical form of the model which does not depend on the best-fit algorithm adopted. A specific solution must be selected by means of some conventionally assigned components of reference liquids, in a way which recalls Drago's calibration procedure for his model equations about the enthalpies of adduct formation in gas-phase or poorly solvating media [39].

Among the infinite and formally equivalent choices, we can define, for instance, a scale based on a realistic ratio of γ^+/γ^- for a reference liquid (as in the Abraham scale, where the water ratio γ^+/γ^- is 5.5); such a scale would also make possible the comparison of acid *and* base components of different solids and liquids. This is, in short, the DVS method proposed by us. Although the procedure seems difficult to carry out, owing to the general disagreement in the values of contact angles for non-polar and basic liquids on common polymers and to an undoubted lack of data about acidic liquids, it provides the most correct way of applying the GvOC theory.

It is interesting to compare the DVS approach with the similar one proposed by Drago. The two methods share some features, such as the resort to best-fit solutions for a concise description of a large amount of experimental data, the non-uniqueness of these best-fit solutions due to the existence of appropriate infinite groups of linear transformations which leave the model equations invariant, or the possibility of defining different scales. However, the Drago theory distinguishes 'acidic' and 'basic' liquids (electron acceptors and donors), each characterized by two variables in such a way that the enthalpy of adduct formation for any acceptor–donor pair is written as

$$-\Delta H = E_A E_B + C_A C_B = D_A^T D_B, \qquad D_A = \begin{pmatrix} E_A \\ C_A \end{pmatrix}, \qquad D_B = \begin{pmatrix} E_B \\ C_B \end{pmatrix}, \quad (6)$$

where the subscripts A and B denote acceptor and donor and E and C represent electrostatic and covalent contributions, respectively (the superscript T stands for 'transpose').

In contrast, no such *a priori* classification of molecules into purely acidic or purely basic occurs in acid–base theories: each liquid or solid is described by the three components γ^{LW}, γ^+, γ^-, which appear in the appropriate expressions (2b)–(3b). These relationships can also be rewritten into the matrix form

$$\gamma_{1,i}^{Tot} = X_i^T R X_i, \qquad \gamma_{s,j}^{Tot} = Y_j^T R Y_j, \qquad \frac{1}{2}(1 + \cos\theta_{ij})\gamma_{1,i}^{Tot} = X_i^T R Y_j \quad (7)$$

if a column vector X_i (or Y_j) of components $\sqrt{\gamma^{LW}}$, $\sqrt{\gamma^+}$, $\sqrt{\gamma^-}$ is associated with each liquid (or solid), and a suitable constant matrix R is introduced; then

$$X_i = \begin{pmatrix} \sqrt{\gamma_{l,i}^{LW}} \\ \sqrt{\gamma_{l,i}^+} \\ \sqrt{\gamma_{l,i}^-} \end{pmatrix}, \qquad Y_j = \begin{pmatrix} \sqrt{\gamma_{s,j}^{LW}} \\ \sqrt{\gamma_{s,j}^+} \\ \sqrt{\gamma_{s,j}^-} \end{pmatrix}, \qquad R = \begin{pmatrix} 1 & 0 & 0 \\ 0 & 0 & 1 \\ 0 & 1 & 0 \end{pmatrix}. \qquad (8)$$

Therefore, although the matrix formalism is applicable to both cases, the situations are quite different. Given any 2×2 real non-singular matrix A, the Drago theory allows one to redefine the acceptor and donor components by means of the linear transformations

$$D_A \longrightarrow A D_A, \qquad D_B \longrightarrow \left(A^{-1}\right)^T D_B \qquad (9)$$

which obviously leave (6) invariant. The use of a different transformation for acidic and basic components makes sense because acceptors and donors are *a priori* recognized and classified, so that there is no reason to impose that the transformation rule must be the same for both.

In contrast, acid–base theories treat any molecule formally in the same way and a unique transformation must be applied to all the substances, by means of an appropriate non-singular 3×3 matrix C, i.e.

$$X_i \longrightarrow C X_i, \qquad Y_j \longrightarrow C Y_j \qquad (10)$$

which must also ensure the invariance of (7). The transformation matrix C satisfies the required condition if and only if

$$C^T R C = R, \qquad (11)$$

and belongs to a subgroup of transformations much smaller than the general group of linear non-singular transformations available for the Drago model. The subgroup, isomorphic to the orthogonal group $O(2, 1; \mathbb{R})$, includes a three-parameter family of non-singular matrices C and provides, therefore, an infinite set of linear transformations (10). The invariance of equations (7) through (10) means invariance of any merit function defined to compute the best-fit solutions and, consequently, applying (10) to a best-fit solution provides a new best-fit solution as well.

4. SOME COMPARISONS WITH AND SOME REFLECTIONS ON THE LITERATURE

4.1. Modelling enthalpy or free energy?

It would be useful to show the relation between the present point of view and that expressed by Fowkes [5]; the relations proposed by GvOC represent the free energy of the interface or, better, the work of the adhesion, which is the difference of the free energies in the formation of the interface. Fowkes expresses the work of adhesion in terms of the enthalpy of the same process; obviously this cannot be done without keeping in mind the thermodynamic relation between these two quantities in an isothermal process:

$$\Delta G = \Delta H - T \Delta S. \tag{12}$$

Fowkes' hypothesis that ΔG can be expressed as the product of the enthalpy by a certain coefficient f is very 'significant', its coherence with the experimental context must be proven and generally speaking is false [34]. Moreover, to assume a value of f close to 1 would correspond to a negligible variation of entropy during the interface formation; this is certainly possible in some cases, but it cannot be considered a general conclusion at all.

Either the Gibbs free energy or the enthalpy of the interface formation process can be expressed by using an LFER equation. Obviously this choice has different consequences on the mathematical form of the remaining quantity.

In particular, if one introduces an LFER equation for the Gibbs free energy, according to GvOC, the expression of enthalpy of adhesion becomes

$$\Delta H = \Delta G + T \Delta S = \Delta G - T \frac{\partial \Delta G}{\partial T} \tag{13a}$$

$$\Delta H_{adh} = W_{adh} - T \frac{\partial W_{adh}}{\partial T} = \gamma_l (1 + \cos\theta) - T \left\{ (1 + \cos\theta) \frac{\partial \gamma_l}{\partial T} + \gamma_l \frac{\partial (\cos\theta)}{\partial T} \right\} \tag{13b}$$

and substituting (3b) in the second term, one obtains

$$\Delta H_{adh} = 2 \left(\sqrt{\gamma_{l,i}^{LW} \gamma_{s,j}^{LW}} + \sqrt{\gamma_{l,i}^{+} \gamma_{s,j}^{-}} + \sqrt{\gamma_{l,i}^{-} \gamma_{s,j}^{+}} \right)$$
$$- 2T \left\{ \frac{\partial \left(\sqrt{\gamma_{l,i}^{LW} \gamma_{s,j}^{LW}} + \sqrt{\gamma_{l,i}^{+} \gamma_{s,j}^{-}} + \sqrt{\gamma_{l,i}^{-} \gamma_{s,j}^{+}} \right)}{\partial T} \right\}, \tag{13c}$$

which can be simplified to:

$$\Delta H_{adh} = 2 \left\{ \left(\sqrt{\gamma_{l,i}^{LW} \gamma_{s,j}^{LW}} - T \frac{\partial \sqrt{\gamma_{l,i}^{LW} \gamma_{s,j}^{LW}}}{\partial T} \right) + \left(\sqrt{\gamma_{l,i}^{+} \gamma_{s,j}^{-}} - T \frac{\partial \sqrt{\gamma_{l,i}^{+} \gamma_{s,j}^{-}}}{\partial T} \right) \right.$$
$$\left. + \left(\sqrt{\gamma_{l,i}^{-} \gamma_{s,j}^{+}} - T \frac{\partial \sqrt{\gamma_{l,i}^{-} \gamma_{s,j}^{+}}}{\partial T} \right) \right\}. \tag{13d}$$

The three terms each inside parentheses can be considered as corresponding, respectively, to the LW and to the acidic and basic components of the total enthalpy of adhesion; so, combining the last two, one obtains

$$\Delta H_{adh} = \{\Delta H_{adh}^{LW} + \Delta H_{adh}^{AB}\}. \tag{14}$$

From this equation one can conclude that by knowing the values of the acid–base components of the work of adhesion and their variation with the temperature it is possible to calculate some corresponding enthalpic quantities.

Note that while from a thermodynamic and a dimensional point of view the following equivalence (or a similar one for the acid–base components) is legitimate

$$\sqrt{\gamma_{l,i}^{LW}\gamma_{s,j}^{LW}} - T\frac{\partial\sqrt{\gamma_{l,i}^{LW}\gamma_{s,j}^{LW}}}{\partial T} = \Delta H_{adh}^{LW}, \tag{15}$$

the situation appears different for terms of the form $\sqrt{\gamma_{s,j}^{LW}} - T\frac{\partial\sqrt{\gamma_{s,j}^{LW}}}{\partial T}$. They are neither enthalpic quantities nor squares of enthalpic quantities, even if their mathematical form looks similar.

However, it still remains still valid that knowing the single acid–base components of the work of adhesion and their variation with the temperature, it is also possible to calculate the quantities as $\gamma_{s,j}^{LW} - T\frac{\partial\gamma_{s,j}^{LW}}{\partial T}$, which can be considered the true acid–base components of the surface enthalpy of a material, but they are not explicitly present in the previous equations.

Vice versa, if one chooses to express the enthalpy by an LFER equation, by proposing the following equivalence (which becomes fully equivalent to the GvOC equation, but in terms of enthalpy):

$$\Delta H_{adh} = 2\left(\sqrt{h_{l,i}^{LW}h_s^{LW}} + \sqrt{h_{l,i}^{+}h_s^{-}} + \sqrt{h_{l,i}^{-}h_s^{+}}\right) \tag{16a}$$

and using the Gibbs–Helmholtz equation, one then obtains

$$\frac{d\Delta W_{adh}/T}{dT} = -\frac{\Delta H_{adh}}{T^2} = -\frac{2\left(\sqrt{h_{l,i}^{LW}h_s^{LW}} + \sqrt{h_{l,i}^{+}h_s^{-}} + \sqrt{h_{l,i}^{-}h_s^{+}}\right)}{T^2}, \tag{16b}$$

where the left-hand side can be estimated from contact angle measurements at different temperatures and represents the known term of an equation in the unknowns h's. An overdetermined set of these equations can be solved with the same methods described for the work of adhesion and its acid–base components.

It is noticeable that again some apparently obvious relationships are not valid; e.g. while it is true that $h^{LW} = \gamma^{LW} - T\frac{\partial\gamma^{LW}}{\partial T}$, the situation is different for the acidic (or the basic) component: $h^{+} \neq \gamma^{+} - T\frac{\partial\gamma^{+}}{\partial T}$.

In both cases, starting from a free energy term or from an enthalpy one, one needs to use measurements of contact angles at different temperatures to express correctly the remaining function, i.e. the acid–base components of surface enthalpy or surface free energy.

By considering correctly the thermodynamic relationships, there is no contradiction between these two approaches. The only problem is to have good experimental data, otherwise the numerical calculation of derivatives is a really cumbersome task.

4.2. Comparison with other proposals to modify the GvOC theory

The paper by Lee [47] in 1996 was the first one where a correlation among the coefficients of solvatochromic scale and the coefficients of the GvOC theory was suggested. However, Della Volpe and Siboni were not aware of Lee's paper at the time they published their paper [11] in 1997, nor did any reviewer note the omission of that paper among the references quoted. Both papers examine the same correlation, but from different points of view.

In Lee's work the correlation is presented as an 'unexpected result', while in the DVS paper this correlation depends on the analysis of the mathematical relations of Taft–Abraham and of the GvOC theory, developed by means of a matrix formalism.

A second conceptual difference is that Lee considered the components γ^+ and γ^- as directly related to the parameters α and β by Taft and others. Now, an LFER equation expresses the interfacial free energy of two materials (or better, the logarithm of an equilibrium constant, proportional to the free energy but actually non-dimensional) as a sum of products where each factor refers to one of the two materials. In the case of the GvOC theory, the work of adhesion or the surface free energy of a liquid or solid depends on the product of the *square roots* of the acidic and basic components, not of the components themselves. In a comparison of the two theories, the correspondence is not between γ^+ and α or γ^- and β, as proposed by Lee, but between $\sqrt{\gamma^+}$ and α or $\sqrt{\gamma^-}$ and β. This changes the ratio between the water acid–base components in a significant way and the value 1.8 suggested by Lee must likely be replaced by $(1.8)^2 = 3.2$.

In ref. [11], two values of the γ^+/γ^- ratio were considered, 5.5 and 6.5. The most recent and experimentally supported is 5.5, calculated from the data of Abraham [41]: the α/β ratio for water was $(0.82/0.35) = 2.34$ in ref. [41] and it must be squared to provide 5.5. The second value 6.5 was only a rather arbitrary guess, although suggested by Taft's work [40]; the α/β ratio proposed in ref. [40] was actually $1.17/0.18 = 6.5$, but the accordingly correct γ^+/γ^- ratio $6.5^2 \sim 42$ seems too high. Moreover, the value of the denominator, 0.18, was considered by Taft as affected by great uncertainty.

Actually, it is not known what is the best γ^+/γ^- ratio for water at 20°C; this will be the subject of the following discussion and of future measurements; it is certainly greater than 1 and the interval 3.2–5.5 can be considered a reasonable guess.

It is worth noting that in common pure water a fast equilibrium (in about 10–20 min) with atmospheric carbon dioxide is established; the final pH is about 5.5–6, which corresponds to a strong excess of hydrogen ions, about 900 times more concentrated than OH^- ions. This can be an explanation of the prevalently acidic character of common pure water. Contact angle measurements of ultra-pure water (at pH 7 and with a very low conductance, about 0.05 μS) on common polymers are not available in the literature, nor would they probably be useful for practical purposes, since water is commonly in equilibrium with carbon dioxide in all the systems of interest, particularly in biological ones. However, for scientific reasons these kinds of measurements would be very interesting, and could even show that ultra-pure water is really a Lewis base before being 'acidified' by carbonic acid formation. This fact could also account for the general law invoked to explain the prevalently basic character of polymer surfaces [24, 25], possibly through the minimization of free energy in the presence of atmospheric water at thermodynamic equilibrium. Although atmospheric water is acidic and ubiquitous, its typical abundance seems too low to play a significant role in the determination of polymer surface acid–base features. Actually, the DVS analysis shows that the basicity of polymer surfaces is presumably overestimated, so that the previous hypothesis does not seem ultimately necessary.

As a third important difference, in Lee's paper no attention is paid to other mathematical aspects of the theory: the ill-conditioning of the set of equations obtained using only three liquids, the limited significance of results calculated only from advancing contact angles, and so forth. In the DVS analysis, it has been widely shown that the use of only three *polar* liquids for the calculation of the LW, acidic and basic components of the surface free energy strongly enhances the ill-conditioning character of the equation set with the worst results. A correct choice is the use of a wide (greater than three) and *proper* (well-balanced) set of monopolar or mainly acidic, monopolar or mainly basic, and mainly dispersive liquids.

Finally, the results by Lee and DVS agree from a qualitative point of view; the general increase of the acidic components and decrease of the basic ones imply that the chemical nature of each liquid is better reflected by its components. Polymer acidic components appear greater than using the GvOC parameters, but never predominant: poly(vinyl chloride) (PVC) is not acidic; poly(vinyl fluoride) (PVF) is a weak acid according to DVS, but not according to Lee.

There is also another set of papers (by Qin and Chang) [48–51] where the problem of acid–base theories has been dealt with by new and original ideas. Qin and Chang introduce as variables the square roots of acid and base components, accept explicitly that they can be positive or negative, and consider that a material is acidic or basic if *both* of its components have a certain sign, positive for acids and negative for bases. They propose, therefore, a new and different set of coefficients for the test liquids, and recalculate the acid–base components of some polymers accordingly.

From our point of view, this approach corresponds to the choice of a particular scale, among the infinite possible ones; the properties of the equation set are not different from those found for the original GvOC model. Indeed, Qin and Chang's expressions for the surface free energy and the work of adhesion of a liquid L and a solid S are [49]

$$\gamma = \frac{1}{2}\left(P_L^d\right)^2 - P_L^a P_L^b,$$

$$\gamma = \frac{1}{2}\left(P_S^d\right)^2 - P_S^a P_S^b W^{adh} = P_S^d P_L^d - \left(P_S^a P_L^b + P_L^a P_S^b\right)$$

(17)

and can be easily reduced to the GvOC form (3a)–(2b) by means of the substitutions:

$$P^d \longrightarrow \sqrt{2\gamma^{LW}}, \qquad P^a \longrightarrow \sqrt{2\gamma^+}, \qquad P^b \longrightarrow -\sqrt{2\gamma^-}.$$

(18)

We tried to apply the procedure by Qin and Chang to the set of data previously described [49], comparing these results with a scale where all the coefficients are chosen as positive; the conclusion is that all the parameters, except for the sign, take the same values.

The problem is the actual acceptability of the values proposed as a reference. Unfortunately, the values for water (and other liquids) chosen by Qin and Chang have been presented at a symposium in 1992 [48] and reported in two subsequent papers in 1995 [49] and 1996 [50], but never fully explained or justified. Thus, we cannot judge the procedure used by the authors, but only the obtained parameters. The latter do not seem coherent with the accepted view of the properties of the test liquids; water appears to be more basic than acid, and methylene iodide and bromonaphthalene are both described as fully basic.

Accordingly, the scale adopted by Qin and Chang shows the same Achilles' heel which affects the original GvOC one; it is not meaningful to compare the acidic and basic properties of the same material, but only the acidic (or basic) parameters of different substances. The scale is arbitrary and cannot be justified.

4.3. The role of receding contact angles

The last point that we would like to discuss in this section is that a wider and proper set of advancing and receding contact angles is necessary, possibly obtained as a result of a round robin, in order to reduce errors.

Although it could seem scandalous, the calculation of surface free energy components of solids from receding contact angles is, in principle, as correct as that carried out from advancing or 'equilibrium' ones, because all of them actually correspond to metastable states.

For the sake of clarity, one could give the name 'apparent' surface free energy components to values calculated from receding angles, to emphasize that the contact angle involved in the calculation is not measured at equilibrium; note, however, that this is always the case. Typically the 'surface science community' uses sessile static contact angles, considering them as equilibrium ones, or simply prefers to base surface free energy calculations on advancing contact angles, but no accepted method exists to measure the true 'equilibrium' value of a contact angle, by distinguishing it from the other infinite number of metastable values [52]. Thus, all calculated values of solid surface free energy available in the literature are 'apparent', unless they stem from (nearly) ideal surfaces, which is very seldom the case. Moreover, in the case of heterogeneous surfaces the important results by Johnson and Dettre [53, 54] entail that both advancing and receding contact angles are representative, the former being better correlated with the low-energy portion of the surface and the latter with the high-energy one. If the two 'apparent' contact angles provide a deeper description of the surface than the 'Young contact angle' does, why not try to extend this reasoning to the calculation of surface free energy?

In two recent papers, Jacobasch and co-workers [13, 55] strongly support the idea that contact angle measurements are not able to evaluate the acidic properties of flame-treated materials. Those papers contain very good contact angle data on flame- and plasma-treated thermoplastic polyolefin (TPO) surfaces for three liquids, measured by the sessile technique. The authors calculated the acid–base surface free energy components by the GvOC theory and compared them with the results of X-ray photoelectron spectroscopy (XPS) and ζ-potential analyses. They found that, differently from the other two techniques, contact angles did not reveal the acidic component.

Their contact angle results are in good agreement with the published data and their presentation is unquestionable, but some criticism can be raised to the calculation procedure in the light of the previous discussion, since:

- the calculations involved only advancing contact angles;

- only three liquids were used;

- the acidic and basic components of each material were directly compared.

Although their procedure constitutes the usual way to apply the GvOC theory, we have already stressed that it is unsatisfactory from several points of view:

- advancing angles are mostly indicative of the low surface free energy portion of a surface, and thus are not necessarily representative of the flame- or plasma-treated portion [52–54];

- the small number of liquids employed enhances the ill-conditioning of the GvOC equation set and produces a significant deviation in the results;

- the use of the liquid coefficients originally proposed by GvOC implies the impossibility of comparing acidic and basic components of the same material.

If the Jacobasch data are treated according to the above criteria, it is possible to show (at least when the advancing contact angles are significantly modified by the flame or plasma treatment) that the acidic component of a flame-treated PP surface is as enhanced as the basic one in comparison with the untreated material. In other cases, however, the advancing contact angles are not sufficiently modified by the process to correctly show the high free energy portion of the surface and the GvOC theory 'seems' to fail.

The use of receding contact angles as a better index of high-energy surfaces and particularly of the acidic (or basic) character of a surface is described in ref. [13]

by the same authors; they consider the general advantages offered by the receding contact angles of poly(propylene–ethylene-co-vinyl acetate) (PP–EVA) blends in water to estimate the acidic character of these surfaces, but do not extend it to the point of calculating the (apparent) surface free energy components from the same angles.

5. THE PROBLEM OF SCALES

5.1. Some new ideas about the choice of a correct scale of acid–base components for solids

We analysed the dependence of the acidic and basic components of solids and liquids on the numerical value of the chosen γ^+/γ^- ratio for water, using the set already introduced in ref. [11]. This is a wide set, including ten liquids and 14 solid polymers; contact angle and surface tension data have been collected from the international literature, whenever possible. It is clear that a large number of measurements would be necessary to provide a more uniform set of contact angle data on standardized materials. In any case, the huge amount of work and the necessity to avoid systematic errors suggest that these kinds of measurements would be better carried out by an international round robin instead of a single laboratory. Looking at the graph of Fig. 5, one can see, for instance, that the acidic component of PVF increases as the water γ^+/γ^- ratio increases. This is generally true for all the acidic components; on the other hand, the basic components become lower, but the variations are not as large as one would expect.

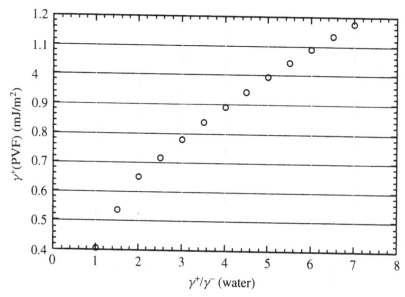

Figure 5. The acidic component of PVF computed by the best-fit method, as a function of the γ^+/γ^- ratio for water. The dispersive component of water is 22 mJ/m^2.

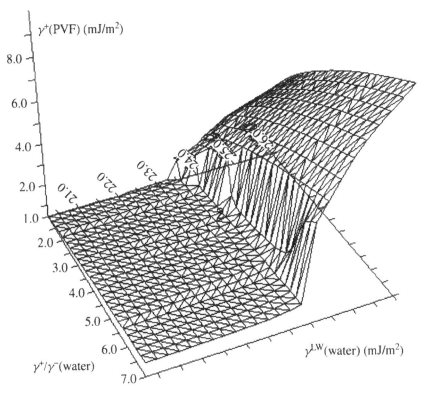

Figure 6. A plot of the acidic component of PVF calculated by the DVS best-fit method vs. the γ^+/γ^- ratio and the value of γ^{LW} for water.

Thus, we decided to explore more extensively the domain of the variables by varying all the elements of the water reference scale; in Figs 6 and 7 the variation of the acidic component of PVF and of the basic component of DMSO vs. the γ^+/γ^- ratio and the value of γ^{LW} for water are plotted. From these graphs it is possible to appreciate an unexpected result. For a value of the dispersive water component of about $23-24$ mJ/m^2, there is a strong non-linear variation of all the components; the data refer to PVF but the situation is similar for other polymers. For the DMSO base component, this non-linear jump occurs at about $26-27$ mJ/m^2; so, the best choice of this parameter is in the interval $23-27$ mJ/m^2.

The global trend of the mathematical function whose minimum gave us the component values is shown in Fig. 8; this function, which we will call henceforth the merit function, is minimized to solve the GvOC overdetermined equation set. It contains all the data and coefficients and depends on 71 unknowns, three of which are the acid–base components of water. By assuming a surface free energy of water of 72.8 mJ/m^2, these components are uniquely fixed by assigning the values of γ^{LW}, γ^+ and γ^- in an appropriate interval each; the minimum of the merit function with respect to the remaining 68 variables, at the given γ^{LW}, γ^+ and γ^- of water, is then calculated and the value of that minimum plotted. The resulting plot is displayed in Fig. 8. As one can see, the minimum of the global merit function becomes slightly

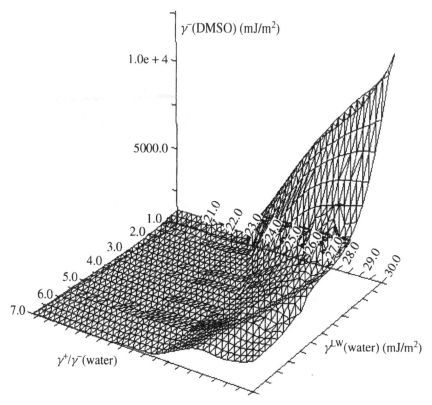

Figure 7. A plot of the basic component of DMSO calculated by the DVS best-fit method vs. the γ^+/γ^- ratio and the value of γ^{LW} for water.

lower with the increase of the water γ^+/γ^- ratio, but the most important point is that the minimum is at a value of the water dispersive component of 26.25 mJ/m^2, completely different from that commonly used of 21.8 mJ/m^2. This prompts us to take a closer look at the origin of this widely accepted value.

The value of 21.8 mJ/m^2 was proposed by Fowkes in 1964 [56]; in that paper he reported measurements of the interfacial energy of water at the interface with completely dispersive liquids, with a very low scatter in the data. In the same paper he also reported the interfacial energy of water on completely dispersive solids, but in this case the numerical values were not reported, but only plotted on a graph (Fig. 6 of ref. [56]) at a magnified scale, which showed that they spread over a wide interval, between 15 and 30 mJ/m^2. It would be very difficult to recalculate the original values from the figure, so we tried to use other literature data to calculate the LW component of water from the interaction of water with a purely dispersive solid whose surface free energy was calculated independently from the contact angle of a purely dispersive liquid on it.

We used equation (3b) in two modified forms, illustrated below: the first one to calculate the LW component of a purely dispersive solid from the contact angle of purely dispersive liquids on the solid, and the second one to calculate the

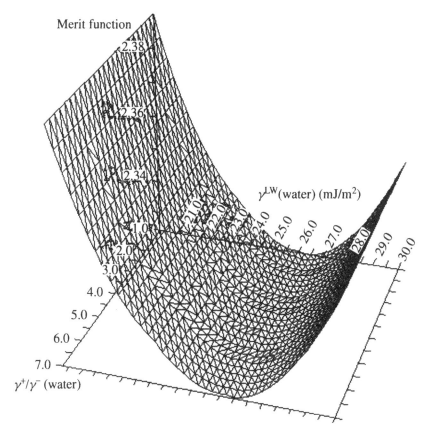

Figure 8. An illustration of the merit function trend. As shown in the discussion, the merit function, used during the calculation of the acid–base components with the DVS method, depends on the acid–base components of water and on the remaining 68 variables, corresponding to the components of the other liquids used and of the polymers. By assuming a surface free energy of 72.8 mJ/m^2, the water components are uniquely fixed by assigning the values of γ^{LW} and γ^+/γ^- in an appropriate interval each; the minimum of the merit function with respect to the remaining 68 variables, at the given γ^{LW}, γ^+ and γ^- of water, is then calculated and the value of that minimum is plotted.

LW component of water from its contact angle on the same solid:

$$\gamma_s^{LW} = \frac{\gamma_1(1+\cos\theta)^2}{4}, \qquad \gamma_1^{LW} = \frac{\gamma_1^2(1+\cos\theta)^2}{4\gamma_s^{LW}}. \tag{19}$$

It is possible to find many useful collections of data in the literature; we found ten such sets, whose sources are cited in the References section [42, 52, 57–59]. The results are shown in Fig. 9; one can see that the estimates are very different, with a large scatter but a mean value greater than that reported for the liquid–liquid interaction, about 27.1 ± 5.7 mJ/m^2. In contrast, the value obtained by Fowkes for liquid–liquid interaction was 21.8 ± 0.2 mJ/m^2.

Is it conceivable that a liquid–liquid interaction is different from a liquid–solid one, even in the case of purely dispersive interactions, namely that the surface free

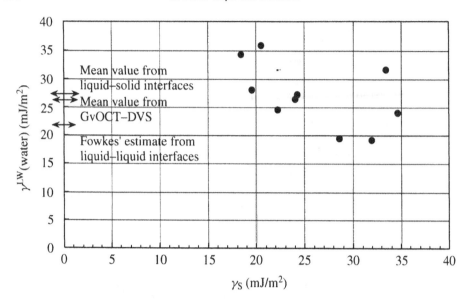

Figure 9. Estimates of the dispersive component of water from data of solid–liquid interfaces. The values on the abscissa are the surface free energies of the solids employed for the calculation. The data sets were collected from refs [41, 51] and [56–58]. The mean values by Fowkes, GvOC, and DVS [11] are shown for comparison.

energy components of a liquid (water) may take different values when it interacts with another liquid or with a solid? The answer is not easy but, in principle, could be affirmative; e.g. one can consider that for a solid polymer at a temperature higher than its T_g the molecular motions required to adapt its structure to the force field of water molecules are much more obstructed than in a liquid. Moreover, the geometrical shape of the interface is different, since even the flattest polymer surface cannot be as flat as a liquid surface can be. A similar result, with a dispersive component for water higher than that commonly accepted, has been obtained by Erbil and Meriç [45] using a non-linear approach to the calculation of the coefficients of a two-component model. Unfortunately, Erbil and Meriç's conclusion, i.e. the presence of a specific contribution in each liquid–solid couple, introduces another *ad hoc* coefficient and is not valid for three-component models, because the corresponding set of equations depends on too many parameters and the number of possible solutions proliferates without control.

These coefficients of water can be reasonably used in the definition of a scale. We can assume a value of 26.25 mJ/m² for the LW component of water, and values of 48.5 and 11.2 mJ/m² for the acidic and basic components, respectively. The γ^+/γ^- ratio is 4.35, exactly in the middle point of the interval 3.2–5.5 previously considered as possibly acceptable. If these new coefficients, different from ref. [11], are employed, we obtain the best-fit results reported in Table 1.

The results are now relatively acceptable; liquids and solids are acidic or basic, and the acidic components are significant and greater than basic ones, consistent with the chemical intuition. PVF and nylon 66 are effectively acidic surfaces, while

Table 1.

Dispersive and acid–base components of some liquids and solids calculated with a reference scale for water: $\gamma^{LW} = 26.25$ mJ/m^2, $\gamma^+ = 48.5$ mJ/m^2 and $\gamma^- = 11.2$ mJ/m^2 by the GvOC theory. The second column contains the square roots of the components, which are the mathematical unknowns, and the third column the components. SD $= \pm2$ mJ/m^2

Material	Square root of the component	Component	Material	Square root of the component	Component
WAlw	5.12	26.2	P3FEt	5.04	25.4
WA+	6.97	48.5	P3FElw	4.80	23.1
WA−	3.34	11.2	P3FE+	1.34	1.79
GLlw	5.92	35.0	P3FE−	0.856	0.733
GL+	5.28	27.8	PVDFt	5.52	30.5
GL−	2.71	7.33	PVDFlw	5.35	28.6
FAlw	5.96	35.5	PVDF+	3.24	10.5
FA+	3.35	11.3	PVDF−	0.290	0.0839
FA−	3.36	11.3	PVFt	6.07	36.8
MIlw	7.13	50.8	PVFlw	5.84	34.1
MI+	0.00	0.00	PVF+	2.87	8.25
MI−	0.00	0.00	PVF−	0.476	0.227
GElw	5.82	33.9	PVDCt	6.35	40.3
GE+	0.983	0.966	PVDClw	6.36	40.4
GE−	7.18	51.6	PVDC+	−0.0431	0.00186
BNlw	6.66	44.4	PVDC−	1.62	2.63
BN+	0.00	0.00	PVCt	6.29	39.5
BN−	0.00	0.00	PVClw	6.25	39.1
DMSOlw	5.68	32.3	PVC+	0.239	0.0572
DMSO+	0.193	0.0373	PVC−	0.940	0.883
DMSO−	27.6	763	PSt	6.18	38.2
BFlw	5.30	28.1	PSlw	6.20	38.4
BF+	4.46	19.9	PS+	−0.162	0.0261
BF−	1.50	2.24	PS−	0.468	0.219
PYlw	5.09	25.9	PMMAt	6.14	37.7
PY+	3.29	10.8	PMMAlw	6.12	37.4
PY−	1.63	2.66	PMMA+	0.104	0.0108
HDlw	5.25	27.6	PMMA−	1.49	2.23
HD+	0.00	0.00	PCTFEt	4.68	21.9
HD−	0.00	0.00	PCTFElw	5.30	28.1
PTFEt	4.49	20.1	PCTFE+	−1.57	2.46
PTFElw	4.49	20.1	PCTFE−	1.97	3.89
PTFE+	0.00	0.00	PA66t	6.54	42.8
PTFE−	0.00	0.00	PA66lw	6.11	37.4
PPt	5.32	28.3	PA66+	1.73	2.99
PPlw	5.32	28.3	PA66−	1.57	2.46
PP+	0.00	0.00	PETt	6.19	38.4
PP−	0.00	0.00	PETlw	6.21	38.5
PEt	5.85	34.2	PET+	−0.0545	0.00297
PElw	5.85	34.2	PET−	1.61	2.58
PE+	0.00	0.00	PBTt	0.774	0.599

Table 1.

(Continued)

Material	Square root of the component	Component	Material	Square root of the component	Component
PE−	0.00	0.00	PBTlw	6.19	38.4
			PBT+	−5.15	26.5
			PBT−	3.67	13.4

WA = water; GL = glycerol; FA = formamide; MI = diiodomethane; GE = ethylene glycol; BN = bromonaphthalene; DMSO = dimethyl sulphoxide; BF = bromoform; PY = pyrrole; HD = hexadecane; PTFE = poly(tetrafluoroethylene); PE = poly(ethylene); PP = poly(propylene); P3FE = poly(trifluoroethylene); PVDF = poly(vinylidene fluoride); PVF = poly(vinyl fluoride); PVDC = poly(vinylidene chloride); PVC = poly(vinyl chloride); PS = poly(styrene); PMMA = poly(methyl methacrylate); PCTFE = poly(chlorotrifluoroethylene); PA66 = poly(hexamethylene adipamide); PET = poly(ethylene terephthalate); PBT = poly(butylene terephthalate).

PMMA remains strongly basic; among liquids, DMSO is the most basic one, while bromoform, pyrrole, water, and glycerol have a strong acid content.

It is interesting to remark that whenever the square root of the acid component is negative, the standard deviation is greater than it, except in one case. The odd behaviour of poly(butylene terephthalate) is probably caused by the quality of the original data; in fact, those data were collected in the laboratory of the authors by analysing the surface of a blood leukodepletion filter made of non-woven fabric, a material whose surface is very hard to analyse by the Wilhelmy microbalance. A systematic error was probably present, so that the data were comparable among them selves but not to other ones.

Once again, however, it is not expected that these calculations will provide 'the correct components'; much work is necessary to obtain good and acceptable contact angles as starting points for the calculation of the acid–base coefficients, work which can only be completed if many laboratories work together in an international research programme. This is also true for the data necessary to set the scale, i.e. to obtain (a) advancing and receding contact angles of dispersive liquids on dispersive solids and (b) advancing and receding contact angles of water on the same solids to set the dispersive component of water from liquid–solid interfaces with a more acceptable standard deviation.

Only if this global work does not produce acceptable results should the 'surface community' reconsider completely its ideas about solid surfaces and the utility of contact angles.

From the above discussion it is easy to suggest that one of the main limits of the data already obtained is the lack of contact angles of monopolar or strong acidic liquids; water is the most powerful acid used up to now. In other terms, in a liquid set where a strong hydrogen-bond acceptor such as the aprotic liquid DMSO is present, we also need a strong hydrogen-bond donor. The main problem is to find liquids with both a high acid or hydrogen-bond donor ability and a high value of the total surface free energy, due to the need to have contact angles different

Table 2.
Monopolar or strong acidic liquids suitable for contact angle measurements on polymers. Data from Abraham [41] and, in parentheses, from Taft and co-workers [40]. Water data are reported for comparison

Liquid	Surface free energy at room temperature (mJ/m^2)	Acid solvatochromic[a]	Basic solvatochromic[a]
H_2SO_4 (98.5%)	55.1	(?)	0 (?)
HNO_3 (98.8%)	42.7	(?)	0 (?)
Formic acid	37.6	0.75	0.38
Mono-chloroacetic acid	35.4	0.74	0.36
Di-chloroacetic acid	35.4	0.90	0.27
Pyrrole	36.6	0.41	0.29
Bromoform	41.53	0.15	0.06
Phenol	40.9	0.60	0.30
Water	72.8	0.82 (1.17)	0.35 (0.18)

[a] The word 'solvatochromic' refers to the particular scale of acid-base strength, developed by the cited authors to be used in their LFER equations.

from zero on common polymers, without significant swelling or reaction effects. Some candidates can be found among those shown in Table 2. Halogen and nitro-derivatives of phenol and cresol can also be suggested.

A very intriguing idea can be the use of hydrogen peroxide; its surface free energy is similar to that of water, about 76 mJ/m^2, while it is estimated to be about 10^6 times less basic than water [60]!

5.2. Some mathematical aspects of the definition of scales

Let us tackle in more detail the problem of scale definition. Suppose that a satisfactory best-fit solution has been found for a given set of donors and acceptors. Our goal is to fix the components of one or more substances in order to remove, at least partially, the invariance property of the GvOC set and select a particular scale. Note that the only requirement is the definition of a scale satisfactory from a physical point of view; thus, the complete removal of invariance does not appear strictly necessary.

The four entries of Drago's matrix A in expression (9) are essentially arbitrary, the only constraint being det $A \neq 0$. Thus, we simply have to compute the parameters E and C for an acceptor and a donor and to give them some conventional values; the set of four equations imposed by the transformation rule will finally determine the four required entries. Some care is needed in the choice of the reference substances, to obtain a set of four independent equations.

The definition of a specific scale for acid-base theories seems a more delicate task. If we suppose the components of a particular liquid to have been calculated, another set of γ^{LW}, γ^+, γ^- for the same liquid can be obtained through a transformation of the invariance group. As $O(2, 1; \mathbb{R})$ essentially depends on a triple of parameters, we expect that whenever three components of some reference liquids are specified, the global best-fit solution is uniquely determined and the scale selected. Of course, we have to check not only that the three imposed conditions are independent, but also that the related matrix C satisfies equation (11). The invariance of $X^T R X$ obviously provides a necessary condition to accomplish the last requirement.

As a less restrictive constraint, we may impose that the acidic and basic components γ^+, γ^- of a liquid are equal, without assigning a specific value to them. In this case, we introduce a unique condition and the scale is not completely fixed (we note here that the GvOC theory in its present form makes this an unsatisfactory choice). It is noticeable, however, that a liquid whose components γ^+, γ^- are both non-zero certainly allows an invariance transformation C leading to $\gamma^+ = \gamma^-$, since, as shown in ref. [11], any diagonal matrix with diagonal elements $(1, e^\zeta, e^{-\zeta})$, $\forall \zeta \in \mathbb{R}$, belongs to $O(2, 1; \mathbb{R})$. For an appropriate value of ζ, such a matrix expresses the condition imposed in the GvOC theory for water, although it remains only a particular case of the most general requirement (11).

We emphasize once again that the reference liquid and its components must be taken 'judiciously', because a wrong choice could be inconsistent with the experimental data. The previous analysis is based on the simple observation that any best-fit algorithm must minimize some (objective) function of the rests $\Delta_{1,i}$, $\Delta_{s,j}$, Δ_{ij} defined below:

$$\Delta_{1,i} = X_i^T R X_i - \gamma_{1,i}^{\text{Tot}}, \qquad \Delta_{s,j} = Y_j^T R Y_j - \gamma_{s,j}^{\text{Tot}}$$

$$\Delta_{ij} = X_i^T R Y_j - \frac{1}{2}(1 + \cos\theta_{ij})\gamma_{1,i}^{\text{Tot}}, \qquad (20)$$

where the column vectors X_i, Y_j and the symmetric orthogonal matrix R have been previously introduced in (8).

The following Theorem 1 and Theorem 2 represent the main mathematical results and ensure the lack of uniqueness for the best-fit solutions. These statements were proven in [11].

Theorem 1. Let A_i, $i = 1, 2, \ldots, L$, and B_j, $j = 1, 2, \ldots, S$, be some (real or complex) 3×3 non-singular matrices such that the substitutions $X_i \rightarrow A_i X_i$ and $Y_j \rightarrow B_j Y_j$ leave the rests (20) invariant. Then

(i) $A_i = B_j = C$, $\forall i = 1, 2, \ldots, L$, $j = 1, 2, \ldots, S$, where C is any 3×3 matrix satisfying $C^T R C = R$;

(ii) the set $\mathbf{G}_3$ of matrices C as above constitutes a group with respect to the usual matrix product. Such a group is isomorphic to the orthogonal group $O(2, 1; \mathbb{R})$.

Theorem 2. The group $\mathbf{G}_3$ includes the three-parameter family of non-singular matrices

$$\pm \exp[\omega_1 E_1 + \omega_2 E_2 + \omega_3 E_3] \qquad \forall \omega_1, \omega_1, \omega_1 \in \mathbb{C}, \qquad (21)$$

where

$$E_1 = \begin{pmatrix} 0 & 1 & 0 \\ 0 & 0 & 0 \\ -1 & 0 & 0 \end{pmatrix}, \quad E_2 = \begin{pmatrix} 0 & 0 & 1 \\ -1 & 0 & 0 \\ 0 & 0 & 0 \end{pmatrix}, \quad E_3 = \begin{pmatrix} 0 & 0 & 0 \\ 0 & 1 & 0 \\ 0 & 0 & -1 \end{pmatrix}. \quad (22)$$

An important by-product of Theorem 2 is the following straightforward corollary.

Corollary 2.1. For any $\omega \in \mathbb{C}$, there holds

$$\exp(\omega E_1) = \begin{pmatrix} 1 & \omega & 0 \\ 0 & 1 & 0 \\ -\omega & -\omega^2/2 & 1 \end{pmatrix}, \quad \exp(\omega E_2) = \begin{pmatrix} 1 & 0 & \omega \\ -\omega & 1 & -\omega^2/2 \\ 0 & 0 & 1 \end{pmatrix}$$

$$\exp(\omega E_3) = \begin{pmatrix} 1 & 0 & 0 \\ 0 & e^\omega & 0 \\ 0 & 0 & e^{-\omega} \end{pmatrix}. \quad (23)$$

This result provides the main tool to prove the statement below, concerning the possibility of redefining acid–base components for a particular liquid or solid.

Theorem 3. Let γ_0^{LW}, γ_0^+, γ_0^- be the non-negative acid–base components of a given material, with $\gamma_0^{LW} + 2\sqrt{\gamma_0^+}\sqrt{\gamma_0^-} > 0$. Consider the new triplet of components γ^{LW}, γ^+, γ^- defined by means of the linear transformation

$$\begin{pmatrix} \sqrt{\gamma^{LW}} \\ \sqrt{\gamma^+} \\ \sqrt{\gamma^-} \end{pmatrix} = C \begin{pmatrix} \sqrt{\gamma_0^{LW}} \\ \sqrt{\gamma_0^+} \\ \sqrt{\gamma_0^-} \end{pmatrix} \quad \text{with} \quad C^T R C = R.$$

Then $\forall \xi \in [0, 1]$ and $\eta \in \mathbb{R}$, a real matrix C as above exists such that

$$\gamma^{LW} = \left[\gamma_0^{LW} + 2\sqrt{\gamma_0^+}\sqrt{\gamma_0^-} \right] \xi$$

$$\gamma^+ = \left[\gamma_0^{LW} + 2\sqrt{\gamma_0^+}\sqrt{\gamma_0^-} \right] e^\eta (1 - \xi)/2 \quad (24)$$

$$\gamma^- = \left[\gamma_0^{LW} + 2\sqrt{\gamma_0^+}\sqrt{\gamma_0^-} \right] e^{-\eta}(1 - \xi)/2.$$

Moreover, C can be chosen in such a way that either $\gamma^+ = 0$ or $\gamma^- = 0$.

The previous Theorem 3 gives us the possibility of assigning the GvOC components to a particular liquid almost arbitrarily. A natural question is whether after this assignment a certain amount of indeterminacy is still available. A simple investigation of the spectral properties of matrices in $\mathbf{G}_3$ provides a satisfactory, positive answer to the problem.

Theorem 4. Let $C \in \mathbf{G}_3$. Then:

(i) there holds $\det(C - \lambda\mathbb{I}) = \det(C^{-1} - \lambda\mathbb{I})$, so that if λ is an eigenvalue, λ^{-1} also is. The spectrum of C always includes $+1$ or -1;

(ii) the eigenvalues of $C = \exp(\omega_1 E_1 + \omega_2 E_2 + \omega_3 E_3)$, $\omega_1, \omega_2, \omega_3 \in \mathbb{C}$, are

$$1 \qquad e^{\sqrt{\omega_3^2 - 2\omega_1\omega_2}} \qquad e^{-\sqrt{\omega_3^2 - 2\omega_1\omega_2}} \tag{25}$$

for $\omega_3^2 - 2\omega_1\omega_2 \neq 0$, whereas in the opposite case there is a unique, simple eigenvalue $+1$. Anyway, for $(\omega_1, \omega_2, \omega_3) \neq (0, 0, 0)$ the eigenspace of $+1$ coincides with the following set:

$$\mathrm{Ker}(C - \mathbb{I}) = \left\{ \mu(\omega_3\omega_2 - \omega_1)^{\mathrm{T}}, \ \mu \in \mathbb{C} \setminus \{0\} \right\}. \tag{26}$$

An immediate consequence of Theorem 4 is the corollary below.

Corollary 4.1. A non-zero vector $(x_1 \ x_2 \ x_3)^{\mathrm{T}} \in \mathbb{C}^3$ is invariant through the linear transformation $C = \exp(\omega_1 E_1 + \omega_2 E_2 + \omega_3 E_3)$, $(\omega_1, \omega_2, \omega_3) \in \mathbb{C}^3 \setminus \{(0, 0, 0)\}$, if and only if

$$C = \exp\left[\mu\left(-x_3 E_1 + x_2 E_2 + x_1 E_3\right)\right] \tag{27}$$

for some $\mu \in \mathbb{C} \setminus \{0\}$.

5.3. Applications to scale definition

We are now ready to apply the previous results to the problem of scale definition in the GvOC theory.

Suppose that the acid–base components $\gamma_{1,0}^{\mathrm{LW}}$, $\gamma_{1,0}^{+}$, $\gamma_{1,0}^{-}$ of a material — referred to as 'primary' reference material from now on — have been estimated in some way. Under the general assumption that $\gamma_{1,0}^{\mathrm{LW}} + 2\sqrt{\gamma_{1,0}^{+}\gamma_{1,0}^{-}} > 0$, we can use Theorem 3 to redefine the components according to

$$\gamma_1^{\mathrm{LW}} = \left[\gamma_{1,0}^{\mathrm{LW}} + 2\sqrt{\gamma_{1,0}^{+}}\sqrt{\gamma_{1,0}^{-}}\right]\xi$$

$$\gamma_1^{+} = \left[\gamma_{1,0}^{\mathrm{LW}} + 2\sqrt{\gamma_{1,0}^{+}}\sqrt{\gamma_{1,0}^{-}}\right]e^{\eta}(1 - \xi)/2 \tag{28}$$

$$\gamma_1^{+} = \left[\gamma_{1,0}^{\mathrm{LW}} + 2\sqrt{\gamma_{1,0}^{+}}\sqrt{\gamma_{1,0}^{-}}\right]e^{-\eta}(1 - \xi)/2$$

with $\xi \in [0, 1]$ and $\eta \in \mathbb{R}$ arbitrarily chosen. By applying the linear transformation (28) to all the materials of the set, we obtain a preliminary definition of the acid–base scale. Although not strictly relevant from a physical point of view, such a definition is obviously not yet complete, since Corollary 4.1 states that there exists a one-parameter group of linear transformations $\mathbb{C}^3 \to \mathbb{C}^3$, with $\mu \in \mathbb{C}$:

$$S(\mu): \quad \begin{pmatrix} \sqrt{\gamma^{LW}} \\ \sqrt{\gamma^{+}} \\ \sqrt{\gamma^{-}} \end{pmatrix}$$

$$\longrightarrow \exp\left[\mu\left(-\sqrt{\gamma_1^{-}}E_1 + \sqrt{\gamma_1^{+}}E_2 + \sqrt{\gamma_1^{LW}}E_3\right)\right] \begin{pmatrix} \sqrt{\gamma^{LW}} \\ \sqrt{\gamma^{+}} \\ \sqrt{\gamma^{-}} \end{pmatrix} \quad (29)$$

through which $(\sqrt{\gamma_1^{LW}}, \sqrt{\gamma_1^{+}}, \sqrt{\gamma_1^{-}})^T$ is invariant. If $\gamma_{2,0}^{LW}$, $\gamma_{2,0}^{+}$, $\gamma_{2,0}^{-}$ are now the components of another material — the 'secondary' reference material — the linear transformation (29) allows us to write

$$\begin{pmatrix} \sqrt{\gamma_2^{LW}} \\ \sqrt{\gamma_2^{+}} \\ \sqrt{\gamma_2^{-}} \end{pmatrix} = \exp\left[\mu\left(-\sqrt{\gamma_1^{-}}E_1 + \sqrt{\gamma_1^{+}}E_2 + \sqrt{\gamma_1^{LW}}E_3\right)\right] \begin{pmatrix} \sqrt{\gamma_{2,0}^{LW}} \\ \sqrt{\gamma_{2,0}^{+}} \\ \sqrt{\gamma_{2,0}^{-}} \end{pmatrix} \quad (30)$$

in such a way that the residual parameter μ can be possibly fixed by assigning one of the numbers $\sqrt{\gamma_2^{LW}}$, $\sqrt{\gamma_2^{+}}$ or $\sqrt{\gamma_2^{-}}$. The choice is not arbitrary, since: (a) the vectors $(\sqrt{\gamma_1^{LW}}, \sqrt{\gamma_1^{+}}, \sqrt{\gamma_1^{-}})$ and $(\sqrt{\gamma_{2,0}^{LW}}, \sqrt{\gamma_{2,0}^{+}}, \sqrt{\gamma_{2,0}^{-}})$ must be linearly independent, otherwise

$$\left(\sqrt{\gamma_{2,0}^{LW}}, \sqrt{\gamma_{2,0}^{+}}, \sqrt{\gamma_{2,0}^{-}}\right)$$

$$\in \text{Ker}\left[\exp\left[\mu\left(-\sqrt{\gamma_1^{-}}E_1 + \sqrt{\gamma_1^{+}}E_2 + \sqrt{\gamma_1^{LW}}E_3\right)\right] - \mathbb{I}\right] \quad (31)$$

and

$$\left(\sqrt{\gamma_2^{LW}}, \sqrt{\gamma_2^{+}}, \sqrt{\gamma_2^{-}}\right) = \left(\sqrt{\gamma_{2,0}^{LW}}, \sqrt{\gamma_{2,0}^{+}}, \sqrt{\gamma_{2,0}^{-}}\right) \quad \forall \mu \in \mathbb{C}. \quad (32)$$

The components γ_2^{LW}, γ_2^{+}, γ_2^{-} do not depend on μ and no full definition of the scale is possible; (b) the transformation $S(\mu)$ can be explicitly written as

$$\sqrt{\gamma_2^{LW}} = c_{11}\sqrt{\gamma_{2,0}^{LW}} + c_{12}\sqrt{\gamma_{2,0}^{+}} + c_{13}\sqrt{\gamma_{2,0}^{-}}$$

$$\sqrt{\gamma_2^{+}} = c_{21}\sqrt{\gamma_{2,0}^{LW}} + c_{22}\sqrt{\gamma_{2,0}^{+}} + c_{23}\sqrt{\gamma_{2,0}^{-}} \quad (33)$$

$$\sqrt{\gamma_2^{-}} = c_{31}\sqrt{\gamma_{2,0}^{LW}} + c_{32}\sqrt{\gamma_{2,0}^{+}} + c_{33}\sqrt{\gamma_{2,0}^{-}}$$

with

$$c_{11} = \cosh(\mu r) - \gamma_1^{LW}\frac{\cosh(\mu r) - 1}{r^2}$$

$$c_{12} = -\sqrt{\gamma_1^-}\sqrt{\gamma_1^{LW}}\frac{\cosh(\mu r) - 1}{r^2} - \sqrt{\gamma_1^-}\frac{\sinh(\mu r)}{r}$$

$$c_{13} = \sqrt{\gamma_1^+}\frac{\sinh(\mu r)}{r} - \sqrt{\gamma_1^+}\sqrt{\gamma_1^{LW}}\frac{\cosh(\mu r) - 1}{r^2}$$

$$c_{21} = -\sqrt{\gamma_1^+}\frac{\sinh(\mu r)}{r} - \sqrt{\gamma_1^+}\sqrt{\gamma_1^{LW}}\frac{\cosh(\mu r) - 1}{r^2}$$

$$c_{22} = \cosh(\mu r) - \sqrt{\gamma_1^-}\sqrt{\gamma_1^+}\frac{\cosh(\mu r) - 1}{r^2} + \sqrt{\gamma_1^+}\frac{\sinh(\mu r)}{r}$$

$$c_{23} = -\gamma_1^+\frac{\cosh(\mu r) - 1}{r^2}$$

$$c_{31} = -\sqrt{\gamma_1^-}\sqrt{\gamma_1^{LW}}\frac{\cosh(\mu r) - 1}{r^2} + \sqrt{\gamma_1^-}\frac{\sinh(\mu r)}{r}$$

$$c_{32} = -\gamma_1^-\frac{\cosh(\mu r) - 1}{r^2}$$

$$c_{33} = \cosh(\mu r) - \sqrt{\gamma_1^-}\sqrt{\gamma_1^+}\frac{\cosh(\mu r) - 1}{r^2} - \sqrt{\gamma_1^{LW}}\frac{\sinh(\mu r)}{r}$$

and $r = \sqrt{\gamma_1^{LW} + 2\sqrt{\gamma_1^+}\sqrt{\gamma_1^-}}$. All the square roots of the new components are given by expressions of the form $a + be^{\mu r} + ce^{-\mu r}$ for suitable constants a, b, c dependent on $\gamma_{2,0}^{LW}, \gamma_{2,0}^+, \gamma_{2,0}^-, \gamma_1^{LW}, \gamma_1^+, \gamma_1^-$ and on the component itself. Constraints on the admissible values of $\mu \in \mathbb{C}$ come from the requirement that all $\sqrt{\gamma_2^{LW}}, \sqrt{\gamma_2^+}, \sqrt{\gamma_2^-}$ be real and non-negative. Typically, μ is a real number.

We expect that whenever the secondary reference material is chosen appropriately, so as to fulfil conditions (a) and (b) for some μ, the parameter takes a uniquely determined value $\mu = \mu^*$ by assigning one of the components $\gamma_2^{LW}, \gamma_2^+, \gamma_2^-$. A mapping through $S(\mu^*)$ of the whole component set will provide the required scale.

It is noticeable that the self-consistency of the method imposes all the square roots of the final components to be non-negative; thus, not every choice will yield satisfactory results. In this sense, both primary and secondary reference materials must be selected 'judiciously', and the conventional values of the corresponding components assigned in a proper way. Of course, our conclusions also remain valid in the case where the square roots of acid–base components are not required to be positive (a constraint in the definition of the scale would simply be removed).

6. CONCLUSIONS

The main problems arising from the application of the GvOC theory to the calculation of Lewis acid–base properties of polymer surfaces from contact angle data can be summarized as follows:

- overwhelming basicity of all surfaces;
- results strongly dependent on the choice of the liquid triplet used for contact angle measurement;
- negative values sometimes obtained for the square roots of the acid–base parameters (which are the mathematical unknowns in the usual application of the GvOC method).

Throughout this paper, it was possible to account for these problems, namely:

- the Lewis base, or electron-donor component, is much greater than the Lewis acid or electron-acceptor component because of the reference values for water chosen in the original GvOC theory. According to this theory, in its usual form, the only meaningful approach is to compare acidic components with other acidic components of different materials, and the same for basic components. A direct comparison of the acidic component with the basic one of the same material has no meaning. This explains some discrepancy between the results obtained from wettability and IGC measurements [6, 12] and, most of all, accounts for the unrealistic (if gauged by common physico-chemical knowledge) complete predominance of monopolar basic surfaces stemming from the application of the GvOC theory. A reference scale for water which is able to overcome this problem has also been suggested, as an example and without any pretension to be right. The possibility of introducing different scales in the GvOC theory has been recognized by rewriting the GvOC equations in a matrix form, able to demonstrate the underlying symmetry properties of the equations themselves. For the calculation of acid–base components, a best-fit approach has been proposed which does not require any starting information about the liquids or polymers and can yield estimates of the acid–base parameters for both the liquids and the polymers involved. Nevertheless, much work is necessary to obtain data suitable for a definitive solution of the 'scale problem';
- the strong dependence of the value of the components on the three liquids employed is due to ill-conditioning of the related set of equations, an intrinsic and purely mathematical feature which cannot be completely cured by any realistic improvement of experimental accuracy. To reduce or eliminate the effect, one only needs a proper set of liquids, representative of all kinds of different liquids;
- as to negative values, whenever a large and proper set of liquids is used and the contact angle data are suitably treated, the number of negative coefficients reduces in a significant way. Moreover, in many cases the standard deviation of a negative component is typically greater than the estimate itself, so that the actual value may be zero. Negative coefficients appear, then, as a simple consequence of measurement uncertainty, combined with the possible ill-conditioning of the equation set. We cannot exclude, however, that in some cases they could have a different origin.

Besides these results, we hope to have sufficiently stressed that many factors, not always considered in everyday practice, come into play in the calculation of surface free energy components from contact angle measurements. Some of them have been frequently invoked and are eminently experimental, namely whether it is better to use the advancing, the receding, or any other contact angle. Some are more subtle and related to more general aspects, such as the effects of the mathematical form of the theory on the final results, the choice of the proper scale of acid–base components, and the definition of scales. No wonder that results achieved neglecting these aspects are tactfully deemed 'theoretically interesting, but not always useful' [61]. There is no hope of obtaining good surface free energy data from contact angle measurements or, more generally, of estimating whether contact angle measurements and related theories can provide useful surface free energy data, unless all of these aspects are taken into due account.

Acknowledgements

This work was supported by the Italian CNR project MSTA II. We are grateful to Dr. M. Morra of NobilBio srl (Asti, Italy) for many scientific suggestions.

REFERENCES

1. A. W. Adamson, *Physical Chemistry of Surfaces*, 5th edn, pp. 107–131. Wiley, New York (1990).
2. J. N. Israelachvili, *Intermolecular and Surface Forces*, 2nd edn, pp. 16–133. Academic Press, London (1992).
3. F. M. Fowkes, in: *Physicochemical Aspects of Polymer Surfaces*, K. L. Mittal (Ed.), Vol. 2, pp. 583–603. Plenum Press, New York (1983).
4. F. M. Fowkes, in: *Surface and Interfacial Aspects of Biomedical Polymers*, J. D. Andrade (Ed.), Vol. 1, Ch. 9, pp. 337–372. Plenum Press, New York (1985).
5. F. M. Fowkes, *J. Adhesion Sci. Technol.* **1**, 7–27 (1987).
6. J. C. Berg, in: *Wettability*, J. C. Berg (Ed.), Ch. 2. Marcel Dekker, New York (1993).
7. R. J. Good and M. K. Chaudhury, in: *Fundamentals of Adhesion*, L. H. Lee (Ed.), Ch. 3. New York, Plenum Press (1991).
8. K. L. Mittal and H. R. Anderson, Jr. (Eds), *Acid–Base Interactions, Relevance to Adhesion Science and Technology*. VSP, Utrecht, The Netherlands (1991).
9. M. Morra, *J. Colloid Interface Sci.* **182**, 312–314 (1996).
10. L. H. Lee, *Langmuir* **12**, 1681–1687 (1996).
11. C. Della Volpe and S. Siboni, *J. Colloid Interface Sci.* **195**, 121–136 (1997).
12. M. L. Tate, Y. K. Kamath, S. P. Wesson and S. B. Ruetsch, *J. Colloid Interface Sci.* **17**, 579–588 (1996).
13. H. J. Jacobasch, K. Grundke, S. Schneider and F. Simon, *J. Adhesion* **48**, 57–73 (1995).
14. G. N. Lewis, *Valence and the Structure of Atoms and Molecules*. The Chemical Catalog Co., New York (1923).
15. W. B. Jensen, *J. Adhesion Sci. Technol.* **5**, 1–7 (1991).
16. S. R. Cain, *J. Adhesion Sci. Technol.* **5**, 71–80 (1991).
17. G. C. Pimentel and A. L. McClellan, *The Hydrogen Bond*. Freeman, San Francisco (1960).
18. N. A. M. Besseling, Ph.D. Thesis, Wageningen Agricultural University, Wageningen, The Netherlands (1993).
19. N. A. M. Besseling and J. M. H. M. Scheutjens, *J. Phys. Chem.* **98**, 11597–11609 (1994).

20. N. A. M. Besseling, *J. Phys. Chem.* **98**, 11610–11622 (1994).
21. N. A. M. Besseling and J. Lyklema, *J. Pure Appl. Chem.* **67**, 881–888 (1995).
22. N. A. M. Besseling, *Langmuir* **13**, 2109–2112 (1997).
23. C. J. van Oss, R. J. Good and M. K. Chaudhury, *J. Protein Chem.* **5**, 385–402 (1986).
24. C. J. van Oss, M. K. Chaudhury and R. J. Good, *Adv. Colloid Interface Sci.* **28**, 35–60 (1987).
25. R. J. Good and C. J. van Oss, in: *Modern Approach to Wettability, Theory and Application*, M. E. Schrader and G. Loeb (Eds), Ch. 1, pp. 1–28. Plenum Press, New York (1991).
26. C. J. van Oss, *Interfacial Forces in Aqueous Media*. Marcel Dekker, New York (1994).
27. M. Sidqi, G. Ligner, J. Jagiello, H. Balard and E. Papirer, *Chromatographia* **28**, 588–596 (1989).
28. V. I. Bogillo and A. Voelkel, *Polymer* **36**, 3503–3510 (1995).
29. A. Voelkel, E. Andrzejewska, R. Maga and M. Andrzejewski, *Polymer* **37**, 455–462 (1996).
30. E. Andrzejewska, A. Voelkel, M. Andrzejewski and R. Maga, *Polymer* **37**, 4333–4344 (1996).
31. T. Young, *Philos. Trans.* **95**, 65 (1805).
32. M. Lampin, R. Warocquier-Clerout, C. Legris, M. Degrange and M. F. Sigot-Luizard, *J. Biomed. Mater. Res.* **36**, 99–108 (1997).
33. J. F. Padday, in: *Surface and Colloid Science*, E. Matijevic (Ed.), Vol. 1, Ch. 2, pp. 39–252. Wiley, New York (1969).
34. J. M. Douillard, *J. Colloid. Interface Sci.* **188**, 511–514 (1997).
35. M. Morra and C. Cassinelli, *J. Biomater. Sci. Polym. Edn* **9**, 55–74 (1997).
36. L. H. Lee, *J. Adhesion* **67**, 1–18 (1998).
37. C. Della Volpe, A. Deimichei and T. Riccò, *J. Adhesion Sci. Technol.* **12**, 1141–1180 (1998).
38. M. H. Adao, B. Saramago and A. Fernandes, *Langmuir* **14**, 4198–4203 (1998).
39. R. S. Drago, *Struct. Bond.* **15**, 73–139 (1973).
40. M. J. Kamlet, J. M. Abboud, M. H. Abraham and R. W. Taft, *J. Org. Chem.* **48**, 2877–2891 (1983).
41. M. H. Abraham, *Chem. Soc. Rev.* **22**, 73–83 (1993).
42. A. Deimichei, Thesis, University of Trento, Trento, Italy (1996).
43. W. H. Press, B. P. Flannery, S. A. Teukolsky and W. T. Vetterling, *Numerical Recipes*. Cambridge University Press, Cambridge (1989).
44. J. K. Spelt, E. Moy, D. Y. Kwok and A. W. Neumann, in: *Applied Surface Thermodynamics*, A. W. Neumann and J. K. Spelt (Eds), Ch. 6, pp. 293–332. Marcel Dekker, New York (1996).
45. H. Y. Erbil and R. A. Meriç, *Colloids Surfaces* **33**, 85–97 (1988).
46. J. Todd, *Basic Numerical Mathematics, Vol. 1. Numerical Analysis*. Academic Press, New York (1978).
47. L.-H. Lee, *Langmuir* **12**, 1681–1687 (1996).
48. W. V. Chang and X. Qin, Paper presented at ACS San Francisco Meeting, San Francisco, California, April 1992.
49. X. Qin and W. V. Chang, *J. Adhesion Sci. Technol.* **9**, 823–841 (1995).
50. X. Qin and W. V. Chang, *J. Adhesion Sci. Technol.* **10**, 963–989 (1996).
51. W. V. Chang and X. Qin, *J. Adhesion Sci. Technol.* (in press).
52. S. Wu, *Polymer Interface and Adhesion*. Marcel Dekker, New York (1982).
53. R. E. Johnson, Jr. and R. H. Dettre, *J. Phys. Chem.* **68**, 1744 (1964).
54. R. E. Johnson, Jr. and R. H. Dettre, in: *Surface and Colloid Science*, E. Matijevic (Ed.), Vol. 2, pp. 85–154. John Wiley, New York (1969).
55. K. Grundke, H. J. Jacobasch, F. Simon and S. Schneider, *J. Adhesion Sci. Technol.* **9**, 327–350 (1995).
56. F. M. Fowkes, *Ind. Eng. Chem.* **56** (12), 40–52 (1964).
57. G. E. H. Hellwig and A. W. Neumann, *Proceedings of the V International Congress on Surface Active Substances*, Sec. B, p. 687. Barcelona (1968).
58. B. Janczuk and T. Bialopiotrowicz, *J. Colloid Interface Sci.* **140**, 362–372 (1990).
59. A. W. Neumann and D. Renzow, *Z. Phys. Chem. (Frankfurt)* **68**, 11 (1969).

60. F. A. Cotton and J. Wilkinson, *Advanced Inorganic Chemistry*. John Wiley, New York (1967).
61. R. E. Baier and A. E. Meyer, in: *Interfacial Phenomena and Bioproducts*, J. L. Brash and P. Wojciechowski (Eds), pp. 85–121. Marcel Dekker, New York (1996).

Apparent and Microscopic Contact Angles, pp. 209–227
J. Drelich, J. S. Laskowski and K. L. Mittal (Eds)
© VSP 2000.

The gap between the measured and calculated liquid–liquid interfacial tensions derived from contact angles

LIENG-HUANG LEE *

*Consultant for Adhesion Science and Polymer Surface Chemistry,
796 John Glenn Blvd., Webster, NY 14580, USA*

Received in final form 18 May 1999

Abstract—We present our new findings about the causes of discrepancies between the measured and calculated liquid–liquid interfacial tensions derived from contact angles. The calculated ones are based on either the equation developed by Fowkes or that by van Oss, Chaudhury and Good (VCG), while the measured ones are based on the sessile drop, weight–volume by Jańzuk *et al.* and the axisymmetric drop shape analysis (ADSA) by Kwok and Neumann. Indeed, there are deviations between the calculated and measured results. For an immiscible liquid–liquid or liquid–solid interface, we prefer to employ Harkins spreading model, which requires the interfacial tension to be constant. However, for the initially immiscible liquid–liquid pairs, we propose an adsorption model, and our model requires the interfacial tension to be varying and the surface tensions of bulk liquids at a distance from the interface to remain unchanged. Thus, the difference between the initial and final interfacial spreading coefficients (S_i) equals the equilibrium interfacial film pressure $(\pi_i)_e$. According to our findings, the calculated interfacial tension represents the initial value $(\gamma_{12})_0$, which differs from the equilibrium value $(\gamma_{12})_e$ obtained experimentally after some time delay. This expected gap at a reasonable time frame is chiefly caused by the equilibrium interfacial film pressure between the two liquids. The initial (or calculated) interfacial tension can be positive or negative, while the equilibrium (or measured) one can reach zero. In fact, the former is shown to have more predictive value than the latter. A negative initial interfacial tension is described to favor miscibility or spontaneous emulsification but it tends to revert to zero instantaneously. Thus, a miscible liquid mixture should have zero interfacial tension. In response to recent papers by Kwok *et al.*, we show that the disagreements between the calculated and measured interfacial tensions are definitely not caused by the failure of the VCG approach. Correct interfacial tensions are calculated for liquid pairs containing formamide or dimethyl sulfoxide (DMSO) by using the dispersion components cited in Fowkes *et al.*'s later publication. With the corrected surface tension components, the equilibrium interfacial film pressures $(\pi_i)_e$'s for at least 34 initially immiscible liquid pairs have been calculated. These values are generally lower than the corresponding spreading pressures π_e's obtained by others using the Harkins model. Recently, we established a relationship between these two film pressures with the Laplace equation and found a new criterion for miscibility to be $(\pi_i)_e = \pi_e$.

---

*Honorary Professor, Chinese Academy of Sciences.

Key words: Axisymmetric drop shape analysis; interfacial film pressure; interfacial tension; Lewis acid–base; sessile drop; spreading coefficient; spreading pressure; surface tension; weight–volume.

1. INTRODUCTION

The determination of liquid–liquid interfacial tension has been used to test various theories for interfacial tension [1]. Recently, Kwok and Neumann [2, 3] have determined the interfacial tensions for a series of liquids with the axisymmetric drop shape analysis (ADSA) method and they found that there were deviations between the calculated and measured results. Consequently, they questioned the validity of the acid–base approach developed by van Oss, Chaudhury and Good (VCG) [4, 5]. The purpose of this paper is to show that these deviations are to be anticipated and that the VCG approach is theoretically sound. In particular, we will demonstrate that for initially immiscible liquid pairs, the equilibrium interfacial film pressure, $(\pi_i)_e$, is the cause of the anticipated deviation [6]; and this surface-chemical factor has never been considered before.

2. THEORETICAL BACKGROUND

2.1. Equilibrium spreading pressure

Bangham and Razouk [7] first observed the effect of the vapor of the liquid adsorbed on the solid on its surface tension. They have indicated that the equilibrium spreading (or film) pressure, π_e, of the adsorbed layer on the solid surface tends to lower the (equilibrium) work of adhesion, W_A, of the solid from the initial work of adhesion, W_{Ao}, which might be determined in the absence of a vapor or *in vacuo*. Thus,

$$\pi_e = W_{Ao} - W_A. \tag{1}$$

π_e can be determined with the Gibbs adsorption isotherm equation [7] by the integration of the surface coverage of the adsorbate (or surface excess) Γ, in moles per unit area, versus the logarithm of its partial pressure:

$$\pi_e = RT \int_0^{p_o} \Gamma \, \mathrm{d}\ln p, \tag{2}$$

where R is the gas constant, T is the temperature, and p_o is the saturated vapor pressure of the liquid at the same temperature. They used the Boltzmann constant κ instead of R for molecular quantities.

2.2. Equilibrium spreading pressure and surface tension of solid

The relationship of the surface tensions of both phases and the interfacial tension affects the wetting and spreading of a liquid on another liquid or a solid. In the beginning, Marangoni's law [8] states that spreading occurs if the surface tension of the underlying liquid surpasses the sum of the tensions of the liquid of the drop and of the interface of two liquids. Later, Harkins and Feldman [9] defined the equilibrium spreading coefficient, S_e, for the spreading of the liquid on a solid surface in equilibrium with the vapor of the liquid:

$$S_e = \gamma_{sv} - (\gamma_{lv} + \gamma_{sl}),\qquad(3)$$

where γ is the surface tension and the subscripts s, l, and v represent the solid, liquid, and vapor phases, respectively. It is important to note that a liquid tends to spread on a solid when S_e is positive, provided that other surface conditions, such as surface roughness, are favorable. In principle, the initial spreading coefficient S_o as defined in the following equation,

$$S_o = \gamma_s - (\gamma_{lv} + \gamma_{sl}),\qquad(4)$$

is always higher than the equilibrium one, and the difference is the equilibrium spreading pressure, π_e, of the vapor on the solid:

$$\pi_e = S_o - S_e = \gamma_s - \gamma_{sv}.\qquad(5)$$

Equation (5) also shows that the surface tension of a clean solid, γ_s, is reduced by the equilibrium spreading pressure to that of the solid in the presence of a vapor, γ_{sv}.

This spreading model by Harkins and Felman has also been applied to immiscible liquid–liquid systems. Some data of spreading coefficients and π_e's for organic liquids versus water (for water, π_w is commonly used) published by Harkins [10], Ottewill [11], Shewmaker *et al.* [12], and Johnson and Dettre [13] have been compiled by Hirasaki [14] (Table 1). We should point out that the calculated interfacial tensions in Table 1 are based on Fowkes equation for a dispersive or non-hydrogen-bonded liquid. Furthermore, the discrepancies between the calculated and measured interfacial tensions do not equal π_w.

2.3. Equilibrium interfacial film pressure and interfacial tension of liquids

Recently, we proposed an adsorption model [6] for the initially immiscible liquid–liquid pair. In that model, the initial and equilibrium interfacial spreading coefficients are defined as

$$(S_i)_o = \gamma_1 - \gamma_2 - (\gamma_{12})_o\qquad(6)$$

and

$$(S_i)_e = \gamma_1 - \gamma_2 - (\gamma_{12})_e,\qquad(7)$$

respectively. Here, γ_1 and γ_2 are the surface tensions of pure liquids at a distance away from the interface, and γ_{12} is assumed to vary during the equilibrium process. In this case, $(\pi_i)_e$, which is the equilibrium interfacial film pressure, equals the

Table 1.

Interfacial tensions, spreading coefficients, and equilibrium spreading pressures for apolar liquid–water pairs (in mN m^{-1}) at 20 °C

Nonpolar liquid	Surface tension		Interfacial tension (calcd.) (exptl.)		Spreading coefficient		Equilibrium spreading pressure*
	γ	γ^d	γ_{12}	γ_{12}	$(S)_o$	$(S)_e$	π_w
Hexane	18.5	18.5	51.2	50.7	3.25	−0.24	3.49 (a)
Heptane	20.3	20.3	51	51.2	2.15	0.39	1.76 (a)
Octane	21.8	21.8	51	51.5	0.15	−0.85	1 (a)
Decane	23.9	23.9	51.1	52	−3.4	−3.9	0.5 (c)
Dodecane	25.1	25.1	51.1	52.8	−5.41	−5.41	< 0 (c)
Tetradecane	25.6	25.6	51	52.2	−7.11	−7.11	< 0 (c)
Cyclohexane	25.5	25.5	51.1	50.2	−2.95	−4.16	1.21 (a)
Benzene	28.85	28.9	34.9	33.9	8.9	−1.63	10.53 (a)
Ethylbenzene	29.2	29.2	38.3	38.4	6.2	0.2	6 (b)
Propylbenzene	28.5	28.5	39.8	39.6	5.1	0	5.1 (b)
n-Butylbenzene	28.7	28.7	40.4	41.4	3.8	0.2	3.6 (b)
Toluene	28.5	28.5	36.1	36.1	8.15	0.17	7.98 (a)
o-Xylene	30.1	30.1	36.1	36.1	6.55	0.9	5.65 (a)
Chloroform	27.15	27.1	31.6	31.6	14	1	13 (a)
Carbon tetrachloride	26.95	27	45	45	0.8	−1.94	2.74 (a)
Carbon disulfide	32.2	32.2	48.4	48.4	−7.85	−9.2	1.36

The majority of the above data were compiled by Hirasaki [14]. The interfacial tensions were calculated using only dispersive liquids with the Fowkes equation. Spreading pressures by (a) Ottewill at 20 °C [11], (b) Shewmaker *et al.* [12], and (c) Johnson and Dettre at 24.5 °C [13].

difference between the initial interfacial tension $(\gamma_{12})_o$ and the equilibrium one $(\gamma_{12})_e$ for the liquid pair:

$$(\pi_i)_e = (S_i)_o - (S_i)_e = (\gamma_{12})_o - (\gamma_{12})_e. \qquad (8)$$

This relation is fundamentally important but not recognized by many. We will illustrate the relevance of this relation to the discrepancies in interfacial tensions

In the laboratory, the interfacial tensions determined without equilibration can be generally described with a curve similar to the one obtained by Jasper *et al.* [15] for the apparent interfacial tension of benzene–water as a function of time (Fig. 1). First, the initial interfacial tension $(\gamma_{12})_o$ decreases with time until it reaches the equilibrium value $(\gamma_{12})_e$ after several minutes to an hour. Thus, the deviation of the calculated (or initial) interfacial tension from the measured (or equilibrium) value is anticipated theoretically, and the deviation has been shown in equation (8) to be related to the equilibrium interfacial film pressure. For an initially immiscible liquid–liquid system, one is dealing with a changing interface. In reality, an interphase instead of an interface is formed at equilibrium, due to mutual dissolution/diffusion between the two phases in contact. Thus, a prolonged equilibration may give rise to erroneous readings on interfacial tension. This discrepancy

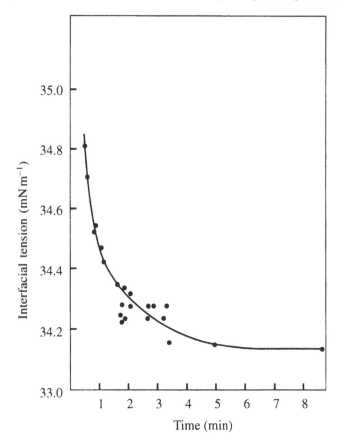

Figure 1. The apparent interfacial tension at the benzene–water interface as a function of time (drop–volume method; 25°C) [15].

may be exemplified if the liquid mixture contains an impurity. In practice, the common effect of an impurity is the suppression of the interfacial tension right from the beginning.

2.4. Calculated and measured liquid–liquid interfacial tensions

The initial interfacial tension can be calculated based on either the Fowkes equation [16] or the VCG methodology [4, 5]. Following Fowkes' formulation, the calculated interfacial tension at a dispersive, or non-hydrogen-bonded, liquid–liquid interface is

$$(\gamma_{12})_0 = \gamma_1 + \gamma_2 - 2(\gamma_1^d \gamma_2^d)^{1/2}. \tag{9}$$

The calculated interfacial tension in equation (9) is identical to the initial interfacial tension prior to the initiation of spreading of the liquid pair. However, for an initially immiscible liquid mixture at equilibrium, the effect of vapor adsorption has to be taken into account, and then the measured (equilibrium) interfacial tension becomes

$$(\gamma_{12})_e = \left[\gamma_1 + \gamma_2 - 2\left(\gamma_1^d \gamma_2^d\right)^{1/2}\right] - (\pi_i)_e. \tag{10}$$

The VCG methodology [4, 5] defines the calculated (initial) interfacial tension $(\gamma_{12})_o$ for a liquid pair capable of hydrogen-bonding interaction as

$$(\gamma_{12})_o = \gamma_1 + \gamma_2 - 2\left(\gamma_1^{LW} \gamma_2^{LW}\right)^{1/2} - 2\left(\gamma_1^+ \gamma_2^-\right)^{1/2} - 2\left(\gamma_1^- \gamma_2^+\right)^{1/2}. \tag{11}$$

Similar to equation (10), the measured (equilibrium) interfacial tension should be

$$(\gamma_{12})_e = \left[\gamma_1 + \gamma_2 - 2\left(\gamma_1^{LW} \gamma_2^{LW}\right)^{1/2} - 2\left(\gamma_1^+ \gamma_2^-\right)^{1/2} - 2\left(\gamma_1^- \gamma_2^+\right)^{1/2}\right] - (\pi_i)_e. \tag{12}$$

In both cases, the deviation between the calculated and measured interfacial tensions is the equilibrium interfacial film pressure. In the presence of vapor adsorption, we prefer to use the dispersion component d instead of the Lifshitz–van der Waals component LW because the former is not affected by the induction and polarization interactions that are now contributing to $(\pi_i)_e$.

2.5. Equilibrium interfacial film pressure and the miscibility of liquids

In principle, the requirements for two liquids to be miscible are similar to those for spontaneous emulsification. In emulsion technology [17], a pressure gradient $\pi_{\bar{g}}$ is assigned to be the film pressure before the curvature [18], in contrast to π_e after the curvature, or on the concave side of the interface. The following equation has been used to predict the effect of $\pi_{\bar{g}}$ on the formation of a micro-emulsion or macro-emulsion at a water–oil interface containing a fraction of alcohol in the oil phase:

$$\gamma_\phi = (\gamma_{o/w})_a - \pi_{\bar{g}}, \tag{13}$$

where $(\gamma_{o/w})_a$ is the interfacial tension at the water–oil interface and γ_ϕ is the total potential (minimum) interfacial tension corresponding to $\pi_{\bar{g}}$ at the water–oil interface in the absence of a stabilizing agent. It is noteworthy that the interfacial tension prior to adsorption at the interface is always larger than the equilibrium one after the adsorption. When $\pi_{\bar{g}} > (\gamma_{o/w})_a$, a negative interfacial tension results and the energy $-\gamma_\phi \, dA$ (A = surface area) would be available to increase the total interfacial area. This has been considered to be a condition for the formation of a stable micro-emulsion with droplet diameters of the order of 8–80 nm. Otherwise, when $\pi_{\bar{g}} < (\gamma_{o/w})_a$, an unstable macro-emulsion forms and the droplet diameters are of the order of magnitude of 1 μm. The interface region corresponding to drop diameters between 1.0 and 0.01 μm is approximately 2.5 nm thick. To restore to equilibrium, a phase separation would soon follow.

Analogously, for the initially immiscible liquid–liquid system, we can reformulate equation (13) for an emulsion as equation (8) for an initially immiscible liquid pair:

$$(\gamma_{12})_e = (\gamma_{12})_o - (\pi_i)_e.$$

Here, $(\pi_i)_e$ is equivalent to $\pi_{\bar{g}}$ in the emulsion case.

Now let us discuss the negative interfacial tension. The lowest interfacial tension at equilibrium is indeed zero. However, the initial (or calculated) interfacial tension can be positive or negative. A negative initial interfacial tension [18] can exist because it is transient in nature and it would lead to a spontaneous increase in the surface area such as with the formation of a micro-emulsion. Here, equilibrium would be attained when the negative interfacial tension reverts to zero or a small positive value by the uncrowding of the interfacial molecules and the resulting loss of pressure at the interface. This could be the reason why at equilibrium, so far, no one has obtained a negative interfacial tension even with the best experimental technique available.

Thus, the initial interfacial tension may have a better predictive value than the equilibrium one because it can be negative. A negative or a very low initial interfacial tension signifies the instability of the system. In two-component systems, zero or negative interfacial tension implies miscibility corresponding to no interfacial tension. The surface area would continue increasing until two phases are mutually dissolved.

It is important to point out that the equilibrium interfacial film pressure, $(\pi_i)_e$, defined in our model [6] is identical to $\pi_{\bar{g}}$ in equation (13). Recently, we found a relationship between the equilibrium spreading pressure π_e and $(\pi_i)_e$ by the use of the Laplace equation:

$$\Delta P = \pi_e - (\pi_i)_e = (\gamma_{12})_e(1/R_1 + 1/R_2),\qquad(14)$$

where ΔP is the gradient between the equilibrium spreading pressure and the equilibrium interfacial film pressure and R_1 and R_2 are principal radii of curvature of the interface. Thus, when ΔP equals zero, $\pi_e = (\pi_i)_e$; as a result, the equilibrium interfacial tension equals zero and the liquids become miscible. This is our newly found criterion for misciblity [6].

In the following sections, we suggest some explanations for the gap between the calculated and measured interfacial tensions of liquids. In particular, we will briefly discuss the effect of $(\pi_i)_e$ on the interfacial tensions of initially immiscible liquid pairs.

3. ANALYSIS OF LITERATURE DATA AND DISCUSSION

3.1. Effect of the impurity of liquids on the equilibrium interfacial tension

In recent years, the experimental technique, namely, ADSA-P (axisymmetric drop shape analysis-profile), for determining liquid–liquid interfacial tension and contact angle developed by Cheng *et al.* [19] has gained acceptance. The accuracy of the experimental interfacial tension is claimed to be better than 0.01 mN m^{-1}. Recently, Kwok *et al.* [3] published interfacial tension results for a number of liquid–liquid

Table 2.
Comparison between the calculated and measured interfacial tensions for liquid–liquid pairs (in mN m^{-1})

Liquid–liquid pair	Calcd. (VCG)	Exptl. (ADSA-P)		Exptl. (sd)		Exptl. (wv)	
	γ_{12}	γ_{12}	%	γ_{12}	%	γ_{12}	%
Water/decane	51.1	51.07	0	51.3	0.4	51.1	0
Water/dodecane	51.1	—	—	50.48	1.2	51.12	0
Water/tetradecane	51.1	—	—	49.95	2.3	51.2	0.2
Water/hexadecane	51.3	52.24	−1.8	51.28	0	51.22	0.2
Water/diio	57.1	—	—	48.62	14.7	49.08	16.3
Water/bromo	55.4	37.95	45	42.9	22.6	41.06	25.9
Water/benzene	34.9	—	—	33.62	3.8	33.51	4.1
Gly/decane	30.9	28.03	10.2	30.96	−0.2	30.32	1.9
Gly/dodecane	30.6	24.74	23.7	30.25	1.2	30.81	−0.7
Gly/tetradecane	30.5	26.4	15.5	30.77	−0.9	31.4	−2.9
Gly/hexadecane	30.3	31.91	−5	31.66	−4.3	31.81	−4.8
Gly/diio	31.7	21.7	46.1	24.26	30.7	24.86	27.5
Gly/bromo	30.7	16.78	83.0	21.15	45.2	21.78	41
Gly/benzene	23.7	—	—	18.91	25.3	19.89	19.2
Form/decane	20.9	27.73	−24.6	28.33	−26.2	27.36	−23.6
Form/dodecane	20	26.1	−23.4	28.22	−29.1	28.4	−29.5
Form/tetradecane	20.2	24.82	−18.6	28.59	−29.3	29.14	−30.7
Form/hexadecane	20	27.25	−26.6	27.4	−27.0	28.7	−30.3
Form/diio	19.8	16.73	18.4	21.75	−7.0	21.02	−5.8
Form/bromo	19.2	16.56	15.9	19.49	−1.5	19.46	−1.3
Form/benzene	14.8	—	—	15.75	−6	15.4	−3.9
Ethylgly/decane	19.3	17.89	7.9	17.92	7.7	16.85	14.5
Ethylgly/dodecane	19.1	17.3	10.4	18	5	17.88	6.8
Ethylgly/tetradecane	17.4	—	—	17.84	−2.5	17.38	−4.1
Ethylgly/hexadecane	19	18.1	5	18.95	10.3	18.52	14.5
Ethylgly/diio	22	10.84	103	13.85	58.8	14.44	52.4
Ethylgly/bromo	20.6	9.72	111.9	10.91	88.8	11.16	84.6
Ethylgly/benzene	14.5	—	—	8.21	77	7.04	106
DMSO/decane	9.3	9.2	1	—	—	—	—
DMSO/dodecane	8.9	11.03	−19	—	—	—	—
DMSO/tetradecane	8.7	11.13	−22	—	—	—	—
DMSO/hexadecane	8.6	11.53	−25	—	—	—	—
DMSO/diio	9.3	0	—	—	—	—	—
DMSO/bromo	8.4	0	—	—	—	—	—

The experimental interfacial tensions were determined at 23°C by Kwok *et al.* [2, 3] with the ADSA-P method, and at 20°C by Jańczuk *et al.* [20] with the sessile drop (sd) and weight–volume (wv) methods. The calculated interfacial tensions were based on van Oss *et al.*'s (VCG) method [4, 5]. The deviation for each measurement is the difference in percent between the measured and calculated interfacial tensions. Liquids: gly, glycerol; form, formamide; ethylgly, ethylene glycol; diio, diiodomethane; bromo, α-bromonaphthalene; and DMSO, dimethyl sulfoxide.

systems. On the surface, their results seriously challenge the validity of the VCG methodology and, perhaps, the surface tension components concept.

First, we compare some of Kwok *et al.*'s results [2, 3] with an earlier parallel work by Jańczuk *et al.* [20]. Strangely, Jańczuk *et al.*'s work has never been cited by Kwok and Neumann. In comparison, Jańczuk *et al.*'s work was carried out at $20 \pm 0.1°C$, while Kwok *et al.*'s experiments were at $23°C$. Unlike the claimed high accuracy for the ADSA-P method, the standard deviations in Jańczuk *et al.*'s work were only ± 0.7 mN m^{-1} for the sessile drop method and ± 0.6 mN m^{-1} for the weight–volume method. In both cases (Kwok *et al.* and Jańczuk *et al.*), the measurements of liquid–liquid interfacial tensions were made after equilibration overnight. We do not know whether this is the correct procedure for achieving equilibration. Generally, it takes several minutes to an hour, as shown in Fig. 1, to achieve equilibrium. Under the long duration of equilibration, a trace amount of impurity can cause a substantial lowering of the liquid–liquid interfacial tension. Furthermore, after equilibration no one will ever be able to observe, if at all possible, any initial negative interfacial tension.

When Jańczuk *et al.*'s data are compared side-by-side with Kwok *et al.*'s data in Table 2, it is apparent that Kwok *et al.*'s data, especially in nonaqueous systems, are very much lower than those obtained by Jańczuk *et al.* with either the sessile drop or the weight–volume method. Through our recent communication, Jańczuk [21] informed us that all of their commercial liquids were further purified. They determined the purity of the liquids, among other things, by chromatographic methods. They found that for the polar liquid–apolar liquid systems the interfacial tensions were in many cases considerably higher than when the commercial liquids without additional purification were used.

Kwok *et al.* did not mention the purification step and appeared to have used all liquids as received; the purity of their liquids was reported according to commercially supplied literature. Also, they determined the density of some liquids saturated by other liquids but did not determine the density of the bare liquids as received; therefore, it is difficult to establish the influence of the saturation of one liquid by another on the density of liquids.

Jańczuk *et al.* who used the apparatus of Krüss made in Germany to measure interfacial tensions by the pendant drop, seriously doubted the accuracy of ADSA-P to be better than 0.01 mN m^{-1}, as claimed by Kwok *et al.* Since Jańczuk *et al.*'s results are generally higher than Kwok *et al.*'s, we believe that Jańczuk *et al.*'s data are more accurate solely based on the purity of their liquids. Of their two methods, Jańczuk [21] believes that the sessile drop method is better than the weight–volume method for interfacial tension measurements, except that it is difficult to choose a plate for liquid pairs for which the contact angle is higher than $90°$.

3.2. Effect of the dispersion component on the calculated interfacial tensions of liquid pairs containing formamide or dimethyl sulfoxide

According to our earlier discussion in this paper on the interfacial tension of liquids, the initial (or calculated) values, except the initial negative interfacial tensions, should be higher than the measured ones and the difference in each set of data is $(\pi_i)_e$. However, after we examined the data in Table 2, we noticed that there were serious problems with formamide and DMSO. Nearly all calculated interfacial tensions with the original γ_l^{LW} components (Table 3) were lower than the corresponding measured values. This is contrary to Gibbs thermodynamics [22]. If it were true, the interfacial film pressure should have been negative, and this is definitely incorrect.

After some calculations, we discovered that the LW components of these two liquids could be the culprit. The original VCG value of γ_l^{LW} for formamide (Table 3) was after Fowkes' original dispersion value γ_l^d [23] of 39 ± 7 mN m^{-1} and it is definitely too high. This higher value apparently resulted in lower calculated interfacial tensions for formamide–other liquid pairs. After we chose the lower dispersion component γ_l^d of 28 mN m^{-1} later used by Fowkes *et al.* [24], it seemed to have solved this particular problem. Similarly, we adopted from the same source the γ_l^d value of 29 mN m^{-1} for DMSO, instead of 36 mN m^{-1} used by VCG. With these lower values of γ_l^d, we recalculated the interfacial tensions for liquid pairs containing formamide or DMSO and the revised values are shown in Table 4. Finally, after the changes, the calculated interfacial tensions for liquid pairs containing formamide or DMSO were actually higher than the measured values obtained by all three methods. We hope that at least this correction might help to resolve one of the controversies surrounding the VCG approach.

Table 3.
Surface tension components for probe liquids (in mN m^{-1}) at 20°C (without equilibrium spreading pressure). (Reference values for water: $\gamma^+ = 34.2$ mN m^{-1}; $\gamma^- = 19$ mN m^{-1})

Liquid	γ	γ^{LW}	γ^{AB}	γ^+	γ^-
Water	72.8	21.8	51	34.2	19
Glycerol	64	34	30	5.3	42.5
Formamide	58	39	19	3.1	29.1
Diiodomethane	50.8	50.8	≈ 0	0	0
Ethylene glycol	48	29	19	2.6	34.8
α-Bromonaphthalene	44.4	43.5	≈ 0	0	0
Dimethyl sulfoxide	44	36	8	0.7	23.8

For the conversion, the ratios for γ^+ (H$_2$O)/γ^+ (Gly), γ^+ (H$_2$O)/γ^+ (Form), γ^+ (H$_2$O)/γ^+ (EG), and γ^+ (H$_2$O)/γ^+ (DMSO) were kept at 6.5, 11, 13, and 51, respectively.

Table 4.
Recalculated and measured interfacial tensions for liquid pairs containing formamide or dimethyl sulfoxide (in mN m^{-1}). (Dispersion components: formamide, 28 mN m^{-1}; DMSO, 29 mN m^{-1})

Liquid–liquid pair	Calcd. (this paper)	Exptl. (ADSA-P) (23°C)		Exptl. (sd) (20°C)		Exptl. (wv) (20°C)	
	γ_{12}	γ_{12}	%	γ_{12}	%	γ_{12}	%
Form/decane	30.1	27.73	8.5	28.33	6.4	27.36	19
Form/dodecane	30	26.1	14.9	28.22	6.3	28.4	5.6
Form/tetradecane	30	24.82	20.9	28.59	4.9	29.14	3
Form/hexadecane	30	27.25	10.1	27.4	9.5	28.7	4.5
Form/diio	33.4	16.73	99.6	21.75	53.6	21.02	37.1
Form/bromo	32.6	16.56	103	19.49	63.3	19.46	67.5
Form/benzene	16.6	—	—	15.75	5.4	15.4	7.8
DMSO/decane	15.3	9.2	66.3	—	—	—	—
DMSO/dodecane	15.1	11.03	37.2	—	—	—	—
DMSO/tetradecane	15.1	11.13	36	—	—	—	—
DMSO/hexadecane	15	11.53	30.4	—	—	—	—
DMSO/diio	17.9	0	—	—	—	—	—
DMSO/bromo	16.4	0	—	—	—	—	—

3.3. Effect of the equilibrium spreading pressure on the surface tension components of probe liquids

VCG selected several probe liquids for the contact angle measurements (Table 3). The surface tension components, e.g. γ^+, hydrogen-bond-donating and γ^-, hydrogen-bond-accepting, were determined from the contact angles of polar liquids on a reference monopolar substrate, e.g. poly(methyl methacrylate) (PMMA). They ignored the lowering of the surface tension of PMMA by vapor adsorption, due to π_e. They also assumed the ratio of γ_w^+ and γ_w^- to be unity for water as the reference; as a result, most polymers appeared to be rather basic. Later we [25–27] found that the correct ratio should be 1.8 (Table 3). On the basis of these data, the surface tension components, in the presence of a vapor, for many polymers and biopolymers have been recalculated [27]. However, in general, most polymers still appear to be more basic than they should be, partially because these surface tensions are γ_{sv} obtained in the presence of a vapor and not the initial surface tensions, γ_s, obtained in the absence of a vapor [6]. One of the typical examples is poly(vinyl chloride). This polymer is expected to be intrinsically acidic, but in the presence of a vapor the γ_{sv} components, γ_{sv}^+ and γ_{sv}^-, signify this polymer to be basic. Our use of the new ratio for water as the reference (Table 3) raised the acidity scale slightly but not enough to render poly(vinyl chloride) acidic. Thus, we suspected that there could be some other problem associated with the VCG approach.

The breakthrough came after we considered the vapor adsorption on PMMA to be substantial and attempted to use the initial work of adhesion relationship by including π_e. There have been several reports claiming that vapors of both

Table 5.
Surface tension components for probe liquids (in $mN\,m^{-1}$) at 20°C (with equilibrium spreading pressure). (Reference values for water: $\gamma^+ = 34.2\,mN\,m^{-1}$; $\gamma^- = 19\,mN\,m^{-1}$)

Liquid	γ	γ^d	γ^h	γ^+	γ^-	$\theta°$ and π_e on PMMA	
Water	72.8	21.8	51	34.2	19	67	23 [31]
Glycerol	64	34	30	17.4	12.9	62	27.3 [31]
Formamide	58	28	30	22.4	10.1	54	31.7 [31]
Diiodomethane	50.8	50.8	≈0	0	0	35.5	31.3 [29]
Ethylene glycol	48	29	19	15.1	6	52	36[a]
α-Bromonaphthalene	44.4	43.5	≈0	0	0	9.5	27.2 [29]
Dimethyl sulfoxide	44	29	15	2	28	16	38 [31]

The dispersion components of surface tension for liquids are based on Fowkes' earlier data. New data on formamide and DMSO are derived from the γ^d values later used by Fowkes *et al.* [24]. The equilibrium spreading pressures were determined with an error of $\pm 12\,mN\,m^{-1}$ by Bellon-Fontaine and Cerf [31]; thus, we used the lower bound of their results as shown in the above table. The data on diiodomethane and α-bromonaphthalene were obtained by Tamai *et al.* [29].

[a] The data on ethylene glycol were obtained from the extrapolation of π_w versus the surface tensions of liquids.

The contact angles for water and DMSO on PTFE were 105° and 78° and the equilibrium spreading pressures of water and DMSO on PTFE were 0.2 and 11 $mN\,m^{-1}$, respectively.

polar [28–31] and apolar liquids [32, 33] can adsorb on PMMA. Furthermore, the adsorption was found to be rather substantial, resulting in relatively high π_e's. Bellon-Fontaine and Cerf [31] examined the adsorption of several liquids on four polymers including PMMA. They indicated that their results had an experimental error of $\pm 12\,mN\,m^{-1}$, partly because they used mercury as the reference for the initial work of adhesion, W_{Ao}. In comparison with the data obtained by other workers [28, 29, 32], their results do appear somewhat high, especially for PMMA and PE; so we used, with caution, the lower bound of their π_e's for our calculations. For example, the lower bound of π_w for PMMA is 23 $mN\,m^{-1}$, which is comparable to the values of 26 $mN\,m^{-1}$ obtained by Busscher *et al.* [30] and 18.6 $mN\,m^{-1}$ by Erbil [32]. For other liquids, the lower bound π_e values are 31.7 $mN\,m^{-1}$ for formamide and 27.3 $mN\,m^{-1}$ for glycerol. The value of 36 $mN\,m^{-1}$ for ethylene glycol was obtained through extrapolation. We also used Tamai *et al.*'s π_e values [29] of 27.2 $mN\,m^{-1}$ for α-bromonaphthalene and 31.3 $mN\,m^{-1}$ for diiodomethane (Table 5), respectively. We believe that so far, all the π_e values listed in Table 5 are rather reasonable.

In order to compensate for the effect of vapor adsorption, we employed equation (15) to determine the ratio of γ^+ of water and γ^+ of the other liquid 1 by including π_e's in the equation:

$$\frac{\gamma_w^+}{\gamma_1^+} = \left[\frac{\gamma_w(1 + \cos\theta_{w/s}) + \pi_w - 2(\gamma_w^d\gamma_s^d)^{1/2}}{\gamma_1(1 + \cos\theta_{1/s}) + \pi_1 - 2(\gamma_1^d\gamma_s^d)^{1/2}} \right]^2. \tag{15}$$

The calculated surface tension components are shown in Table 5. Since DMSO is a base, a similar equation was used to calculate the ratio of γ_w^-(water)$/\gamma_l^-$ of DMSO on PTFE, not on PMMA, based on contact angles and π_e reported in ref. [31]. The calculated results for DMSO are also included in Table 5. More refined values based on reasonably accurate spreading pressure data are needed for future studies.

In comparison with Table 3, the surface tension components of the liquids in Table 5 are listed as d for dispersion, h for hydrogen-bonding, instead of AB for the broader Lewis acid–base, γ^+ and γ^-, and π_e for the equilibrium spreading pressure resulting from the combined induction and polarization interactions. It is important to note that π_e, not γ^d as defined, depends on polar interactions (polarization and induction) between a solid and the vapor of a liquid, and π_e for a solid varies from liquid to liquid. We have used the dispersion components (not affected by π_e) cited by Fowkes *et al.* in their later paper [24]. In general, after recalculation, the liquids, e.g. glycerol, formamide, and ethylene glycol, appear to be rather acidic, and γ_l^+ and γ_l^- for these liquids are now also comparable to the α and β parameters of the linear free energy relationship [6].

It is important to note that by using the surface tension components in Table 5, for the first time we are able to show that poly(vinyl chloride) in the absence of vapor adsorption is rather acidic. The initial surface tension γ_s for PVC is 50.6 mN m^{-1} instead of γ_{sv} of 43.8 mN m^{-1}; the γ_s^+ value is calculated to be 7.3 mN m^{-1}, instead of γ_s^+ of 0.04 mN m^{-1}; and γ_s^- is 2 mN m^{-1}, instead of γ_{sv}^- of 3.5 mN m^{-1}. Here, γ_s for PVC can also be obtained by the sum of π_w (6.5 mN m^{-1}) and γ_{sv} (43.8 mN m^{-1}) to give a value of 50.3 mN m^{-1}. Both values are in excellent agreement. Thus, it is clear that PVC is intrinsically acidic, and there is nothing wrong with the VCG methodology. Since the polarity of PVC has been used as a test for the VCG methodology, we are now convinced that the VCG methodology appears sound after modifications. We hope that with our proposed minor modifications, by including π_e the VCG methodology can be broadly used in conjunction with other methods for the wetting thermodynamics. For further demonstration, we will use the values in Table 5 for our calculations in the following sections.

3.4. Derivation of the equilibrium interfacial film pressure from the liquid–liquid interfacial tension

In Table 1, the equilibrium spreading pressure of immiscible liquids in the presence of water vapor is expressed as π_w. For alkanes, the π_w values are low; however, for aromatics the π_w values are high for *o*-xylene (5.65 mN m^{-1}), toluene (7.98 mN m^{-1}), and benzene (10.39 mN m^{-1}). For halogenated compounds, the π_w values are somewhat high for carbon tetrachloride (2.74 mN m^{-1}) and very high for chloroform (13 mN m^{-1}). The interfacial tensions are calculated based on the original Fowkes equation for a dispersive liquid. It is important to point out that in Table 1 the differences between the calculated and the measured interfacial tensions for some immiscible liquids do not equal the equilibrium spreading pressures. The discrepancies are especially large for aromatics and halogenated compounds.

Table 6.

Equilibrium interfacial film pressures derived from interfacial tensions for liquid–liquid pairs (in mN m^{-1})

Liquid–liquid pair	Calcd. γ_{12}	Exptl. (ADSA-P) $(\pi_i)_e$	Exptl. (sd) $(\pi_i)_e$	Exptl. (wv) $(\pi_i)_e$
Water/decane	51.1	0	−0.2	0
Water/dodecane	51.1	—	0.6	0
Water/tetradecane	51.1	—	1.2	−0.1
Water/hexadecane	51.3	−0.9	0	0.1
Water/diio	57.1	—	8.5	8
Water/bromo	55.4	17.6	12.5	14.3
Water/benzene	34.9	—	1.3	1.4
Gly/decane	30.9	2.9	−0.1	0.6
Gly/dodecane	30.6	5.9	0.4	−0.2
Gly/tetradecane	30.5	4.1	−0.3	−0.9
Gly/hexadecane	30.3	−1.6	−1.4	−1.5
Gly/diio	31.7	10	7.4	6.8
Gly/bromo	30.7	13.9	9.6	8.9
Gly/benzene	23.7	—	4.7	3.8
Form/decane	30.1	2.4	1.8	2.7
Form/dodecane	30	3.9	1.8	1.6
Form/tetradecane	30	5.2	1.4	0.9
Form/hexadecane	30	2.8	2.6	1.3
Form/diio	33.4	16.7	11.7	12.4
Form/bromo	32.6	16	13.1	13.1
Form/benzene	16.6	—	0.9	1.2
Ethylgly/decane	19.3	1.4	1.4	2.5
Ethylgly/dodecane	19.1	1.8	1.1	1.2
Ethylgly/tetradecane	17.4	—	−0.4	0
Ethylgly/hexadecane	19	0.9	0.1	0.5
Ethylgly/diio	22	11.2	8.2	7.6
Ethylgly/bromo	20.6	10.9	9.7	9.4
Ethylgly/benzene	14.5	—	6.3	7.5
DMSO/decane	15.3	6.1	—	—
DMSO/dodecane	15.1	4.1	—	—
DMSO/tetradecane	15.1	4	—	—
DMSO/hexadecane	15	3.5	—	—
DMSO/diio	17.9	17.9	—	—
DMSO/bromo	16.4	16.4	—	—

The experimental interfacial tensions were determined by Kwok et al. [2, 3] with the ADSA-P method, and by Jańczuk et al. [20] with the sessile drop (sd) and weight–volume (wv) methods. The calculated equilibrium interfacial film pressure is the difference between the calculated and experimental interfacial tensions. Liquids: gly, glycerol; form, formamide; ethylgly, ethylene glycol; diio, diiodomethane; bromo, α-bromonaphthalene; and DMSO, dimethyl sulfoxide.

For immiscible liquid pairs or liquid–solid pairs, besides Harkins' spreading technique, there are other methods for the determination of π_e, such as vapor adsorption [28, 29], ellipsometry [30], etc. Previously, Xu *et al.* [34] estimated π_w of water on mercury from the interfacial tension to be 32 mN m^{-1} *vis-à-vis* some experimental value [10] of 25 mN m^{-1} obtained by other means. In this case, water–mercury is a typical immiscible liquid pair.

However, for the initially immiscible liquid pairs, the discrepancy in interfacial tensions is caused by the equilibrium interfacial film pressure $(\pi_i)_e$, not by the equilibrium spreading pressure π_e in the above immiscible liquid systems. Furthermore, when the calculated interfacial tensions are obtained with the revised surface tension components in Table 5, we can estimate $(\pi_i)_e$ from the difference between the correctly calculated and the accurately measured interfacial tensions.

For our calculations, the interfacial tension for a miscible liquid pair at equilibrium is assigned to be zero. Furthermore, all negative interfacial tensions were treated as transitory and reverting to zero. We calculated the $(\pi_i)_e$ values for at least 34 liquid pairs from all three methods and the results are listed in Table 6. It is apparent that the data obtained from the sessile and the weight–volume methods agree very well with each other. As recommended by Jańczuk, the sessile drop method appears to be more reliable.

In general, the magnitude of $(\pi_i)_e$'s in liquid–liquid pairs are similar to that for the equilibrium spreading pressures, π_w, listed in Table 1. In general, the π_w values are low for alkanes but high for aromatics and halogenated liquids. However, some of the $(\pi_i)_e$ values are smaller than the corresponding π_e's; for example, benzene has a $(\pi_i)_e$ of 1.3 mN m^{-1} in comparison with π_e of 10.53 mN m^{-1}. Since these two film pressures are far apart, benzene does not mix with water but forms a lens instead. Both diiodomethane and α-bromonaphthalene yield high $(\pi_i)_e$'s to all five opposing polar liquids selected for this study. In Table 6, there are several slightly negative $(\pi_i)_e$'s resulting from calculations which should be treated as zero.

3.5. Effect of the equilibrium interfacial film pressure on the miscibility of liquids

Several miscible and immiscible liquid pairs were used by Kwok *et al.* [2, 3] to test the VCG methodology. The calculated interfacial tensions based on the VCG methodology are compared with the recalculated values on the basis of our parameters for the probe liquids (Table 5). Kwok *et al.* determined interfacial tensions for immiscible liquid pairs with the ADSA-P methods. Our recalculated results (Table 7) show more miscible liquid pairs to have negative initial interfacial tensions. Our recalculated data for the water–glycerol, water–formamide, and formamide–ethylene glycol pairs yielded small but positive initial interfacial tensions. The signs for the interfacial tension for the first two pairs were reversed after the recalculation. As we have discussed, these results are for the initial interfacial tensions and a small positive value can also signify miscibility. This should not cause any concern because $(\pi_i)_e$ can make a difference, as shown in equation (8).

Table 7.

Liquid-pair miscibility versus interfacial tensions and equilibrium interfacial film pressures (in mN m^{-1}) at 20°C

Liquid–liquid pair	Calcd. (VCG) γ_{12}	Calcd. (this paper) γ_{12}	Exptl. (ADSA) γ_{12}	Miscibility	Calcd. equilibrium inter. film pre. $(\pi_i)_e$
Water/gly	−14.5	3.8	0	misc.	3.8
Water/form	−4.6	3.1	0	misc.	3.1
Water/ethylgly	−12.7	−4.5	0	misc.	0
Water/DMSO	−3.5	−7.7	0	misc.	0
Gly/form	1.4	−0.1	0	misc.	0
Gly/ethylgly	1.1	−0.9	0	misc.	0
Gly/DMSO	4.9	−9.2	0	misc.	0
Form/ethylgly	0.6	1.1	0	misc.	1.1
Form/DMSO	1.1	−14.1	0	misc.	0
DMSO/diio	9.3	16.2	0	misc.	16.2
DMSO/ethylgly	2	−14	0	misc.	0
DMSO/bromo	8.4	16.4	0	misc.	16.4
Diio/bromo	0.2	0.3	0	misc.	0.3
Bromo/decane	3.2	3.8	0	misc.	3.8
Bromo/dodecane	2.6	3.4	0	misc.	3.4
Bromo/tetradecane	2.3	3.3	0	misc.	3.3
Bromo/hexadecane	2	2.9	0	misc.	2.9
Diio/water	57	57	48.6[a]	immisc.	8.4
Diio/gly	31.7	31.7	21.7	immisc.	10
Diio/ethylgly	22	22.2	10.84	immisc.	11.4
Diio/form	33	33	16.73	immisc.	16.3
Bromo/water	55	54.6	37.95	immisc.	16.7
Bromo/gly	30.7	30.7	16.78	immisc.	13.9
Bromo/ethylgly	20.6	20.4	9.74	immisc.	10.7
Bromo/form	19.2	31.7	16.56	immisc.	15.1
Water/*trans*-decalin	51.6	51.6	37.62	immisc.	14
Water/*cis*-decalin	51.6	51.6	31.48	immisc.	20

These ADSA results were obtained by Kwok and Neumann [2, 3].

[a] Data of Jańczuk et al. [20].

Equilibrium interfacial film pressures were calculated from the difference between the calculated (this paper) and measured interfacial tensions. Liquids: gly, glycerol; form, formamide; ethylgly, ethylene glycol; DMSO, dimethyl sulfoxide; bromo, α-bromonaphthalene; and diio, diiodomethane.

In fact, both $(\pi_i)_e$ and hydrogen-bonding can drive interfacial tension to zero or negative; thus, miscibility or spontaneous emulsification of liquids results.

It should be noted that halogenated compounds have low $(\pi_i)_e$ values against apolar liquids including another halogenated one, e.g. diiodomethane versus α-bromonaphthalene. However, it is different against a polar medium. The recalculated interfacial tensions for the diiodomethane–DMSO and α-bromonaphthalene–DMSO pairs appear to be rather high, 16 mN m^{-1}. We do not know the reason for the high interfacial tensions except to speculate that DMSO itself has the highest dipole moment of 4.0 Debye [35], which could lead to significant dipole–dipole interactions with dipolar halogenated compounds. Eventually, with the effect of $(\pi_i)_e$ the initial interfacial tensions were reduced to zero, and these liquid pairs finally became miscible.

The deviations of the calculated interfacial tensions from the measured ones are rather high for the immiscible water–organic liquid pairs containing halogenated compounds and decalins [35]. As we discussed previously, all these results are within anticipation, provided that we take $(\pi_i)_e$'s into account. For these liquid pairs, the $(\pi_i)_e$ values were not high enough to lower the interfacial tensions to zero; thus, they remained immiscible.

As we have shown in equation (15), one of the criteria for miscibility [6] is the gradient of film pressures, ΔP. When ΔP is zero, $(\pi_i)_e = \pi_e$. As a result, the equilibrium interfacial tension is zero and the liquid pair dissolves. Thus, some initially immiscible liquids become miscible later.

4. CONCLUSIONS

The experimental technique ADSA-P employed by Kwok *et al.* [2, 3] was accurate without question, but the equilibration of liquid pairs overnight might not be proper, especially if an impurity is present in the liquid mixture. In comparison with Jańczuk' *et al.*'s data, Kwok *et al.*'s interfacial tensions for apolar liquid pairs appeared to be too low, presumably due to the presence of some impurities. Since Jańczuk *et al.* used purified liquids [21], their data appear to be superior to Kwok *et al.*'s results. It is important to note that equilibration also removes the opportunity for observing the initial interfacial tension.

For an immiscible liquid–liquid interface or liquid–solid interface, we employed Harkins' spreading model and we included π_e in our formulations. However, for initially immiscible liquids, we needed to apply our adsorption model. In general, the calculated (or initial) interfacial tension should be higher than the measured (or equilibrium) one and the difference is $(\pi_i)_e$. In comparison, the initial interfacial tension, which can be positive or negative, has a better predictive value because a negative initial interfacial tension can lead to miscibility or spontaneous emulsification. However, after the equilibrium is reached, the lowest interfacial tension should always be zero. This may explain why the fleeting negative interfacial tension has never been observed experimentally.

The original γ_l^{LW} components for formamide and DMSO used by VCG were found to be too high and the revised γ_l^d values for both liquids later used by Fowkes *et al.* yielded correct initial interfacial tensions for the respective liquid pairs. Indeed, there is no need to keep using the term Lifshitz–van der Waals component, especially when the equilibrium spreading pressure is taken into account. Otherwise, the VCG approach with modifications is theoretically sound.

Our study on the surface tension components finally led to the successful calculation of $(\pi_i)_e$ values for at least 34 liquid pairs and to an explanation for the miscibility of liquids with $(\pi_i)_e$ and negative interfacial tension. These results reaffirm the belief that with a correct formulation and accurate experimental data we can prove that for an initially immiscible liquid pair the gap between the calculated and measured interfacial tensions is actually the equilibrium interfacial film pressure.

Acknowledgements

I gratefully acknowledge the correspondence on experimental details with Professor Bronislaw Jańczuk, Department of Chemistry, Maria-Curie Sklodowska University, Lublin, Poland. I also thank the three reviewers for their constructive comments.

REFERENCES

1. R. E. Johnson, Jr. and R. H. Dettre, *Langmuir* **5**, 295 (1989).
2. D. Y. Kwok and A. W. Neumann, *Can. J. Chem. Eng.* **74**, 551 (1996).
3. D. Y. Kwok, Y. Lee and A. W. Neumann, *Langmuir* **14**, 1548 (1998).
4. C. J. van Oss, M. K Chaudhury and R. J. Good, *Adv. Colloid Interface Sci.* **28**, 35 (1987).
5. R. J. Good, M. K. Chaudhury and C. J. van Oss, in: *Fundamentals of Adhesion*, L. H. Lee (Ed.), p. 153. Plenum Press, New York (1991).
6. L. H. Lee, *J. Colloid Interface Sci.* **214**, 64 (1999).
7. D. H. Bangham and R. I. Razouk, *Trans. Faraday Soc.* **33**, 1459, 1463 (1937).
8. C. Marangoni, *Tipographia dei fratelli Fusi, Pavia* (1865); also *Ann. Phys. Chem. (Poggendorf)* **143**, 337 (1871).
9. W. D. Harkins and A. Feldman, *J. Am. Chem. Soc.* **44**, 2665 (1922).
10. W. D. Harkins, *The Physical Chemistry of Surface Films*. Reinhold, New York (1952).
11. R. H. Ottewill, Ph.D. Thesis, University of London (1951).
12. J. E. Shewmaker, C. E. Vogler and E. R. Washburn, *J. Phys. Chem.* **58**, 945 (1954).
13. R. E. Johnson, Jr. and R. H. Dettre, *J. Colloid Interface Sci.* **21**, 610 (1966).
14. G. J. Hirasaki, in: *Contact Angle, Wettability and Adhesion*, K. L. Mittal (Ed.), p. 183. VSP, Utrecht, The Netherlands (1993).
15. J. J. Jasper, M. Nakonecznyj, C. S. Swingley and H. K. Livingston, *J. Phys. Chem.* **74**, 1535 (1970).
16. F. M. Fowkes, *J. Adhesion Sci. Technol.* **1**, 7 (1987).
17. J. H. Schulman, *J. Phys. Chem.* **63**, 1677 (1959).
18. L. M. Prince, *J. Colloid Interface Sci.* **23**, 165 (1967).
19. P. Cheng, D. Li, L. Boruvka, Y. Rotenberg and A. W. Neumann, *Colloids Surfaces* **3**, 151 (1990).
20. B. Jańczuk, W. Wójcik and A. Zdziennicka, *J. Colloid Interface Sci.* **157**, 384 (1993).
21. B. Jańczuk, private communication (5 August 1998).
22. J. T. Davies and E. K. Rideal, *Interfacial Phenomena*. Academic Press, New York (1963).
23. F. M. Fowkes, *Ind. Eng. Chem.* **56**, 40 (1964).
24. F. M. Fowkes, F. L. Riddle, Jr., W. E. Pastore and A. A. Weber, *Colloids Surfaces* **43**, 367 (1990).
25. L. H. Lee, *Langmuir* **12** 1681 (1996); also *Langmuir* **12**, 5972 (1996).
26. L. H. Lee, *J. Adhesion* **63**, 187 (1997).
27. L. H. Lee, *J. Adhesion* **67**, 1 (1998).
28. Y. Tamai, *Prog. Colloid Polym. Sci.* **61**, 93 (1976).
29. Y. Tamai, T. Matsunaga and K. Horiuchi, *J. Colloid Interface Sci.* **60**, 112 (1977).

30. H. J. Busscher, G. A. M. Kip, A. Van Silfhout and J. Arends, *J. Colloid Interface Sci.* **114**, 307 (1986).
31. M.-H. Bellon-Fontaine and O. Cerf, *J. Adhesion Sci. Technol.* **4**, 475 (1990).
32. H. Y. Erbil, *Langmuir* **10**, 2006 (1994).
33. J. Berg and P. N. Jacob, *J. Adhesion* **54**, 115 (1995).
34. Z. Xu, Q. Liu and J. Ling, *Langmuir* **1**, 1044 (1995).
35. R. J. Good and E. Ebling, in: *Chemistry and Physics of Interfaces — II*, p. 71. American Chemical Society, Washington, DC (1971).

Apparent and Microscopic Contact Angles, pp. 229–243
J. Drelich, J. S. Laskowski and K. L. Mittal (Eds)
© VSP 2000.

Effect of an external radiofrequency electric field on the surface free energy components of calcium carbonate in the presence of cationic and anionic surfactants

EMIL CHIBOWSKI*, LUCYNA HOŁYSZ and MONIKA LUBOMSKA

Department of Physical Chemistry, Faculty of Chemistry, Maria Curie-Skłodowska University, 20-031 Lublin, Poland

Received in final form 5 February 1999

Abstract—The surface free energy components of calcium carbonate for bare and surfactant-treated surfaces were studied by the thin-layer wicking technique. The surface precoverage from a 10^{-3} M aqueous solution of cetyltrimethylammonium bromide (CTMABr) and sodium dodecyl sulfate (SDS) caused a decrease in the electron donor component and an increase in the electron acceptor component. The changes were more pronounced in the case of CTMABr than for SDS. Consequently, in the presence of the cationic surfactant, the surface becomes hydrophobic while the SDS-precovered surface is still slightly hydrophilic. The presence of a radiofrequency (RF) electric field (44 MHz, 60 V peak-to-peak no-load amplitude) during wicking experiments perpendicularly to the wicking direction also decreases the electron donor component and increases the electron acceptor component. The increase is relatively higher for the surfactant-precovered surfaces. The RF treatment does not change the hydrophobicity of the bare surface but changes the hydrophobic CTMABr-precovered surface to slightly hydrophilic. On the contrary, the slightly hydrophilic SDS-preadsorbed surface becomes slightly hydrophobic in the presence of RF. It is concluded that the free energy changes appearing in the presence of RF are due to reorientation of the adsorbed surfactant molecules and/or changes in the structure of the water molecules hydrating the surface.

Key words: Calcium carbonate; surfactants; radiofrequency field; surface free energy components.

1. INTRODUCTION

Although the determination of the surface free energy of solids, as well as that of liquids, is still an open problem, the approach of van Oss, Good, and Chaudhury (vGC) [1–3] is commonly accepted at present. The main problem is with the choice of an absolute scale of the acid–base parameters for the liquids used for the determination of contact angles. Up to now, it has been based on the assumption

---

*To whom correspondence should be addressed. E-mail: emil@hermes.umcs.lublin.pl

of van Oss and co-workers that for water the electron donor parameter, γ_1^-, is equal to the electron acceptor one, γ_1^+. As the experimental value of the acid–base component for water equals 51.0 mJ/m^2 (at 20°C) [4], both parameters have to be equal to 25.5 mJ/m^2, if one accepts the combining rule for the electron donor–acceptor interactions to be appropriate. The rule is applied for the dispersion interaction (or Lifshitz–van der Waals interaction, γ_1^{LW}, in the vGC approach). A consequence of this equality is a particular set of these parameters for other liquids used as probes for contact angle or thin-layer wicking methods. However, very recently, different values of γ_1^- and γ_1^+ for water have been suggested [5, 6]. On the basis of solvatochromic parameters, Lee [5] suggests that for water at room temperature the ratio of the electron donor and electron acceptor parameters equals 1.8. As a result, the acidic parameter for water $\gamma_1^+ = 34.2$ mJ/m^2 and the basic one $\gamma_1^- = 19$ mJ/m^2. Moreover, Della Volpe and Siboni [6], while discussing the mathematical problem involved in solving the equations with three unknowns to determine the acid–base parameters, offer another set of these parameters for water, namely $\gamma_1^+ = 65$ mJ/m^2 and $\gamma_1^- = 10$ mJ/m^2. As can be seen, in the latter two approaches a higher acidic character (proton-donating) rather than basic is preferred for water. Nevertheless, irrespective of which set of parameters for the interfacial interactions is accepted as the correct one, which at present cannot be judged, the relative changes in the acid and base parameters of a given surface can still give some insight into the interfacial process studied.

In this investigation we were interested in studying the effect of a radiofrequency electric field (RF; 44 MHz, 60 V peak-to-peak no-load amplitude) on the wicking process of a thin porous layer of $CaCO_3$ by probe liquids and, as a result, on its surface free energy interactions.

The investigations of the external field effects (magnetic, high frequency electric RF, microwave) on the properties of dispersed systems have become of increasing interest to many researchers [7–23]. It became clear that both magnetic and RF fields also affected the behavior of non-magnetic materials, such as calcium carbonate [9–14, 16, 17, 23], polystyrene latex [15], and some oxides [9, 18–23]. The so-called 'memory effect' cannot be fully explained yet by applying the known theories. Such 'non-chemical' treatment of the dispersed systems may have applications in many practical processes, such as waste water treatment, to diminish the hard-scale deposition in industrial pipes and to enhance aggregation of colloidal particles, as well in biological systems, where long-lasting changes were observed in the cells [24, 25].

Calcium carbonate ($CaCO_3$) is a principal component of the hard scale depositing inside pipes as well as of many precipitates present in waste waters. Higashitani et al. [12–14] have found a distinct effect of the magnetic field treatment of $CaCl_2$ and Na_2CO_3 on precipitated calcium carbonate, even though it was precipitated about 2 h after the field had ceased. From the magnetically treated solutions a mixture of calcite and aragonite precipitated with a larger size than from untreated solutions, from which only calcite precipitated. Therefore, it seemed interesting to us to investigate whether a RF field would affect the surface properties of calcium carbonate [16, 17]. One of the important characterizations of a surface is the determination of the surface free energy interactions, which are expressed by the energy components, apolar γ_s^{LW}, and polar acid and base, γ_s^+ and γ_s^-, respectively. The field was applied vertically to the wicking direction. Having determined the penetration rates for nonane, water, and formamide in thin bare layers of calcium carbonate and those equilibrated with the liquid vapors, it was possible to calculate the surface free energy components from a general form of Washburn's equation [26–28]. Because surfactants are used to aggregate colloidal particles, we also investigated the influence of preadsorbed cetyltrimethylammonium bromide (CTMABr) and sodium dodecyl sulfate (SDS) on the calcium carbonate surface on its surface free energy components. The effect of the RF field irradiation was evaluated. The surface free energy components were calculated using the two sets of surface tension components for water and formamide mentioned above.

2. EXPERIMENTAL

2.1. Materials and reagents

The CaCO₃ used for the studies was from Fluka ($> 99.0\%$) and was used as received. The specific surface area of the sample determined with an ASAP 2405N V1.01 instrument (Micromeritics Instr. Co., USA) from nitrogen adsorption was BET, 0.99 m^2/g; Langmuir, 1.45 m^2/g; and single point at $p/p_0 = 0.199–0.88$ m^2/g. Thus, the average value was 1.11 m^2/g. SDS was from Koch-Light Laboratories Ltd., UK (p.a.), and CTMABr was from Merck ($> 99\%$). These surfactants were used for adsorption experiments on the calcium carbonate. For the thin-layer wicking experiments, the following liquids were used: formamide (p.a., R.C.B. Brussels), n-nonane (pure, Reachim, Russia), and water (doubly distilled and deionized, Milli-Q system).

2.2. Adsorption experiments

The adsorption of CTMABr and SDS was conducted from a 10^{-3} M aqueous solution (natural pH). A 15 g sample of CaCO₃ was poured into 100 ml of the solution and left for 24 h, while shaking by hand several times. The slurry was then filtered and the solid was dried at 50°C for 4 h. The solution was analyzed spectrophotometrically by Few's method [29] to determine the concentration of CTMABr with Orange II, and by the Jones method [29] with methylene blue to determine the concentration of SDS. The complexes were extracted into the chloroform phase and the concentrations of CTMABr and SDS were determined from the respective calibration curves. The wavelength, λ, used for CTMABr was 485 nm and those for SDS, 570 and 600 nm. The extinction was measured with a Specol 11 spectrometer (Carl Zeiss, Jena, Germany). The resulting adsorption thus determined amounted to 5.1 $\mu mol/g$ for CTMABr and 3.9 $\mu mol/g$ for SDS.

2.3. Thin the layer wicking experiments

Thin layers of calcium carbonate, untreated and with preadsorbed CTMABr or SDS (from 10^{-3} M) were deposited on 2.5×10 cm glass plates which had earlier been very carefully cleaned with chromic acid, then rinsed repeatedly with water, and washed in an ultrasonic bath. Five millilitres of the $CaCO_3$ suspension (6% w/w, homogenized with ultrasound for 15 min) was poured onto each plate, left overnight at room temperature to evaporate water, and then dried for 1 h at 110°C and 2 h at 145°C and kept in a desiccator (denoted as bare plates). Some plates were equilibrated in a closed vessel with nonane (24 h), water (20 h), and formamide (48 h) vapors to obtain an equilibrium duplex film on the calcium carbonate surface (denoted as precontacted plates) and then used for the penetration of the liquids. To determine the surface free energy components, wicking experiments were also conducted with plates that were not contacted with the vapors (bare plates). The wicking experiments were conducted in a sandwich chamber (Chromdes, Lublin) used for thin-layer chromatography. Details of the method have been described elsewhere [26–28]. Each wicking experiment was repeated for at least three plates and the average time t vs. distance x^2 was plotted to find whether the obtained relationship obeyed a modified [26–28] Washburn's equation:

$$x^2 = \frac{Rt}{2\eta}\Delta G, \qquad (1)$$

where R is the effective radius of the inter-particle capillaries in the porous layer, η is the liquid viscosity, and ΔG is the specific change in the free energy accompanying the process. Depending on the liquid/solid system studied, the value of ΔG assumes different values [26–28]. In this notation, a positive value of ΔG means that the process is spontaneous. For n-alkane penetrating the plate precontacted with its vapor, the change in the specific free energy accompanying the process, denoted here as ΔG_p, is equal to γ_l (surface tension of the liquid), which allows determination of the effective radius R, which results in the original Washburn equation. However, for the same n-alkane penetrating the bare surface of the porous layer, $\Delta G_b = W_A - W_C$, i.e. the difference between the work of adhesion and the work of cohesion of the liquid. In the case of a higher surface tension liquid (water, formamide), from some theoretical considerations and experimental data it resulted that $\Delta G_b - \Delta G_p = W_A - W_C$. Having determined ΔG_b and ΔG_p for nonane, water, and formamide, it is possible to calculate the surface free energy components of the solid, if the liquid surface tension components are known, by simultaneously solving the three equations with three unknowns. Applying van Oss et al.'s approach to the surface free energy formulation [1–3], the work of adhesion reads:

$$W_A = 2\left[\left(\gamma_s^{LW}\gamma_l^{LW}\right)^{1/2} + \left(\gamma_s^+\gamma_l^-\right)^{1/2} + \left(\gamma_s^-\gamma_l^+\right)^{1/2}\right], \qquad (2)$$

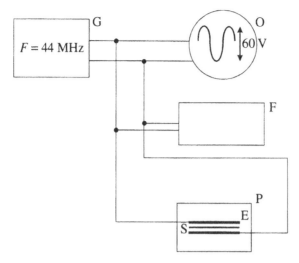

Figure 1. Scheme of the experimental set-up for the radiofrequency (RF) field treatment: G — 44 MHz generator; O — oscilloscope; F — frequency meter; P — Teflon sandwich chamber; E — silver electrodes; S — sample.

where the subscripts s and l mean solid and liquid, respectively. For an n-alkane, the last two terms in equation (2) are zero.

2.4. Thin-layer wicking in the presence of a RF field

To find out whether the RF field affected the free energy of interactions, the same set of experiments as those described above was carried out, but in the presence of the field. It was applied vertically relative to the direction of the liquid penetration via two silver electrodes separated by 7 mm. The plate was placed between the electrodes and the upper electrode was made of a silver net (with 0.1 mm holes) to observe the penetrating front. A schematic diagram of the measuring system is shown in Fig. 1. To ensure the same conditions (ambient temperature, humidity, etc.), the wicking experiments were conducted simultaneously in two sandwich chambers, to one of which the RF field was applied. As mentioned above, for a particular system tested, at least three plates were wicked in each chamber and the average times of the wicking distances were used for the calculations.

3. RESULTS AND DISCUSSION

From Washburn's equation [equation (1)] it results that at constant temperature the relationship $x^2 = f(t)$ should be linear, if ΔG is constant along the penetrated distance. Several examples of this relationship are presented in Figs 2 and 3. Figure 2 presents the wicking results for n-nonane, water, and formamide on CaCO₃, for both bare plates and plates equilibrated with the liquid vapors (precontacted plates); Fig. 3 shows the same relationships in the presence of a RF field. For practical

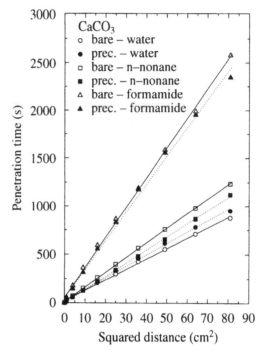

Figure 2. Penetration times of the probe liquids (n-nonane, water and formamide) into thin layers of CaCO₃, for bare plates and plates precontacted with the liquid vapors.

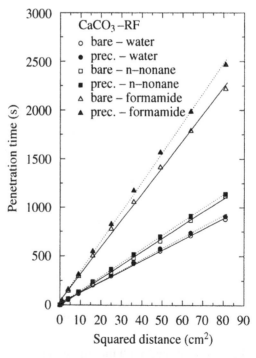

Figure 3. The same relationships as in Fig. 2, but in the presence of a RF field.

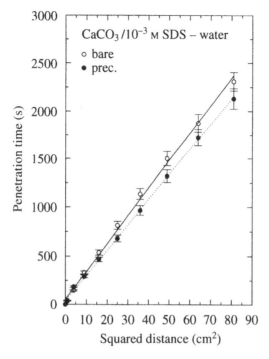

Figure 4. Penetration times of water for bare and vapor precontacted plates for calcium carbonate with preadsorbed SDS.

reasons (in fact, the wicking time is recorded for particular distances 1 cm apart each), the reversed function is plotted in the figures. As can be seen, Washburn's equation works quite well because the $x^2 = f(t)$ dependences are linear. In general, the wicking rates of the probe liquids are different for bare and precontacted plates. Whether the wicking rate is higher or lower for a bare or precontacted plate depends on the work of spreading W_S, which is the difference between the work of adhesion W_A and the work of cohesion W_C of the liquid [26–28].

In most cases, the RF field affects the penetration rates of the probe liquids used. The extent of the changes depends on the kind of probe liquid used (i.e. the magnitude of the interactions present), as well the state of the CaCO$_3$ surface, i.e. bare or precontacted. The results presented in Figs 2 and 3 allow calculations of the surface free energy components of CaCO$_3$ in the absence and presence of a RF field. To show the reproducibility of the wicking experiments and the RF effect, some examples are given in Figs 4–7, now for water on calcium carbonate with preadsorbed CTMABr or SDS surfactant. In Fig. 4, the results of water penetration for calcium carbonate with preadsorbed SDS both on bare and on water vapor equilibrated CaCO$_3$ surfaces are presented; Fig. 5 shows the same relationships but in the presence of a field. There are distinct differences in the penetration rate of water between the bare and precontacted surfaces and the differences are even larger when the RF field interacted (Fig. 5). As mentioned in Section 2, the

E. Chibowski et al.

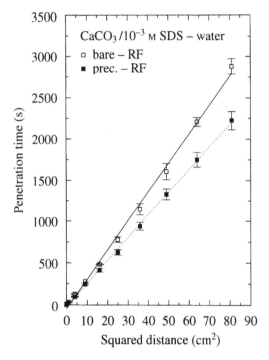

Figure 5. The same relationships as in Fig. 4, but in the presence of a RF field.

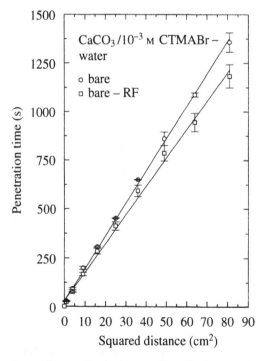

Figure 6. Penetration times of water for the bare surface of calcium carbonate with preadsorbed CTMABr in the absence and presence of a RF field.

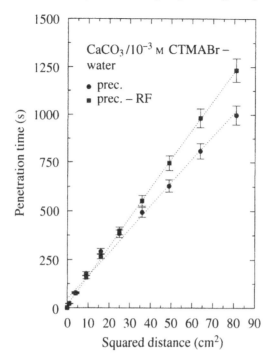

Figure 7. Penetration times of water for the precontacted surface of calcium carbonate with preadsorbed CTMABr in the absence and presence of a RF field.

plotted relationships in Figs 4 and 5 are the average of 3–4 wicking experiments. Similar relationships for the CTMABr-treated CaCO$_3$ surface are shown in Figs 6 and 7. To depict the RF field effect better, this time the penetration rates of water are shown for a bare surface (Fig. 6) and a surface equilibrated with water vapor (Fig. 7) with and without a RF field. The differences are clearly beyond the standard deviations shown and can be considered statistically significant. Based on the results presented in Figs 2–7 and others not presented here, and using the probe liquids' surface tension components (Table 1), the free energy components for the unmodified and the surfactant-modified CaCO$_3$ surfaces both in the absence and in the presence of a RF field were calculated according to the procedure described above [26–28]. Thus, the calculated components for the CaCO$_3$ surface by applying the vGC parameters are listed in Table 2, and those using Lee's values are shown in Table 3. For comparison, in Table 2 the values determined in our previous paper [16] are also given. There the plates with the thin layer of CaCO$_3$ were first treated with the RF field for 15 or 25 min and then the surface free energy components were determined from thin-layer wicking experiments, but in the absence of a field.

As can be seen from Table 2, the determined value of the Lifshitz–van der Waals component γ_s^{LW} is practically the same and the preadsorbed surfactants do not affect it, except for a small decrease for the SDS-treated surface. One would expect, for

E. Chibowski et al.

Table 1.
Surface free energy components of the probe liquids, in mJ/m^2 at $20°C$

Liquids	η	γ_l^{LW}	γ_l^-	γ_l^+	γ_l^-	γ_l^+
			van Oss and Good [30]		Lee [5]	
n-Nonane	22.91	22.91	0	0	0	0
Water	72.8	21.8	25.5	25.5	19.0	34.2
Formamide	58.0	39.0	39.6	2.28	29.1	3.1

Table 2.
Surface free energy components of $CaCO_3$ calculated from the data of van Oss and Good [30], and work of water spreading, W_S, in mJ/m^2

$CaCO_3$	Without RF field				With RF field			
	γ_s^{LW}	γ_s^-	γ_s^+	W_S	γ_s^{LW}	γ_s^-	γ_s^+	W_S
1. Bare	48.5	68.2	0.01	3.9	51.6	52.6	0.7	3.4
2. 10^{-3} M CTMABr	48.4	33.5	1.5	−9.8	50.8	36.6	3.5	1.0
3. 10^{-3} M SDS	46.8	53.1	1.1	2.5	44.0	30.1	5.8	−3.9
4. Bare [16]	48.0	79.0	0	8.9	48.0^a	40.7^a	1.34^a	−4.8
					48.0^b	63.9^b	1.77^b	13.2

[a] 15 min RF treated prior to wicking experiments.
[b] 25 min RF treated prior to wicking experiments.

Table 3.
Surface free energy components of $CaCO_3$ calculated from Lee's data [5], in mJ/m^2

$CaCO_3$	Without a RF field			With a RF field		
	γ_s^{LW}	γ_s^-	γ_s^+	γ_s^{LW}	γ_s^-	γ_s^+
1. Bare	48.5	51.3	0.01	51.6	39.3	1.0
2. 10^{-3} M CTMABr	48.4	25.0	2.0	50.8	27.2	4.7
3. 10^{-3} M SDS	46.8	39.6	1.5	44.0	22.4	7.9

a surfactant-treated surface, a much lower apolar component γ_s^{LW}, similar to the n-alkane surface energy, i.e. about $25–30$ mJ/m^2. Taking the specific surface of the sample 1.11 m^2/g, the surface areas per CTMABr and SDS molecule are 36 and 47 $Å^2$, respectively. This shows that in the case of SDS, the surface coverage is about half a statistical monolayer, while about one statistical monolayer is present in the case of CTMABr. The reason why the latter surfactant does not decrease the apolar component might be due to nonuniform adsorption of its molecules, possibly on electron donor sites which appears in the essentially decreased γ_s^- component (Table 2). On the other hand, polar polymers, such as polyethylene glycol or poly(methyl methacrylate), also show an apolar γ_s^{LW} component of the order of 43 mJ/m² [30]. Therefore, in the light of this it seems understandable that no essential changes in the apolar component were observed for the surfactant-treated calcium carbonate surface. In the presence of a RF field, a small increase in γ_s^{LW} was observed for the untreated surface and with preadsorbed CTMABr, while for the SDS-preadsorbed surface there was a decrease of 2.8 mJ/m² (Table 2).

Of course, it is possible to speculate about the mechanisms by which the field affects the penetration rates of a liquid. However, a simple heating effect can rather be excluded, especially at the applied field strength. The unbound water molecules show a maximum dielectric loss in the range of microwaves, i.e. 1–10 GHz. The dipoles of bound water exhibit a maximum dielectric loss around 100 MHz [31] and it is much lower than that of free water molecules. The dielectric losses are due to the loss current and appear as dissipated energy as heat [32]. No visible effects of heating were observed in the wicking experiments. In the light of Washburn's equation, the observed changes in the penetration rates can be interpreted as changes in the solid/liquid free energy interactions irrespective of the mechanism. These changes are included in the ΔG parameter of equation (1). They may solely result from changes in the liquid surface tension components, the solid surface free energy components, or both. Because, at present, we are not able to 'correct' the liquid surface tension components for the RF field effect, the observed changes in wicking rates (Figs 2–7) were totally ascribed to the changes in the solid surface, probably via adsorbed water and the surfactant molecules. Very interesting results on electromagnetic treatment were very recently published by Colic and Morse [22, 33]. They concluded that the observed effects were connected with some perturbations at the gas/liquid (nanobubbles) interface, which takes a long time to equilibrate ('memory effect'). Small amounts of ozone, superoxide, hydrogen peroxide, and atomic hydrogen are produced during the field action. These moieties can influence the structure of water and its reactivity. However, more work has to be done to explain the observed effects.

A more distinctive RF field effect is observed in the case of a polar (acid–base) interaction (Table 2). The electron donor component, γ_s^-, for an untreated surface determined in these experiments is lower than that obtained previously for the same original sample. This may be due to some differences in the thin-layer preparation. The presence of CTMABr or SDS diminishes the interaction and the change is much more pronounced for the cationic surfactant, which is understandable keeping in mind the presence of a nitrogen atom in the molecule which may interact as the electron acceptor with the calcium carbonate surface, and thus may more effectively decrease the γ_s^- component than SDS does. The RF field effect on the electron donor parameter of CaCO₃ depends greatly on the sample surface. For a surfactant-free surface as well as one precovered with SDS, the field decreases the γ_s^- interaction, while for a CTMABr-precovered surface a small increase is observed. However, in all cases the electron acceptor component, γ_s^+, is increased and the increase is relatively high, especially for the SDS-treated CaCO₃ surface.

The same was observed for RF-pretreated plates [16], but there both acid and base parameters depended on the field treatment time, which also corresponded with zeta potential changes of calcium carbonate [17] and other solids [18, 19]. These parameters changed in an oscillatory manner as a function of the treatment time. Oscillatory changes in the zeta potential were also observed by Colic and Morse [22] for rutile. In the case of the zeta potential, this 'memory effect' also varied in an oscillatory manner as a function of time after the field had ceased [18, 19]. We tried to explain the observed effects as a result of the field energy absorption. It might cause a shift in the adsorption–desorption equilibrium of H^+/OH^- ions and changes in the structure of water dipoles hydrating the solid surface.

It seemed interesting to calculate the work of spreading of water, W_S, to evaluate the surfactant and the field effects on the hydrophobicity of the calcium carbonate surface. The results are shown in Table 2. As can be seen, the untreated surface of $CaCO_3$ is slightly hydrophilic, because the work of spreading is positive. The adsorption of the cationic CTMABr surfactant makes it hydrophobic because the work of spreading is negative. However, the preadsorbed SDS practically does not change the hydrophobicity, although it lowers the electron donor component and simultaneously increases the electron acceptor component as well. As a result, the total interactions are of the same order. It is worth noting that the decrease in the electron donor parameter for SDS is much lower than for CTMABr (Table 2), presumably because the polar sulfate group of SDS exposes the electron donor component. The RF field affects the hydrophobicity of the surfactant-treated $CaCO_3$ surface, while for the bare surface it is practically the same in presence of the field. What is interesting is that the hydrophobic CTMABr-treated surface becomes very slightly hydrophilic and that the slightly hydrophilic SDS-treated surface becomes slightly hydrophobic in the RF field. As can be seen (Table 2 and previous study [16]), 15 min RF pretreatment caused hydrophobization of the surface; however, after 25 min RF treatment the surface became more hydrophilic than the untreated one. Based on the present knowledge, these somewhat surprising changes can be attributed to some rearrangements of the adsorbed surfactant molecules as well as of water dipoles hydrating the surface. However, these are only speculations.

As mentioned in the Introduction, there is a problem with the absolute values of acid–base parameters. Lee [5] suggested for water at room temperature $\gamma_s^- = 19$ mJ/m^2 and $\gamma_s^+ = 34.2$ mJ/m^2, which also affects the parameters for the other liquids used (Table 1). Using Lee's values, we present in Table 3 the recalculated components of the calcium carbonate surface. As it was not the principal purpose of this paper, we would only like to state that the conclusions which can be drawn from these values are essentially the same. As was also stated by Lee [5], the use of Lee's parameters causes a decrease in γ_s^- and an increase in γ_s^+, but which parameters are more real is an open problem for further studies. However, it should be stressed that because for a particular liquid (or solid) γ_s^- and γ_s^+ are not independent parameters, the calculated work of spreading, W_S, is the same regardless of which set of parameters is used for the calculations. Also, it is worth mentioning that using another set of probe liquids the calculated components would be somewhat different, but very probably their relative changes would lead to the same conclusions (see also ref. [6]). There would be a problem, however, if one were to eliminate water as one of the probe liquids. This is because if the magnitudes of γ_s^- and γ_s^+ of the two polar probe liquids used are similar (and low, as usually happens for γ_s^+), then the experimental error (resulting from the contact angle measurements or wicking and the following calculations) can be extremely large. The general problem dealing with the determination of components is the film pressure π_e resulting from the presence of the film behind the drop settled on the solid surface, or ahead of the visible front in the wicking experiments (precursor film). It decreases (or in some systems may increase) the solid surface free energy. π_e will depend on the probe liquid and, therefore, actually three new unknowns should be inserted into the equations, thus giving six unknowns in the three equations. Therefore, they cannot be solved. Hence, the film pressures are commonly neglected in the calculations. This is probably the reason why the components determined with the various probe liquids differ markedly for some solids.

4. CONCLUSIONS

The adsorbed surfactants do not affect the Lifshitz–van der Waals component γ_s^{LW}, which is practically the same as for untreated surface. The bare CaCO₃ surface is slightly hydrophilic, because the work of water spreading W_S is positive. But the adsorbed cationic CTMABr surfactant makes it hydrophobic, because W_S is negative.

The adsorbed CTMABr or SDS diminishes the electron donor component γ_s^- and the change is more pronounced for the cationic surfactant. This may result from the presence of a nitrogen atom in the molecule, which can interact as the electron acceptor with the calcium carbonate surface, and thus more effectively decrease the γ_s^- component than the SDS does. The adsorbed SDS practically does not change the hydrophobicity, although it lowers the electron donor component, but simultaneously the electron acceptor component increases. As a result, the total water–surface interactions are of the same order.

The RF field slightly affects the γ_s^{LW} interactions of CaCO₃. An increase is observed for the bare and the CTMABr-precovered surface and a small decrease appears for the SDS-treated surface.

The field effect on the electron donor parameter of CaCO₃ depends greatly on the state of the sample surface. For the surfactant-free surface and the SDS-precovered surface, the field decreases the γ_s^- interaction, while for the CTMABr-precovered surface a small increase is observed. For all the CaCO₃ surfaces studied, the RF field causes a relatively high increase in the electron acceptor component γ_s^+. This increase is especially distinct for the SDS-treated surface, presumably because the polar sulfate group of SDS exposes the electron donor component.

The RF field affects the hydrophobicity of the surfactant-treated $CaCO_3$ surface, while for the bare surface, it is practically the same in the presence of the field.

The observed RF field effects can probably be attributed to some rearrangements of the adsorbed surfactant molecules as well as of water dipoles hydrating the surface.

REFERENCES

1. C. J. van Oss, M. K. Chaudhury and R. J. Good, *Chem. Rev.* **88**, 927 (1988).
2. C. J. van Oss, R. J. Good and M. K. Chaudhury, *Langmuir* **4**, 884 (1988).
3. C. J. van Oss, *Interfacial Forces in Aqueous Media*. Marcel Dekker, New York (1994).
4. F. M. Fowkes, *J. Phys. Chem.* **66**, 382 (1962).
5. L.-H. Lee, *Langmuir* **12**, 1861 (1996).
6. C. Della Volpe and S. Siboni, *J. Colloid Interface Sci.* **195**, 121 (1997).
7. E. Chibowski, S. Gopalakrishanan, M. A. Busch and K. W. Busch, *J. Colloid Interface Sci.* **139**, 43 (1990).
8. K. W. Busch, M. A. Busch, R. E. Darling, S. Maggard and S. W. Kubala, *Trans. Inst. Chem. Eng.* **75B**, 105 (1997).
9. K. W. Busch and M. A. Busch, *Desalination* **109**, 131 (1997).
10. M. Ozaki, H. Suzuki, K. Takahashi and E. Matijević, *J. Colloid Interface Sci.* **113**, 76 (1986).
11. N. Kallay and E. Matijević, *Colloids Surfaces* **39**, 161 (1986).
12. K. Higashitani, K. Okuhara and S. Hatade, *J. Colloid Interface Sci.* **152**, 125 (1992).
13. K. Higashitani, A. Kage, A. Katamura, K. Imai and S. Hatade, *J. Colloid Interface Sci.* **156**, 90 (1993).
14. K. Higashitani, H. Iseri, K. Okuhara, A. Kage and S. Hatade, *J. Colloid Interface Sci.* **172**, 383 (1995).
15. K. W. Busch, S. Gopalakrishnan, M. A. Busch and E. Tombácz, *J. Colloid Interface Sci.* **183**, 528 (1996).
16. E. Chibowski and L. Hołysz, *J. Colloid Interface Sci.* **164**, 245 (1994).
17. E. Chibowski, L. Hołysz and W. Wójcik, *Colloids Surfaces A* **92**, 79 (1994).
18. E. Chibowski and L. Hołysz, *Colloids Surfaces A* **101**, 99 (1995).
19. E. Chibowski and L. Hołysz, *Colloids Surfaces A* **105**, 211 (1995).
20. K. W. Busch, M. A. Busch, S. Gopalakrishnan and E. Chibowski, *Colloid Polym. Sci.* **273**, 1186 (1995).
21. E. Chibowski and A. Sołtys, in: *Adhesion Science and Technology. Festschrift in honor of Dr. K. L. Mittal on the occasion of his 50th Birthday*, W. J. van Ooij and H. R. Anderson, Jr. (Eds), pp. 831–855. VSP, Utrecht, The Netherlands (1998).
22. M. Colic and D. Morse, *Langmuir* **14**, 783 (1998).
23. L. Yezek, R. L. Rowell, M. Larwa and E. Chibowski, *Colloids Surfaces A* **141**, 67 (1998).
24. A. Ontiveros, J. D. G. Durán, F. González-Caballero and E. Chibowski, *J. Adhesion Sci. Technol.* **10**, 999 (1996).
25. S. B. Sisken and J. Walker, in: *Electromagnetic Fields; Biological Interactions and Mechanisms*, M. Blank (Ed.). American Chemical Society, Washington, DC (1995), in ref. 22.
26. E. Chibowski and L. Hołysz, *Langmuir* **8**, 710 (1992).
27. E. Chibowski and F. González-Caballero, *Langmuir* **9**, 1069 (1993).
28. E. Chibowski and L. Hołysz, *J. Adhesion Sci. Technol.* **11**, 1289 (1997).

29. M. J. Rosen and H. A. Goldsmith, *Systematic Analysis of Surface-Active Agents*, P. J. Elving and I. M. Kolthoff (Eds), Vol. 12, pp. 402–407. Wiley, New York (1972).
30. C. J. van Oss and R. J. Good, *J. Macromol. Sci. — Chem.* **A26**, 1183 (1989).
31. F. Henry, C. Pichor, A. Kamel and M. S. El-Aasser, *Colloid Polym. Sci.* **267**, 48 (1989).
32. C. P. Smyth, *Dielectric Behavior and Structure*, Ch. 2. McGraw-Hill, New York (1955).
33. M. Colic and D. Morse, *Phys. Rev. Lett.* **80**, 2465 (1998).

Apparent and Microscopic Contact Angles, pp. 245–259
J. Drelich, J. S. Laskowski and K. L. Mittal (Eds)
© VSP 2000.

AFM measurements of hydrophobic forces between a polyethylene sphere and silanated silica plates — the significance of surface roughness

J. NALASKOWSKI [1,*], S. VEERAMASUNENI [1,†], J. HUPKA [2]
and J. D. MILLER [1,‡]

[1] *Department of Metallurgical Engineering, University of Utah, Salt Lake City, UT 84112, USA*
[2] *Department of Chemical Technology, Technical University of Gdansk, 80-952 Gdansk, Poland*

Received in final form 20 May 1999

Abstract—Recently, substantial research effort has been devoted to the study of non-DLVO forces between hydrophobic surfaces. However, the significance of surface roughness in the analysis of these hydrophobic attractive forces has not been given sufficient consideration and research is now in progress to attend to this issue. Fused silica plates covered with adsorbed octadecyltrichlorosilane (OTS) were characterized by water contact angle measurements and atomic force microscopy (AFM). Surfaces with different surface coverages and different contact angles were obtained by variation of the adsorption time. OTS formed patches on the silica surfaces, the lateral size and height of which depended on the adsorption time. Such surfaces exhibit differences in roughness at the sub-nanometer level. Using the AFM colloidal probe technique, forces between a polyethylene sphere and silanated silica surfaces were measured in water. Long-range attractive forces were found, usually referred to as hydrophobic forces. The resulting force vs. distance curves were fitted with a double exponential function. The magnitude of the short-range part of the force curves seems to correlate with water contact angles at silanated silica surfaces. On the other hand, the range of the long-range force correlates with the roughness of the silanated silica surface. These results with silanated silica surfaces were compared with the AFM results for polyethylene and graphite surfaces and on the basis of these experimental efforts, it appears that the nature of these hydrophobic attractive forces is related to surface roughness.

Key words: Silanation of silica; octadecyltrichlorosilane; atomic force microscopy (AFM); hydrophobic forces; roughness.

---

*On leave from Department of Chemical Technology, Technical University of Gdansk, 80-952 Gdansk, Poland.

†Present address: USG Corporation, Research Center, 700 North Highway 45, Libertyville, IL 60048-1296, USA.

‡To whom correspondence should be addressed. E-mail: jdmiller@mines.utah.edu

1. INTRODUCTION

The existence of non-DLVO attractive interactions between hydrophobic surfaces, frequently much stronger than van der Waals forces and with a much longer range, has been reported in the literature [1–3]. Variation in the magnitude and range of these attractive forces is considerable and not well explained. Little attention has been given to the effect of surface roughness, which should be of considerable importance.

The so-called hydrophobic force is important in many areas of particle separation technology, especially flotation, where differences in interaction between hydrophobic air bubbles and hydrophobic particle surfaces are responsible for the separation process. Despite many experimental studies during the last 10 years and some theoretical analyses, there is no theory which can explain all experimental observations. Some authors explain the origin of hydrophobic forces as an effect of interfacial water structure [4], or electrostatic interaction [5], or an effect due to the formation of cavities or nanobubbles in the interfacial region between hydrophobic surfaces [6, 7]. The last mechanism seems to be the most probable and popular in the recent literature. Regardless of the phenomenological details, the hydrophobic force (F_h) can be represented empirically as

$$F_h/R = C_0 \exp(-H/D_0), \tag{1}$$

where R is the radius of curvature, C_0 is a pre-exponential factor, H is the separation distance, and D_0 is the decay length.

More often, especially for stronger hydrophobic interactions, a double exponential function is used, i.e.

$$F_h/R = C_0 \exp(-H/D_0) + C_1 \exp(-H/D_1), \tag{2}$$

where it has been suggested that the parameters C_0 and C_1 are related to the interfacial tensions between the interacting solids with liquid between them, while D_0 and D_1 are referred to as decay lengths [8]. It was reported that even a third exponential term is sometimes necessary to fit the experimental data [2]. The hydrophobic force can also be described by a simple power law [9]:

$$F_h/R = -K/6H^2, \tag{3}$$

which is similar to the equation used for describing dispersion forces. Such a form is commonly used due to its simplicity, with only one parameter, K, to fit. This parameter K can be compared with the Hamaker constant and treated in the same manner by using a combining rule for obtaining K for asymmetric interactions.

Many investigators have reported direct hydrophobic force measurements between solid surfaces using the surface force apparatus (SFA) [10]. These experiments were mostly limited to mica surfaces, hydrophobized by self-assembled or Langmuir–Blodgett deposited cationic surfactants (single- or double-chain amines). Also, more recently, atomic force microscopy (AFM) [11] was used for hydrophobic force measurements by means of the colloidal probe technique. For colloidal probes, most researchers have used hydrophobized (silanated) glass [2] or silica spheres, while others have used a polypropylene hemisphere [7]. Interactions with flat surfaces, such as silanated silica, contaminated stainless steel, and polymer films, have been considered. Silanated silica is an especially good material for study, due to the broad range of hydrophobicity (measured by water contact angle) obtained by variation of the silanation time. Numerous works regarding silanation and silane layer characterization have been reported in the literature [12–17]. However, adsorbed surfactants, even strongly bonded to the surface, may always be questioned due to the possible rearrangement or desorption of adsorbed molecules.

Interestingly, these attractive forces were not observed for all hydrophobic surfaces [18]. In addition, the reported range of this hydrophobic force, despite the examination of surfaces having a similar degree of hydrophobicity, varies from less than 10 nm to 300 nm or more [1–3, 6, 7, 19–22]. Strikingly, most SFA studies, where surfaces were atomically smooth, reported a shorter range of hydrophobic forces (often below 20–30 nm). On the other hand, AFM studies, usually with not so well-defined surfaces, give longer ranges of attraction (even up to 300 nm). Although the issue of surface quality has not been discussed in detail in the literature, it is expected that surface preparation will have a significant effect on interaction force measurements. Differently prepared surfaces exhibit different degrees of heterogeneity and roughness and these differences can lead to variation in the force measurement. In addition, the presence and the amount of dissolved gas in the system seem to play an important role in most systems [7]. This effect is, to some extent, evidence for the theory that hydrophobic attraction originates from the cavitation effect. However, the issue is still discussed in the literature. How strong is this effect and for what kind of surfaces is it important?

In this study, using AFM, silanated silica surfaces with different water contact angles were characterized and subsequently interaction forces with a polyethylene colloid probe were measured. These results with silanated silica surfaces were compared with the AFM results for polyethylene and graphite surfaces and on the basis of these experimental efforts, it appears that the nature of these hydrophobic attractive forces is related to surface roughness.

2. EXPERIMENTAL

2.1. Chemicals

Octadecyltrichlorosilane (OTS, 95% purity) was obtained from Aldrich. Cyclohexane (spectrophotometric grade, Mallinckrodt) was dried over freshly activated 3 Å molecular sieves (Mallinckrodt). Fused silica (optical grade) plates were obtained from Harrick, Inc. The polyethylene (PE) powder was a low-density polyethylene (Scientific Polymer Product, Inc.) with a molecular weight of 1800 and a melting point of 117°C. Other reagents included chloroform (spectrophotometric grade, J. T. Baker), glycerol (certified ACS grade from Fisher Scientific), ammonium hydroxide (reagent grade, Fisher Scientific), hydrogen peroxide (reagent grade, Fisher Scientific), nitrogen (reagent grade N_2, Mountain Airgas), highly ordered pyrolytic graphite (HOPG; Union Carbide), and deionized water ($18\ M\Omega/cm$) obtained using a Milli-Q System (Millipore).

2.2. Silanation procedure

Prior to each experiment, silica plates were rinsed with acetone, methanol, and water, and then immersed in a mixture of $5\ H_2O : 1\ H_2O_2 : 1\ NH_4OH$ for 25 min at 80°C. After wet cleaning, the silica plates were exposed to a cold argon plasma for 30 min in Plasmod cleaner. Such cleaned silica discs were placed into Millipore water for eventual rebuilding of hydroxyl groups at the silica surface for several hours. Before silanation, the silica discs were dried using a stream of nitrogen. Such cleaned silica surfaces were fully wettable with water (contact angle less than 7°) and no contamination was detected using optical microscopy and AFM.

OTS solution preparation and the complete silanation process were conducted in a glove box under a slight nitrogen pressure in order to avoid the presence of atmospheric water vapor in the system. A 1.01×10^{-3} M OTS solution was prepared in dried cyclohexane and placed in a glass beaker (the beaker had been exposed to the OTS solution before the experiment in order to saturate the glass surface with OTS molecules and the eliminate reaction with the walls during the silanation experiments). The beaker was placed in an ultrasonic bath filled with hexadecane (to avoid water exposure). The ultrasonic bath was cooled during silanation; a constant temperature of 24°C was maintained. The silica plates were placed on a Teflon rack and immersed in the OTS solution for a specified time. Then the rack with silanated plates was removed from the OTS solution and immersed in dried cyclohexane and subsequently in a dried chloroform bath to remove excess OTS from the surface. After that, the plates were removed from the glove box, rinsed with copious amounts of chloroform, and dried in a nitrogen stream. Each silica plate was silanated using fresh OTS solution.

2.3. Contact angle measurements

Contact angles of water on fresh and silanated silica plates were measured by a Ramé-Hart goniometer using the sessile drop technique. The drop volume was increased and decreased until the three-phase boundary moved over the material surface, and advancing and receding contact angles were measured as described in a previous contribution [23]. Reported values are the average values from measurements for drop diameters between 3 and 7 mm.

2.4. Topography measurements using AFM

Topography features of the prepared silica surfaces before and after treatment with OTS were measured using a Digital Instruments Nanoscope III E AFM. Measurements were done in the contact mode (in air) by means of triangular oxygen-sharpened silicon nitride cantilevers (spring constant 0.12 N/m). Topography images were obtained from the 'high' channel with relatively high values of integral and proportional gains. Images were planarized using a third-order planarization algorithm. Section profiles through the asperities were taken. Roughness measurements were performed using Digital Instruments Nanoscope III E software for 0.25 μm^2 areas of the sample surfaces.

2.5. Force measurements using AFM

Spherical polyethylene (PE) particles were obtained by the melting of PE dispersion in glycerol. After appropriate rinsing and drying, this procedure was found not to change the surface properties of PE, which retained a high degree of hydrophobicity [24].

A spherical PE particle was glued to the AFM cantilever using a procedure described elsewhere [25, 26]. Tipless rectangular cantilevers (Digital Instruments, Inc.) were used. Particle interaction forces were measured with a Digital Instruments Nanoscope III E using a liquid cell. Deionized Milli-Q water was outgassed before measurements by subsequent boiling/freezing under vacuum several times. The same cantilever and PE sphere were used for all measurements. Deflection vs. distance curves in pure water were obtained for differently treated silica plates using the PE sphere as a colloidal probe. At least six measurements were taken at different locations on each sample surface. After the force measurements, SEM photographs of the cantilevers were taken in order to measure the diameter of the PE particle and the dimensions of the cantilever. The spring constant was calculated from the dimensions of the cantilever and was found to be 30 N/m. The spherical PE particle was measured to be 14 μm in diameter. Conversion of deflection curves to force vs. separation distance plots was done by the AFM analysis software [27] and average force curves were plotted.

3. RESULTS AND DISCUSSION

A plot of the water contact angle as a function of the silanation time is shown in Fig. 1. The untreated silica sample was fully wetted by water (advancing contact angle below 7°). The largest water advancing contact angle of 108° was found after 5400 s of treatment and corresponded quite well to what would be expected for a densely packed hydrocarbon monolayer. Despite precautions taken during silanation with regard to the presence of water in the system and maintaining a constant temperature below the reported transition temperature for OTS [28, 29], contact angle hysteresis [difference between the advancing (θ_A) and receding (θ_R) contact angles] was significant. The value of contact angle hysteresis was the largest for a short time of silanation and decreased with the silanation time from 32° for a 20 s treatment time to 12° for 5400 s.

Atomic force microscope images of the silica plates before and after different times of silanation are shown in Figs 2–6. The clean silica surface was found

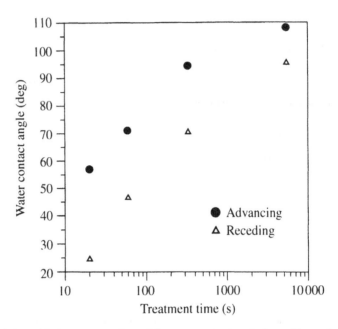

Figure 1. Variation with the treatment time of the water contact angle for a silica surface treated with 1.01×10^{-3} M OTS.

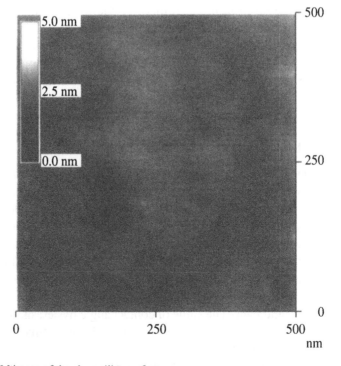

Figure 2. AFM image of the clean silica surface.

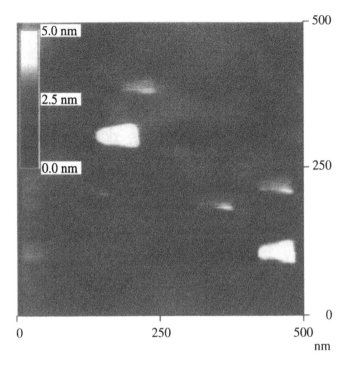

Figure 3. AFM image of the silanated silica surface after 20 s of silanation with 1.01×10^{-3} M OTS.

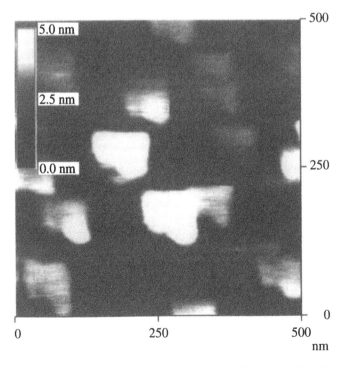

Figure 4. AFM image of the silanated silica surface after 60 s of silanation with 1.01×10^{-3} M OTS.

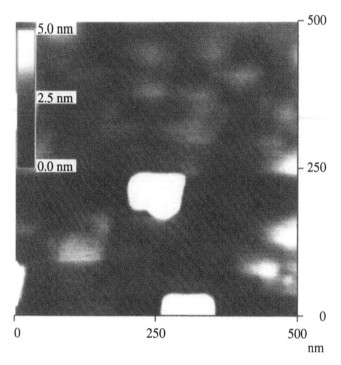

Figure 5. AFM image of the silanated silica surface after 330 s of silanation with 1.01×10^{-3} M OTS.

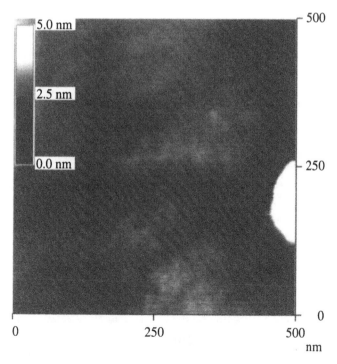

Figure 6. AFM image of the silanated silica surface after 5400 s of silanation with 1.01×10^{-3} M OTS.

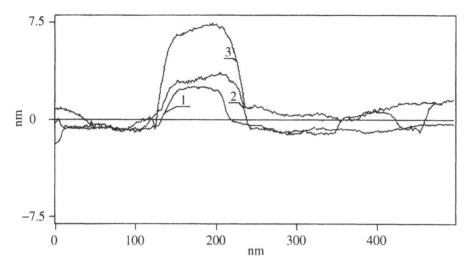

Figure 7. Profile of a section through the OTS patches after different silanation times; (1) 20 s; (2) 60 s; (3) 330 s.

to be homogeneous (without features on the surface) and to have a mean surface roughness of 0.21 nm. The images after silanation show a structure associated with the OTS layer. Below monolayer coverage, OTS forms patches on the silica surface. The patches are more or less rectangular and have average dimensions of 100 × 60 nm. A similar form for OTS layers has already been reported in the literature [30, 31]. Although the predominant mechanism of adsorption seems to rely on increasing the number of patches on the surface, we also found differences in the lateral size of the patches after different silanation times. In addition, the height of the OTS patches increases with the reaction time (see Fig. 7). On the surface of silica samples, patches with different heights were found and for different patches the height varied from 2 to 8 nm. Because the height of an OTS molecule adsorbed at the surface is approximately 2 nm (slightly less than the length of an OTS molecule, due to the angle formed with the surface), some of the patches seem to be composed of more than one layer of OTS molecules. With the silanation time, the number of patches on the surface increases and for longer times, coverage becomes more uniform. Finally, after 5400 s, the AFM image shows a homogeneous surface of OTS. However, from the AFM images we are not able to say if OTS covered the silica surface in the form of a monolayer or as a multilayer. Also, we can only speculate about the origin of such specific, regularly shaped OTS patches. Such geometry may be related to the hydrophobic chain arrangement on the surface, the adsorption of OTS aggregates formed in the solution, or may be due to the existence of some microcrystals at the silica surface, not visible with AFM. Such microcrystals, if they exist, should have a higher surface energy, which would lead to the preferential adsorption of OTS. The possibility of such preferential adsorption at silica grain boundaries has been reported in the literature [30]. Such grains or microcrystals were not clearly visible in the

Table 1.
Characterization of OTS/silica surfaces with respect to the extent of silanation

No.	Silanation time (s)	Advancing water contact angle (degrees)	Hysteresis (degrees)	Fractional surface coverage (%)	Mean roughness (nm)	C_1 (mN/m)	D_1 (nm)
0	0	< 7	< 4	0	0.217	—	—
1	20	57	32	0.04	0.296	−7.4	31
2	60	71	25	0.30	0.806	−50	88
3	330	94	24	0.62	0.687	−61	45
4	5400	108	12	1	0.284	−60	28

AFM images of our silica plates; however, X-ray diffraction analysis of the silica plate showed slight traces of crystallinity. The shape of the OTS patches may also be affected, to some extent, by the shape of the AFM tip. Additional experiments should be performed to determine the origin of the shape of OTS patches. The fractional coverage was roughly calculated from the AFM images and is given in Table 1. The change in fractional coverage and height of the patches leads to a difference in the mean roughness of the samples at the sub-nanometer level. The values of the mean roughness measured from the AFM images for a 500 × 500 nm area together with other measured properties of the silica surfaces are given in Table 1. The roughness values (in this case, closely related to the chemical heterogeneity of the silica surface) correspond well with the values of contact angle hysteresis.

Interaction forces between the silanated silica plates and the PE sphere in water were measured using AFM. PE exhibits a high degree of hydrophobicity ($\theta_A = 92°$, $\theta_R = 67°$) and a surface roughness of 2.59 nm for a 0.25 μm^2 area, as established in a previous contribution [24]. Attractive forces were found between PE and all silica plates except the untreated one. The average force/radius vs. separation distance curves are shown in Fig. 8. An arrow indicates the point where the gradient of the attractive force exceeded the spring constant of the cantilever. At this point, the PE sphere jumps into contact and at smaller distances the force curve may be considered to be unstable. Despite such a consideration, this part of the curve at short separation distances is repeatable and gives some information about the magnitude of the short-range attractive force. The interaction force between PE and the untreated (hydrophilic) silica was exclusively repulsive. This force arises from electrostatic repulsion between the surfaces, because both PE and silica are negatively charged in water. The observed attractive interactions between PE and the silanated silica surfaces exhibit the nature of hydrophobic interactions. Although the net force is observed in these experiments, the attractive van der Waals force is much smaller in range and magnitude than the observed forces. The electrostatic interactions are repulsive and negligibly small when compared with the observed forces.

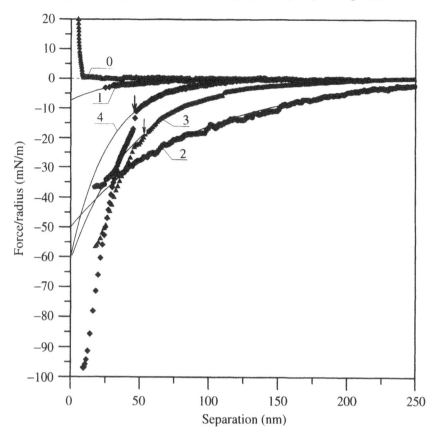

Figure 8. Force/radius vs. separation curves between the PE colloidal probe and the silanated silica surfaces in water. Arrows indicate the jump-to-contact distance; the solid line represents the fitted exponential function. For key to surface characterization see Table 1.

These force curves can be fitted with a double exponential function or power function; however, we found that the exponential function gave a much better fit for these data. Because the first term of this function is related to the short-range hydrophobic attraction, which cannot be measured due to the jump to contact in some cases, we used only the second term of the double exponential function, which is responsible for the long-range hydrophobic force. The fitting was done for distances beyond the jump-to-contact distance and the exponential function parameters are given in Table 1.

Very long-range distances for attraction were observed, even up to 250 nm, and interestingly no direct correlation between the water contact angle on silanated silica and the range of forces was found. The largest range of attraction was measured for sample 2, with a water contact angle of 71°. However, the magnitude of the short-range attraction (after jump to contact) increases monotonically with the water contact angle (or fractional surface coverage). The decay length (range of forces) D_1 seems to be related to surface roughness and increases with an increase in the

Table 2.

Comparison of different hydrophobic surfaces and their interactions with a hydrophobic PE colloidal probe

Surface	Advancing water contact angle (degrees)	Hysteresis (degrees)	Mean roughness (nm)	C_1 (mN/m)	D_1 (nm)
HOPG	88	26	0.169	−72	6
OTS (3)	94	24	0.687	−61	45
PE	92	15	1.278	−40	64

mean surface roughness. The C_1 parameter (magnitude of long-range attraction) does not correlate with roughness, but rather with the contact angle or fractional surface coverage; however, it is almost the same for samples 3 and 4 (see Table 1). Previous reports [2, 31] suggested that C_1 was related to the surface coverage, but found that D_1 was related to the lateral size of the OTS domain or the distances between them, and was almost constant for different surface coverages (with the same size of domains). But in our present study we found large differences in D_1 and very small differences in the lateral dimension of the OTS patches during the silanation process.

A theoretical mechanism of the hydrophobic force origin, discussed in the literature, involves pair electrostatic interactions between hypothetical heterogeneous surfaces, with rectangular domains on the surface [32, 33], which may be the case with silanated surfaces. Also, the origin of the hydrophobic interactions and differences in their range, the lateral diffusion, and the rearrangement of adsorbed compounds was proposed [34]. Such mechanisms are possible with physisorbed surfactants and even with silane layers to some extent, but should not take place with bulk hydrophobic materials like polymers. In the last case, long-range attraction should not exist, but such attraction has also been reported for polymers [7, 35].

In this regard, we performed additional measurements between a PE sphere and a PE flat surface (obtained by melting PE on a mica surface), and between a PE sphere and a freshly cleaved HOPG graphite surface in water. The results were compared with the PE–OTS silanated silica surface and system 3 was chosen because of the similarity in water contact angle. The parameters for these systems are given in Table 2. Experimental force curves and exponential function fitting are shown in Fig. 9. Despite the similar water contact angle and hysteresis for all samples, the range for the attractive force is different. The PE–HOPG system shows no long-range attraction, whereas the PE–PE system had a slightly longer range for attraction than the PE–OTS system. The roughness of high-quality HOPG is very low (0.169 nm) compared with the PE surface, which was quite rough (1.278 nm). In some regard, the nature of these systems is different. PE and HOPG are chemically homogeneous, whereas OTS forms adsorbed patches on the silica surface. On the other hand, PE and OTS have a similar chemical composition (both are composed of hydrocarbon chains), whereas HOPG is, of course, elemental

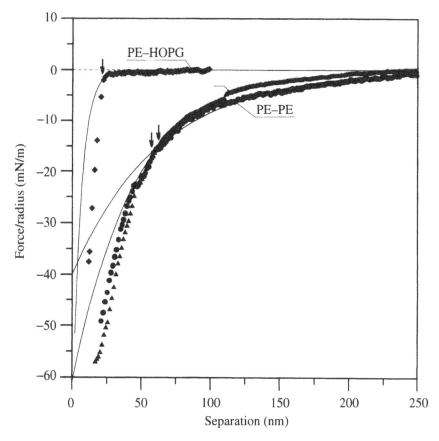

Figure 9. Force/radius vs. separation curves between the PE colloidal probe and the silanated silica surface (No. 3), the HOPG surface, and the PE surface in water. Arrows indicate the jump-to-contact distance; the solid line represents the fitted exponential function. For key to surface characterization see Tables 1 and 2.

carbon. Nevertheless, in all three cases, D_1 was found to correlate with surface roughness (see Table 2).

In summary, there is no chemical heterogeneity (domains) in the case of PE and HOPG, and lateral rearrangement of the hydrophobic groups is not possible in these cases. We cannot prove how the range of hydrophobic attraction may be affected by the roughness of the surface, but it is possible that surface roughness can promote the formation of cavities before approach or when the two surfaces approach each other. These cavities may be responsible for the strong, long-range hydrophobic attraction, which has been reported in the literature [6, 7, 36]. In the same way, it is expected that surface roughness may also be an important factor when the issue of dissolved gas is considered. The absence of cavitation or bubble formation at very smooth hydrophobic surfaces in supersaturated gas solutions has been reported in the literature [37]. Nanometric irregularities at a hydrophobic surface would facilitate the fixation of nanobubbles and promote the cavitation phenomenon. In view of the foregoing, the significance of surface roughness on the measurement of hydrophobic forces must be given appropriate consideration in future research.

4. CONCLUSIONS

We have found that OTS adsorbs as quite uniform, rectangular patches on silica surfaces below monolayer coverage. The lateral size of these patches was found to be 60×100 nm and was almost constant during the silanation process. The height of these patches increased with the silanation time from 2 to 8 nm. Such surfaces exhibit different degrees of hydrophobicity (water contact angle from 57° to 108°) and different roughnesses on the nanometer level.

The forces between a PE sphere and silanated silica in water were measured using AFM. A strong, long-range attraction was found and fitted using an exponential function. The magnitude of the short-range hydrophobic force correlated with water contact angles on the silanated silica surfaces. However, the range (or decay length D_1) was not related to the contact angle on the silanated surface or fractional surface coverage. The range of the long-range force was related to surface roughness and increased with the roughness of the surface.

In addition, force measurements between a PE sphere and a HOPG surface and a PE sphere and a PE surface were performed in water and compared with force measurements for the PE–OTS system. Despite similar water contact angles and the chemical homogeneity of HOPG and PE, the range of attraction was much shorter for the very smooth HOPG surface ($Ra = 0.169$ nm), whereas for the rough PE surface ($Ra = 1.278$ nm) the range was slightly greater than for the silanated silica surface. It is concluded that surface roughness can be a very important factor in influencing hydrophobic force measurements. It appears that the surface roughness can alter the range of these forces by promoting the formation of vapor or gas cavities between interacting surfaces.

Acknowledgements

Financial support from the US Environmental Protection Agency, National Center for Environmental Research and Quality is gratefully appreciated. Although the research described in this article has been funded partially by the US Environmental Protection Agency through grant No. R825306-01-0 to the University of Utah, it has not been subjected to the Agency's required peer and policy review and, therefore, does not necessarily reflect the views of the Agency and no official endorsement should be inferred. In addition, support from the Basic Energy Science Division of DOE (DOE Grant No. 93-ER14315) and from the US–Japan Center of Utah is gratefully appreciated.

REFERENCES

1. J. N. Israelachvili and R. Pashley, *Nature* **300**, 341 (1982).

2. Y. I. Rabinovich and R. H. Yoon, *Langmuir* **10**, 1903 (1994).
3. R. H. Yoon and S. A. Ravishankar, *J. Colloid Interface Sci.* **176**, 391 (1996).
4. J. C. Eriksson, S. Ljunggren and P. M. Claesson, *J. Chem. Soc., Faraday Trans.* **285**, 162 (1989).
5. E. Ruckenstein and N. V. Churaev, *J. Colloid Interface Sci.* **147**, 535 (1991).
6. H. K. Christenson and P. M. Claesson, *Science* **239**, 390 (1988).
7. L. Meagher and V. S. J. Craig, *Langmuir* **10**, 2736 (1994).
8. R. M. Pashley, P. M. McGuiggan, B. W. Ninham and D. F. Evans, *Science* **229**, 1088 (1985).
9. P. M. Claesson, C. E. Blom, P. C. Herder and B. W. Ninham, *J. Colloid Interface Sci.* **114**, 234 (1986).
10. J. N. Israelachvili and G. E. Adams, *J. Chem. Soc., Faraday Trans.* **74**, 975 (1978).
11. G. Binnig, C. F. Quate and C. Gerber, *Phys. Rev. Lett.* **56**, 930 (1986).
12. J. Sagiv, *J. Am. Chem. Soc.* **102**, 92 (1980).
13. P. Silberzan, L. Leger, D. Ausserre and J. J. Benattar, *Langmuir* **7**, 1647 (1991).
14. M. Wei, R. S. Bowman, J. L. Wilson and N. R. Morrow, *J. Colloid Interface Sci.* **157**, 154 (1993).
15. S. Biggs and F. Grieser, *J. Colloid Interface Sci.* **165**, 425 (1994).
16. C. P. Tripp and M. L. Hair, *Langmuir* **7**, 923 (1991).
17. J. D. Le Grange and J. L. Markham, *Langmuir* **9**, 1746 (1993).
18. J. L. Parker, P. M. Claesson, J. H. Wang and H. K. Yasuda, *Langmuir* **10**, 2766 (1994).
19. Y. Tsao, S. X. Yang, D. F. Evans and H. Wennerstrom, *Langmuir* **7**, 3154 (1991).
20. Y. Tsao, D. F. Evans and H. Wennerstrom, *Langmuir* **9**, 779 (1993).
21. J. Wood and R. Sharma, *Langmuir* **11**, 4797 (1995).
22. K. Kurihara, S. Kato and T. Kunitake, *Chem. Lett.* 1555 (1990).
23. J. Drelich, J. D. Miller and R. J. Good, *J. Colloid Interface Sci.* **179**, 37 (1996).
24. J. Nalaskowski, J. Drelich, J. Hupka and J. D. Miller, *J. Adhesion Sci. Technol.* **13**, 1 (1999).
25. W. A. Ducker, T. J. Senden and R. M. Pashley, *Nature* **353**, 239 (1991).
26. S. Veeramasuneni, M. R. Yalamanchili and J. D. Miller, *J. Colloid Interface Sci.* **184**, 594 (1996).
27. L. Ip and D. Y. Chan, AFM Analysis 1.0.3. University of Melbourne, Victoria, Australia (1994).
28. J. B. Brzoska, I. B. Azouz and F. Rondelez, *Langmuir* **10**, 4367 (1994).
29. A. N. Parkih, D. L. Allara, I. B. Azouz and F. Rondelez, *J. Phys. Chem.* **98**, 7577 (1994).
30. D. H. Flinn, D. A. Guzonas and R. H. Yoon, *Colloids Surfaces* **87**, 163 (1994).
31. Y. I. Rabinovich and R. H. Yoon, *Colloids Surfaces* **93**, 273 (1994).
32. W. J. C. Holt and D. Y. C. Chan, *Langmuir* **13**, 1577 (1997).
33. S. J. Miklavic, D. Y. C. Chan, L. R. White and T. W. Healy, *J. Phys. Chem.* **98**, 9022 (1994).
34. H. K. Christenson and V. V. Yaminsky, *Colloids Surfaces* **129–130**, 67 (1997).
35. M. E. Karaman, L. Meagher and R. M. Pashley, *Langmuir* **9**, 1220 (1993).
36. V. V. Yaminsky and B. W. Ninham, *Langmuir* **12**, 4969 (1996).
37. W. L. Ryan and E. A. Hemmingsen, *J. Colloid Interface Sci.* **157**, 312 (1993).

Apparent and Microscopic Contact Angles, pp. 261–282
J. Drelich, J. S. Laskowski and K. L. Mittal (Eds)
© VSP 2000.

Self-organization in unstable thin liquid films: dynamics and patterns in systems displaying a secondary minimum

JAYANT SINGH and ASHUTOSH SHARMA *

Department of Chemical Engineering, Indian Institute of Technology, Kanpur 208 016, India

Received in final form 3 May 1999

Abstract—Dynamics, stability, morphology, and dewetting of a thin film (< 100 nm) under the influence of a long-range van der Waals attraction combined with a short-range repulsion are studied based on numerical solutions of the nonlinear two-dimensional (2-D) thin film equation. Area and connectivity measures are used to analyze the morphology and the distinct pathways of evolution of the surface instability. The initial disturbance resolves into an undulating structure of uneven 'hills and valleys'. Thereafter, the morphology depends on the mean film thickness relative to the minimum of the force curve. Relatively thin films to the left of the minimum transform directly into an array of droplets via the fragmentation of ridges. At long times, the droplets merge due to ripening. In contrast, relatively thick films are dewetted by the formation and growth of isolated, circular holes. Coalescence of holes eventually leads to the formation of ridges and drops. Films of intermediate thickness display a rich combination of different morphologies. Thus, the morphology and the sequence of evolution depend crucially on the form of the potential and the film thickness relative to the location of the minimum in the force vs. thickness curve. Different types of patterns can, therefore, even co-exist on a heterogeneous surface.

Key words: Thin films; dewetting; pattern formation; spinodal decomposition.

1. INTRODUCTION

Stability, dynamics, morphology, and adhesion failure by dewetting in thin films (< 100 nm) are of increasing concern in a host of technological and scientific settings such as coatings, flotation, emulsions, and physical and biological thin film phenomena (e.g. wetting, adhesion, colloid stability, biomembrane morphology).

It is now well understood [1–4] that the free surface of a thin film is unstable and deforms spontaneously whenever its disjoining pressure, engendered by the excess intermolecular interaction, increases with increasing local film thickness. In other words, instability by the 'spinodal decomposition' mechanism results whenever the

---

*To whom correspondence should be addressed. E-mail: ashutos@iitk.ac.in

spinodal parameter, defined as the second derivative of the free energy per unit area, is negative. The total free energy is the sum of its many possible components, the most prominent of which are the Lifshitz–van der Waals interactions (which include the dispersion and dipolar interactions), electrostatic interaction, and shorter-range acid–base, entropic, and structural interactions. In our nomenclature, an 'attractive' ('repulsive') component is one which engenders attraction (repulsion) between the two surfaces (interfaces) of the film.

Experiments [5–18] have shown that surface instability leads to a variety of microstructures in thin films, ranging from micro-droplets to holes, as well as a spectrum of bicontinuous undulating structures. However, theoretical understanding of self-organized thin film patterns and their relationship to intermolecular interactions have been rudimentary, and even deceptive, in so much as they are based largely either on the linear stability analysis [1, 2, 4, 19–23] or on one-dimensional (1-D) nonlinear analysis [19–22]. While these approaches have uncovered much of the underlying physics, they cannot be used to resolve the actual three-dimensional (3-D) morphologies and pathways of dewetting. The problem of 3-D nonlinear pattern selection has only recently been addressed based on the full 2-D nonlinear equation of evolution for thin films [24–26]. In particular, two distinct types of thin film systems have been studied, both of which display a lone primary minimum in the free energy close to the point of dewetting or pseudo-dewetting: (a) thin films subject to the Lifshitz–van der Waals attraction [24] and (b) thin films subject to a long-range van der Waals repulsion combined with a shorter range attraction [25, 26]. Films of system (a), regardless of their thickness, as well as relatively thick films of system (b), were found to dewet the substrate by the formation of isolated circular holes. Interestingly, in films of system (b), the morphological pathway of dewetting changed as the film thickness was reduced closer to the location of the minimum in the spinodal parameter. A decrease in the film thickness leads first to dewetting by an undulating bicontinuous pattern (rather than holes) and then by an array of isolated circular droplets. Thus, while a bicontinuous undulating pattern has usually been considered to be a hallmark of spinodal processes [1, 11–13, 27], nonlinear interactions can lead to a variety of other self-organizing patterns.

The main objective of this investigation was to uncover the variety of morphological patterns *on a homogeneous substrate* which can form spontaneously in an unstable film subject to the long-range van der Waals attraction combined with a short-range repulsion (these will be referred to as 'Type II' systems) and the conditions for the selection of a particular pattern. The combination of short- and long-range repulsive and attractive forces in thin films (as in flocculating colloids) usually leads to the existence of a secondary minimum in the free energy, disjoining pressure, and spinodal parameter isotherms. The long-range attraction is usually due to the Lifshitz–van der Waals interaction on nonwettable substrates [4, 20, 28, 29]. The shorter-range repulsion can be due to one or more of the following: van der Waals interaction with a thin wettable solid coating of the substrate [22], hydrophilic repulsion in aqueous films [28], electrical double-layer repulsion, and entropic effects in adsorbed or grafted polymer films [14–16, 30, 31]. In other classes of interesting systems, the presence of a significant secondary (and even tertiary) minimum is engendered by the layering or ordering of repeat units, e.g. liquid crystals [18], films containing colloidal particles or surfactants [32, 33], block polymers, and organic multilayers [34].

The basic physics and 1-D nonlinear simulations for type II thin films are reported elsewhere [4, 20]. The growth of surface instability leads to (pseudo-) dewetting, and eventually to a quasistable structure consisting of a periodic array of cylindrical micro-drops in (quasi-) equilibrium with the intervening thin films. True dewetting or drying of the substrate does not occur since a relatively thick adsorbed flat film of thickness corresponding to the *secondary* minimum in the free energy is left behind. When the free energy per unit area displays both a primary minimum (molecularly close to the substrate) and a secondary minimum (at relatively large thickness), we use the term 'pseudo-dewetting' to signify that an equilibrium film of thickness close to the secondary minimum is left on the surface. The usual term 'dewetting', which conveys the idea of an almost dry surface, can be used when the equilibrium film left on the substrate corresponds to the primary minimum. Although both types of dewetting are seen in experiments on liquid-crystal films [18], spinodal dewetting of relatively thick films can lead only to pseudo-dewetting. True dewetting requires the overcoming of an energy barrier separating the primary and secondary minima, which may be possible by 'nucleation' induced by heterogeneities [18].

In this paper we address the problem of pattern selection, complete 3-D morphology, and morphological pathways of pseudo-dewetting for unstable type II films displaying a secondary minimum in the spinodal region. These issues are investigated based on numerical solutions of the nonlinear 2-D thin film equation. The simulations reported here provide a formalism for correlating the film morphology with the interfacial interactions and the film thickness, and will make it possible to compare theory and experiments directly in the future. Such a formalism will also help us to address the inverse problem of characterization of surface interactions from the observed morphology.

2. THEORY

We consider a thin (< 100 nm) Newtonian fluid film on a smooth, chemically homogeneous solid surface ($x - y$ plane). Evaporation and condensation are neglected, as is the case for high viscosity, low vapor pressure polymer films. The following nondimensional thin film equation [20, 24–26], derived from the Navier–Stokes equations, governs the stability and spatio-temporal evolution of a thin film subject to excess intermolecular interactions (where $\Phi_H = [2\pi h^2/|A|][\partial^2 \Delta G/\partial H^2]$, and A is the effective Hamaker constant):

$$\partial H/\partial T + \nabla \cdot \left[H^3 \nabla (\nabla^2 H)\right] - \nabla \cdot \left[H^3 \Phi_H \nabla H\right] = 0, \tag{1}$$

where $H(X, Y, T)$ is the nondimensional local film thickness scaled with the mean thickness, h. The spatial coordinates, X, Y, in the plane of the substrate are scaled with the characteristic length scale for the van der Waals case [20], $(2\pi\gamma/|A|)^{1/2}h^2$. Nondimensional time, T, is scaled with $(12\pi^2\mu\gamma h^5/A^2)$, where γ and μ refer to the film surface or interfacial tension and viscosity, respectively. A renormalized real time, $t_N = t(A^2/12\pi^2\mu\gamma) = Th^5$, can also be defined to remove the influence of the mean film thickness. Finally, $\Phi_H = (2\pi h^4/|A|)(\partial^2\Delta G/\partial h^2)$ is the nondimensional spinodal parameter. ∇ is the usual 2-D gradient operator, $i(\partial/\partial X) + j(\partial/\partial Y)$. The rationale for this type of nondimensionalization is given elsewhere [20, 22].

The first term of the thin film equation denotes unsteady force (containing viscosity), which merely retards the growth of instability. The second term denotes the effect of surface tension for a curved surface, which in a 3-D geometry may be stabilizing (due to 'in-plane' curvature as in 2-D cases) or destabilizing (due to transverse curvature as in Rayleigh instability of circular cylinders). The third term describes the effect of excess intermolecular interactions, which engender instability by causing flow from thinner to thicker regions in the case of negative 'diffusivity', namely when the spinodal parameter $\Phi_H < 0$.

For simulations, we consider a fairly general excess intermolecular interaction free energy composed of a long-range (algebraic) attraction and a short-range (exponential) repulsion [4, 20–25, 29, 35], i.e.

$$\Delta G = -(A/12\pi h^2) + S^P \exp\left[(d_0 - h)/l_p\right], \tag{2}$$

where d_0 (= 0.158 nm) is a molecular cut-off thickness [29], l_p is the decay length for the short-range repulsion and h is the film thickness. A is the effective Hamaker constant, which is positive (signifying long-range attraction) when the surface tensions for the substrate and the bounding medium are both either larger or smaller than that of the film material [4, 20, 22, 29]. S^P (> 0) measures the strength of the shorter-range non-van der Waals repulsion at the substrate surface at $h = d_0$. Qualitative variation of the spinodal parameter $\partial^2\Delta G/\partial h^2$ is as shown in Fig. 1 for two different sets of parameters. A different qualitative variation of the spinodal parameter results for type IV systems where the net van der Waals force is repulsive ($A < 0$) and the shorter-range non-van der Waals forces are attractive ($S^P < 0$). The morphology of type IV systems is considered elsewhere [25, 26].

While a convenient analytical representation for repulsion is chosen in equation (2) and realistic sets of parameters chosen for illustration, the key features of morphology (e.g. whether a circular hole forms, etc.) and the pathways of evolution were found to be independent of these details. Based on our previous work [24–26] and a large number of additional simulations, we have verified that the distinct qualitative pathways of pattern evolution depend only on the qualitative *form* of the potential, rather than on its precise analytical representation. The key feature of the potential being investigated here is the presence of an unstable secondary spinodal region, which is separated from the primary spinodal region by

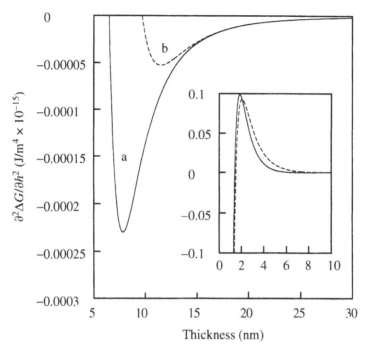

Figure 1. Variations of the spinodal parameter (force per unit volume, $[\partial^2 \Delta G / \partial h^2] \times 10^{-15}$) with the film thickness. Curve a: $A = 9.41 \times 10^{-21}$ J, $S^P = 1.2$ mJ/m^2, $l_p = 0.83$ nm. Curve b: $A = 9.41 \times 10^{-21}$ J, $S^P = 1.2$ mJ/m^2, $l_p = 1.1$ nm. The inset shows details for small thickness. Instability occurs for 1.39 nm $> h > 6.49$ nm for curve a and for 1.5 nm $> h > 9.7$ nm for curve b, where $\partial^2 \Delta G / \partial h^2 < 0$.

an intervening region where films are stable. The problem of pattern formation in the primary spinodal region has already been addressed [24–26]. Hence, we will confine attention to the self-organization around the secondary minimum.

The linear stability analysis [1–4, 20, 24] of the 3-D thin film equation (1) predicts a dominant nondimensional characteristic length scale of the instability $\lambda = 4\pi/\sqrt{-\Phi_H}$, which is the diagonal length of a unit square cell of length $L = \lambda/\sqrt{2}$.

In order to address the problem of pattern selection, we directly solved the nonlinear thin film equation numerically over an area of $9L^2$, starting with an initial small amplitude (≈ 1 Å) random perturbation. An ADI (alternating-direction implicit) technique was implemented, which combines the accuracy of an implicit integration scheme with the efficiency of an explicit scheme. Details of this numerical method can be found elsewhere [36, 37]. Briefly, over a single time step, all derivatives in one direction (say X) are discretized by an implicit scheme, and all derivatives in the other direction (Y), as well as the mixed derivatives, are discretized by an explicit scheme. At the next time step, the discretizing schemes in each direction are reversed. A 45 × 45 grid was usually found to be sufficient for convergence when central differencing in space with half-node interpolation was combined with Gear's algorithm for time marching. The latter is a variable-order accurate method with automatic adjustment of time steps, which is especially suitable for stiff equations. Several simulations were also repeated for a 60×60 grid and some simulations were repeated over an area of $16L^2$ to rule out the influence of the grid density and the finite domain size on the result.

3. RESULTS AND DISCUSSION

We found two completely different sequences of evolution by which (pseudo) dewetting could occur, depending on the distance between the mean film thickness and the location of the minimum in the spinodal parameter. All of the simulations reported below are for the system represented by curve a of Fig. 1, unless otherwise noted.

Figure 2 summarizes the major events in the time evolution of patterns in a 7 nm thick film (which lies to the left of the minimum of the spinodal curve a in Fig. 1). Figure 2A presents the gray scale images of the evolution at increasing times (images A1–A10 from left to right), and Fig. 2B displays selected 3-D images of the same evolution (images B1–B4). The hatched areas in Fig. 2A and all of the subsequent figures denote a largely flat film of equilibrium thickness corresponding to the secondary minimum in the free energy per unit area curve (pseudo-dewetting). The initial random disturbance first transforms into a bicontinuous structure composed of uneven long hills and valleys, which then evolves into small-amplitude circular depressions and liquid ridges (images A2 and B1 in Fig. 2). Fragmentation of ridges directly produces increasingly circular droplets which become increasingly isolated by the coalescence and thinning of valleys surrounding them (image A3). The first signs of pseudo-dewetting appear *after* the formation of droplets (hatched areas in images A4 and B2). At this stage, droplets grow both by continued thinning of valleys, which increases the dewetted area, and by merger or ripening, which is driven by the flow from smaller to larger droplets due to Laplace pressure gradients (images A5–A7). After dewetting is complete, further evolution occurs only by the ripening of the structure in which larger drops grow at the expense of smaller ones, thus decreasing the number density of drops (images A7–A10 and images B2–B4). Eventually, a truly thermodynamic stable state is reached, which is represented by the co-existence of a single drop (image A10) with its surrounding equilibrium flat thin film, which may be called the situation of 'pseudo-partial wetting'. Thus, for films thinner than the secondary minimum in the spinodal parameter, the onset of dewetting by the formation of a contact line occurs *after* the formation of isolated, mature circular droplets. Droplets are formed directly by the fragmentation of an initially bicontinuously undulating structure.

Figure 3 shows the fractional area covered by the liquid regions above the mean thickness (curve a) and above a cut-off equilibrium thickness (curve b) for the 7 nm film of Fig. 2 (the equilibrium cut-off thickness was defined as 15% thicker than the final minimum equilibrium thickness at long times). The fractional dewetted

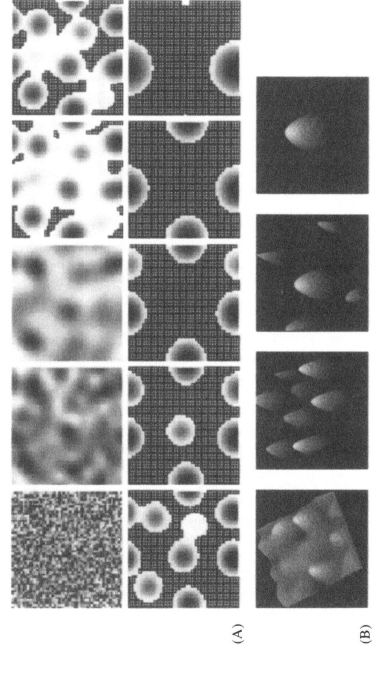

(A)

(B)

Figure 2. Different stages of the evolution of a 7 nm thick film, which is thinner than the location of the minimum (at 7.82 nm) in the spinodal parameter vs. thickness curve a of Fig. 1. The area of each cell is $9L^2$ (nondimensional wavelength $L = 17.24$). Series A: continuous shading is employed between the minimum and maximum thicknesses in each image. The hatched regions represent the pseudo-dewetted areas. The onset of dewetting is clearly seen in image A4 (from left to right, top). Gray scale images in the two rows correspond to (from left to right): $T = 0$, 8, 91, 206, 241, 362, 859, 1455, 2004, and 7124, respectively. The corresponding nondimensional maximum and minimum thicknesses (H) in images 1 to 10 are (1.099, 0.9), (1.03, 0.97), (1.13, 0.94), (1.98, 0.82), (2.1, 0.81), (2.3, 0.79), (2.73, 0.78), (3.03, 0.78), (3.26, 0.77), and (4.04, 0.77), respectively. Series B: 3-D morphologies of the film are shown at $T = 92$, 242, 4453 and 7123, respectively.

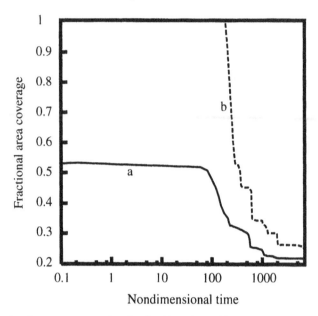

Figure 3. Fractional area coverage by the liquid regions thicker than the threshold thickness. Variations of the fractional coverage above the mean thickness (curve a) and above 15% of the equilibrium thickness (curve b) with the nondimensional time are shown for a 7 nm film.

area is simply 1– fractional area coverage in curve b. Inspection of curve b shows that after the onset of dewetting (at nondimensional time $T \approx 200$), the dewetted area increases almost linearly until the process of ripening becomes important (at $T \approx 300$); thereafter, the dewetted area increases in discrete steps which mirror the disappearance of smaller drops by ripening. Similarly, curve a shows a rapid decline in the area occupied by the liquid ridges/droplets above the mean film thickness until the onset of dewetting, and a slower decline thereafter. The initial rapid decline is due to the fragmentation of ridges to produce more compact circular droplets of increased heights. After the onset of dewetting, slower changes occur due to retraction of droplets making them more circular and compact ($T < 600$), and thereafter by ripening.

Figure 4 for the 7 nm film of Fig. 2 quantifies the lateral extent and amplitude of depressions (valleys or regions below the mean thickness), as well as the amplitude of ridges (regions above the mean thickness). A convenient way to quantify the lateral extent is the effective mean radius, defined as twice the area covered by depressions divided by the total perimeter of depressions. In the first phase ($T < 80$), there is only a rearrangement leading to a bicontinuous pattern. Depressions merely widen slowly without becoming deeper. The minimum thickness then declines, leading to the onset of dewetting ($T \approx 200$), by which time droplets have emerged (the maximum nondimensional thickness increases to about 2). After dewetting begins, the effective radius of depressions grows with time almost as $\log(T)$. Figure 4 also clearly points out that in relatively thin films, a significant

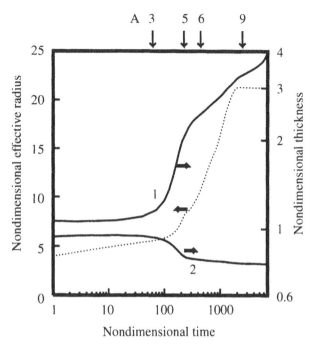

Figure 4. Variations of the nondimensional effective radius of depressions (below the mean thickness), maximum thickness (curve 1), and minimum thickness (curve 2) for a 7 nm film. The arrows numbered 3, 5, 6, and 9 on top indicate the times corresponding to images 3, 5, 6, and 9 of series A in Fig. 2.

growth of depressions leading to dewetting is concluded only after the formation of droplets, which continue to grow further by ripening, but with different kinetics.

The same qualitative pathway of morphological evolution was seen in a large number of simulations (not shown) for all thicknesses to the left of the minimum in the $\partial^2 \Delta G / \partial h^2$ vs. thickness curve, regardless of the numerical values of the parameters A, S^P and l_p which characterize the potential.

In contrast to the above scenario, Fig. 5 shows a different pattern of self-organization and a different pathway of dewetting for a thick film (25 nm), which is much thicker than the thickness at which the minimum in the spinodal parameter occurs (Fig. 1). The initial random disturbances (image A1 in Fig. 5) are again first organized into a small-amplitude bicontinuous pattern on a length scale close to the dominant wavelength calculated from the linear stability analysis (image A2). Thereafter, largely circular depressions emerge and quickly grow into full thickness isolated holes which initiate dewetting (images A3, A4, B1, and B2). New holes continue to form, expand, and coalesce thereafter (images A5–A8, B3, and B4). The axisymmetry of the holes is gradually lost because of the interaction with neighboring holes. A large-scale structure of liquid ridges and dewetted areas forms by repeated coalescence of holes (image A9), which then transforms into droplets by fragmentation of ridges, which is reminiscent of the Rayleigh instability (images A10 and B5).

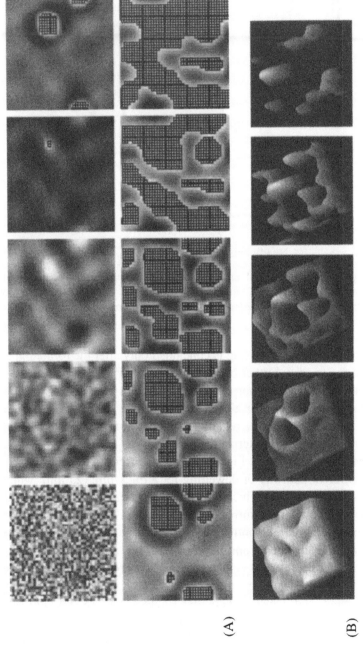

Figure 5. Different stages of evolution in a 25 nm thick film, which is much thicker than the location of the minimum (at 7.82 nm) in the spinodal parameter vs. thickness curve a of Fig. 1. The area of each box is $9L^2$ ($L = 8.9$). Series A: gray scale images correspond to (from left to right) $T = 0$, 0.06, 5.16, 8.8, 10.2, 12.0, 13.5, 15.5, 20.4, and 1687, respectively. The onset of dewetting can be clearly seen in image A4. The corresponding nondimensional maximum and minimum thicknesses (H) in images 1–10 are (1.09, 0.9), (1.05, 0.94), (1.09, 0.89), (1.17, 0.22), (1.44, 0.206), (2.08, 0.205), (2.64, 0.204), (2.83, 0.204), (4.07, 2.05), and (5.39, 0.205), respectively. Series B: 3-D morphologies are shown at $T = 8.8$, 11.7, 13.5, 15.5, and 1687.

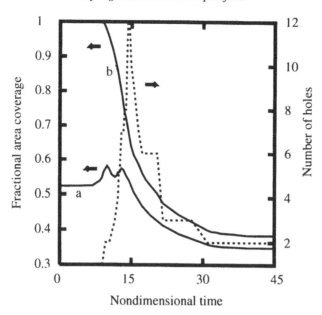

Figure 6. Fractional area coverage by the liquid regions thicker than the threshold thickness. Variations of the fractional coverage above the mean thickness (curve a) and above 15% of the equilibrium thickness (curve b) with the nondimensional time are shown for a 25 nm film. The number of holes is shown by the dotted curve.

Figure 6 shows the fractional area covered by the liquid regions above the mean thickness (curve a), and above the cut-off equilibrium thickness (curve b), during the course of evolution for the 25 nm thick film. The same figure also gives information on the temporal evolution of the number of holes, which initially increase and then decline due to coalescence. The area still covered by the liquid declines sharply during the hole formation and growth ($T < 15$) and more slowly thereafter. The area covered by ridges above the mean thickness (curve a) increases slightly after the formation of rims around the holes, but starts to decline after the hole coalescence dominates ($T > 15$). A slight intermediate oscillation is due to a composite effect of the two concurrent processes of hole growth and hole coalescence.

Figure 7 (to be contrasted with Fig. 4) presents the effective radius of depressions and the minimum and maximum thicknesses for the 25 nm film. In the first phase ($T < 7$), there is a significant widening of depressions without a concurrent change in amplitudes. Rupture by holes occurs rather explosively, without a significant increase in the maximum thickness during the hole formation. The appearance of new holes decreases the radius slightly. The continued hole expansion and dewetting, however, lead to a sustained growth of the effective radius at long times.

A very similar sequence of morphological evolution (bicontinuous pattern → circular holes with rims → hole expansion → hole coalescence) was also shown [24–26] by simulations in the following two general classes of systems: (a) films subject only to a long-range van der Waals attraction, regardless of their mean thickness; and (b) relatively thick, type IV films close to their upper spinodal

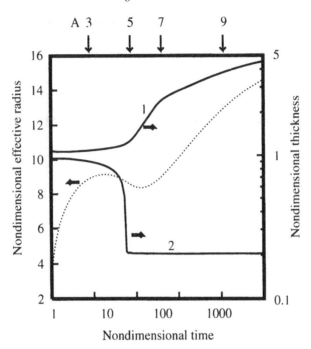

Figure 7. Variations of the nondimensional effective radius of depressions (below the mean thickness), the maximum thickness (curve 1), and the minimum thickness (curve 2) with time for a 25 nm thick film. The arrows numbered 3, 5, 7, and 9 on top indicate the times corresponding to images 3, 5, 7, and 9 of series A in Fig. 5.

boundary. In the latter system, the spinodal (unstable) region occurs in the vicinity of the primary minimum, which is different from the case considered here. The present study thus reinforces the belief in a general pathway of pattern evolution and dewetting by the formation of isolated holes whenever the initial mean thickness is sufficiently high so that the repulsive interactions at the minimum thickness are encountered only after a considerable growth of the instability. This conclusion appears to be independent even of the general classes of the potential, namely it seems to hold for all combinations of attractive and repulsive, and of short- and long-range forces. The formation of circular holes is therefore not always indicative of 'nucleation' by large dust particles, defects, etc., but can also occur by a spontaneous growth of surface instability as the film thickness increases towards the upper spinodal boundary (for the type II system, the upper spinodal is at infinity).

Figure 8 shows the evolution for an intermediate thickness film (10 nm in Fig. 1). The initial bicontinuous pattern resolves into a mixture of both drops and depressions displaying less axisymmetry in varying proportions (image 1). Dewetting still occurs by the formation of holes, which now all appear within a narrower window of time compared with thicker films. Coalescence of holes produces a bicontinuous dewetted structure made up of long ridges and flat valleys. Ridges disintegrate to produce the droplets.

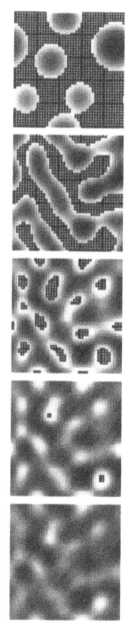

Figure 8. Evolution of the pattern in an intermediate thickness film (10 nm). The area of each box is $16L^2$ ($L = 9.27$). The onset of dewetting can be clearly seen in the second image. The gray scale images correspond to (from left to right) $T = 9, 15, 19, 35$, and 435, respectively. The corresponding nondimensional maximum and minimum thicknesses (H) are $(1.15, 0.81)$, $(1.4, 0.57)$, $(1.69, 0.53)$, and $(3.78, 0.52)$, respectively.

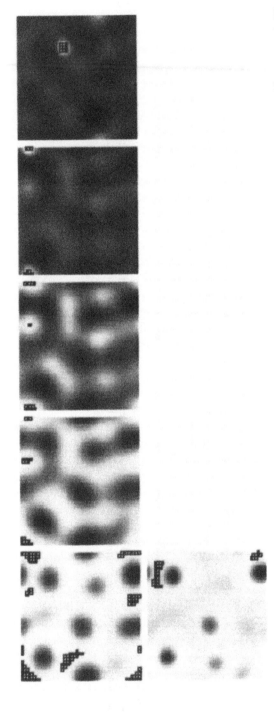

Figure 9. Morphology at the onset of dewetting in 7 nm, 8 nm, 10 nm, 15 nm, and 25 nm thick films of the system represented by curve a in Fig. 1. For a different potential (curve b), the lone image in the next row represents the pattern at the onset of dewetting in a 10 nm film. The areas shown are $9L^2$. A continuous linear gray scale between the minimum and the maximum thicknesses in each image is used.

Thus, different pathways of pattern evolution can produce strikingly different morphologies at the onset of dewetting in films of different thicknesses as summarized clearly in Fig. 9. For a 7 nm film, dewetting occurs by the formation of circular droplets. For intermediate thickness, slightly to the right of the spinodal minimum, a bicontinuous pattern of ridges and valleys exists at the onset of dewetting, the structure being more uniform and connected for thicker films. At still higher thickness (15 and 25 nm), dewetting occurs by isolated holes which become increasingly circular for thicker films.

The morphological pathway of dewetting thus depends only on the relative location of the film thickness *vis-à-vis* the location of the minimum in the spinodal parameter. Figure 9 further illustrates this point in the last image, which shows the pattern at the time of dewetting for a 10 nm film of a different system represented by curve b of Fig. 1, where the minimum occurs at 12 nm. Dewetting indeed occurs by droplets, rather than by a bicontinuous structure. Thus, as far as the qualitative morphological features are concerned, a 10 nm thick film of a system displaying the spinodal minimum at 12 nm is similar to a 7 nm film of a system with the spinodal minimum at 8 nm. Thus, the pattern is governed not by the film thickness *per se*, but by its positioning *vis-à-vis* the location of the spinodal minimum, which is different for different systems.

Recent experiments of Herminghaus *et al.* [18] with thin liquid crystal films show the same qualitative variation of the free energy as used in the simulations here. The most prominent feature is the existence of a primary and a secondary minimum. Interestingly, films close to the secondary minimum showed two distinct, but co-existing, patterns and pathways of dewetting [18]: (a) pseudo-dewetting in the secondary minimum by a correlated undulating structure similar to the one seen in simulations for relatively thin to moderate thickness films (Figs 4, 8, and 11) before the long-term formation of circular drops, and (b) true dewetting in the primary minimum by the formation of randomly distributed circular holes. Based on our simulations, we hypothesize that the latter structure results from the heterogeneity of the substrate in conjunction, possibly, with the heterogeneity of the molecular ordering close to the substrate surface, which leads to a weaker short-range repulsion and therefore a shift of the secondary minimum to a smaller thickness closer to the primary minimum. On such randomly distributed patches, pseudo-dewetting should occur by the formation of circular holes since the distance between the (secondary) spinodal minimum and the film thickness increases. The equilibrium film left at the base also becomes much thinner so that 'nucleation' to the primary minimum now becomes easier. Indeed, in another set of experiments with molten gold films [18], which do not experience a soft repulsion, correlated holes could be witnessed. This is in conformity with the simulations reported here and those published previously for purely attractive long-range forces [24]. In experiments on aqueous films on graphite [17] and on low-molecular-weight polystyrene films [11] on silicon (both of which are type IV systems considered elsewhere [25, 26]), holes due to both nucleation and spinodal dewetting seem to co-exist. For aqueous films, the population of holes produced by the spinodal mechanism is expected to increase with increased rate of evaporation [35], as evidenced also in the experiments [17].

Simulations show that the magnitude of the characteristic length scale of the instability is well predicted by the linear analysis of the dominant wavelength *until the onset of dewetting*. For example, a maximum of eight droplets and 12 holes appear in Figs 2 and 5 (see also Fig. 6), respectively. Since these simulations are carried out over an area of $9L^2$, the linear theory would anticipate nine structures. The hole density was almost always found to be slightly larger, since additional holes at late times continue to form in the interstitial spaces whenever the space available is larger than the critical length. It is well known that the critical wavelength for the instability is smaller by a factor of $\sqrt{2}$ than the dominant wavelength of the linear theory. Of course, at late times, the structure coarsens due to hole coalescence or by ripening of drops, and the length scale of the structure can no longer be predicted by the linear theory.

Finally, we consider the Minkowski functional for the connectivity of patterns [38, 39], which is a useful topological measure used for the analysis of thin films [13, 18] and other patterns [38]. To analyze the connectivity of a pattern, a two-color representation of the film morphology is first formed by dividing the regions above (black) and below (gray) a threshold thickness. We chose the mean film thickness to threshold the image in order to facilitate convenient comparisons with future experimental work. The connectivity is simply a measure of the difference between the number of isolated (disconnected) gray domains (G) and the number of isolated black domains (B). The normalized connectivity is $[(G - B)/(G + B)]$. Connectivity is positive (negative) if the number of isolated depressions is larger (smaller) than the number of distinct ridges. The connectivity measure is also an indicator of the mean curvature of the pattern [37], positive (negative) connectivity implying the dominance of concave (convex) black regions. Thus, the presence of isolated holes (droplets) leads to a positive (negative) connectivity. Connectivity vanishes when the numbers of black and gray domains are equal. The evolution of connectivity with time also gives a clue regarding whether the pattern is correlated (e.g. arising through a spinodal mechanism) or random (e.g. a Poissonian process) [13]. Further details regarding computations of connectivity from the gray scale images (or profiles) are given elsewhere [38, 39].

Figure 10 shows the temporal evolution of connectivity for 7 nm, 10 nm, and 25 nm thick films. Typical two-color representations of the morphology for 7 nm (droplet forming) and 25 nm (hole-forming) films are summarized in Figs 11 and 12, respectively. For the 7 nm film, connectivity is initially negative since the initial rearrangement produces a structure in which the number of isolated ridges exceeds the number of isolated valleys (image 1 in Fig. 11). The isolated hills coalesce to form longer interconnected ridges, some of which trap isolated depressions (image 2 in Fig. 11). Connectivity now becomes positive. Further evolution produces a bicontinuous pattern where both the ridges and the valleys are equally connected

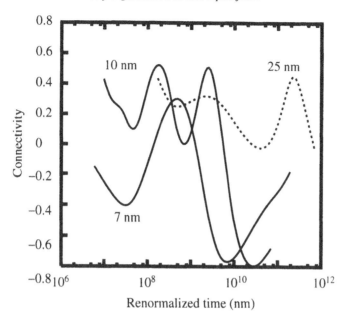

Figure 10. Connectivity of the pattern at a threshold thickness equal to the mean film thickness for 7 nm, 10 nm, and 25 nm thick films. A scaled renormalized time, t_N (in nm) $= Th(h/d_0)^4$, as defined in the text, is used ($d_0 = 0.158$ nm).

(image 3 in Fig. 11), and the connectivity thus once again passes through zero. Fragmentation of ridges into droplets leads to negative connectivity (image 4), and finally, at late times, ripening (image 5) increases connectivity due to decreased drop density.

In contrast to the above scenario, connectivity for a thick hole forming 25 nm thick film remains positive at all stages of evolution and dewetting except at very long times after the conclusion of coalescence leading to droplet formation (not shown). Connectivity initially declines (image 1 in Fig. 12) until a bicontinuous pattern is formed (image 2 in Fig. 12). The formation of isolated depressions developing into holes (image 3 in Fig. 12) increases the connectivity until a maximum is reached after some coalescence (image 4). Thereafter, the fragmentation of ridges starts to produce isolated (noncircular) droplets, which decreases the connectivity. Further evolution would have produced an array of droplets (not shown), leading to negative connectivity.

The connectivity features for intermediate thickness (e.g. curve for 10 nm in Fig. 10) are more complex, but are clearly derived from a combination of the characteristics of thin and thick films. In general, if one considers the evolution *until dewetting*, the regions showing positive (negative) connectivity enlarge (shrink) as the film gets thicker. In fact, for films thicker than the critical thickness, connectivity always remains positive. Thus, the connectivity measure appears to be a useful tool for characterizing and discriminating the thin film patterns and their pathways of

Figure 11. Two-color representation of the film morphology in a 7 nm thick film. Black and gray represent the regions above and below the mean thickness, respectively. The images correspond to (from left to right) $t_N(\times 10^{-7}) = 4.12, 23, 159, 816,$ and 3230 nm, respectively. t_N is as defined in Fig. 10.

Figure 12. Two-color representation of the film morphology in a 25 nm thick film. Black and gray represent the regions above and below the mean thickness, respectively. The images correspond to (from left to right) $t_N (\times 10^{-9}) = 1.6, 100, 180, 280,$ and 600 nm, respectively. t_N is as defined in Fig. 10.

evolution, especially from experiments. The theory presented here will be useful for this purpose.

Interestingly, it may be noted that two-color representations of the morphology (e.g. Figs 11 and 12), which are often used to depict experimental morphologies, carry much less information than their gray scale counterparts. In fact, mere visual observation of Figs 11 and 12 does not readily suggest the rather drastic differences in the pathways of evolution evidenced in the detailed gray scale images (the formation of droplets in Fig. 12 at long times is not shown, the addition of which would make it almost identical to Fig. 11). Moreover, patterns can be made to appear different by choosing a different threshold thickness in the two-color representations. A more complete analysis of an experimental pattern requires computations of connectivity at different threshold thicknesses [26, 38]. This would facilitate a more meaningful comparison between simulations and experiments.

4. CONCLUSIONS

The morphological pathways of spontaneous pattern evolution and dewetting have been studied in type II films in the spinodal region around the secondary minimum of the spinodal parameter, which is the negative curvature of the excess free energy per unit area. Relatively thick films some distance away to the right of the spinodal minimum are dewetted by the formation of isolated circular holes, which grow and coalesce to form a giant structure of liquid ridges, which in turn decays into droplets. In thick films, not all holes appear within a narrow window of time, as may be concluded from the linear analysis. Droplets are formed at an increasingly earlier stage of evolution as the film thickness decreases. Films thinner than the spinodal minimum decay into droplets via a bicontinuous structure before the onset of dewetting. An important conclusion is that different morphologies (isolated circular holes, droplets, bicontinuous) and their combinations can all be produced depending on the film thickness *vis-à-vis* the location of the minimum in the force vs. thickness curve. Moreover, all three patterns (holes and drops of varying sizes, bicontinuous ridges, and droplets) can co-exist in varying proportions at a given time even on homogeneous surfaces, especially for intermediate thickness films. Since most real (and even experimental) surfaces are heterogeneous, they can support a wide variety of patterns resulting from the spinodal dewetting alone. The results presented here are also germane to films displaying many local spinodal minima due to layering, as the evolution in each local minimum is governed by the same considerations.

Regardless of the morphology, however, the structure is correlated, and its length scale is adequately predicted by the linear analysis *until the onset of dewetting*. Hole coalescence or droplet ripening produces larger-scale structures at late times after dewetting. However, it can be noted that on a heterogeneous surface, large-scale correlation of the structure may be lost, but the local morphology will still be governed by the local potential.

The above results should aid our understanding, design, and interpretation of thin film experiments, where the linear analysis has hitherto been used as the guide.

Acknowledgements

Discussions with S. Herminghaus, K. R. Mecke, G. Reiter, and K. Jacobs are gratefully acknowledged. This work was supported by a grant from the Indo-French Centre for the Promotion of Advanced Research/Centre Franco-Indien Pour la Promotion de la Recherche.

REFERENCES

1. A. Vrij, *Discuss. Faraday Soc.* **42**, 23 (1966).
2. E. Ruckenstein and R. K. Jain, *J. Chem. Soc., Faraday Trans. 2* **70**, 132 (1974).
3. B. V. Derjaguin, *Theory of Stability of Thin Films and Colloids.* Consultants Bureau/Plenum Press, New York (1989).
4. A. Sharma, *Langmuir* **9**, 861 (1993).
5. G. Reiter, *Phys. Rev. Lett.* **68**, 75 (1992).
6. G. Reiter, *Langmuir* **9**, 1344 (1993).
7. G. Reiter, P. Auroy and L. Auvray, *Macromolecules* **29**, 2150 (1996).
8. W. Zhao, M. H. Rafailovich, J. Sokolov, L. J. Fetters, R. Plano, M. K. Sanyal, S. K. Sinha and B. B. Sauer, *Phys. Rev. Lett.* **70**, 1453 (1993).
9. J. M. Guerra, M. Srinivasrao and R. S. Stein, *Science* **262**, 1395 (1993).
10. L. Sung, A. Karim, J. F. Douglas and C. C. Han, *Phys. Rev. Lett.* **76**, 4368 (1996).
11. R. Xie, A. Karim, J. F. Douglas, C. C. Han and R. A. Weiss, *Phys. Rev. Lett.* **81**, 1251 (1998).
12. J. Bischof, D. Scherer, S. Herminghaus and P. Leiderer, *Phys. Rev. Lett.* **77**, 1536 (1996).
13. K. Jacobs, S. Herminghaus and K. R. Mecke, *Langmuir* **14**, 965 (1998).
14. G. Reiter, A. Sharma, R. Khanna, A. Casoli and M.-O. David, *J. Colloid Interface Sci.* **214**, 126–128 (1999).
15. G. Reiter, A. Sharma, R. Khanna, A. Casoli and M.-O. David, *Europhys. Lett.* **46**, 512–518 (1999).
16. G. Reiter, A. Sharma, A. Casoli, M.-O. David, R. Khanna and P. Auroy, *Langmuir* **15**, 2551–2558 (1999).
17. U. Thiele, M. Mertig and W. Pompe, *Phys. Rev. Lett.* **80**, 2869 (1998).
18. S. Herminghaus, K. Jacobs, K. Mecke, J. Bischof, A. Fery, M. Ibn-Elhaj and S. Schlagowski, *Science* **282**, 916 (1998).
19. M. B. Williams and S. H. Davis, *J. Colloid Interface Sci.* **90**, 220 (1982).
20. A. Sharma and A. T. Jameel, *J. Colloid Interface Sci.* **161**, 190 (1993).
21. A. Sharma and A. T. Jameel, *J. Chem. Soc., Faraday Trans.* **90**, 625 (1994).
22. R. Khanna, A. T. Jameel and A. Sharma, *Ind. Eng. Chem. Res.* **35**, 3108 (1996).
23. A. Sharma and G. Reiter, *J. Colloid Interface Sci.* **178**, 383 (1996).
24. R. Khanna and A. Sharma, *J. Colloid Interface Sci.* **195**, 42 (1997).
25. A. Sharma and R. Khanna, *Phys. Rev. Lett.* **81**, 3463 (1998).
26. A. Sharma and R. Khanna, *J. Chem. Phys.* **110**, 4929 (1999).
27. J. W. Cahn, *J. Chem. Phys.* **42**, 93 (1965).
28. J. N. Israelachvili, *Intermolecular and Surface Forces.* Academic Press, London (1992).
29. C. J. van Oss, M. K. Chaudhury and R. J. Good, *Chem. Rev.* **88**, 927 (1988).
30. K. R. Shull, *J. Chem. Phys.* **94**, 5723 (1991).

31. K. R. Shull, *Faraday Discuss.* **98**, 203 (1994).
32. X. L. Chu, A. D. Nikolov and D. T. Wasan, *J. Chem. Phys.* **103**, 6653 (1995).
33. V. Bergeron and C. J. Radke, *Langmuir* **8**, 3020 (1992).
34. H. Riegler, A. Asmussen, M. Christoph and A. Davydov, in: *Proceedings of 30th Rencontres de Moriond, Short and Long Chains at Interfaces*, J. Daillant, P. Guenoun, C. Marques, P. Muller and J. T. T. Van (Eds), pp. 307–312. Editions Frontieres, Gif-sur-Yvette, France (1995).
35. A. Sharma, *Langmuir* **14**, 4915 (1998).
36. N. N. Yanenko, *The Method of Fractional Steps*. Springer-Verlag, New York (1971).
37. D. A. Anderson, J. C. Tannehill and R. H. Pletcher, *Computational Fluid Mechanics and Heat Transfer*. Hemisphere, Washington, DC (1984).
38. K. R. Mecke, *Phys. Rev. E* **53**, 4794 (1996).
39. A. Rosenfield and A. C. Kak, *Digital Picture Processing*. Academic Press, New York (1976).

Part 3

Wetting of Heterogeneous, Rough and Curved Surfaces

Apparent and Microscopic Contact Angles, pp. 285–300
J. Drelich, J. S. Laskowski and K. L. Mittal (Eds)
© VSP 2000.

Underwater captive bubble technique on curved surfaces of blended polydimethylsiloxanes

ERIC A. ANGU [1], RYAN C. CAMERON [2], STEVE K. POLLACK [2]
and WILLIAM E. COLLINS [1,*]

[1] *Department of Chemical Engineering, Howard University, Washington, DC 20059, USA*
[2] *Department of Chemistry, Howard University, Washington, DC 20059, USA*

Received in final form 27 April 1999

Abstract—A modified underwater captive bubble technique was used to determine the air–water contact angle of water on a series of polymer biomaterials, including polydimethylsiloxane blends. These consisted of unfunctionalized polydimethylsiloxane (PDMS) blended with small percentages of either ethylene oxide (EO)- or phosphorylcholine (PC) ester-terminated polydimethylsiloxanes to increase surface hydrophilicity. The latter was formed by the hydrosilation of a dimethylsilane-terminated polydimethylsiloxane with ω-vinylphosphorylcholine ester. The ester was synthesized from 9-hydroxy-1-decene and 2-chlorotrimethylamine incorporating the phosphoryl functionality. The blends were cast from a dilute solution of chloroform onto low-density polyethylene (LDPE). A comparison of methods to implement the underwater captive bubble (UWCB) technique, the curved UWCB technique or the modified UWCB technique, a convenient method for determining θ_{AW} on curved surfaces, demonstrated that both methods were accurate and precise. As the percentage of functionalized polydimethylsiloxane in a family of blends increased, θ_{AW} decreased. The θ_{AW} value on the EO-siloxane blends was nonetheless typically 30° less than that on the PC-siloxane blends. The values of θ_{AW} on blends determined with the modified UWCB technique demonstrated that too little functionalized polydimethylsiloxane was present to produce a hydrophilic surface promising hemocompatibility.

Key words: Phosphoryl choline ester; captive bubble technique; air–water contact angle; polysiloxanes; poly(ethylene oxide); polymer blends; hemocompatibility.

1. INTRODUCTION

The surface properties of biomaterials often determine their hemocompatibility. The air–water contact angle θ_{AW}, an important surface property, is directly related to the surface free energy of a biomaterial. Contact angle measurement is considered a

---

*To whom correspondence should be addressed. E-mail: wcollins@scs.howard.edu

valuable technique to characterize biomaterial surfaces [1] because θ_{AW} is sensitive to the outermost 3–20 Å of a surface and its chemistry.

A number of techniques can be used to measure θ_{AW} accurately, among them the capillary rise, Wilhelmy plate, and underwater captive bubble (UWCB) techniques. The UWCB technique is not as versatile as other techniques and measures only the receding contact angle [2]. Accordingly, the UWCB technique is more sensitive to the polar rather than the dispersive component of the surface free energy [3]. The UWCB technique is nonetheless useful because it can measure θ_{AW} on curved surfaces with considerable precision [4]. Thus, the UWCB technique can be applied to measure the θ_{AW} values of biomaterials without refabricating them to disturb the prospective nuances of surface structure. These nuances include roughness, heterogeneous or nonequilibrium crystal structure, microphase separation, additives, chemical impurities, and surface modifications such as grafting or plasma treatment, nuances likely to determine the biological response to biomaterials. Replacing the conventional flat plate with a biomaterial tubing, the Wilhelmy plate technique is transformed into a 'Wilhelmy tube' technique [3]. However, the Wilhelmy tube technique furnishes useful results only in the unlikely case that the inside and outside tubing surfaces possess the same chemistry and structure. The capillary rise technique is confounded by contact angles near 90°, measuring them only with substantial difficulty.

Circular tubings are often used in biomedical implants. Such tubing is used in catheters, vascular prostheses, pacemaker wire insulation, connecting tubing for dialysis, and so forth [4]. When θ_{AW} is measured on surfaces that are not flat, a sufficiently small drop or bubble size is required to neglect all effects of curvature. While it is possible, though difficult, to make θ_{AW} measurements on highly convoluted surfaces, the surface of circular tubing facilitates calculations of the effect of curvature on bubble dimensions. Such a technique enables θ_{AW} to be measured directly on the fabricated material. Device fabrications have been shown to affect the surface properties of the device material. Accordingly, it is incorrect to characterize a device fabricated with injection molding by measuring surface properties on a sample fabricated by another technique, like solvent casting. To overcome this problem, one must develop a technique that will measure θ_{AW} on the surface of a material, as fabricated.

This report presents a UWCB technique procedure to measure θ_{AW} on curved surfaces with more convenience, called 'the modified UWCB technique'. The procedure can be applied to other fluids, e.g. octane or dimethylsulfoxide, with equal success. Accordingly, the procedure will furnish data that can be employed to calculate the surface free energies of biomedical polymers in the form of circular tubing [2]. The procedure utilizes automated goniometry to apply previously developed procedures for the UWCB technique applied to curved surfaces more efficiently. This has particular application to biomaterials research. Polymer surface groups are sufficiently mobile to orient depending on the chemical environment (hydrophobic or hydrophilic) [1, 5, 6]. To demonstrate the applicability of the modified UWCB tech-

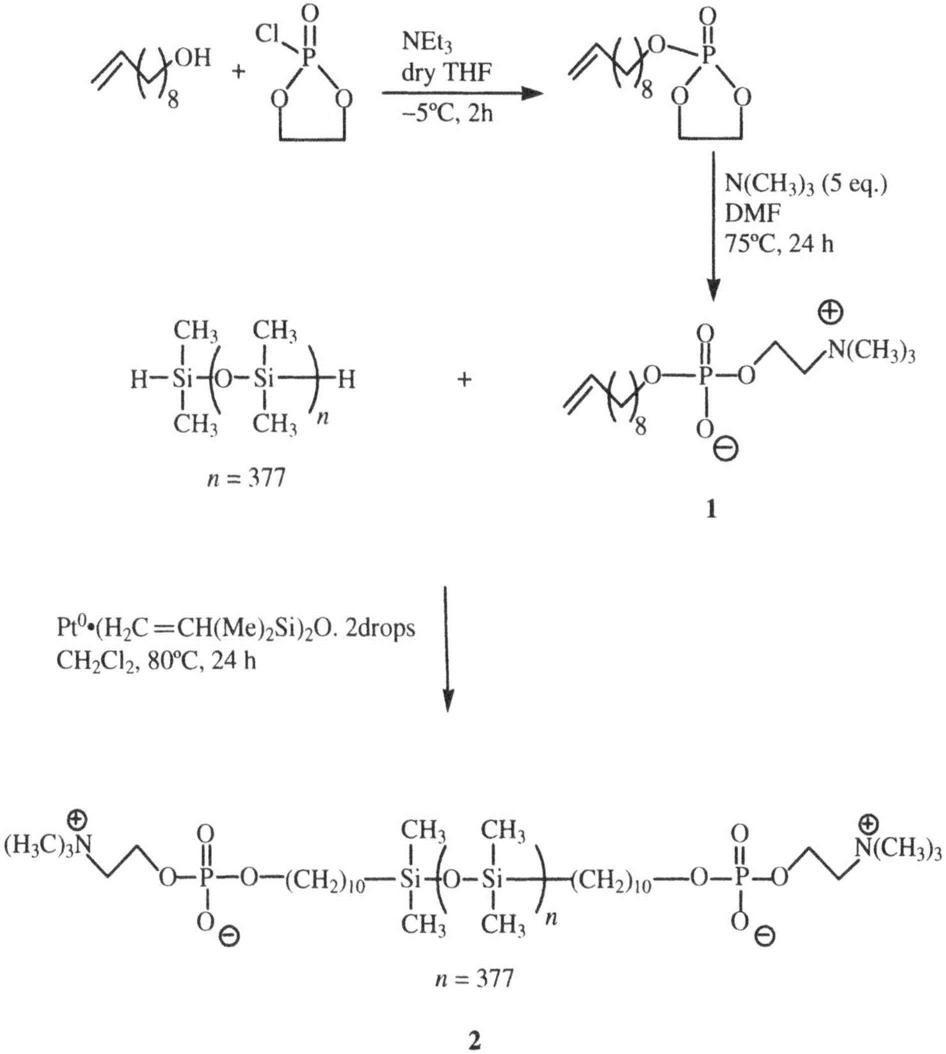

$n = 377$

1

$Pt^0 \bullet (H_2C=CH(Me)_2Si)_2O$. 2drops
CH_2Cl_2, 80°C, 24 h

$n = 377$

2

Figure 1. Synthesis of the ω-vinylphosphorylcholine ester, which was grafted onto the termini of diethylsilane-terminated PDMS using hydrosilation.

nique, it was used to measure the θ_{AW} values on commercial biomaterial tubings: polypropylene (PP), low-density polyethylene (LDPE), solution grade Biomer™ (SB), and Teflon fluorinated ethylene propylene copolymer (FEP). These air–water contact angles were compared with θ_{AW} on the same curved biomaterial surfaces as measured with the method of Lelah *et al.* [4] (the 'curved UWCB technique').

This new procedure used to measure θ_{AW} on curved surfaces with the UWCB technique was subsequently applied to determine θ_{AW} for blends of functionalized and unfunctionalized polydimethylsiloxanes. Primarily investigated were blends including telechelic siloxane precursors terminated with phosphorylcholine esters (PC-siloxanes) (Fig. 1). The PC-siloxanes were blended with unfunctionalized polydimethylsiloxane to form 'PC-siloxane blends'. These esters behave similarly to natural phospholipids in many respects [7, 8]. Hypothetically, a phospholipid surface would be hemocompatible since cells and platelets fail to adhere to one another under normal circumstances. Accordingly, biomaterials functionalized with phospholipid or phosphorylcholine esters would display promising hemocompatibility as the biomaterials literature suggests [9].

The PC-siloxanes studied were blended up to 10% with uncrosslinked polydimethylsiloxane under the conjecture that the phosphorylcholine ester groups would prefer a hydrophilic environment. These groups would accordingly preponderate at the interface between the polymer coating and aqueous solution. A hydrophilic surface would presumably result. In support of this hypothesis, it was recently demonstrated that the end groups of functionalized siloxanes dominated the surface energetics of these materials [10].

The modified UWCB technique demonstrated that the PC-siloxane blends contained too little PC-siloxane to be hydrophilic, though the results remained auspicious. Corroboration of these results was obtained analogously using ethylene oxide-terminated polydimethylsiloxane (EO-siloxanes) blended with uncrosslinked polydimethylsiloxane (EO-siloxane blends).

2. MATERIALS AND METHODS

2.1. Preparation of PC-siloxane blends

Polydimethylsiloxane functionalized with phosphorylcholine ester chain ends was synthesized to form PC-siloxanes of a molecular weight around 20 000 [11]. These were produced from commercial dimethylsilane-terminated PDMS (Gelest, Tullytown, PA) and the 9-hydroxy-1-decenyl ester of phosphorylcholine. The phosphorylcholine ester was synthesized beforehand with the reaction of 9-hydroxy-1-decene with 2-chlorotrimethylamine to form **1** (Fig. 1). The 2-chlorotrimethylamine was obtained by reacting 2-oxo-1,3,2-dioxamphospholoane with trimethylamine. The ω-vinylphosphorylcholine ester was grafted onto the termini of the PDMS chain using hydrosilation (**2** in Fig. 1). The phosphate ester, which is chemically quite sensitive to acidic or basic conditions, was retained in this process. Blends of 1, 3, or 10% PC-siloxane in trimethylsiloxy-terminated PDMS of molecular weight 90 000 (Gelest) formed the PC-siloxane blends. The resultant PC-siloxane blends were subsequently cast from chloroform onto the lumen of LDPE tubing to furnish suitable mechanical properties.

2.2. Preparation of other polymer surfaces

The EO-siloxane blends, 1 or 10% EO-siloxane, molecular weight about 5000 (Gelest), blended with trimethylsiloxy-terminated PDMS, molecular weight 90 000, were analogously cast from chloroform onto LDPE. For purposes of comparison, pure EO-siloxane was also examined. All coating procedures were conducted at room temperature. Dilute solutions containing 3% wt/vol. of the blends in chloroform were stirred for at least 24 h. The blends were subsequently coated onto 0.125 in inner diameter (ID) tubing of LDPE (Intramedic® PE-350, Clay Adams, Parsippany, NJ) for 2 h under static conditions. The coated tubing was next dried in a nitrogen atmosphere for a week, followed by further drying under vacuum for another 10 days to ensure complete removal of the solvent.

The remaining polymers were used as received: LDPE, 0.125 in ID Teflon FEP (Cole Parmer, Niles, IL), and polypropylene (Cole Parmer). As received solutions of 30% wt/vol. of SB (Ethicon, Somerville, NJ) in dimethylacetamide were diluted to 2% for coating, stirred to homogeneity, and coated onto LDPE, using a procedure analogous to that used to coat the PC-siloxane blends. The tubings were washed in 0.125% Ivory detergent; rinsed with over 1000 tubing volumes of deionized, distilled water, and dried under vacuum for 24 h or more.

2.3. Contact angle measurements

A VCA 2500 video contact angle system (AST Products, Billerica, MA) was used to measure θ_{AW} using the UWCB technique. A sample holder, attached to a stand and clamp, contained the submerged sample in a glass container (Fig. 2). The sample consists of polymer tubing that has been sectioned in half longitudinally. The clamp has an internal diameter approximately equal to the external diameter of the tubing. This prevents the development of crimps and stresses at the tubing surface.

A disadvantage of any UWCB technique is the scrupulous care that must be taken to ensure that the large volume of water required for this technique remains pure

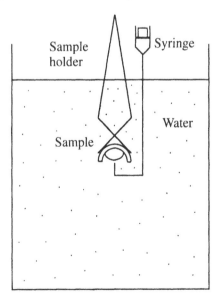

Figure 2. Apparatus to determine the contact angles on circular tubing using the UWCB technique. Neither the camera nor the automated goniometer is displayed.

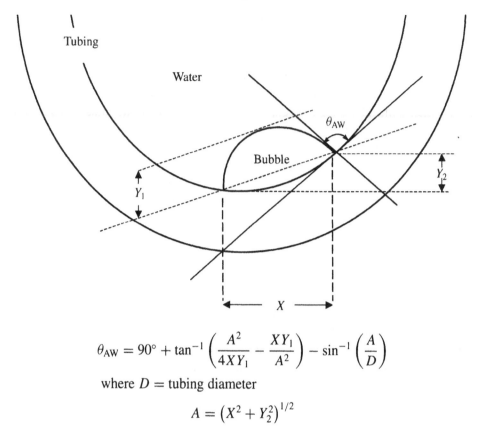

$$\theta_{AW} = 90° + \tan^{-1}\left(\frac{A^2}{4XY_1} - \frac{XY_1}{A^2}\right) - \sin^{-1}\left(\frac{A}{D}\right)$$

where D = tubing diameter

$$A = \left(X^2 + Y_2^2\right)^{1/2}$$

Figure 3. Geometric representation of the determination of θ_{AW} on circular tubing using the UWCB technique.

and does not become contaminated during the course of the measurements. (Such care was taken, to furnish a procedure that was feasible.) A very small (0.1–0.5 μl) bubble was introduced to the sample from a hand held 'J' type syringe (Hamilton Co., Reno, NV). Gentle tapping of the syringe released the bubble so that it floated to the tubing surface. The image of the bubble adherent to the tubing sample was photographed and transferred to a computer screen. The angle measured in the water phase between tangents to the curve surfaces at their point of intersection is defined as θ_{AW}, as shown in Fig. 3. The bubble dimensions Y_1, Y_2, and X, as defined in Fig. 3, were measured from enlarged photographs of the bubble. According to the curved UWCB technique [2], one calculates θ_{AW} according to

$$\theta_{AW} = 90° + \tan^{-1}\left(\frac{A^2}{4XY_1} - \frac{XY_1}{A^2}\right) - \sin^{-1}\left(\frac{A}{D}\right), \tag{1}$$

where

$$A = \left(X^2 + Y_2^2\right)^{1/2} \tag{2}$$

and D is the inner diameter of the test tubing.

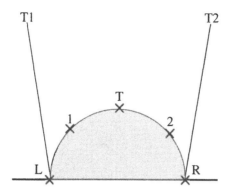

Figure 4. Implementation of the UWCB technique using the operating software of a video contact angle goniometer to mark with a cross the right (R) and left (L) points of contact of the drop with the surface, the top of the drop (T), and two intermediate positions on the surface of the drop.

The curved UWCB technique assumes that the captive bubble is microscopic and spherical. Thus, alternatively one can draw a circle through the bubble circumference and draw tangents to the bubble and tubing at their point of intersection. To diminish human error in the modified UWCB technique, automated goniometry was used to perform these drawings. A flat surface facilitates demonstration because an automated video contact angle goniometer ordinarily captures a digital image of a liquid drop on a horizontal, planar surface. Once this image is stored, the standard software allows the operator to mark the right and left points of contact of the drop with the surface, the top of the drop, and two intermediate positions on the surface of the drop (Fig. 4). The software then fits an arc to the five points, assuming points L and R lie on a plane. The five points can be independently adjusted to obtain the best visual fit of the arc to the air–liquid interface. Finally, the software determines the angle T1-L-R (or T2-R-L). These angles ordinarily correspond to the air–liquid contact angle.

Accordingly, the modified UWCB technique uses the VCA 2500 system to define the points L and R, which represent the points where the bubble contacts the lumen of the tubing. The complement of the angles K' is defined by placing point T at the apex of the bubble and points 1 and 2 in intermediate positions to optimize the fit to the air–liquid interface. The air–solid interface is subsequently defined by placing point T on the apex of the arc. The fit to this arc defines the complements of K'' (Fig. 5). The desired air–liquid contact angle is provided by the intersecting tangents to the interfaces of the two curves (Fig. 6) as

$$\theta_{AW} = 180° - \left(K' + K''\right). \tag{3}$$

The contact angle given in equation (3) is the same θ_{AW} as that calculated with the curved UWCB technique, but with more speed and convenience. Both methods were used to determine θ_{AW} for most surfaces. However, only contact angles measured with the modified UWCB technique were determined for the EO-siloxane blends.

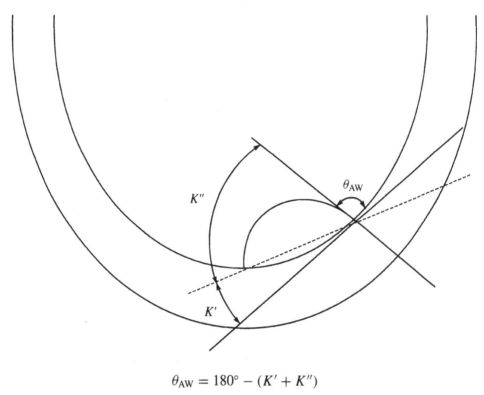

$$\theta_{AW} = 180° - (K' + K'')$$

Figure 5. Implementation of the modified UWCB technique.

3. RESULTS AND DISCUSSION

Table 1 compares θ_{AW} values obtained with the curved UWCB technique and modified UWCB technique. Both procedures have low typical standard deviations. Both procedures predict the same value of θ_{AW} statistically for a variety of polymer surfaces, from the relatively hydrophilic SB to the highly hydrophobic FEP. However, the values of θ_{AW} obtained by the curved UWCB technique were typically 2–4° lower than the values of θ_{AW} obtained for the modified UWCB technique. This is possible evidence of systematic error, corroborated by trends in the values of θ_{AW} differences. For example, the difference in θ_{AW} between the two methods diminishes with the value of θ_{AW}. Also, there are some differences in applying the curved UWCB technique in this report as compared with Lelah's original studies. Our values of θ_{AW} on SB determined with the curved UWCB technique differed from Lelah's values at a statistically significant level of $p < 0.001$ using Student's unpaired, two-tailed t-test [12]. There are plausible reasons for these differences in θ_{AW} on solution grade Biomer. Previously the θ_{AW} value on SB was determined as 56°; θ_{AW} on extruded-grade Biomer™ (EB) was 47° [13]. Furthermore, the influence of fabrication type on polymer surface properties was demonstrated by suspending extruded-grade Biomer™ in dimethylacetamide to make a 2% homogeneous solution of the polyetherurethane. This was solvent-

Table 1.
Various methods to determine θ_{AW} (in degrees)

Polymer	Lelah[a]	Curved UWCB technique[b]	Modified UWCB technique[c]
SB	60.9 ± 2.5	67.4 ± 2.9	68.2 ± 1.5
LDPE	88.4 ± 3.6	85.7 ± 5.4	88.4 ± 4.1
PP	—	89.2 ± 5.0	93.1 ± 2.3
Teflon FEP	103 ± 1	101.5 ± 4.4	107.3 ± 2.7
10% PC[d]	—	101.8 ± 1.4	105.0 ± 5.0
3% PC	—	108.1 ± 2.5	111.6 ± 4.4
1% PC	—	110.6 ± 3.1	114.7 ± 1.3

Averages given along with their standard deviations, where n is the number of data points taken for a given sample (technique and surface).

[a] The values determined with the curved UWCB technique for θ_{AW} on SB and LDPE (NIH reference tubing) [2]; n was not given. The value for θ_{AW} on Teflon FEP (flat sheet) [14] was taken from the literature.

[b] $n = 5$.

[c] $n = 11$.

[d] The PC-siloxane blend containing 10% PC-siloxane.

cast onto LDPE tubing as described above. The θ_{AW} value on this surface is $55°$, far closer to that on SB than to that on EB. Unfortunately no standard deviation is provided with these data, although they imply that scatter in the values of SB θ_{AW} was observed. The values for SB θ_{AW} in Table 1 are about $7°$ higher than the values that Lelah determined for the SB θ_{AW}, but this falls within the outer limits of the scatter in his data ($60.9° + 2.5° - 56° = 7.4°$). Furthermore, the receding θ_{AW} on SB has been determined as $77°$ using a technique other than the UWCB technique. The particular SB lot can affect the value of θ_{AW}. It has been shown that SB BSUA001 lot was surface-contaminated with poly(diisopropyl aminoethyl methacrylate), an additive that could not be extracted [15]. The SB lot used was EQ203-L180B in ref. [13] (though unknown in ref. [4]) and in this study it was BSQXXI2. These factors demand consideration when appraising the difference between the SB θ_{AW} measured with Lelah's application of the curved UWCB technique and our applications of either method. We emphasize that we determined the receding air–water contact angle, which can differ widely from values of θ_{AW} determined as advancing air–water contact angles.

The value for θ_{AW} on polyethylene determined with the curved UWCB technique is identical to the value of θ_{AW} on LDPE determined with the modified UWCB technique. Our application of the curved UWCB technique measured a θ_{AW} that also agrees well with a previous determination of the θ_{AW} on LDPE tubing (Intramedic® PE-350 0.125 in ID) with the curved UWCB technique, a value of $85°$, no standard deviation cited [13]. Materials examined with the curved UWCB technique were fabricated differently to the conditions used in this report. For the National Institutes of Health (NIH) standard polyethylene as tubing, a flat sheet, or a flat sheet rolled

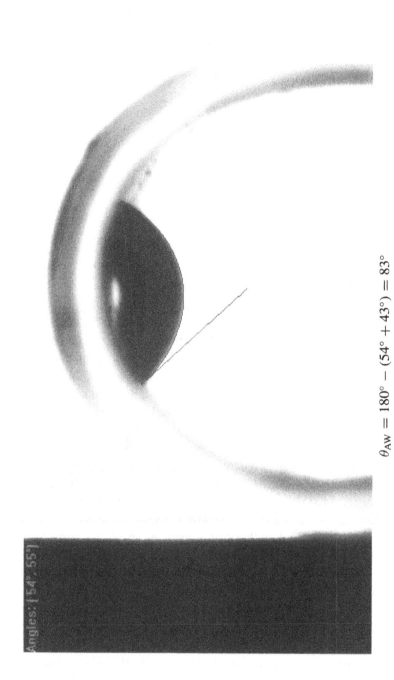

$$\theta_{AW} = 180^\circ - (54^\circ + 43^\circ) = 83^\circ$$

Figure 6. Geometric representation of the determination of θ_{AW} on circular tubing using the modified UWCB technique (1% EO-siloxane). Angle K.

Figure 6. (Continued). Angle K'.

Table 2.
Role of functionalized end groups in the determination of θ_{AW}

Polymer	Air–water contact angle (degrees)	Significance of the null hypothesis[a]
Pure EO-Siloxane[b]	77.4 ± 3.0	—
10% EO[b,c]	79.2 ± 2.0	Not significant
1% EO[b]	83.0 ± 1.4	< 0.001
10% PC[d]	105.0 ± 5.0	< 0.001
3% PC[d]	111.6 ± 4.4	< 0.01
1% PC[d]	114.7 ± 1.3	Not significant

Averages given along with standard deviations.
[a] Probability that the null hypothesis is significant between this and the preceding value of θ_{AW}.
[b] $n = 18$.
[c] EO-siloxane blend containing 10% ethylene oxide-terminated polydimethylsiloxane.
[d] $n = 11$.

into a cylinder, θ_{AW} determined with the curved UWCB technique was the same as that determined on flat surfaces. When SB corresponded to solution grade Biomer™ solvent cast on a flat sheet, it was demonstrated that for SB as a flat sheet or a flat sheet rolled into a cylinder, θ_{AW} was the same. Our preliminary work measuring θ_{AW} on flat surfaces coated with the blended polydimethylsiloxanes gives the same values as those obtained with the modified UWCB technique for curved surfaces.

In Table 1, the PP θ_{AW} is about $4°$ greater than the LDPE θ_{AW}, perhaps reflecting greater oxidation of LDPE by the ambient environment [16]. Perhaps the pendant methyl groups of PP provide more resistance to oxidation; a high percentage of PP crystallinity can increase its θ_{AW} [17]. FEP exhibited the largest difference in average contact angles in Table 1, but this difference does not seem significant.

This study showed that θ_{AW} depended on the properties of the PC-siloxane blends as expected. For instance, as the percentage of PC-siloxane in the blend increased, θ_{AW} decreased for any of the methods given in Table 1. For a given method, the θ_{AW} value for 1% PC differed from that on 10% PC at excellent levels of significance (Table 2). For a given method, the θ_{AW} value for 3% PC differed from that on 10% PC at high levels of significance ($p < 0.01$), perhaps reflecting the likelihood that the blends began to present significant amounts of PC at the blend–aqueous solution interface for compositions above 10% PC-siloxane. Appropriately, θ_{AW} was statistically indistinguishable between methods for a given surface. Any differences seem to suggest that the tangent drawing protocols of the automated goniometer can be refined. A refinement of the software controlling the VCA 2500 instrument would presumably furnish the same average θ_{AW} for either method. The blend θ_{AW} was high regardless of the composition. It is somewhat unusual for the θ_{AW} value on any PC-siloxane blend to exceed the θ_{AW} value on Teflon FEP, a very hydrophobic material. The values of θ_{AW} that we determined for FEP were nonetheless plausible

(Table 1). Other investigators have determined $102 \pm 4°$ as the θ_{AW} value of Teflon FEP film, applying the sessile drop method [18]. The static θ_{AW} value on Silastic™ medical-grade flat sheet was measured as $105 \pm 1.3°$ [19] using this method. (This resembles blend contact angles with low percentages of PC-siloxanes considering the methodology used.) This comparison must be qualified because the blends are fabricated far differently than Silastic™, a crosslinked, air-permeable surface.

Table 2 compares the θ_{AW} values determined with the modified UWCB technique for the different siloxane blends. As the percentage of ethylene oxide-terminated polydimethlysiloxane increases, θ_{AW} decreases. For the EO-siloxane blends, θ_{AW} is significantly less than that for the PC-siloxane blends and exhibits significant hydrophilicity due to ethylene oxide. The θ_{AW} value for 1% EO differed from that on 10% EO or on pure EO-siloxane at excellent levels of significance ($p < 0.001$). The similarity in θ_{AW} for 10% EO or pure EO-siloxane possibly suggests a preponderance of EO-siloxane at the blend–aqueous solution interface as hypothesized.

Pure EO-siloxane was considerably less hydrophobic than polyethylene oxide (PEO). For example, the advancing contact angles on polyethylene oxide grafted to poly(ethylene terephthalate) are as low as $25°$ [20], falling with increasing PEO molecular weight. Presumably the receding contact angles on PEO grafted onto poly(ethylene terephthalate) would be still lower. In order to maximize protein rejection, our EO-siloxane blends would need a higher percentage of EO-siloxane in the blend and perhaps a much higher molecular weight in the ethylene oxide block. The Wilhelmy plate technique shows that blending 10 weight% of PEO-polypropylene oxide-PEO (PEO-PPO-PEO) of four PEO molecular weights with Pellethane, a commercial polyurethane, decreases the receding θ_{AW} but slightly [21]. The blend with an average PEO molecular weight of 1950 displayed a receding θ_{AW} value of $61°$, but that with an average PEO molecular weight of 8750 displayed a receding θ_{AW} value of $49°$.

It must be noted that contact angle measurement, whether by the UWCB technique or by pendant drop techniques, can display hysteresis. This is related to the 'stick–slip' phenomenon, which in turn results from the role of metastable states and local minima in contact angle in these measurements [22]. Small changes in the volume of the captive bubble can induce appreciable differences in the θ_{AW} value measured, even on smooth surfaces, as long as they are heterogeneous. Accordingly, contact angles can vary more than $10°$ on the same sample as the drop or bubble volume varies.

It was recently shown that the θ_{AW} value on polyurethanes functionalized with phosphatidylcholine analogous moieties was sensitive to the polymer surface [23]. Stable placement of such moieties at an interface is thus encouraged. Our simple coating procedure represents a strategy that conveniently places biocompatible functionalities at an interface with good specificity. This specificity is established by the sensitivity of θ_{AW} measurement to the chemistry and outermost characteristics of a surface, but can be further investigated with X-ray photoelectron spectroscopy or secondary ion mass spectrometry. The θ_{AW} value on both blend families differed significantly from that on the bare LDPE substrate, suggesting that our coating strategy was successful. The θ_{AW} values on our blends displayed low standard deviations, substantiating coating stability.

PC-siloxane surfaces in contact with aqueous solutions will presumably strive to maximize the PC contact with the aqueous phase. When blended at low percentages, this would form a heterogeneous surface. The θ_{AW} trends suggest that the PC-siloxane blends lack hemocompatibility. They incorporate too little PC-siloxane to decrease θ_{AW} and form a hydrophilic surface. Previous investigations have shown that biomembrane structures which consist of phospholipids display good hemocompatibility. Thromboblastographic studies [8, 16] have shown that polymeric lipids and polyesters having a phosphorylcholine group were relatively nonthrombogenic. Moreover, they showed that it was not the total phospholipid structure with lipid chains which was essential for hemocompatibility but the phosphorylcholine group itself. Phosphorylcholines form a monolayer when spread at an air–water interface. This forms a self-assembled monolayer which can then be transferred onto a solid support; the process can be repeated if a multilayer is desired. Thus, phosphorylcholine polar groups have been attached in a number of different ways to a variety of surfaces [8]. Physisorbable phosphorylcholine-containing polymer films have been synthesized [7] to coat hydrophobic surfaces such as LDPE, polypropylene, and unplasticized poly(vinyl chloride). Furthermore, copolymers of 2-methacryloyloxyethyl phosphorylcholine (MPC) [24] with hydrophobic alkyl methacrylates were used to produce polymers with excellent hemocompatibility. For example, a copolymer of poly(MPC-co-n-butyl methacrylate) [poly-(MPC-co-BMA)] diminished platelet adherence and aggregation with increasing MPC composition. Using 0.32 mole fraction of MPC in this copolymer completely suppressed the activation of platelets and the formation of fibrin.

Further study copolymerized MPC with either poly(hydroxyethylmethacrylate) (HEMA) or poly(n-butyl methacrylate) [25]. Compared with the MPC-containing copolymers, the unfunctionalized hydrogels were significantly more thrombogenic when exposed to whole human blood containing no anticoagulants. The hemocompatibility of MPC-containing copolymers has been examined with respect to mechanism. A quartz crystal microbalance was used to quantify the adsorption of dipalmitoylphosphatidylcholine (DPPC) liposomes [26] to poly(hydroxyethyl methacrylate) either unfunctionalized or copolymerized with an ω-methacryloyloxyalkylphosphorylcholine (MAPC) of three different possible chain lengths. More liposomes adsorbed onto unfunctionalized HEMA. The morphology of liposomes adsorbed onto functionalized or unfunctionalized HEMA was characterized with atomic force microscopy. Liposomes adsorbed onto the MAPC-containing HEMA copolymers retain their original, spherical shape. According to the minimum interfacial free energy hypothesis [2], this implies a very low interfacial tension between the surfaces of a MAPC-containing copolymer and the adherent liposome. However, DPPC liposomes adsorbed onto unfunctionalized HEMA distort, penetrate the hydration layer, and diminish less favorable surface energetics. Furthermore, MPC functionalization enhances protein desorption. The measurement of hysteresis on protein-preadsorbed HEMA, poly-(MPC-co-BMA), or poly(ethylene terephthalate) with dynamic contact angle demonstrates that poly-(MPC-co-BMA) behaves nearly like a bare surface after repeated cycles [27]. Ueda *et al.* [27] did not measure the air–water contact angle for any of their copolymerized or unfunctionalized hydrogels, though it is known that for HEMA, θ_{AW} is typically 15°, but for alkyl methacrylate polymers, typically 60° [2]. Presumably copolymerization of the hydrogels with an MAPC decreases θ_{AW}.

The hemocompatibility of any biomaterial functionalized with phosphorylcholine depends on the likelihood that the phosphorylcholine head groups preponderate at the interface between the biomaterial and the aqueous solution. This report shows that blending small amounts of PC-siloxanes with PDMS decreases θ_{AW}, but not to the extent likely to eliminate protein adsorption and enhance hemocompatibility. Future studies would accordingly increase the amount of PC-siloxanes in the blends or would use the pure functionalized PDMS. These studies would presumably incorporate a more precise software package to minimize the errors experienced in θ_{AW} measurements. This package should serve to realize the promise of the new procedure to implement the UWCB technique on curved surfaces.

4. CONCLUSIONS

A modified UWCB technique was used to determine the air–water contact angle on a series of polymer biomaterials, including polydimethylsiloxane blends. These consisted of unfunctionalized polydimethylsiloxane blended with small percentages of either ethylene oxide- or phosphorylcholine ester-terminated polydimethylsiloxanes to increase surface hydrophilicity. The blends were cast from a dilute solution of chloroform onto LDPE. Comparison of the curved UWCB technique with the modified UWCB technique, a convenient, automated method of determining θ_{AW} on curved surfaces, demonstrated the accuracy and precision of the latter method. The values of blend θ_{AW} determined with the modified UWCB technique demonstrated that too little functionalized polydimethylsiloxane was present to produce a hydrophilic surface. The EO-siloxane blends were nonetheless more hydrophilic than the PC-siloxane blends or unfunctionalized PDMS. As the percentage of functionalized polydimethylsiloxane in a family of blends increased, θ_{AW} decreased. It was suggested that a refinement in the software driving the automated goniometer would improve the accuracy of the modified UWCB technique to some extent.

Acknowledgements

Mr. Kabelo Masiane is thanked for his contact angle work. This research was supported, in part, by the Joint Advisory Research Council of Howard University.

300   *E. A. Angu* et al.

REFERENCES

1. B. D. Ratner, *Cardiovasc. Pathol.* **2**, 87S (1993).
2. J. D. Andrade, S. M. Ma, R. N. King and D. E. Gregonis, *J. Colloid Interface Sci.* **72**, 488 (1979).
3. W. G. Pitt, B. R. Young and S. L. Cooper, *Colloids Surfaces* **27**, 345 (1987).
4. M. D. Lelah, T. G. Grasel, J. A. Pierce and S. L. Cooper, *J. Biomed. Mater. Res.* **19**, 1011 (1985).
5. S. R. Holmes-Farley and G. M. Whitesides, *Langmuir* **3**, 62 (1987).
6. S. R. Holmes-Farley, R. H. Reamey, T. J. McCarthy, J. Deutch and G. M. Whitesides, *Langmuir* **1**, 725 (1985).
7. D. Chapman, *Langmuir* **9**, 39 (1993).
8. N. Nakabayashi and K. Ishihara, in: *Proteins at Interfaces II*, T. A. Horbett and J. L. Brash (Eds), pp. 385–394. American Chemical Society, Washington, DC (1995).
9. K. Ishihara, T. Tsuji, T. Kurosaki and N. Nakabayashi, *J. Biomed. Mater. Res.* **28**, 225 (1994).
10. C. Jalbert, J. T. Koberstein, A. Hariharan and S. K. Kumar, *Macromolecules* **30**, 4481 (1997).
11. R. C. Cameron and S. K. Pollack, *Polym. Mater. Sci. Eng.* **77**, 576 (1997).
12. I. Miller and J. E. Freund, *Probability and Statistics for Engineers*. Prentice-Hall, Upper Saddle River, NJ (1985).
13. M. D. Lelah, L. K. Lambrecht, B. R. Young and S. L. Cooper, *J. Biomed. Mater. Res.* **17**, 1 (1983).
14. D. K. Pettit, T. A. Horbett and A. S. Hoffman, *J. Biomed. Mater. Res.* **26**, 1259 (1992).
15. J. R. Rasmussen, E. R. Stedronsky and G. M. Whitesides, *J. Am. Chem. Soc.* **99**, 4736 (1977).
16. B. J. Tyler, B. D. Ratner, D. G. Castner and D. Briggs, *J. Biomed. Mater. Res.* **26**, 273 (1992).
17. R. F. Brady and S. J. Bonafede, *Polym. Mater. Sci. Eng.* **78**, 180 (1998).
18. Y. Tomada and Y. Ikada, *J. Biomed. Mater. Res.* **28**, 783 (1994).
19. T. Okada and Y. Ikada, *J. Biomed. Mater. Res.* **27**, 1509 (1993).
20. W. R. Gombotz, W. Guanghui, T. A. Horbett and A. S. Hoffman, *J. Biomed. Mater. Res.* **25**, 1547 (1991).
21. J. H. Lee, Y. M. Ju, W. K. Lee, K. D. Park and Y. H. Kim, *J. Biomed. Mater. Res.* **40**, 314 (1998).
22. A. Marmur, *Colloids Surfaces A* **186**, 209 (1998).
23. Y. Li, T. Yokawa, K. H. Matthews, T. Chen, Y. Wang, M. Kodama and T. Nakaya, *Biomaterials* **17**, 2179 (1996).
24. K. Ishihara, R. Aragaki, T. Ueda, A. Watenabe and N. Nakabayashi, *J. Biomed. Mater. Res.* **24**, 1069 (1990).
25. K. Ishihara, H. Oshida, Y. Endo, T. Ueda, A. Watenabe and N. Nakabayashi, *J. Biomed. Mater. Res.* **26**, 1543 (1992).
26. Y. Iwasaki, S. Tanaka, M. Hara, K. Ishihara and N. Nakabayashi, *J. Colloid Interface Sci.* **192**, 432 (1997).
27. T. Ueda, K. Ishihara and N. Nakabayashi, *J. Biomed. Mater. Res.* **29**, 381 (1995).

Apparent and Microscopic Contact Angles, pp. 301–318
J. Drelich, J. S. Laskowski and K. L. Mittal (Eds)
© VSP 2000.

Thermodynamic equilibrium and stability of liquid films and droplets on fibers

ALEXANDER V. NEIMARK *

TRI/Princeton, 601 Prospect Avenue, Princeton, NJ 08542-0625, USA

Received in final form 10 February 1999

Abstract—The modeling of liquid spreading and penetration into fibrous materials requires a better understanding of the interactions of thin liquid films and small droplets with single fibers. The wetting properties of fibers may differ significantly from those of plane solid surfaces. Convex surfaces of fibers imply a positive Laplace pressure acting on the liquid–gas interface. This effect causes liquid film instability and hinders droplet spreading. Liquid films on fibers are stable when the destabilizing action of the Laplace pressure is balanced by liquid–solid adhesion. Equilibrium configurations of liquid droplets and films are determined by the competition between capillary and adhesion forces. A general analytical solution is presented for the equilibrium profile of the transition zone between a film and a droplet residing on a cylindrical fiber. A new equation for apparent contact angles on fibers is derived. Adhesion forces, including van der Waals and polar interactions, are expressed in terms of disjoining pressure. Explicit formulae for calculations of equilibrium droplet profiles, film thicknesses, apparent contact angles, and stability factors are presented in the form of expressions which include both the measurable geometrical parameters and the presumably known parameters of liquid–solid interactions, such as apolar and polar spreading coefficients. The method developed is applicable for analyses of apparent contact angles and film stability on fibers and other cylindrical surfaces, particularly nanofibers. It is shown that a transition from partial wetting to non-wetting may occur as the fiber diameter decreases. Depending on the fiber diameter, contact angles of water on hydrophobic carbon fibers may vary from 75° (plane graphite surface) to 100–130° (carbon nanotubes).

Key words: Fibers; liquid films; liquid droplets; contact angle; adhesion; nanotubes; nanocomposites.

1. INTRODUCTION

The modeling of practical problems of liquid spreading and penetration in fibrous materials requires a better understanding of the interactions of thin liquid films and small droplets with single fibers. Despite several decades of intensive experimental and theoretical studies of fiber wetting phenomena, the fact that the wetting

---

*To whom correspondence should be addressed. E-mail: aneimark@triprinceton.org

properties of fibers may differ significantly from those of plane solid surfaces has not been appreciated by experimentalists until recently [1]. This effect becomes more prominent as the fiber diameter decreases and may lead to qualitatively different behaviors (non-wetting versus partial wetting) in the case of nanofibers. The specifics of fluid–fiber interactions are of special technological importance in view of the recent discovery of nanotubes composed of various materials such as carbon, boron nitride, silica, alumina, and titanium oxides, and the rapidly growing research interest in fiber-reinforced nanocomposites and other nanotechnologies [2–4].

The convex surfaces of fibers imply positive curvatures of coating liquid films and, therefore, a positive Laplace pressure, P_c, acting on the film–air surface:

$$P_c = \gamma / (a + h). \tag{1}$$

Here, γ is the liquid–gas surface tension, a is the fiber radius, and h is the liquid film thickness. The Laplace pressure, when positive, tends to squeeze the liquid out of the film. This effect causes a well-known phenomenon of capillarity-driven instability of films on cylindrical surfaces of wires and fibers. Goren [5] and then other researchers [6, 7] studied the stability of liquid films on cylindrical wires. They concluded that a uniform liquid coating was always unstable and collapsed into a periodic array of droplets, independent of the film thickness, the solid surface curvature, and liquid properties, such as viscosity, surface tension, and density. The above-mentioned parameters determine the rate of film destruction.

In making this conclusion, the authors ignored the liquid–solid interactions, or adhesion. Adhesion is a stabilizing factor. When the film–solid attraction dominates the surface-tension forces, the films are stable. Derjaguin [8] introduced the disjoining pressure, $\Pi(h)$, to account for fluid–solid interactions. He suggested that the thermodynamic equilibrium and stability of thin liquid films on curved solid surfaces be studied as a competition between capillarity and adhesion. While the curvature-dependent Laplace pressure favors the smearing of the liquid over the solid surface, the thickness-dependent disjoining pressure, $\Pi(h)$, opposes film thinning. The equilibrium liquid configurations are observed when these forces are in balance.

The analyses of thermodynamic equilibrium and the stability of thin liquid films on curved solid surfaces in the Derjaguin approximation [8, 9] are based on the assumption that the Laplace and disjoining pressures make additive contributions to the mechanical balance equation. This means that in the absence of external forces, the Laplace pressure and the disjoining pressure may vary along the equilibrium film surface, due to variations in the film thickness and substrate curvature; however, their sum must be constant:

$$P_{cap} = P_l - P_g = \gamma \left(\frac{1}{R_1} + \frac{1}{R_2} \right) - \Pi(h). \tag{2}$$

Here, R_1 and R_2 are the two principal radii of curvature of the outer surface of the film. The capillary pressure, P_{cap}, is introduced as the difference between the liquid pressure, P_1, and the gas pressure, P_g, and is therefore positive for films on convex surfaces. P_{cap} is equal to the capillary pressure in the bulk liquid that is in equilibrium with the film.

It is worth noting that the Derjaguin approach [equation (2)] adopted in this paper is a 'first-order' approximation that takes into account the solid–liquid interactions in terms of the disjoining pressure. There are obvious and well-known shortcomings of this approach. It is assumed that the Laplace pressure and the disjoining pressure in curved films are not entirely coupled. The Laplace pressure is determined by the curvature of the outer film surface and does not depend on the film thickness. The surface tension, γ_{lg}, is the liquid–gas surface tension for the bulk phases. The disjoining pressure is determined by the film thickness and does not depend on the film curvature. Adsorption films are considered liquid-like and are characterized by the same density as the bulk liquid. This raises a legitimate concern regarding the applicability of the disjoining pressure concept to systems with large contact angles when droplets are in contact with monolayer and sub-monolayer adsorption films. Such films can also be described by using the disjoining pressure with certain reservations and understanding that the disjoining pressure in this case is a formal quantity related to the chemical potential [9–11].

The Derjaguin equation (2) has been successfully applied for studies of the equilibrium and stability of thin liquid films in various pore systems, colloids, foams, and other dispersed media [9–12].

In the case of a uniform liquid film on a cylindrical fiber, $R_1 = \infty$, $R_2 = a + h$, equation (2) reduces to

$$P_{cap} = \frac{\gamma}{a + h} - \Pi(h). \tag{3}$$

The disjoining pressure stabilizes the film when its magnitude increases with film thinning, i.e. when its first derivative is negative, $\Pi'(h) < 0$ [9]. The Derjaguin approximation for curved films implies that the film is stable when the capillary pressure, equation (3), increases with film thinning, i.e. the stability condition of films on fibers can be expressed as $\delta P_{cap}/\delta h < 0$ [13], or through a dimensionless stability factor, Δ:

$$\Delta = 1 + \frac{(a + h)^2}{\gamma} \Pi'(h) < 0. \tag{4}$$

When the stability factor Δ is negative, any infinitesimal disturbance on the film surface is damped due to the action of the fluid–solid interactions.

Brochard-Wyart [14] analyzed the thermodynamic equilibrium and stability of liquid films on fibers in terms of competition between the Laplace and disjoining pressures for the van der Waals fluid–solid interactions. Spreading conditions for small droplets have been analyzed by Novakowski and Ruckenstein [15]. Recently, a nonlinear stability analysis of van der Waals films has been done by Chen and Hwang [16].

In the present paper we develop an analytical model to analyze equilibrium configurations of liquid droplets and films on cylindrical surfaces. The solution to this problem provides a better understanding of the physical mechanisms determining the equilibrium distributions of liquid in fiber structures and the rates of liquid penetration and spreading. The equilibrium configurations of droplets on fibers were studied in the literature earlier [17–19] ignoring the fluid–solid interactions. The authors assumed that the droplet shape was determined exclusively by the Laplace pressure on the liquid–gas interface. Therefore, the droplet outer surface was regarded as a surface of constant curvature. However, it is well documented in the literature (see refs [20, 21] and refs therein) that typically advancing menisci are gradually transformed into so-called precursor films and do not form the contact angle that they would form on plane surfaces. There exists a transition zone between the film and the apparent droplet edge. Similar observations were made for liquid menisci on fibers contacting the free surface of a bulk liquid.

We present below an analytical solution for the equilibrium profile of the transition zone between a film and a droplet residing on a cylindrical fiber under the action of Laplace forces on the liquid–gas interface and adhesion forces expressed in terms of the disjoining pressure.

Similar problems of the equilibrium liquid distribution in slit-shaped, wedge-shaped, and cylindrical pores have been examined earlier by Derjaguin et al. [22], Renk et al. [23], Neimark and Kheifets [24], and in several later publications [24–26]. Neimark and Kheifets [24] have derived general analytical solutions for these geometries for an arbitrary form of the disjoining pressure isotherm. They concluded that the characteristic scale of the transition zone in pores was of the order of the geometrical mean of the pore width and film thickness. We use below the same approach to study the liquid equilibrium on fibers.

2. EQUILIBRIUM SHAPES OF DROPLETS ON FIBERS. GENERAL EQUATIONS AND BOUNDARY CONDITIONS

Consider a droplet of a wetting fluid residing on a fiber (Fig. 1). Outside the droplet, the fiber is coated by a film. The droplet and film are in equilibrium and represent a continuous liquid layer of varying thickness and curvature. The equilibrium shape of the liquid layer is determined by equation (2), which represents the balance of the Laplace and disjoining pressures. Because the gas pressure, P_g, is constant, the condition of mechanical equilibrium in the liquid implies a constant liquid pressure in both the droplet and the film. Thus, the right-hand side of equation (2) must be constant along the liquid profile. At the top of the droplet, the disjoining pressure is negligible and the layer shape is adequately approximated by an unduloid of

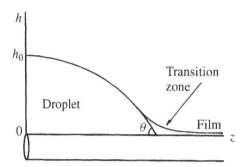

Figure 1. Schematic profile of a droplet adhering to a fiber. θ is the apparent contact angle defined via the equicurvature approximation [18].

constant curvature:

$$\left(\frac{1}{R_1} + \frac{1}{R_2}\right)_{z=0} = \frac{1}{R_0} + \frac{1}{a + h_0}.$$

Here, h_0 is the droplet height and R_0 is the radius of curvature of the layer profile at the droplet top, $z = 0$ (z is the coordinate along the fiber axis, Fig. 1). The capillary pressure, P_{cap}, introduced in equation (2) as the constant pressure difference across the gas–liquid interface is determined by the droplet curvature at the top of the droplet:

$$P_{cap} = \frac{\gamma}{R_0} + \frac{\gamma}{a + h_0}. \tag{5}$$

Sufficiently far from the droplet top, the layer is uniform. The equilibrium uniform film thickness, h_∞, is determined by equation (3):

$$P_{cap} = \frac{\gamma}{a + h_\infty} - \Pi(h_\infty). \tag{6}$$

We are mostly interested in a transition zone where the layer curvature continuously varies from the unduloid curvature equation (5) to the film curvature equation (6). In the transition region, the contributions of the Laplace and disjoining pressures are comparable.

Equation (2) for mechanical equilibrium of the liquid layer of varying thickness (droplet passing into film) can be expressed as a second-order differential equation with regard to the layer thickness $h(z)$:

$$\gamma\left[-\frac{h''}{(1 + h'^2)^{3/2}} + \frac{1}{(a + h)(1 + h'^2)^{1/2}}\right] = P_{cap} + \Pi(h). \tag{7}$$

The boundary conditions are

$$h(0) = h_0, \qquad h'(0) = h'(\infty) = 0. \tag{8}$$

Solution of equation (7) with the boundary conditions (8) gives the equilibrium profile of the droplet and transition zone between the droplet and the film.

To derive an analytical solution, let us introduce a new variable:

$$y(h) = (1 + h'^2)^{-1/2}. \qquad (9)$$

Note that this new variable is related to the inclination of the profile to the fiber axis. It is equal to the cosine of angle θ between the profile tangent and fiber axis:

$$h'^2 = \tan^2 \theta \Rightarrow y(h) = (1 + h'^2)^{-1/2} = \cos \theta. \qquad (10)$$

In the capillarity theory, this angle at the droplet edge is associated with the contact angle. Thus, equation (7) transforms into

$$\gamma \left(\frac{dy}{dh} + \frac{y}{a+h} \right) = P_{cap} + \Pi(h). \qquad (11)$$

The boundary conditions (8) transform into

$$y(h_0) = y(h_\infty) = 1. \qquad (12)$$

3. EQUILIBRIUM SHAPES OF DROPLETS ON FIBERS. GENERAL SOLUTION

Equation (11) can be integrated to yield

$$\gamma \, d\big[y(a+h)\big] = (a+h)\big(P_{cap} + \Pi(h)\big) \, dh \qquad (13)$$

and

$$y = \frac{(a+h_0)}{(a+h)} - \frac{1}{\gamma(a+h)} \int_h^{h_0} (a+h_1)\big(P_{cap} + \Pi(h_1)\big) \, dh_1. \qquad (14)$$

From equation (14) at $h \to h_\infty$ and $y \to 1$, we obtain the following relations for the capillary pressure, P_{cap}, of the droplet of height h_0:

$$P_{cap} = \frac{\gamma(h_0 - h_\infty) - \int_{h_\infty}^{h_0}(a+h)\Pi(h) \, dh}{\frac{1}{2}\big[(a+h_0)^2 - (a+h_\infty)^2\big]} = \frac{\gamma}{a + (h_0 + h_\infty)/2}$$

$$- \frac{\int_{h_\infty}^{h_0}(a+h)\Pi(h) \, dh}{(h_0 - h_\infty)\big(a + (h_0 + h_\infty)/2\big)}. \qquad (15)$$

Equation (15) together with condition (6) of the equilibrium film constitute the closed system of algebraic equations with respect to two unknowns, P_{cap} and h_∞. With P_{cap} and h_∞ obtained at given h_0, the equilibrium profile is calculated directly from equation (14) rewritten in the form

$$\frac{dh}{dz} = -\sqrt{\frac{(a+h)^2}{\big[(a+h_0) - \frac{1}{\gamma}\int_h^{h_0}(a+h_1)\big(P_{cap} + \Pi(h_1)\big) \, dh_1\big]^2} - 1}. \qquad (16)$$

The general solution valid for any type of disjoining isotherm reads

$$z(h) = \int_h^{h_0} \left[\frac{(a + h_1)^2}{\left[(a + h_0) - \frac{1}{\gamma} \int_{h_1}^{h_0}(a + h_2)\left(P_{\text{cap}} + \Pi(h_2)\right) dh_2\right]^2} - 1 \right]^{-1/2} dh_1.$$

(17)

4. APPARENT CONTACT ANGLE

Equations (14) and (15) shed light on the physical meaning of apparent contact angles formed by droplets residing on cylindrical fibers. Because of the non-spherical geometry of droplets adhering to fibers, the contact angle has to be defined from geometrical considerations somewhat differently from the known sessile drop method. This problem was studied by Carroll [18, 19], who derived an equation for the profile of a droplet residing on a fiber under the action of the Laplace pressure. Below, first we show that our general solution reduces to the Carroll equation when the action of the disjoining pressure is neglected, and second, we derive an explicit relation between the geometrical or apparent contact angle involved in the Carroll approximate approach and the adhesion properties of the liquid–fiber system expressed in terms of the disjoining pressure.

Carroll [18] considered the droplet profile as a figure of constant curvature, which forms a given angle, θ, with the fiber surface (Fig. 1). Carroll's equation for the capillary pressure reads

$$P_{\text{cap}} = \frac{2\gamma\left((a + h_0)/a - \cos\theta\right)}{a\left[\left((a + h_0)/a\right)^2 - 1\right]}.$$

(18)

Equation (18) follows from equation (14) at $y = \cos\theta$, $\Pi(h) = 0$, and $h_\infty = 0$, which can be rewritten as

$$\cos\theta = \frac{(a + h_0)}{a} - \frac{1}{\gamma a} \int_0^{h_0} (a + h_1) P_{\text{cap}} \, dh_1$$

$$= \frac{(a + h_0)}{a} - \frac{1}{2\gamma a}\left[(a + h_0)^2 - a^2\right] P_{\text{cap}}.$$

(19)

In fact, equation (19) should be regarded as a definition of the apparent contact angle, which can be calculated from measurable dimensions of the droplet: its height, h_0, and the radius of curvature at the top, R_0 (5):

$$\cos\theta = \frac{(a + h_0)}{a} - \frac{1}{2a}\left[(a + h_0)^2 - a^2\right]\left(\frac{1}{a + h_0} + \frac{1}{R_0}\right)$$

$$= 1 - \frac{h_0(a + h_0)(2a + h_0) - R_0 h_0^2}{2a R_0(a + h_0)}.$$

(20)

The condition $\cos\theta < 1$ implies an upper limit of the droplet radius of curvature, i.e.

$$R_0 < (a + h_0)(2a + h_0)/h_0. \tag{21}$$

The contact angle, defined via equation (19), is not a microscopic contact angle at the droplet edge. The existence of the transition zone, which would be revealed by zooming-in on the droplet edge, implies that the tangent angle gradually decreases to zero. Correspondingly, the microscopic $\cos\theta$ varies within the transition zone, according to equation (14), approaching unity in the region of the film. At $y = \cos\theta = 1$, equation (14) can be rewritten in a form close to equation (18) as

$$1 + \frac{1}{\gamma(a + h_\infty)} \int_{h_\infty}^{h_0} (a + h)\Pi(h)\,dh$$
$$= \frac{(a + h_0)}{(a + h_\infty)} - \frac{1}{2\gamma(a + h_\infty)}\left[(a + h_0)^2 - (a + h_\infty)^2\right]P_{\text{cap}}. \tag{22}$$

Comparing equations (18) and (22), we conclude that the left-hand side of equation (22) represents the cosine of the apparent contact angle of the droplet residing on the fiber covered by the film of thickness h_∞. Thus, we arrive at the desired relation between the apparent contact angle and the disjoining pressure integral:

$$\cos\theta = 1 + \frac{1}{\gamma(a + h_\infty)} \int_{h_\infty}^{h_0} (a + h)\Pi(h)\,dh. \tag{23}$$

The equilibrium film thickness, h_∞, is determined by equation (6).

In the limit of $a \to \infty$ and $P_{\text{cap}} \to 0$, equation (23) reduces to the Frumkin–Derjaguin equation [8, 9] for the cosine of an equilibrium contact angle for plane substrates:

$$\cos\theta_e = 1 + S_e/\gamma = 1 + \frac{1}{\gamma} \int_{h_e}^{\infty} \Pi(h)\,dh, \tag{24}$$

where

$$S_e = \int_{h_e}^{\infty} \Pi(h)\,dh \tag{25}$$

is the equilibrium spreading coefficient, which determines the conditions of spreading for a droplet on a pre-wetted surface covered by an equilibrium film (it can be a multilayer, monolayer, or sub-monolayer film depending on the given liquid–solid system). The equilibrium film thickness, h_e, is determined from the condition $\Pi(h_e) = 0$.

The equilibrium contact angle [equation (24)] can be defined and measured, for example by the sessile drop method, provided the spreading coefficient $S_e < 0$. The equilibrium spreading coefficient is equal to the integral of the disjoining pressure over the film thickness from the equilibrium film thickness h_e to infinity. In the case of positive spreading coefficients, $S_e > 0$, we deal with a phenomenon of absolute wetting, i.e. when the equilibrium droplet does not exist: any droplet placed on a plane substrate would spread out and form a thin 'pancake' [27].

As follows from the augmented Frumkin–Derjaguin equation (23), the condition of absolute wetting for fiber surfaces differs from that for plane substrates ($S_e > 0$) and reads

$$\int_{h_\infty}^{\infty} (1 + h/a)\Pi(h)\, dh > 0. \qquad (26)$$

That is, an absolute wetting liquid with a positive spreading coefficient would not necessarily spread over a fiber but would form a finite contact angle when inequality (26) does not hold. This phenomenon for van der Waals (apolar) films has been discussed by Brochard-Wyart [14]. In Section 5, we derive explicit expressions for spreading conditions for fibers, taking into account short-range (acid–base) and long-range (dispersion) interactions.

For most practical applications, the film thickness h_∞ is much smaller than both the droplet height h_0 and the fiber radius a. When these strong inequalities hold, equation (24) reduces to the form

$$\cos\theta = 1 + \frac{1}{\gamma} \int_{h_\infty}^{\infty} \Pi(h)\, dh, \qquad (27)$$

which is identical to the Frumkin–Derjaguin equation (24) for plane surfaces. The difference between the above equation for cylindrical fibers and the Frumkin–Derjaguin equation for plane surfaces is in the lower limit of integration. The equilibrium film thickness h_∞ depends on the fiber curvature and is defined by equation (6).

In Section 6, we demonstrate that in the case of nanofibers, the wetting conditions depend critically on the fiber curvature.

5. INFLECTION ANGLE

Direct optical measurement of the contact angles on fibers is a difficult and inaccurate procedure. Carroll's approach [18], which determines the contact angle indirectly from measurements of the droplet dimensions, gives the value for an apparent contact angle.

Another approach has been suggested recently by McHale *et al.* [1]. The droplet profile $h(z)$ changes the sign of its curvature dy/dh from positive (convex) to negative (concave) at the inflection point, which can be visualized and identified in direct observations. McHale *et al.* [1] introduced the inflection angle, θ_{inf}, from the slope of the droplet profile at the inflection point. The cosine of the inflection angle θ_{inf} is determined by equation (19) at $dy/dh = 0$ (Fig. 2):

$$\gamma \frac{\cos\theta_{inf}}{a + h_{inf}} = P_{cap} + \Pi(h_{inf}). \qquad (28)$$

A. V. Neimark

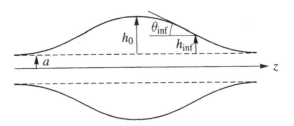

Figure 2. Definition of the inflection angle [1].

Introduction and experimental determination of the inflection angle are advantageous only when the inflection point is at an appreciable distance from the droplet edge. In this case, the elevation of the inflection point, h_{inf}, should be considerably greater than the equilibrium film thickness. For such thick films, the disjoining pressure is negligible and can be omitted in equation (28). Taking into account equation (18), we exclude P_{cap} from equation (28) and arrive at the following relation between the cosines of the inflection and apparent contact angles:

$$\cos \theta_{inf} = \frac{2(a + h_{inf})\left[(a + h_0)/a - \cos \theta\right]}{a\left[(a + h_0)^2/a^2 - 1\right]} \tag{29}$$

and vice versa,

$$\cos \theta = 1 + \frac{h_0}{a} - \frac{a}{2(a + h_{inf})}\left[(a + h_0)^2/a^2 - 1\right]\cos \theta_{inf}. \tag{30}$$

Equation (30) seems to be useful for the determination of the contact angle on fibers, especially when the droplet edge cannot be well defined due to uncertainty caused by the transition zone. Equation (30) can be rewritten as

$$\cos \theta = 1 + \frac{h_0}{a} - \frac{2a + h_0}{2(a + h_{inf})}\frac{h_0}{a}\cos \theta_{inf} = 1 - \frac{h_0}{a}\left(\frac{2a + h_0}{2(a + h_{inf})}\cos \theta_{inf} - 1\right). \tag{31}$$

6. ACCOUNTING FOR APOLAR AND POLAR INTERACTIONS. EXPLICIT EQUATIONS FOR THE CONTACT ANGLE, EQUILIBRIUM FILM THICKNESS, AND STABILITY FACTOR

In the Derjaguin approach adopted here, adhesion forces caused by liquid–solid interactions are expressed in terms of the disjoining pressure, $\Pi(h)$, which depends on the film thickness. Different types of liquid–solid intermolecular interactions make contributions to the resulting disjoining pressure. They are divided into long-range and short-range interactions [9, 28–34]. The short-range interactions include various polar interactions, such as acid–base interactions and hydrophobic interactions, and effects of the electrical double layer. The long-range dispersion (or van der Waals) forces are dominant in the case of thick films (> several nanometers).

In general, the disjoining pressure $\Pi(h)$ can be represented as a sum of the apolar (van der Waals) and polar (acid–base) components:

$$\Pi(h) = \Pi_d(h) + \Pi_p(h). \tag{32}$$

For thick films, the major contribution to the disjoining pressure is due to the van der Waals interactions. For the van der Waals interactions [9, 28],

$$\Pi_d(h) = 2d_0^2 S_d/h^3. \tag{33}$$

Here, d_0 is a lower cut-off distance of fluid–solid intermolecular forces where the Born repulsion may be replaced by the hard wall (vertical rise in the potential to infinity prohibiting overlap of fluid and solid molecules). d_0 is a molecular scale of the order of 0.1–0.2 nm. S_d is the dispersion component of the spreading coefficient S. For wetting liquids, $S_d > 0$, and the dispersion component of the disjoining pressure (33) is positive.

The polar component of the disjoining pressure is assumed to decay exponentially with a correlation length, l_0 [9, 28, 32–34]:

$$\Pi_p(h) = S_p l_0^{-1} \exp\left((d_0 - h)/l_0\right). \tag{34}$$

The correlation length for water interaction with solid substrates is of the order of 0.6 nm or even larger [32]. The exponential term [equation (34)] in the disjoining pressure isotherm is introduced to account effectively for all short-range interactions including, but not restricted to, the acid–base interactions. Derjaguin *et al.* [9] referred to this term as a structural component of the disjoining pressure.

The sum of apolar and polar components expressed through equations (33) and (34) gives the following expression for the disjoining pressure [32–34]:

$$\Pi(h) = 2d_0^2 S_d/h^3 + S_p l_0^{-1} \exp\left((d_0 - h)/l_0\right), \tag{35}$$

which can be used for quantitative estimates of the fluid–solid interactions in practical systems. The spreading coefficients S_d and S_p for a particular liquid–solid pair can be calculated from the apolar and polar components of the liquid and solid surface tensions and the contact angle on a plane solid [28]. The spreading coefficients S_d and S_p may be positive or negative depending on the system. By varying the values of S_d and S_p, it is possible to describe different types of the disjoining pressure isotherm presented schematically in Fig. 3 [10].

The isotherms in Figs 3a, 3b, and 3c correspond to the conditions of absolute wetting (the horizontal lines BF in Figs 3b and 3c are drawn from the condition of equal areas BCD and DEF, by the Maxwell rule (see ref. [10] for details). The isotherms in Figs 3d, 3e, 3f, and 3g are characterized by a positive spreading coefficient [equation (25)] and correspond to partial wetting and correspondingly to finite contact angles.

The sum

$$S_i = S_d + S_p \tag{36}$$

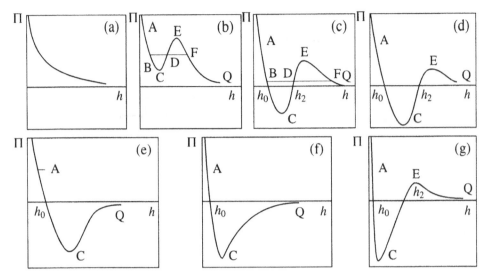

Figure 3. Different types of disjoining pressure isotherm [10].

is equal to the so-called initial spreading coefficient [30], which determines the spreading of a droplet placed on a clean plane surface. The initial spreading coefficient is equal to the integral of the disjoining pressure:

$$S_i = \int_{d_0}^{\infty} \Pi(h)\, dh. \tag{37}$$

The initial spreading coefficient determines the conditions for spreading of a droplet placed on a dry solid surface. When $S_i < 0$, an initial contact angle can be introduced:

$$\cos \theta_i = 1 + S_i/\gamma. \tag{38}$$

The initial spreading coefficient [equation (37)] differs from the equilibrium spreading coefficient S_e, defined by equation (25). Correspondingly, the initial contact angle [equation (38)] differs from the equilibrium contact angle defined by the Frumkin–Derjaguin equation (25).

General equations [(3), (4), (23), (24), (37)] are applicable for any type of fluid–solid interaction expressed in terms of the disjoining pressure. In the vast majority of fluid–solid pairs which are of practical interest to fiber technologies, adhesion is determined by the interplay of dispersion and acid–base interactions. Therefore, for further estimates it is useful to present the explicit equations for the disjoining pressure isotherm in the form of equation (35).

In the case of flat surfaces, the corresponding equations for the equilibrium and stability of liquid films have been derived in refs [32–34]. The sign of the disjoining pressure derivative determines the conditions of film stability on flat surfaces. Films are thermodynamically stable when

$$\Pi'(h) = -6d_0^2 S_d/h^4 - S_p l_0^{-2} \exp\left((d_0 - h)/l_0\right) < 0. \tag{39}$$

The thickness, h_e, of the film which is in equilibrium with a macroscopic drop is determined by the disjoining pressure isotherms from the condition $\Pi(h_e) = 0$.

$$\Pi(h)|_{h=h_e} = 0 \Rightarrow 2d_0^2 S_d/h_e^3 = -S_p l_0^{-1} \exp\left((d_0 - h_e)/l_0\right)$$

or

$$\frac{d_0}{h_e} = \left(-\frac{S_d}{S_p}\frac{2d_0}{l_0}\exp\left((d_0 - h_e)/l_0\right)\right)^{1/3}. \tag{40}$$

Depending on the ratio of the apolar and polar components of the spreading coefficient, S_d and S_p, the algebraic equation (40) may have two, one or zero roots. The smallest root corresponds to the equilibrium film thickness h_e which appears in the Frumkin–Derjaguin equation (24).

Substitution of the disjoining isotherm [equation (35)] in the Frumkin–Derjaguin equation (24) gives an expression for the cosine of the contact angle in plane geometries through the parameters of apolar and polar interactions:

$$\cos\theta_e = 1 + \frac{1}{\gamma}\int_{h_e}^{\infty}\Pi(h)\,\mathrm{d}h = 1 + \frac{1}{\gamma}\left[S_d\left(\frac{d_0}{h_e}\right)^2 + S_p\exp\left((d_0 - h_e)/l_0\right)\right]$$

$$= 1 + \frac{S_d}{\gamma}\left[\left(\frac{d_0}{h_e}\right)^2 - \frac{2l_0}{d_0}\left(\frac{d_0}{h_e}\right)^3\right] = 1 + \frac{S_d}{\gamma}\left(\frac{d_0}{h_e}\right)^2\left[1 - \frac{2l_0}{h_e}\right]. \tag{41}$$

A necessary condition of the finite contact angle is $2l_0 > h_e$, provided that $S_d > 0$.

To analyze the influence of liquid–solid interaction parameters on fiber wetting, it is convenient to use the dimensionless variables

$$\begin{aligned}\tilde{h} &= h/d_0;\\ \tilde{l} &= l/d_0;\\ \tilde{a} &= a/d_0;\\ \tilde{\Pi} &= \Pi(h)/\left(\frac{\gamma}{a}\right)\end{aligned} \tag{42}$$

and to use the liquid surface tension γ and the initial spreading coefficient S_i or the initial contact angle, $\cos\theta_i = 1 + S_i/\gamma$, as the primary parameters for accounting for liquid–solid interactions. Therewith, the apolar and polar components of the spreading coefficient are reduced to the liquid surface tension:

$$S_d = \lambda_d\gamma \tag{43}$$

$$S_p = \lambda_p\gamma; \quad \lambda_p = -(1 + \lambda_d - \cos\theta_i). \tag{44}$$

The above equations are rewritten in the dimensionless form. The disjoining pressure is expressed as

$$\tilde{\Pi}(\tilde{h}) = \frac{a}{d_0}\left[2\lambda_d/\tilde{h}^3 - (1 + \lambda_d - \cos\theta_i)\tilde{l}_0^{-1}\exp\left((1 - \tilde{h})/\tilde{l}_0\right)\right]. \tag{45}$$

The disjoining pressure derivative is expressed as

$$\tilde{\Pi}'(\tilde{h}) = \tilde{a}\left[-6\lambda_d/\tilde{h}^4 + (1 + \lambda_d - \cos\theta_i)\tilde{l}_0^{-2}\exp\left((1 - \tilde{h})/\tilde{l}_0\right)\right]. \tag{46}$$

The films on fibers are thermodynamically stable when the stability factor Δ is negative [equation (4)]. This condition differs substantially from that for plane geometries [equation (38)], due to the destabilizing action of the Laplace pressure. The stability factor [equation (4)] for films on fibers reads

$$\Delta = 1 + \frac{(a+h)^2}{\gamma}\Pi'(h) = 1 + \tilde{a}(1 + \tilde{h}/\tilde{a})^2\Pi'(\tilde{h})$$

$$= 1 - \tilde{a}^2(1 + \tilde{h}/\tilde{a})^2\left[6\lambda_d/\tilde{h}^4 - (1 + \lambda_d - \cos\theta_i)\tilde{l}_0^{-2}\exp\left((1 - \tilde{h})/\tilde{l}_0\right)\right]. \tag{47}$$

The contact angle on plane surfaces [equation (41)] is expressed as

$$\cos\theta_e = 1 + \frac{1}{\gamma}\int_{h_e}^{\infty}\Pi(h)\,dh = 1 + \lambda_d/\tilde{h}_e^2 - (1 + \lambda_d - \cos\theta_i)\exp\left((1 - \tilde{h}_e)/\tilde{l}_0\right). \tag{48}$$

The contact angle on fiber surfaces [equation (23)] is expressed as

$$\cos\theta_e = 1 + \frac{1}{\gamma(a + h_\infty)}\int_{h_\infty}^{\infty}(a+h)\Pi(h)\,dh$$

$$= 1 + \frac{1}{1 + \tilde{h}_\infty/\tilde{a}}\left[\lambda_d/\tilde{h}_\infty^2 - (1 + \lambda_d - \cos\theta_i)\exp\left((1 - \tilde{h}_\infty)/\tilde{l}_0\right)\right]$$

$$+ \frac{1/\tilde{a}}{1 + \tilde{h}_\infty/\tilde{a}}\left[\lambda_d/\tilde{h}_\infty - (1 + \lambda_d - \cos\theta_i)\tilde{l}_0\left(1 + \tilde{h}_\infty/\tilde{l}_0\right)\right.$$

$$\left. \times \exp\left((1 - \tilde{h}_\infty)/\tilde{l}_0\right)\right]. \tag{49}$$

For most practical applications, the strong inequality $h_\infty \ll a$ holds. In this case, equation (49) reduces to equation (48); however, the equilibrium film thickness h_∞ which appears in this equation is determined by the fiber curvature by condition (6), which implies that

$$2d_0^2 S_d/h_e^3 = -S_p l_0^{-1}\exp\left((d_0 - h_e)/l_0\right) + \frac{\gamma}{a + h_\infty} - P_{cap}, \tag{50}$$

or in dimensionless form

$$2\lambda_d/\tilde{h}_\infty^3 = -(1 + \lambda_d - \cos\theta_i)\tilde{l}_0^{-1}\exp\left((1 - \tilde{h}_\infty)/\tilde{l}_0\right) + \frac{1}{\tilde{a}}\left(\frac{1}{1 + \tilde{h}_\infty/\tilde{a}} - \tilde{P}_{cap}\right). \tag{51}$$

The initial contact angle on fiber surfaces is determined by equation (49) with d_0 as the low limit of integration:

$$\cos\theta_i = 1 + \frac{1}{\gamma(a + d_0)}\int_{d_0}^{\infty}(a + h)\Pi(h)\,dh. \tag{52}$$

7. WETTING OF NANOFIBERS

It is anticipated that newly synthesized nanofibers will soon be used in numerous technologies, replacing currently available commercial fibers, in the first turn in nanocomposite processing. As a prominent example, we consider the effects of fiber curvature on the wetting of carbon nanotubes.

Carbon nanotubes represent new nanofibers with unique mechanical and electro-physical properties. A variety of single- and multi-wall carbon nanotubes have been synthesized with fiber diameters starting from *ca.* 1 nm [2, 35–38]. The outer sur-face of carbon nanotubes is graphite-like; therefore for modeling the interaction of fluids with nanotubes, it is justified to use intermolecular potentials introduced in the literature to model graphite surfaces.

The disjoining pressure isotherm of water on graphite in the form of equation (35) has been used by Thiele *et al.* [39] with the following parameters: apolar component of the spreading coefficient $S_d = 106 \times 10^{-3}$ N/m, polar component of the spreading coefficient $S_p = -159 \times 10^{-3}$ N/m, Born repulsion length $d_0 = 0.158$ nm, and correlation length $l = 0.6$ nm. The dimensionless parameters $\lambda_d = S_d/\gamma = 1.468$ and $l_0/d_0 = 3.8$. It is worth noting that the above value of the polar component of the spreading coefficient is smaller than the minimum possible value of -102×10^{-3} N/m that would be predicted in the framework of the acid–base interaction model for water [28]. This apparent inconsistency may be attributed to some specific structural interactions of water with graphite beyond the acid–base interactions.

This set of parameters implies a finite initial contact angle $\theta_i = 74.5°$, as calculated by equation (38) with the initial spreading coefficient $S_i = -53 \times 10^{-3}$ N/m and the water surface tension $\gamma = 72.2 \times 10^{-3}$ N/m. However, the equilibrium contact angle on the plane graphite surface, θ_e, calculated from the Frumkin–Derjaguin equation (48), where the equilibrium film thickness h_e is determined by equation (40), turns out to be $\theta_e = 109°$. That is, the graphite surface characterized by the parameters recommended in ref. [39] is non-wetting. After the formation of the equilibrium adsorption layer of effective thickness $h_e = 0.29$ nm, the equilibrium contact angle exceeds 90° (we adopt the commonly used classification of wettability into three classes: absolute wetting for $\theta = 0°$, partial wetting for $\theta < 90°$, and non-wetting for $\theta > 90°$ [40]). At the same time, the acute initial contact angle implies that when a dry solid is dipped into a liquid (drop or bulk), one observes spontaneous spreading along the dry surface until the equilibrium angle is established. Spontaneous spreading is possible only in the case of partial wetting.

In the case of fibers, we expect that due to the positive Laplace pressure, the non-wetting effects will be more prominent. As the fiber diameter decreases, the initial contact angle increases and exceeds 90°. Under these conditions, partial wetting of the dry surface ceases. The results of calculations of the initial contact angle [equation (52)] are presented in Fig. 4. Transition from partial wetting of the dry surface to non-wetting, i.e. initial contact angle $\theta_I = 90°$, is observed at the fiber

A. V. Neimark

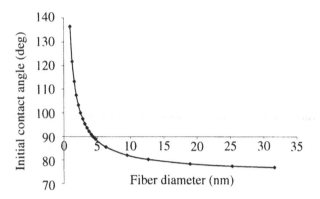

Figure 4. Dependence of the initial contact angle on the fiber diameter. Water on carbon nanotubes with a graphite-like surface characterized by the parameters recommended in ref. [39]. Initial contact angle 75°, equilibrium contact angle 109°, $\lambda_d = S_d/\gamma = 1.468$, $l_0/d_0 = 3.8$, $d_0 = 0.158$ nm. Transition from partial wetting of the dry surface to non-wetting (90° initial contact angle) is observed at the fiber diameter $2a = 4.1$ nm.

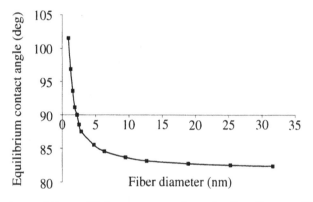

Figure 5. Dependence of the equilibrium contact angle on the fiber diameter. Wetting of carbon nanotubes with a graphite-like surface characterized by initial contact angle 75°, equilibrium contact angle 82°, $\lambda_d = S_d/\gamma = 0.5$, $l_0/d_0 = 3.8$, and $d_0 = 0.158$ nm. Transition from partial wetting to non-wetting (90° contact angle) is observed at the fiber diameter $2a = 2.2$ nm.

diameter $2a = 4.1$ nm. For fibers of diameter characteristic of single-wall carbon nanotubes, *ca.* 1 nm, the water contact angles achieve values typical for mercury.

To demonstrate the effect of fiber curvature on partial wetting, consider a system with the same parameters as above (the initial contact angle 75°, $\lambda_d = S_d/\gamma = 0.5$, $l_0/d_0 = 3.8$, $d_0 = 0.158$ nm), but with reduced energy characteristics, $\lambda_d = S_d/\gamma = 0.5$ instead of 1.468. Note that in this case, the polar component of the spreading coefficient $S_p = -89.1 \times 10^{-3}$ N/m, is not in conflict with predictions of the acid–base interaction model [28]. This set of parameters implies an acute equilibrium contact angle equal to 82°. We expect that due to the positive Laplace pressure, partial wetting will be hindered. As the fiber diameter decreases, the equilibrium contact angle increases and exceeds 90°. The results of calculations of the equilibrium contact angle [equation (49)] are presented in Fig. 5. Transition from partial wetting to non-wetting (90° contact angle) is observed at the fiber diameter $2a = 2.2$ nm.

These estimates show that the wetting properties of nanofibers may differ significantly from those of plane surfaces.

8. CONCLUSIONS

We have presented a general analytical solution for the equilibrium profile of the transition zone between a film and a droplet residing on a cylindrical fiber. Adhesion forces, including van der Waals and polar interactions, are expressed in terms of the disjoining pressure. The method developed is applicable for analyses of apparent contact angles and film stability on fibers and other cylindrical surfaces. A new equation for apparent contact angles formed by droplets on fibers has been derived. Explicit formulae for calculations of the equilibrium droplet profiles, film thicknesses, apparent contact angles, and stability factors have been presented in the form of expressions which include measurable geometrical parameters and presumably known parameters of liquid–solid interactions, such as apolar and polar spreading coefficients. The method developed has been applied for studies of the wetting of nanofibers, in particular carbon nanotubes. It is shown that a transition from partial wetting to non-wetting may occur as the fiber diameter decreases. Depending on the fiber diameter, the contact angles of water on hydrophobic carbon fibers may vary from 75° (plane graphite surface) to 100–130° (carbon nanotubes). The estimates presented demonstrate that the fiber curvature has to be considered an important factor affecting fiber wettability.

Acknowledgement

This work was supported by a group of TRI corporate participants.

REFERENCES

1. G. McHale, N. A. Kab, M. I. Newton and S. M. Rowan, *J. Colloid Interface Sci.* **186**, 453–461 (1997).
2. T. W. Ebbesen (Ed.), *Carbon Nanotubes: Preparation and Properties*. CRC Press, Boca Raton, FL (1996).
3. B. C. Satishkumar, A. Govindaraj, E. M. Voli, L. Basumallic and C. N. R. Rao, *J. Mater. Res.* **12**, 604–609 (1997).
4. T. Kasuga, M. Hiramatsu, A. Hoson, T. Sekino and K. Niihara, *Langmuir* **14**, 3160–3163 (1998).
5. S. L. Goren, *J. Fluid Mech.* **12**, 309–319 (1962).
6. M. Johnson, R. D. Kamm, L. W. Ho, A. Shapiro and T. J. Pec, *J. Fluid Mech.* **233**, 141–149 (1991).
7. A. L. Yarin, A. Oron and P. Roseneau, *Phys. Fluids A* **5**, 91–98 (1993).
8. B. V. Derjaguin, *Acta Phys.-Chim. USSR* **12**, 181–200 (1940); *J. Colloid Interface Sci.* **49**, 249–258 (1974).

9. B. V. Derjaguin, N. V. Churaev and V. M. Muller, *Surface Forces*. Plenum Press, New York (1987).

10. L. I. Kheifets and A. V. Neimark, *Multiphase Processes in Porous Media*. Khimia, Moscow (1982).

11. D. Li and A. W. Neumann, in: *Applied Surface Thermodynamics*, A. W. Neumann and J. K. Spelt (Eds), pp. 109–168. Marcel Dekker, New York (1996).

12. A. V. Neimark and M. Vignes-Adler, *Phys. Rev. E* **51**, 788–791 (1995).

13. V. M. Starov and N. V. Churaev, *Colloid J. USSR* **40**, 909–914 (1978).

14. F. Brochard-Wyart, *C.R. Acad. Sci. Paris, Ser. II* **303**, 1077–1080 (1986).

15. B. Novakowski and E. Ruckenstein, *J. Colloid Interface Sci.* **148**, 273–279 (1992).

16. J.-L. Chen and C.-C. Hwang, *J. Colloid Interface Sci.* **182**, 564–569 (1996).

17. D. A. White and J. A. Tallmadge, *J. Fluid Mech.* **23**, 325–339 (1965).

18. B. J. Carroll, *J. Colloid Interface Sci.* **57**, 488–495 (1976).

19. B. J. Carroll, *Langmuir* **2**, 248–250 (1986).

20. D. Quere, J.-M. Di Meglio and F. Brochard-Wyart, *Rev. Phys. Appl.* **23**, 1023–1030 (1988).

21. J. C. Bacri, C. Frenois, R. Perzynski and D. Salin, *Rev. Phys. Appl.* **23**, 1017–1022 (1988).

22. B. V. Derjaguin, V. M. Starov and N. V. Churaev, *Colloid J. USSR* **38**, 875–879 (1976).

23. F. Renk, P. C. Wayner and G. M. Homsy, *J. Colloid Interface Sci.* **67**, 408–416 (1978).

24. A. V. Neimark and L. I. Kheifets, *Colloid J. USSR* **43**, 402–407 (1981).

25. M. Kagan, W. V. Pinczewski and P. E. Oren, *J. Colloid Interface Sci.* **170**, 426–431 (1995).

26. M. Kagan and W. V. Pinczewski, *J. Colloid Interface Sci.* **180**, 293–295 (1996).

27. J. F. Joanny and P. G. de Gennes, *C.R. Acad. Sci. Paris, Ser. II* **299**, 279–282 (1984).

28. C. J. van Oss, *Interfacial Forces in Aqueous Media*. Marcel Dekker, New York (1994).

29. G. Hirasaki, in: *Interfacial Phenomena in Petroleum Recovery*, N. R. Morrow (Ed.), pp. 23–99. Marcel Dekker, New York (1991).

30. G. Hirasaki, in: *Contact Angle, Wettability and Adhesion*, K. L. Mittal (Ed.), pp. 183–220. VSP, Utrecht, The Netherlands (1993).

31. G. Hirasaki, *J. Adhesion Sci. Technol.* **7**, 285–322 (1993).

32. A. Sharma, *Langmuir* **9**, 861–871 (1993).

33. A. Sharma and A. T. Jameel, *J. Colloid Interface Sci.* **161**, 190–208 (1993).

34. A. Sharma, *Langmuir* **14**, 4915–4928 (1998).

35. S. Iijima, *Nature* **354**, 56–58 (1991).

36. T. W. Ebbesen and P. M. Ajayan, *Nature* **363**, 220–222 (1992).

37. S. Iijima and T. Ichihashi, *Nature* **363**, 603–605 (1993).

38. R. S. Smalley, *Rev. Mod. Phys.* **69**, 723–730 (1997).

39. U. Thiele, M. Mertig and W. Pompe, *Phys. Rev. Lett.* **80**, 2869–2872 (1998).

40. C. A. Miller and P. Neogi, *Interfacial Phenomena*. Marcel Dekker, New York (1985).

Apparent and Microscopic Contact Angles, pp. 319–331
J. Drelich, J. S. Laskowski and K. L. Mittal (Eds)
© VSP 2000.

Estimation of contact angles on fibers

G. McHALE*, S. M. ROWAN, M. I. NEWTON and N. A. KÄB

*Department of Chemistry and Physics, The Nottingham Trent University,
Clifton Lane, Nottingham NG11 8NS, UK*

Received in final form 20 March 1999

Abstract—A droplet of liquid placed on a flat high-energy solid surface spreads to give a thin film so that no macroscopic droplet shape exists. On a chemically identical solid surface with only the geometry changed to a cylinder, the same droplet can have an equilibrium conformation. When the equilibrium conformation is of a barrel type, the profile of the droplet changes rapidly in curvature as the three-phase contact line is approached and the direct measurement of the contact angle is difficult. This work considers the theoretical profile for barrel-type droplets on cylinders and discusses how the inflection angle in the profile depends on droplet parameters. Experimental results are reported for poly(dimethylsiloxane) oils on a range of fiber surfaces and these are used to estimate the equilibrium contact angle from the inflection angle. The drop radius and volume dependence of the inflection angle is confirmed.

Key words: Wetting; fibers; contact angles.

1. INTRODUCTION

A knowledge of the contact angle of a liquid on a solid surface is important in understanding wetting and adhesion [1, 2]. Changes in the equilibrium contact angle are used as indications of changes in the wettability of surfaces and are related by Young's law, $\cos \theta = (\gamma_{SV} - \gamma_{SL})/\gamma_{LV}$, where the γ_{ij} are the interfacial tensions. Optical observation of the profile of a droplet on a surface is an important technique for accurate determination of the contact angle. However, materials often arise in the form of a fiber and this presents quite specific problems in measuring the contact angle from the observed profile. On a flat surface, a droplet of liquid characterized by a spreading power $S = \gamma_{SV} - \gamma_{SL} - \gamma_{LV} \geqslant 0$ will spread to form a thin film so that no macroscopic shape exists. On a fiber of the same material, a droplet may

---

*To whom correspondence should be addressed. E-mail: glen.mchale@ntu.ac.uk

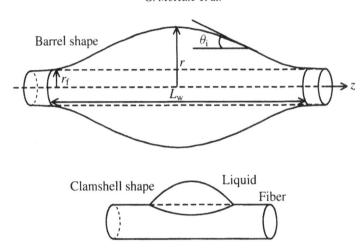

Figure 1. The two types of equilibrium conformation possible for droplets on fibers. The four directly accessible measurable profile parameters are the fiber radius r_f, the droplet radius r, the wetted length L_W, and the inflection angle θ_i.

exhibit either a clam-shell or a barrel-type conformation depending on its volume (Fig. 1) [3, 4]. Moreover, it is consistent with a vanishing equilibrium contact angle for the barrel-type droplet on a fiber to have a macroscopic shape rather than simply forming a thin sheathing film [2, 5–7]. This type of droplet requires an inversion of curvature in the profile close to the fiber surface. The excess pressure across a curved surface is given by the Laplace law $\Delta P = \gamma_{LV}(1/R_1 + 1/R_2)$, where R_1 and R_2 are the two principal radii of curvature. This excess pressure can be reduced by either increasing the two radii of curvature or introducing a relative sign between the radii. Film formation on a flat surface requires both radii to become large and this is consistent with conservation of the liquid volume. However, for barrel-type droplets on a fiber, volume conservation necessarily means that increasing one radius of curvature involves a reduction in the other. A limit is set on the reduction in the radius of curvature by the radius of the fiber. Thus, a reduction in the excess pressure can be achieved whilst maintaining finite absolute values for the radii of curvature provided that there is a change in curvature, i.e. an inflection, in the droplet profile.

The point of inflection in the profile of a barrel-type droplet on a fiber can occur close to the fiber surface and this complicates the measurement of the contact angle. A system with a vanishing equilibrium contact angle will have an inflection angle that depends on droplet volume and this angle may be in excess of 30°. A simple tangent method for determining the contact angle will almost certainly produce an estimate somewhere between the true value and the inflection angle. To enable a simple optical determination, Carroll [8, 9] therefore developed a method using the maximal droplet radius and wetted length. A more sophisticated technique using a numerical fit to the complete optical profile has also been reported [10, 11]. In this work, we report on a variation of the method used by Carroll [8]. This involves measuring the inflection angle, in addition to the maximal droplet radius and the fiber radius. The contact angle can then be estimated using a numerical minimization procedure as previously reported [7]. An alternative scheme based on analytical approximations for the inflection angle, which is straightforward to implement, is developed in this paper. Finally, experimental results are presented for barrel-type droplets on fibers of copper, polyester, polyethylene, and polypropylene, and the consistency of contact angle determination using the inflection angle, rather than the wetted length, is discussed.

2. THEORY

2.1. Droplet profile and the inflection angle

The change in curvature in the profile of a barrel-type droplet makes the direct estimation of contact angle difficult even when the contact angles are relatively large. The problem becomes increasingly difficult as the fiber diameter reduces. This motivated Carroll [8] to develop a method for estimating the contact angle from measurements of the wetted droplet length along the fiber and the maximal droplet radius in the plane normal to the fiber axis (also see Roe [12]). This work was a generalization of earlier derivations of equations for the barrel profile, which imposed a vanishing contact angle at the fiber surface [13, 14]. Subsequently, we reviewed Carroll's derivation and developed programs to implement his equations [7]. The defining equation for the profile of a barrel-type droplet on a fiber whose axis is in the z-direction is

$$\frac{d\bar{z}}{d\bar{x}} = \frac{-(\bar{x}^2 + an)}{\sqrt{(n^2 - \bar{x}^2)(\bar{x}^2 - a^2)}},$$ (1)

where $\bar{z} = z/r_f$ and $\bar{x} = x/r_f$ are the reduced co-ordinates and $n = r/r_f$ is the maximum radius of the droplet in reduced co-ordinates. The parameter a depends on both the contact angle and the droplet radius and is given by

$$a = \frac{n\cos\theta - 1}{n - \cos\theta}.$$ (2)

The point of inflection in the profile is given by the maximum in equation (1) and can be found by setting the differential to zero. The non-zero solution is

$$\bar{x} = \pm\sqrt{an}$$ (3)

and substituting equation (3) into equation (1) gives the formula for the inflection angle, θ_i:

$$\tan\theta_i = \pm\left(\frac{n - a}{2\sqrt{an}}\right).$$ (4)

Thus, the inflection angle depends on both the contact angle and the reduced radius of the droplet. Since droplets of differing volume have different values of reduced radius and wetted length, the inflection angle will not be the same for any two arbitrary droplets of the same liquid on fibers of the same material. Physically, the position of the inflection must lie between the fiber surface and the maximal droplet radius. For a given equilibrium contact angle less than 90°, applying this requirement to equation (3) gives

$$n_{\min} = \frac{1 + \sin\theta}{\cos\theta}, \tag{5}$$

which sets a minimum value of the maximal droplet radius, n, possible for any given liquid–fiber system. Any solution of equation (4) must be consistent with the requirement of $n > n_{\min}$.

2.2. Analytical approximations for $\theta(\theta_i, n)$

Numerically, the contact angle can be deduced from measurements of the fiber radius r_f and two other parameters chosen from (r, L, θ_i, V). Carroll [8] has produced tables for this purpose covering the physical range of $n = 1$–10 and θ up to 90°. We have also previously described a simple computational minimization scheme that enables this to be done [7]. An alternative numerical scheme has been described by Wagner [15] (see also refs [16] and [17]). However, the tables are based on measurements of the reduced length and radius, rather than the inflection angle. Although our numerical method does allow the contact angle to be estimated from the inflection angle, it is convenient to provide a simple approximate analytical method based on the inflection angle.

In many contact angle problems, the small angle approximation $\cos\theta \approx 1 - \theta^2/2$ proves remarkably accurate, even for apparently quite large contact angles. This is due to the next term in the series being of quartic order. For example, at 40° the error is only 1.3% and at 50° it is still a moderate 3.7%. To obtain an analytical approximation for the contact angle in terms of the inflection angle and the reduced radius, we therefore expand $\cos\theta \approx 1 - \theta^2/2$ in equation (2) and substitute it into equation (4). Rearranging the resulting expression for the contact angle gives (in radians)

$$\theta \approx \theta_0 = \frac{2(n-1)}{n+1}\sqrt{\frac{2\sqrt{n}\tan\theta_i}{n-1} - 1} \tag{6}$$

and this is accurate to 0.1° for $\theta < 12°$ and to 1° for $\theta < 24°$. Taking the expansion in $\cos\theta$ to the fourth power gives the improved approximation

$$\theta \approx \theta_1 = \sqrt{\frac{12}{7}\left(-1 + \sqrt{1 + \frac{7}{6}\theta_0^2}\right)} \tag{7}$$

and equation (7) extends the range of contact angles for accuracies of 0.1° and 1° to $\theta < 18°$ and $\theta < 34°$, respectively. Both equation (6) and equation (7) tend to overestimate the contact angle. A further improvement can be found by using an expansion of $\cos \theta$ about θ_1, obtained from equation (7), and this predicts

$$\theta \approx \theta_2 = \theta_1 + \frac{(n - a_1)(n \cos \theta_1 - 1)(n - \cos \theta_1)(\tan \theta_i - \tan \theta_i^1)}{(n + a_1)(n^2 - 1) \sin \theta_1}, \qquad (8)$$

where θ_i^1 is the inflection angle calculated from equation (4) using the measured value of n and θ_1 from equation (7). Similarly, a_1 is computed from equation (2) using n and θ_1. Equations (6)–(8) are all based on small angle approximations and cover the range of droplet parameters that are likely to be observed. However, whilst equation (8) is accurate to within $0.1°$ for $\theta < 20°$ and to within $1°$ for $\theta < 40°$, for the larger angles it becomes increasingly inaccurate, e.g. for $\theta = 60°$ equation (8) predicts an angle of $61.8°$. In practice, equations (6)–(8) are likely to be more accurate than stated since the maximum theoretical error occurs for drops with reduced radii close to the minimum allowed and these are less likely to be observed.

Contact angles of greater than $60°$ are unlikely to be observed, since at such high angles droplets on fibers tend to change conformation from the barrel shape to the clam-shell shape [4, 8]. However, for completeness it is worth considering these higher angles. In such extreme cases, it is better to expand $\cos \theta$ about $\theta = 90°$, using $\phi' = (\pi/2 - \theta)$ as the small parameter. It should be noted that some difficulty occurs in identifying leading order terms as $n\phi'$ cannot be assumed small, even though ϕ' is small. One such expansion, which is accurate at high contact angles, is

$$\theta_e \approx \theta_{high} = \pi/2 - \frac{1}{n}\left[1 + \left(\frac{1 + n^2}{2n \tan \theta_i}\right)^2\right]. \qquad (9)$$

Equation (9) tends to underestimate the actual contact angle. The error using equation (9) is maximum at n_{min} and within $0.1°$ for $\theta = 75°$, and becomes greater as the contact angle reduces. Equation (9) outperforms equation (8) for angles of $65°$ or greater. It should be noted that for $\theta > 78.6°$ the reduced radius exceeds the maximum value of $n = 10$ considered by Carroll [8] and by the approximation scheme presented in this work.

The analytical approximation scheme can be summarized as follows. To estimate the contact angle from measured values of n and θ_i using equations (6)–(9), equation (8) should be used initially. If the predicted value of θ is greater than $68.5°$, which corresponds to the worst case n for a true contact angle of $64°$, then equation (9) will give a better estimate. The worst case values in this procedure will be in error by no more than $3.5°$ and will be better than $1.8°$ for $\theta \leqslant 60°$. The contact angles in equations (6)–(9) are in radians and have been checked against the numerical results for $n = 1$–10 and $\theta = 1$–$78°$. The consistency of a predicted contact angle with the minimum permitted value of n should be verified using equation (5).

3. MATERIALS AND METHODS

In this work, droplets of 1000 cSt viscosity poly(dimethylsiloxane) (PDMS) oil were deposited onto a range of fibers of different materials. These fibers included electrochemically etched copper wires with diameters in the range 40–500 μm and polyester fibers with nominal diameters 40, 80, and 120 μm. Polypropylene (PP) fibers with diameters between 20 and 35 μm and poly(ethylene terephthalate) (PET) fibers with diameters between 40 and 80 μm were also examined. The electrochemical preparation of the copper wires ensures that a controlled range of fiber diameters with chemically similar surfaces can be produced. Moreover, a flat copper surface with PDMS provides a high-energy system, thus providing a large contrast between the intrinsically low contact angle and the expected droplet size-dependent inflection angle. The in-house preparation of these copper wires has been described elsewhere [7]. Since the primary aim in the present work was to examine the measurement technique using the inflection angle, rather than to provide accurate contact angles for specific materials, the commercial fibers were used as received.

The fibers were mounted in a system with two cameras equipped with microscope barrels and objective lenses, thus enabling both a top view and a side view of the droplet on the fiber. Droplets were deposited from the tip of a wire onto the fibers by lightly bringing them into contact. It is, therefore, likely that droplets were effectively deposited onto a fiber that already had a light coating of PDMS. Indeed, in a number of cases, chains of droplets were observed on the commercial fibers, indicating that a film had been deposited and had subsequently broken up into individual droplets. The images observed from above were captured into a personal computer (PC) using a DataTranslation DT3512 framegrabber card. This card has a one-to-one aspect ratio, which is a significant advantage since the measurements involved either angles or reduced radii and reduced lengths. A calibration piece with 100 divisions each separated by 10 μm was used to verify the aspect ratio and to check for optical distortions. A single fixed magnification was used for all measurements and this had a conversion of 352 pixels per mm; the images had a pixel resolution of 758 × 576 and used 256 gray levels. Images were not captured using the side-view camera, which was used only to give an approximate indication of whether the droplets were symmetric both above and below the fiber. Capillary, rather than gravity, forces are expected to mould the shape of the droplets on the fibers provided that the Goucher number, $Go = r_f/\kappa^{-1}$, is small [7]; κ^{-1} is the capillary length for the liquid. For PDMS $\kappa^{-1} = 1.5$ mm, which gives a fiber diameter of 300 μm before the Goucher number exceeds 0.1 and we would expect gravity forces to be important. Indeed, our previous work for PDMS on copper has shown that symmetric droplets are obtained once the fiber diameter decreases below 0.3 mm [7].

Measurements were obtained from the captured images of droplets using the PC-based ImageTool software. A simple algorithm involving a median filter, thresholding, and then a 5 × 5 convolution filter (HAT) was used to identify the edges of both the droplet and the fiber. The fiber diameter was estimated from the average measured at either side of the droplet. Similarly, the wetted length was the average measured in the upper and lower halves of the image. The inflection angle was taken as the average of the maximum slopes in the four quarters of a droplet image. In each case, a simple tangent was aligned on the profile by eye. The final measurement taken from the droplet profile was of the maximal droplet diameter. An alternative edge detection algorithm based on a 5×5 gradient magnitude and thinning routine was also used to detect the edges, but this did not produce significantly different values for the measured parameters. The largest error is expected to exist in the identification and measurement of the diameter of the smallest fibers since they correspond to around 10 pixels. Analysis of the images showed that it was not possible to measure the contact angle directly at the three-phase contact line.

4. RESULTS AND DISCUSSION

In Carroll's approach [8], the contact angle can be deduced from measurements of the reduced droplet radius and the reduced wetted length and this we did using our previously developed programs [7]. The contact angle can then be used with the reduced droplet radius in equations (2) and (4) to estimate the inflection angle. Figure 2 shows the correlation between the inflection angle thus estimated and the directly measured inflection angle. Droplets of all sizes and for all the different fiber materials are shown (circles). The previously published data for PDMS on etched

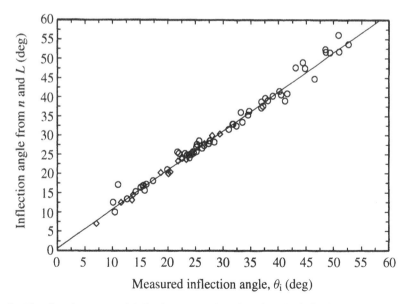

Figure 2. The directly measured inflection angle plotted against the inflection angle deduced from measurements of reduced radius, n, and reduced wetted length, L, for PDMS on copper, polyester, polypropylene, and poly(ethylene terephthalate) fibers. The straight-line fit has a slope of 1.020 and an intercept of $0.585°$. Data from ref. [8] have been included (diamond symbols).

G. *McHale* et al.

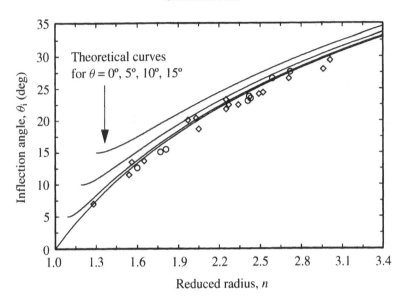

Figure 3. The data for inflection angle and reduced radius for barrel-type droplets of PDMS on copper lie on a curve parallel to, but slightly below, that predicted by a vanishing contact angle. The trend of the data towards the upper right in this figure shows that the inflection angle increases substantially with reduced volume. The lowest of the solid curves indicates the theoretical curve for $0°$.

copper wires have also been included in Fig. 2 (diamond symbols). The best-fit straight line through the data has a slope of 1.020 and an intercept of $0.585°$ with $R^2 = 0.992$. The agreement in the directly measured and deduced inflection angles over the large angular range from below $10°$ to over $50°$ is impressive.

The consistency of the current analysis with the previous work can be verified by examining the data for PDMS on copper over a similar range of reduced radii and wetted lengths. Figures 3 and 4 show the directly measured inflection angle against the reduced radius and reduced wetted length, respectively. The solid curves are the values computed using Carroll's formulae for contact angles of $0°$, $5°$, $10°$, and $15°$. In each of the diagrams, the trend in data follows the expected theoretical curves. In both cases, moving from the lower left to the upper right corresponds to increasing the reduced volume, volume/r_f^3, of the droplets. Slightly more scatter is apparent in Fig. 4, which uses the reduced wetted length of the droplet. This is to be expected since for a system with a small contact angle, it is very difficult to identify the precise contact between fiber and liquid. Consequently, the wetted length measurements tend to have an uncertainty of around 10%. In contrast, the reduced radius is easier to measure, to within typically 2%, once the profile has been identified. Although this clearly reduces the scatter in the data in Fig. 3, it does not necessarily mean that the reduced radius is known more accurately. The estimation of the reduced radius may be more susceptible to a small systematic error arising, for example, from the method used to identify the edge of either the fiber or the droplet. The data plotted in Fig. 4 suggest that the contact angle is around $3–4°$, whereas the data from the same droplets, but plotted in the form of Fig. 3 lie slightly

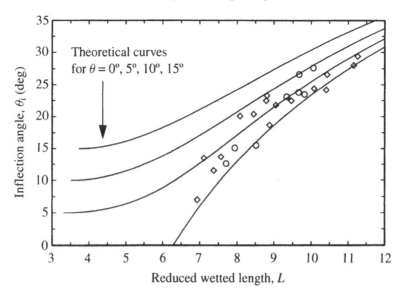

Figure 4. The trend towards an increasing inflection angle with increasing reduced volume, seen in Fig. 3, is apparent when the dependence on reduced wetted length is examined. A slightly higher contact angle of around 3–4° is indicated by this graph compared with Fig. 3.

below the 0° curve. The equivalent data can also be plotted using the reduced radius and reduced wetted length, as suggested by Carroll [8], and this indicates a contact angle of around 6°. Since the optical system used has a one-to-one aspect ratio, the difference in predicted contact angles is probably not due to a systematic error in the inflection angle. The simplest systematic error that would bring the three estimates of contact angle into agreement is an overestimate of the reduced radius. For example, if the image processing definition of the edges produced a slight underestimate of the separation of horizontal lines, then the effect on the fiber radius would tend to be greater than that on the droplet radius. This would reduce n and shift the experimental points to the left in Fig. 3. Such a mechanism would also marginally increase the reduced wetted length, so shifting the experimental points to the right in Fig. 4. Such a small systematic error would be consistent with Fig. 5 and increase the agreement between the three estimates of contact angle.

Figures 6–8 show equivalent data for droplets of PDMS on the three different sizes of polyester fibers. Although the data set is more limited than for copper, a wider parameter range is encompassed. In each graph, the data follow the expected trends with increasing inflection angle as the droplet size increases. The measured inflection angle–reduced radius plot, Fig. 6, indicates a contact angle of around 0°, although a precise estimate is difficult because of the proximity of the theoretical curves for 0° and 5°. Measured inflection angle considered against reduced wetted length (Fig. 7) suggests a contact angle of around 5–6°, whereas using the reduced radius and the reduced wetted length (Fig. 8) indicates that the contact angle is slightly higher, at around 7–8°. Comparing these three estimates

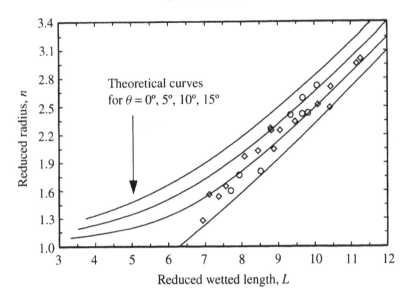

Figure 5. The data in Fig. 3 plotted as reduced radius against reduced wetted length. The contact angle can be estimated from the solid curves as originally suggested by Carroll [8].

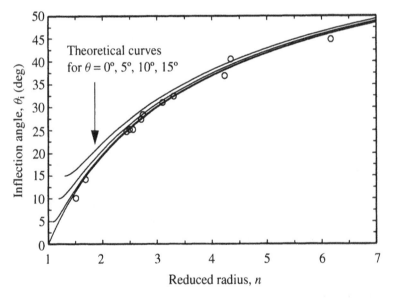

Figure 6. A graph of the measured inflection angle against the measured reduced radius for droplets of PDMS on polyester fibers (nominal radii of 40, 80, and 120 μm). The inflection angle increases as the reduced radius increases.

with those obtained for copper, we observe the same trend in contact angle with the smallest value deduced from θ_i and n and the largest from n and L.

The consistency of the contact angle estimates suggests that the measurement of inflection angle is a useful approach. The inflection angle can be combined with measurements of the reduced radius and length to provide a set of three estimates of

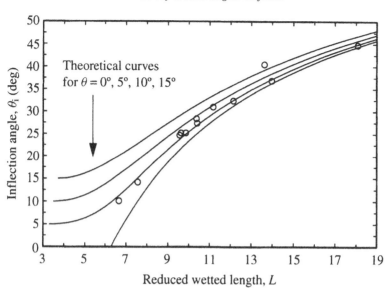

Figure 7. The inflection angle plotted against the reduced wetted length for the same droplets of PDMS on polyester fibers as shown in Fig. 6. The contact angle deduced from this graph is around 5–6°, which is slightly higher than that deduced using Fig. 6.

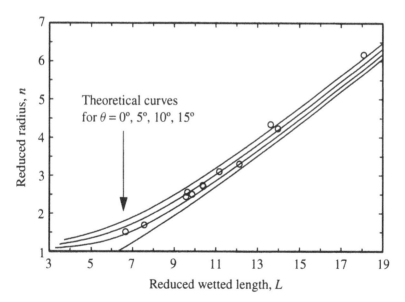

Figure 8. A graph of the reduced radius against the reduced wetted length for the data of PDMS on polyester fibers shown in Fig. 6.

the contact angle and so provide a self-consistency check. Alternatively, measurements of θ_i and n can be used with the analytical approximations, equations (6)–(9), to provide a simple method of estimating θ.

5. CONCLUSIONS

The theoretical profile for barrel-type droplets on fibers has been considered. It has been suggested that measurements of the inflection angle in the droplet profile provide additional information, in addition to the droplet radius and wetted length, for estimating the contact angle of the system. The inflection angle has been shown to depend strongly on the droplet size. An analytical approximation has been developed to allow the contact angle to be calculated from measurements of the inflection angle and the droplet radius. Droplets of poly(dimethylsiloxane) on copper, polyester, polypropylene, and poly(ethylene terephthalate) have been observed and the consistency of the measured inflection angle with the value predicted from the droplet radius, and wetted length has been confirmed for inflection angles between 10° and 55°. Independent measurements of the inflection angle, droplet radius, and wetted length allow three estimates of the contact angle to be obtained. Moreover, these three estimates provide a consistency check on the estimate of the contact angle and may allow any systematic error within the measurements to be identified.

Acknowledgements

We acknowledge B. J. Carroll (Unilever Research Laboratory, Port Sunlight) for providing some of the fibers used in this work and for invaluable discussions. Image processing was performed using the UTHSCSA ImageTool program (developed at the University of Texas Health Science Center at San Antonio, TX, and available from the Internet by anonymous ftp from maxrad6.uthscsa.edu).

REFERENCES

1. A. W. Adamson, *Physical Chemistry of Surfaces*. Wiley, New York (1976).
2. L. Léger and J. F. Joanny, *Rep. Prog. Phys.* **55**, 431–486 (1992).
3. B. J. Carroll, *J. Colloid Interface Sci.* **9**, 195–200 (1984).
4. B. J. Carroll, *Langmuir* **2**, 248–250 (1986).
5. D. Quéré, J.-M. Di Meglio and F. Brochard, *Rev. Phys. Appl.* **23**, 1023–1030 (1988).
6. F. Brochard-Wyart, J.-N. Di Meglio and D. Quéré, *J. Phys. France* **51**, 293–306 (1990).
7. G. McHale, N. A. Käb, M. I. Newton and S. M. Rowan, *J. Colloid Interface Sci.* **186**, 453–461 (1997).
8. B. J. Carroll, *J. Colloid Interface Sci.* **57**, 488–495 (1976).
9. B. J. Carroll, *J. Adhesion Sci. Technol.* **6**, 983–994 (1992).
10. A. Kumar and S. Hartland, *J. Colloid Interface Sci.* **124**, 67–76 (1988).
11. A. Kumar and S. Hartland, *J. Colloid Interface Sci.* **136**, 455–469 (1990).
12. R.-J. Roe, *J. Colloid Interface Sci.* **50**, 70–79 (1975).
13. G. Mozzo and R. Chabord, *Proc. 23rd Annual Technical Conference*, The Society of the Plastics Industry, Section 9-C, pp. 1–8 (1968).
14. H. M. Princen, in: *Surface and Colloid Science*, E. Matijevic (Ed.), Vol. 2, p. 1. Wiley-Interscience, New York (1969).
15. H. D. Wagner, *J. Appl. Phys.* **67**, 1352–1355 (1990).

16. B. J. Carroll, *J. Appl. Phys.* **70**, 493–494 (1991).
17. H. D. Wagner, *J. Appl. Phys.* **70**, 495 (1991).

Apparent and Microscopic Contact Angles, pp. 333–348
J. Drelich, J. S. Laskowski and K. L. Mittal (Eds)
© VSP 2000.

Evolution of surface chemistry and physical properties during thin film polymerization of thermotropic liquid crystalline polymers

KUI-XIANG MA [1,2], TAI-SHUNG CHUNG [1,2,*], PRAMODA K. PALLATHADKA [1] and SI-SHEN FENG [1,2]

[1] *Institute of Materials Research and Engineering, 10 Kent Ridge Crescent, Singapore 119260*
[2] *Department of Chemical Engineering, National University of Singapore, 10 Kent Ridge Crescent, Singapore 119260*

Received in final form 13 February 1999

Abstract—By applying a novel thin film polymerization technique, X-ray photoelectron spectroscopy (XPS), and the Lifshitz–van der Waals acid–base (LWAB) theory, we have determined the time evolution of surface chemistry and surface free energy during the polymerization of liquid crystalline poly(p-oxybenzoate/2,6-oxynaphthoate) at a molar ratio of 50/50. The surface free energy components of these main-chain liquid crystalline copolyesters were calculated from contact angle measurements using a Ramé-Hart goniometer and a three-liquid procedure (water, glycerol, and diiodomethane). The experimental data suggest that the Lewis base parameter (γ^-) during thin film polymerization decreases rapidly with the progress of polymerization, while the Lewis acid parameter (γ^+) and the Lifshitz–van der Waals parameter (γ^{LW}) are almost invariant. The surface roughness data measured by atomic force microscopy (AFM) suggested that the increase in water contact angle (or the decrease in γ^-) was not caused by the change in surface roughness, but by the change in surface chemistry, i.e. due to the reaction of acetoxy and carboxy groups to release acetic acid during the polymerization reaction. In addition, the XPS results coincide with our previous Fourier transform infrared spectroscopy results showing that the condensation polymerization is much faster in the beginning than in the later stages. Consequently, the decrease in γ^- in the early stages of the polymerization is well explained.

Key words: Contact angle; surface free energy; acid–base interactions; roughness; liquid crystalline polymers; thin film polymerization.

---

*To whom correspondence should be addressed. E-mail: chencts@nus.edu.sg

1. INTRODUCTION

Because of their unique anisotropic mechanical properties, excellent chemical resistance, and high thermal stability, thermotropic liquid crystalline polymers (LCPs) have attracted interest from both academia and industry. Several publications have described the scientific progress on this subject [1–4]. Currently, LCPs have two major applications: one is for electronic devices and the other is for structural composites.

Using the monomer composition of Hoechst Celanese's liquid crystalline polyester Vectra™ A-950 [a random copolymer of 73 mol% 4-hydroxybenzoic acid (HBA) and 27 mol% 6-hydroxy-2-naphthoic acid (HNA)] as an example, Geil and co-workers [5–7] were the pioneers to investigate the microstructure of LCPs during polymerization using optical microscopy, transmission electron microscopy, X-ray diffraction, differential scanning calorimetry, and Fourier transform infrared spectroscopy (FTIR). They developed a so-called 'constrained thin film polymerization', in which a thin layer of monomers in specific ratios was sandwiched between two glass slides and wrapped with an aluminum foil. The thin film polymerization was then conducted on a temperature-controlled hot plate. A series of homopolymers and copolymers were synthesized using 4-acetoxybenzoic acid (ABA) and 2,6-acetoxynaphthoic acid (ANA), which are obtained by acetylation from HBA and HNA, respectively.

Our research group also invented a novel thin film polymerization technique to examine *in situ* and analyze polycondensation reactions and liquid-crystal phase growth mechanisms for wholly aromatic LCPs, such as poly(oxybenzoic acid (OBA)/oxynaphthoic acid (ONA)) copolymers, as well as poly(OBA) and poly(ONA) homopolymers [8]. In our study, a thin layer of monomers was sandwiched between two glass slides on a hot plate and the thin film polymerization occurred at the top glass plate. It was found that there was a sublimation–recrystallization–melting–polymerization–LC domain formation–crystallization process during the preparation of the copolymer.

However, the success of using LCPs for multilayer and miniature electronic devices depends on the surface free energy of the LCPs and their adhesion properties. To our surprise, only a few papers have been published on the surface free energy of LCPs [9–14] and most of them studied side-chain liquid-crystalline polymers [9–11]. For example, Uzman et al. [9] investigated the surface tension (γ) of branched polyethylene melts, and two side-chain liquid-crystalline polyacrylate (LCPA) melts in isotropic, nematic, and smectic states, using the pendant drop method. Correia et al. [10] reported the surface tension of a side-chain liquid crystalline polyacrylate (poly({3-[4-(4-cyanophenyl)phenoxy] propyloxycarbonyl)}ethylene)) provided by Merck.

Chung and co-workers [12–14] were the first to study the surface free energies of main-chain liquid crystalline polyesters [Vectra™ A-950 (Hoechst Celanese, USA) and Xydar™ SRT-900 (Amoco, USA)] and polyester-amide [Vectra™ B-950 (Hoechst Celanese, USA)]. Various surface free energy calculation models were employed to analyze the contact angle data. They found that the surface free energy values did match between two-liquid geometric-mean and three-liquid Lifshitz–van der Waals acid–base (LWAB) approaches if the correct combinations of test liquids were used. However, the three-liquid LWAB approach provided more information, on, for example, the acidity and basicity of LCP surfaces [13]. They also reported that these LCP surface free energies could be theoretically estimated through addition of the group contribution of their repeat units [12].

As atomic force microscopy (AFM) has become a very important tool for imaging surfaces [15–18] in various fields, especially in investigating surface roughness, researchers have used this instrumentation to study the influence of surface roughness on wetting behavior. Miller and co-workers [16, 17] found that a PTFE surface with an RMS roughness of 83 nm exhibited an advancing water contact angle of 146°, while another surface with an RMS roughness of 8.3 nm showed a contact angle of 94°. Hence, they demonstrated that nanometer-size surface roughness also affected the wetting behavior of their PTFE thin films. Busscher *et al.* [19], Hitchcock *et al.* [20], and Onda *et al.* [21] also reported similar effects of roughness on the wettability of polymers and ceramics. It was believed that one way to control the wettability of solid surfaces was to alter the geometrical structure rather than by chemical modification of the solid surfaces.

In this paper, we intend to investigate the surface free energy changes during the polymerization reaction for wholly aromatic copolymers prepared from 4-acet-oxybenzoic acid (ABA) and 2,6-acetoxynaphthoic acid (ANA), and to examine the surface morphology, roughness, and compositions of the thin film-polymerized liquid crystalline polymers, after different reaction times, by AFM and XPS.

2. MODERN ACID–BASE THEORY

In the modern theory of surface science, Fowkes was the first to propose the theory of acid–base interfacial interactions [22]. The surface free energy consists of two components: one is apolar (γ^d) and the other is acid–base (γ^{AB}). In order to determine the strengths of the acidic and basic sites of a polymer, Fowkes suggested using spectroscopic or calorimetric methods.

Van Oss and Good [23] created two new parameters to define the acid–base (γ^{AB}) interaction and to quantify the strength of acid–base interactions:

$\gamma^+ \equiv$ (Lewis) acid parameter of surface free energy;

$\gamma^- \equiv$ (Lewis) base parameter of surface free energy.

$$\gamma^{AB} = 2\sqrt{\gamma^+ \gamma^-}. \tag{1}$$

They also developed a 'three-liquid procedure' and the following equation to determine γ_S using contact angle measurements plus a traditional matrix scheme:

$$\gamma_{Li}(1 + \cos\theta_1) = 2\left(\sqrt{\gamma_S^{LW}\gamma_{Li}^{LW}} + \sqrt{\gamma_S^+\gamma_{Li}^-} + \sqrt{\gamma_S^-\gamma_{Li}^+}\right)$$

$$\gamma_{L2}(1 + \cos\theta_2) = 2\left(\sqrt{\gamma_S^{LW}\gamma_{L2}^{LW}} + \sqrt{\gamma_S^+\gamma_{L2}^-} + \sqrt{\gamma_S^-\gamma_{L2}^+}\right) \qquad (2)$$

$$\gamma_{L3}(1 + \cos\theta_3) = 2\left(\sqrt{\gamma_S^{LW}\gamma_{L3}^{LW}} + \sqrt{\gamma_S^+\gamma_{L3}^-} + \sqrt{\gamma_S^-\gamma_{L3}^+}\right).$$

In short, to determine the surface free energy, γ_S, of a polymer solid, it was recommended [23] to select three or more liquids from the reference liquids table, with two of them being polar and the other one apolar. Since the LW, Lewis acid (γ^+), and Lewis base (γ^-) parameters of γ_{L1}, γ_{L2}, and γ_{L3} in equation (2) are available [23], one can determine the LW, Lewis acid, and Lewis base parameters of γ_S by solving these three equations simultaneously using the experimental contact angle data. Hence, γ_S of the polymer solid can be obtained.

3. EXPERIMENTAL

3.1. Preparation of monomers: 4-acetoxybenzoic acid (ABA) and 2,6-acetoxynaphthoic acid (ANA)

The ABA and ANA monomers were prepared by acetylation, respectively, of 4-hydroxybenzoic acid (HBA) and 2,6-hydroxynaphthoic acid (HNA), with acetic anhydride in the presence of a catalytic amount of pyridine. The reaction took 4 h to complete, followed by recrystallization of ABA in butyl acetate and ANA in methanol, respectively. NMR confirmed the success of monomer acetylation.

3.2. Thin film polymerization

ABA and ANA monomers with specific mole ratios were dissolved in acetone. The solution was then deposited onto micro-glass slides and acetone evaporated in a few minutes. Another micro-glass slide was used to cover the one with monomers and was spaced about 0.5 mm, using an aluminum foil. The whole set was then wrapped with aluminum foil, followed by heating using a digital hot plate controlled at $270°C \pm 1\%$. Samples were removed from the thermostat hot stage immediately after 4, 8, 15, and 30 min, 1 h and 2 h heating times and cooled down under ambient environment conditions ($\sim 25°C$).

Since the monomer mixture was deposited on the bottom slide of the sample package, the monomers sublimed from the bottom slide and condensed on the top one. Microscopic study on the surface of the top slide suggested that this sublimation–condensation on a glass slide yielded much better monomer packing and eliminated defects better than conventional approaches [5–7]. Once the monomers had sublimed at the top glass slide, the polymerization reaction occurred immediately on the top slide. A polarizing light microscope (Olympus BX50) was utilized to examine the sample morphology.

3.3. Test liquids

Deionized water (prepared in this laboratory), glycerol (from BDH), and di-iodomethane (from Nacalai Tesque, Japan) were chosen as the test liquids because relevant data are available for these liquids [23]. All were reagent grade and were used as received. Table 1 presents their basic surface tension parameters (in mJ/m^2).

3.4. Contact angle measurements

The contact angle measurements were made on a Ramé-Hart Contact Angle Go-niometer (Model 100-22) by the sessile drop method at 25°C using an environmental chamber. A built-in image system provided by Ramé-Hart was able to acquire the images and transmit them to a computer, and to perform the image analysis. Liquid droplets were placed by a Gilmont micro-syringe onto the surfaces of ABA/ANA thin copolymer film samples (the top glass slide).

Advancing and receding contact angles were obtained with a tilting base using the basic technique and the principle developed elsewhere [24, 25]. When a liquid drop resting on a solid surface is inclined, it deforms. The contact angle at the advancing edge of the drop increases while the angle at the receding edge decreases, as shown in Fig. 1. The drop remains firmly adhered to the surface and stationary until the advancing (θ_a) and receding (θ_r) angles exceed certain critical values.

Normally five droplets in different regions of the same piece of film were deposited, and two pieces of sample were used, to obtain reliable contact angle data. Thus, ten contact angle values were averaged for each kind of sample as well as for each test liquid.

Table 1.
Surface tension parameters (in mJ/m^2) of the test liquids

	Water	Glycerol	Diiodomethane
γ^+	25.5	3.92	0.0
γ^-	25.5	57.4	0.0
γ^{AB}	51.0	30.0	0.0
γ^{LW}	21.8	34.0	50.8
γ	72.8	64.0	50.8

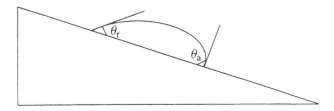

Figure 1. A liquid drop on the tilting base showing the advancing and receding contact angles.

3.5. Surface roughness determination by AFM

Atomic force microscopy (AFM) has become a very important tool for imaging surfaces [16–18] in various fields. The atomic force microscope has a cantilever equipped with a tip, which extends vertically from the free end of the cantilever, and the whole cantilever–tip unit is attached to the cantilever holder of the AFM. While scanning, a sharp tip at the end of a cantilever scans over a surface, and the tip and cantilever are deflected by different surface features. The topographic image of the surface is obtained from the deflection of the cantilever. In the present study, an Explorer 2000 AFM, from Topometrix, was used in the contact mode, and silicon nitride 1520 cantilever tips with a force constant of 0.064 N/m were chosen to investigate the surface roughness of the thin copolymer films after different reaction times. The surface topographical images were processed using TMX2000 (Topometrix) image software.

The RMS roughness (R_q) is defined as the standard deviation of the height values within the given area:

$$R_q = \sqrt{\Sigma(Z_i - Z_{ave})^2/N}, \tag{3}$$

where Z_{ave} is the average of the Z values within the given area, Z_i is the Z value for a given point, and N is the number of points within the given area. The thin film samples were imaged over a scan size of 50 μm × 50 μm; at least three different locations were scanned; and the RMS roughnesses were averaged to give the sample roughness.

3.6. X-ray photoelectron spectroscopy (XPS)

The thin copolymer film surfaces were also characterized using XPS in this study. The XPS measurements were preformed on a VG ESCA Lab MK II spectrometer with a magnesium anode source producing MgK_α (125.3 nm) X-rays at 12 keV. The XPS data analysis was done using software from VG ESCA Lab. Atomic concentrations were determined from the peak areas using the sensitivity factors provided by the manufacturer.

4. RESULTS AND DISCUSSION

4.1. Evolution of the contact angle and surface free energy of 50/50 ABA/ANA copolymers

All the measured advancing and receding contact angles on the 50/50 ABA/ANA thin copolymer films along with their standard deviations are listed in Table 2 for the three test liquids and various reaction times.

Since the advancing contact angles represent the equilibrium contact angles more closely than the receding angles [26], only the advancing contact angles will be discussed hereafter. Table 2 shows the contact angles with the most polar liquid,

Table 2.
Contact angles of the test liquids on 50/50 ABA/ANA thin copolymer films

Reaction time (min)	Contact angle	Water	Glycerol	Diiodomethane
4	Advancing	67.7 ± 1.61	66.8 ± 0.48	35.9 ± 2.16
	Receding	63.4 ± 1.69	62.0 ± 0.80	29.4 ± 1.33
8	Advancing	74.8 ± 1.33	66.8 ± 1.44	36.4 ± 1.07
	Receding	70.4 ± 0.51	61.3 ± 1.02	27.4 ± 0.90
15	Advancing	79.3 + 1.21	66.9 ± 1.97	38.2 ± 0.18
	Receding	74.0 ± 1.19	59.5 ± 1.53	32.8 ± 0.59
30	Advancing	81.5 ± 1.05	70.5 ± 0.44	34.8 ± 1.34
	Receding	76.3 ± 1.29	63.4 ± 0.21	28.0 ± 0.98
60	Advancing	83.5 ± 1.00	69.0 ± 2.36	36.0 ± 2.80
	Receding	77.3 ± 1.24	61.8 ± 2.35	29.2 ± 3.05
120	Advancing	83.9 ± 1.57	67.0 ± 1.62	39.0 ± 0.39
	Receding	76.8 ± 1.86	58.5 ± 2.54	34.6 ± 0.71

water; the less polar liquid, glycerol; and the apolar liquid, diiodomethane, on 50/50 ABA/ANA thin copolymer films after different reaction times. As expected, the polar liquids give higher contact angles than the apolar liquid. Moreover, it is quite obvious that the water contact angle increases with an increase in the reaction time, e.g. from 67° at 4 min to 83° at 2 h of reaction time. However, it is interesting to see that neither the glycerol nor the diiodomethane contact angle varies significantly compared with water. Clearly, with an increase in the copolymerization time, the polarity (in terms of Lewis acid–base interactions) of the copolymer surfaces decreases.

In order to obtain the surface free energies of the 50/50 ABA/ANA thin copolymer films, van Oss and Good's Lifshitz–van der Waals acid–base approach (three-liquid LWAB method) was utilized. Table 3 shows the calculated results using equation (2). In Table 3, the Lewis acid parameter (γ^+) of the ABA/ANA copolymers (50/50 ratio) is very small, varying from zero to 0.2 mJ/m^2, hence resulting in quite small acid–base interactions ($\gamma^{AB} = 2\sqrt{\gamma^+\gamma^-}$). Moreover, the Lifshitz–van der Waals parameter (γ^{LW}) does not vary much with the variation of reaction time, e.g. γ^{LW} is in the range of 40.1–42.1 mJ/m^2 for the 50/50 ABA/ANA copolymers.

However, the Lewis base parameter (γ^-) decreases monotonically with an increase in the reaction time for the 50/50 ABA/ANA copolymers. γ^- decreases from 17.2 mJ/m^2 at 4 min of reaction time to 2.60 mJ/m^2 at 2 h of reaction time. In other words, the Lewis base parameter decreases with the progress of

K.-X. *Ma* et al.

Table 3.
Surface free energy components (in mJ/m^2) of 50/50 ABA/ANA thin copolymer films (calculated with the three-liquid LWAB method)

Reaction time (min)	γ^+	γ^-	γ^{AB}	γ^{LW}	γ
4	0.00	17.2	0.00	41.6	41.6
8	0.02	9.02	0.89	41.4	42.2
15	0.15	5.18	1.78	40.5	42.3
30	0.00	4.92	0.24	42.1	42.4
60	0.10	3.04	1.12	41.6	42.7
120	0.26	2.60	1.64	40.1	41.8

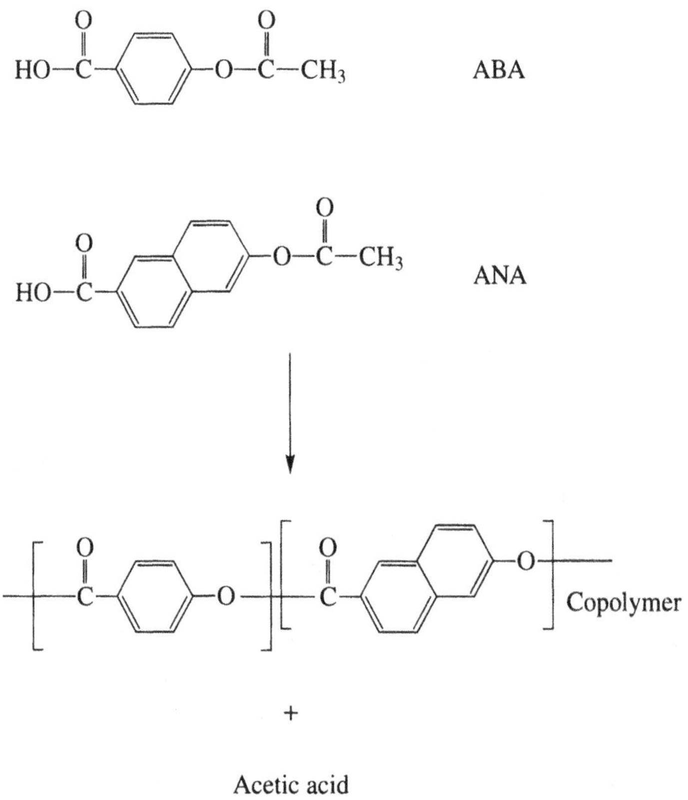

ABA

ANA

Copolymer

+

Acetic acid

Figure 2. Monomer structures and the polycondensation reaction.

ABA/ANA copolymerization. This decreasing phenomenon may be explained from the monomers' structures as well as the copolymerization formulation, as shown in Fig. 2. With the progress of copolymerization, more acetoxy and carboxyl groups combine to release acetic acid and therefore the decrease in γ^- of the surface is understandable.

4.2. RMS roughness of 50/50 ABA/ANA thin copolymer films

The thin film surfaces of 50/50 ABA/ANA copolymers with varying reaction time were examined by contact mode AFM, so as to study the influence of surface roughness on the wetting characteristics.

The AFM images of the thin film surfaces with 4, 8, 15, and 30 min, and 1 and 2 h reaction times are presented in Figs 3a–3f and the corresponding RMS roughness values are presented in Table 4. The topographical images of the thin film surfaces were recorded at three different locations, and one of them is shown in Fig. 3. The RMS surface roughness from various reaction times, R_q, listed in Table 4, was obtained from equation (3).

It can be seen from Table 4 that the RMS roughness values after different reaction times are comparable, except for 42 nm at 4 min and 164 nm at 2 h, although there is a trend that the surface roughness increases with increasing reaction time. For example, the roughness increases slowly from 11 nm to 21 nm from 8 min through 1 h. Moreover, it is clearly shown in Fig. 3 that after 4 min reaction time, the oligomer is randomly distributed and the LC texture appeared after 8 min and develops throughout the whole reaction period.

It has been reported elsewhere that the LC morphology changes in this novel thin-film copolymerization are sequential [8, 14]: generation of an anisotropic phase and coalescence of nematic droplets at the early stage (e.g. from the beginning to 4 min), formation of a schlieren texture and a stripe texture (e.g. from 8 min through 30 min), and then the development of a stripe or banded texture (e.g. from 1 h to 2 h). On comparing the AFM images shown in Fig. 3, it is understandable that at 4 min reaction time, the oligomer (pre-polymer) is undergoing formation and coalescence of nematic droplets when melting. This pre-polymer recrystallizes (Fig. 3a) after cooling to room temperature. The recrystallized morphology is the likely reason for the rougher surface and higher R_q. With the progress of copolymerization, a low-molecular-weight LC copolymer is being formed, as well as the stripe texture. Hence, the surface images from 8 min through 30 min are similar (Figs 3b–3d) and they have close R_q values. With further development of the stripe texture through annealing or solid-state polymerization, as shown in the images of 1 h

Table 4.
RMS roughness values of 50/50 ABA/ANA thin copolymer films (as determined by AFM)

Sample (with different reaction time)	RMS roughness (nm)
4 min	42
8 min	11
15 min	12
30 min	18
1 h	22
2 h	165

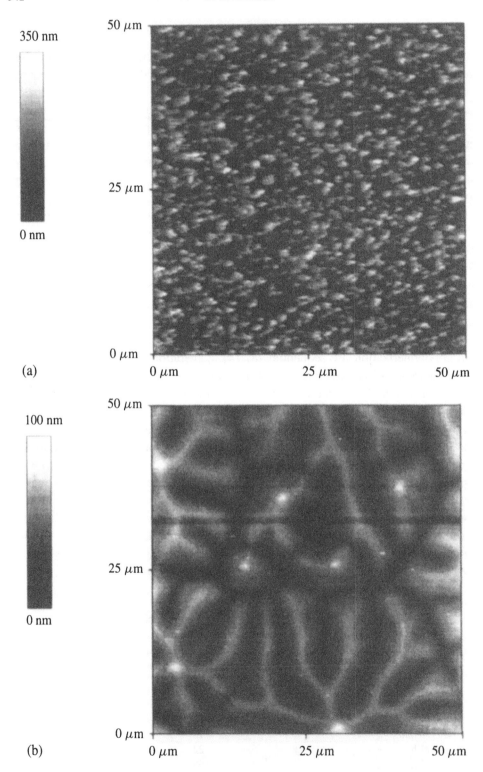

Figure 3. AFM images of 50/50 ABA/ANA thin copolymer films at different reaction times: (a) 4 min; (b) 8 min; (c) 15 min; (d) 30 min; (e) 1 h; (f) 2 h.

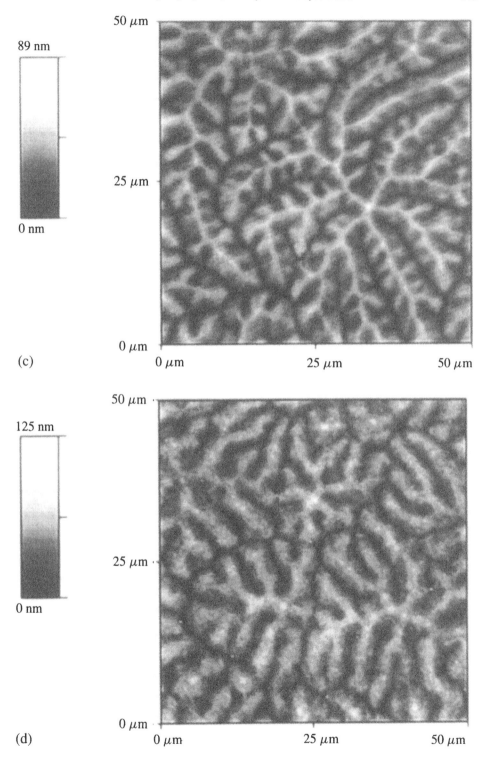

(c)

(d)

Figure 3. (Continued).

K.-X. *Ma* et al.

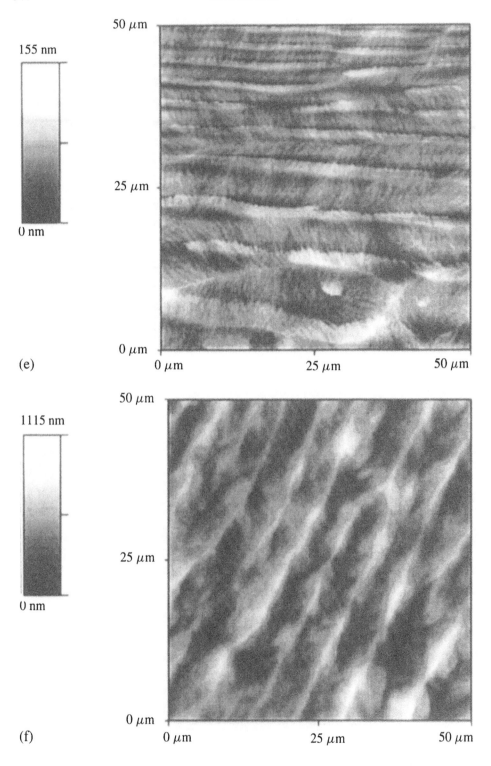

(e)

(f)

Figure 3. (Continued).

and 2 h (Figs 3e and 3f), we observe the formation of a banded structure and/or crystallization. R_q increases up to 164 nm at 2 h, indicating that the LC texture may also develop vertically because of chain folding and packing.

As shown in Table 2, the advancing water contact angles on the thin copolymer film increases consistently from 67° at 4 min through 83° at 2 h reaction time. For instance, R_q of the thin copolymer film decreases from 41 nm to 11 nm from 4 min to 8 min, while the water contact angle still increases from 67° to 74°. In particular, R_q increased from 21 nm to 164 nm from 1 h to 2 h, which is almost 8 times different; however, the water contact angles are nearly equal (83°). In contrast, the data from Miller and co-workers [16, 17], in which the surface roughness decreased the PTFE wettability, showed the opposite trend.

Our previous work [14] had explained why the water contact angles increased and hence γ^- decreased with increasing reaction time of ABA/ANA copolymers. Based on the FTIR spectra, it was shown that both carboxyl and acetoxy groups from the monomers (ABA and ANA, respectively) decreased to below the detectable limits after 2 h, while the polyester group increased with increasing reaction time. Therefore, due to the combination of acetoxy with carboxyl groups to release acetic acid during the copolymerization, the polarity of the thin copolymer film surface decreases, and hence the water contact angle increased and γ^- decreased.

So our results clearly indicate that the decrease of wettability of the ABA/ANA thin copolymer film with varying reaction time is not due to the surface roughness change, and the dominating force is the surface chemistry change through polycondensation.

4.3. XPS data on 50/50 ABA/ANA thin copolymer films

Figure 4 shows the XPS spectra of a 50/50 ABA/ANA thin copolymer film after 2 h of reaction time with the widescan (a) as well as deconvoluted XPS spectra for C 1*s* (b) and O 1*s* (c) as an example. From the spectra it is clear that there are no impurities except the accepted C, O, and their Auger peaks present in the samples. Moreover, the XPS data did not show any peak shift as the polymerization progressed.

The calculated atomic concentrations of C and O are shown in Table 5. It is apparent that the atomic % of C is different from that of O; specifically, the atomic C% increases and O% decreases as the copolymerization reaction proceeds. It is therefore assumed that the ABA/ANA polycondensation reaction progresses with an increase in the atomic concentration of C and a decrease in that of O.

The theoretical atomic concentrations of these two elements can be calculated by counting their respective number of atoms from the monomer structure and the ideally fully reacted copolymer, respectively. The atomic C% and O% of the monomer mixture are estimated as 71.4 and 28.6, and those of the 50/50 ABA/ANA copolymer are 85.7 and 14.3, respectively. Based on these calculations, the atomic C% should increase and O% should decrease with the progress of

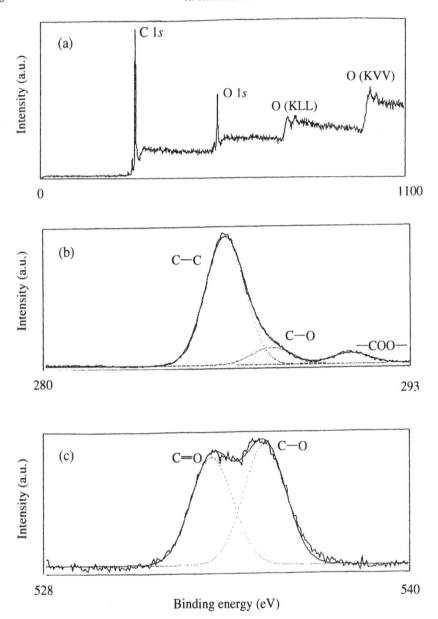

Figure 4. XPS spectra of a 50/50 ABA/ANA thin copolymer film. (a) Widescan; (b) C 1s; (c) O 1s.

the ABA/ANA copolymerization. This expectation is consistent with our XPS experimental results.

Moreover, it is obvious that the atomic C% increases more rapidly at the beginning of copolymerization, i.e. from 4 min to 15 min, than in the later stage, i.e. from 30 min to 2 h, where the atomic C% is almost constant. These XPS results coincide with the FTIR results in our previous paper [14], in which the condensation polymerization was much faster in the beginning stage than in the later stage.

Table 5.
Elemental composition of 50/50 ABA/ANA thin copolymer films at varying reaction time (experimental data from XPS)

Reaction time (min)	Atom% C	Atom% O
4	77.2	22.8
8	81.8	18.2
15	84.0	16.0
30	85.6	14.4
60	85.6	14.4
120	83.5	16.5

Therefore, it is well explained that γ^- decreases by a much greater magnitude during 4–15 min of reaction time than for 30 min to 2 h.

5. CONCLUSION

A novel thin-film polymerization technique and the Lifshitz–van der Waals acid–base theory were utilized to investigate the acid–base interactions of wholly aromatic liquid crystalline poly(ABA/ANA) with a 50/50 mole ratio as a function of the polymerization time. The surface free energy components of these liquid crystalline copolymers were calculated from contact angle measurements using a Ramé-Hart goniometer with the three-liquid approach, while their surface chemistry was analyzed by XPS.

The experimental data showed that the Lewis acid parameter (γ^+) and the Lifshitz–van der Waals parameter (γ^{LW}) of the copolymers were almost invariant with the progress of the polycondensation reaction, while the Lewis base parameter (γ^-) decreased rapidly. The surface roughness data measured by AFM suggested that the increase in the water contact angle (or the decrease in γ^-) was not caused by the change in surface roughness, but by the change in surface chemistry, i.e. by the reaction of acetoxy and carboxy groups to release acetic acid during the polymerization reaction. The XPS results also confirmed that the condensation polymerization did occur on the 50/50 ABA/ANA thin copolymer film surface, and the speed of reaction was much faster in the early stage than in the later stage.

Acknowledgements

We express our gratitude to the Institute of Materials Research and Engineering (IMRE), Singapore for financial support and the National University of Singapore (NUS) for research funding (No. RP 950696). Special thanks go to Dr. K. L. Mittal for his careful corrections and helpful comments, Miss Sixue Cheng for her useful technical help, and Dr. M. Sanjoy for providing the monomers. Thanks are also due to Dr. J. S. Pan at IMRE and Mr. G. H. Zhu at NUS for their kindness and skills in obtaining XPS and AFM data. Special thanks are given to Professor R. J. Good at State University of New York at Buffalo and Dr. J. Drelich at Michigan Technological University for their useful comments when one of us presented parts of this work at the 216th Annual Meeting of the American Chemical Society held in Boston in 1998.

REFERENCES

1. G. W. Calundann and M. Jaffe, in: *Proceedings of the Robert A. Welch Conferences on Chemical Research, XXVI, Synthetic Polymers*, 15–17 November, p. 247. Robert A. Welch Foundation, Houston, TX (1982).
2. T. S. Chung, G. W. Calundann and A. J. East, in: *Encyclopedia of Engineering Materials*, N. P. Cheremisinoff (Ed.), Vol. 2, p. 625. Marcel Dekker, New York (1989).
3. M. Jaffe, G. W. Calundann and H. N. Yoon, in: *Handbook of Fibre Science and Technology*, M. Lewin and J. Preston (Eds), Vol. 3, p. 83. Marcel Dekker, New York (1989).
4. A. A. Handlos and D. G. Baird, *J. Macromol. Sci.: Rev. Macromol. Chem. Phys.* **C35**, 183 (1995).
5. F. Rybnikar, B. L. Yuan and P. H. Geil, *Polymer* **35**, 1863 (1994).
6. F. Rybnikar, B. L. Yuan and P. H. Geil, *Polymer* **35**, 1831 (1994).
7. J. Liu, F. Rybnikar and P. H. Geil, *J. Macromol. Sci. — Phys.* **B35**, 375 (1996).
8. S. X. Cheng, T. S. Chung and S. Mullick, *Chem. Eng. Sci.* **54**, 663–674 (1999).
9. M. Uzman, B. Song, T. Runke, H. Cackovic and J. Springer, *Makromol. Chem.* **192**, 1129 (1991).
10. N. T. Correia, J. J. M. Ramos, B. J. V. Saramago and J. C. G. Calado, *J. Colloid Interface Sci.* **189**, 361 (1997).
11. J. Wang, G. Mao, C. K. Ober and E. J. Kramer, *Macromolecules* **30**, 1906 (1997).
12. T. S. Chung, K. X. Ma and M. Jaffe, *Macromol. Chem. Phys.* **199**, 1013 (1998).
13. K. X. Ma, T. S. Chung and R. J. Good, *J. Polym. Sci.: Polym. Phys.* **36**, 2327 (1998).
14. T. S. Chung and K. X. Ma, *J. Phys. Chem. B* **103**, 108 (1999).
15. G. Binning, C. F. Quate and Ch. Gerber, *Phys. Rev. Lett.* **56**, 930 (1986).
16. J. D. Miller, S. Veeramasuneni, J. Drelich and M. R. Yalamanchili, *Polym. Eng. Sci.* **36**, 1849 (1996).
17. S. Veeramasuneni, J. D. Miller, J. Drelich and M. R. Yalamanchili, *Prog. Org. Coat.* **31**, 265 (1997).
18. Y. L. Zhang and G. M. Spinks, *J. Adhesion Sci. Technol.* **11**, 207 (1997).
19. H. J. Busscher, A. W. J. Vanpelt, P. De Boer, H. P. De Jong and J. Arends, *Colloids Surfaces* **9**, 319 (1984).
20. S. J. Hitchcock, N. T. Carroll and M. G. Nicholas, *J. Mater. Sci.* **16**, 714 (1981).
21. T. Onda, S. Shibuichi, N. Satoh and K. Tsujii, *Langmuir* **12**, 2125 (1996).
22. F. M. Fowkes, *J. Adhesion Sci. Technol.* **1**, 7 (1987).
23. R. J. Good, in: *Contact Angle, Wettability, and Adhesion*, K. L. Mittal (Ed.), p. 3. VSP, Utrecht, The Netherlands (1993).
24. G. MacDougall and C. Okrent, *Proc. R. Soc. London Ser. A* **180**, 151 (1942).
25. C. W. Extrand and Y. Kumagai, *J. Colloids Interface Sci.* **184**, 191 (1996).
26. R. J. Good, in: *Surface Colloid and Science. Vol. II: Experimental Methods*, R. J. Good and R. R. Stromberg (Eds), p. 1. Plenum Press, New York (1979).

Apparent and Microscopic Contact Angles, pp. 349–375
J. Drelich, J. S. Laskowski and K. L. Mittal (Eds)
© VSP 2000.

The combined effect of roughness and heterogeneity on contact angles: the case of polymer coating for stone protection

C. DELLA VOLPE [1,*], A. PENATI [1], R. PERUZZI [2], S. SIBONI [1], L. TONIOLO [2] and C. COLOMBO [2]

[1] *Department of Materials Engineering, University of Trento, Via Mesiano 77, 38050 Trento, Italy*
[2] *CNR Research Center 'Gino Bozza' per lo studio delle cause di deperimento e dei metodi di conservazione delle opere d'arte, Piazza L. da Vinci 32, 20133, Milan, Italy*

Received in final form 30 June 1999

Abstract—The individual effects of heterogeneity and roughness on contact angles have been repeatedly analysed in the literature, but the application of the accepted models to practical situations is often not correctly performed. In the present paper the combined effects of roughness and heterogeneity on the contact angles of water on stone surfaces protected by a hydrophobic polymer coating are considered. Two different kinds of calcareous stone with different surface roughnesses and porosities were protected against the effect of water absorption by two different polymer coatings. The contact angles of water on the protected stone surfaces were measured by the Wilhelmy and the sessile drop techniques. A comparison of the results obtained shows not only the limits of the static sessile drop technique, but also the combined effect of roughness and heterogeneity. Some considerations are developed on the application of commonly accepted models to surfaces with a combination of roughness and heterogeneity. Some other results obtained with techniques such as roughness measurements, mercury porosimetry, energy dispersive X-ray spectroscopy (EDXS), thermogravimetric analysis (TGA), water absorption by capillarity experiments (WAC), all able to show the structure and properties of the obtained films, are also compared with those obtained from contact angle measurements. It is concluded that the static contact angle is not well correlated with the degree of protection; on the contrary, the receding contact angles are well correlated with the degree of protection actually obtained. An ideal protecting agent should have a receding contact angle greater than $90°$.

Key words: Contact angle; roughness; heterogeneity; monument protection.

*To whom correspondence should be addressed. E-mail: devol@devolmac.ing.unitn.it

1. INTRODUCTION

Ancient monuments are an important economic resource; millions of people travel around the world to visit them in many countries, such as Italy. But their existence and protection are also a challenge for surface science and technology.

Time, acid rain, smog, microbes, UV rays, wind, and also simple contact with human hands are some of the many factors that produce an increasing degradation of the materials of these monuments. In many cases, however, these factors work 'through' water, so that the problem of their protection has as a major task the control of the absorption (and, in some cases, of the release) of water. It is also important to distinguish between the consolidation and the protection of a monument; in this paper, only the protection aspect is analysed, by considering it as 'an operation whose purpose is to slow down the processes of deterioration' [1].

A common method of protection consists in the formation of a polymeric film at the interface between the monument and the environment; this film should reduce the absorption of water and, thus, the effect of practically all the causes of degradation.

For obvious reasons these films are made of hydrophobic materials; the hydrophobic polymers generally used should be able to adhere to the monument surface and thus prevent or reduce water absorption. It is very important to find the exact conditions (e.g. concentration of polymer in the protecting agent, time of treatment, etc.) to maximize the protection with the minimum quantity of polymer. In fact, the 'golden rule' of art protection is that the protecting agent must be removable, in the future, when better protection methods are hopefully available. Moreover, in some cases, particularly for buildings, water must also be eliminated 'from' the material, which can absorb it from the foundations.

The contact angle seems to be an ideal method to test the protection obtained by these treatments, due to its speed and ease of execution; however, the variability of natural stones, and therefore the combination of roughness and heterogeneity of their surfaces, makes the evaluation and the meaning of the obtained values of contact angles a challenging task.

Considerable research effort has been devoted in this field [2–10]; however, all the general problems of contact angle theory and practice, for example, the low importance attached to the receding values of contact angles or the false idea that the commonly measured static values of contact angles can be approximated to the equilibrium or Young contact angle, or again the neglect of roughness effects on contact angle values and of polymer surface mobility, leave an enormous possibility for new research efforts.

Finally, in some countries such as Italy, there is great interest to develop a complete set of techniques and methods to solve the problems of the protection of monuments, so it is absolutely necessary to clarify the role that common techniques such as contact angle measurement can really play in this context.

The present paper concerns the analysis of the protecting effects of two different polymers, commercially named Paraloid B72 [a copolymer of ethyl methacrylate

(EMA) and methyl acrylate (MA), with a monomer composition EMA/MA = 70/30] and Paraloid B67 (polyisobutyl methacrylate), on two calcareous stones with very different porous and roughness properties: Candoglia marble and Noto stone (used, respectively, in many ancient monuments such as the Milan cathedral and in the baroque buildings of Noto, a town of Sicily).

The advancing and receding contact angles of water on these surfaces and their comparison with the static angles show the extreme importance of the receding values for the evaluation of the actual protection obtained using a 'hydrophobic' film.

The contact angle data have been compared with the results of other techniques such as roughness measurement, mercury porosimetry, energy dispersive X-ray spectroscopy (EDXS), thermo-gravimetric analysis (TGA), all able to show the structure and the properties of the coating films. Water absorption by capillarity experiments (WAC) on some samples is also discussed. Some considerations are presented about the combined effects of roughness and heterogeneity in affecting the experimental values of the contact angles, starting from the models proposed in the literature [11–19].

2. MATERIALS AND METHODS

2.1. Materials

The samples tested measured $5 \times 2 \times 0.5$ cm; they were obtained from fresh quarry stone by a wire metal saw without any lubricating agent. Before every treatment and/or test, they were washed with deionized water and then dried as described in the following.

Candoglia marble is a crystalline calcareous stone extracted from the dioritic-kinzigitic formation in the south of the Alps; Noto stone is a bio-calcareous Miocenic fossiliferous stone from the Palazzolo formation.

The chemical structures of Paraloid B72 (B72) and Paraloid B67 (B67) (by Rohm & Haas) are shown in Fig. 1.

2.2. Treatment of the samples

The samples were dried (24 h at 40°C and then 24 h at 300 mTorr) and then treated by a B72 or B67 solution at a concentration of 0.5, 1.5, 3, or 4.5% by weight in ethyl acetate and 6% in acetone. The absorption of the solution was carried out by capillary absorption using the same procedure as that described for water absorption in the next paragraph. The time of absorption was 7 h for Candoglia marble and 6 h for Noto stone. After the treatment, the samples were left to dry, placing the treated face upwards for 24 h, and then dried as before.

The samples were also weighed before and after the treatment; the weight difference was about 0.005–0.007% for Candoglia marble (6–10 mg per sample)

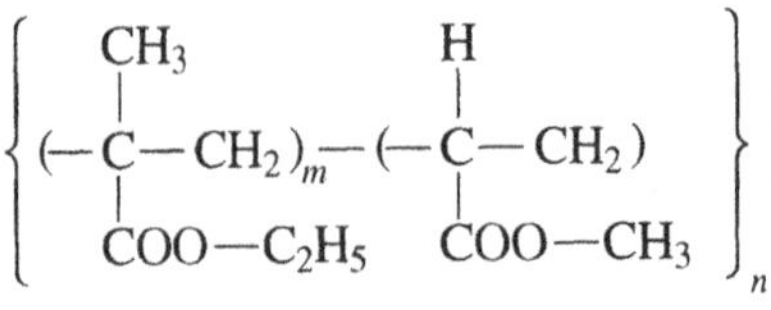

$$\left\{ \begin{array}{cc} CH_3 & H \\ | & | \\ (-C-CH_2)_m - (-C-CH_2) \\ | & | \\ COO-C_2H_5 & COO-CH_3 \end{array} \right\}_n$$ Ethyl methacrylate–methyl acrylate copolymer

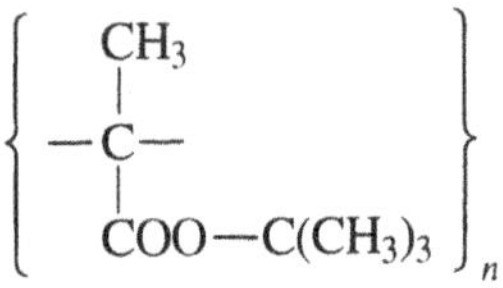

$$\left\{ \begin{array}{c} CH_3 \\ | \\ -C- \\ | \\ COO-C(CH_3)_3 \end{array} \right\}_n$$ Poly(isobutyl methacrylate)

Figure 1. The molecular structures of Paraloid B72 [a copolymer of ethyl methacrylate (EMA) and methyl acrylate (MA), with a monomer composition EMA/MA = 70/30] and Paraloid B67 (polyisobutyl methacrylate).

and about 0.7–1.6% for Noto stone (0.5–1 g per sample) and the quantity increased with the concentration of the treating solution. The mean ratio between the amount of polymer absorbed by the two stones at the same concentration of protecting polymer was of the order of 1 : 100 between Candoglia and Noto stones. However, a more precise indication of the quantity of material absorbed in the surface zone of the samples is reported in the EDXS analyses.

2.3. Water absorption by capillarity (WAC)

This test was executed as recommended in NORMAL 11/85 [20] (NORMAL is a collection of procedures published by the Italian CNR). Dried samples were weighed and separately placed in vessels containing a pile of filter papers immersed in water up to half of their height, with the face of the sample on the paper; in this way, the absorption occurred only through the base by vertical suction. The vessels containing samples were closed and kept at a constant temperature of 20°C. At regular intervals of time, the samples were removed and weighed after sponging them with a wet cloth to eliminate excess water drops. For every sample the results have been reported as $mg\,cm^{-2}$ of water versus the square of time in s (Darcy's law).

2.4. Static contact angles (SCA)

Static contact angles of water (HPLC quality, Merck Bracco, Italy) on the samples were measured by the sessile drop method using a Lorentzen & Wettre Surface Wettability Tester (Lorentzen & Wettre, Stockholm) (modified by fitting it to a horizontally pivoted table) in accordance with NORMAL 33/89 [21]. Each value is the mean of the measurements performed on 10–20 single drops of 5 μl, by determining their geometrical dimensions and applying the so-called 'spherical approximation'.

2.5. Dynamic contact angles (DCA)

Dynamic contact angles of water (HPLC quality, Merck Bracco, Italy) on the described samples, before and after protection by the polymers, were measured using a new, currently developed model of the Wilhelmy microbalance by Gibertini (Milan, Italy) at a speed of $105 \ \mu m \, s^{-1}$. The liquid was contained in a vessel with a large diameter (about 16 cm) to reduce or eliminate the variation of the reference level during the immersion of the sample due to the significant volume of the sample or to the water absorption. This phenomenon is commonly neglected, but it has been showen that it can be significant [22].

The speed chosen was a compromise between the reliability of the results and the necessity to reduce the water absorption, if present. Note that in this investigation the dynamic contact angle and the corresponding meniscus were considered fully equivalent to those obtainable during static contact angle measurements. This non-trivial assumption turns out to be completely justified in the present analysis, owing to the very small values of the capillary numbers considered (about 1.5×10^{-6} for an immersion speed of $1 \times 10^{-4} \ m \, s^{-1}$, taking the water viscosity as $1.2 \times 10^{-3} \ Pa \, s$ and surface tension as $72 \times 10^{-3} \ N \, m^{-1}$ [22]. Some further comments will be made in the Discussion.

The water absorbed by the sample can, in principle, modify the force value and so the calculated contact angle; this is not the case, except in a few cases, as will be clear from the final discussion.

It is, however, worthy of note that the amount of water absorbed during the Wilhelmy experiment is very difficult to calculate; the sample can absorb water by vertical suction [2] from the base of the sample, whose area is about $1 \ cm^2$ and this is true throughout the immersion, with an increasing effect due to the increasing hydrostatic pressure. Only when the meniscus touches the vertical walls of the sample and begins to move along them does a second kind of mechanism, horizontal suction, which has the greatest effect [2], occur.

However, due to the meniscus shape, if the contact angle is higher than 90°, during the very first part of immersion only vertical suction from the base of the sample is present; e.g. if $\theta = 120°$, the first 2.7 mm of the immersion is free from horizontal suction [12].

The technique of repeated immersion shows very well the amount of water effectively absorbed and allows one to easily evaluate its relative importance. Moreover, consider that for contact angle values between 90° and 120°, as is the case for our sample, the $\cos^{-1}$ function changes slowly with the decrease of the negative measured force; e.g. $\cos^{-1}(-0.1) \cong 96°$, while doubling the negative force $\cos^{-1}(-0.2) \cong 101°$, with an effect of only a few degrees. Thus, in many cases the absorption effect can be, and effectively is, lower than the standard deviation of the contact angles obtained.

The treatment of the data obtained has been done by the method described in the literature [23] using programs such as Kaleidagraph 3.05 for Macintosh©. The following description can be useful for those interested in monument protection, but are not sufficiently expert regarding the Wilhelmy technique. Advancing and receding contact angle values were obtained by immersing a sample of constant geometrical cross section, suspended from a Wilhelmy microbalance, in the test liquid. The forces acting on the sample are the sample weight, the buoyancy, and the surface tension along the immersion perimeter:

$$F^1 = F - mg = P\gamma \cos\theta - \rho g V,$$

where F^1 is the total force measured on the sample, m and P are the mass and perimeter of the sample, V is the volume of sample immersed in the liquid of density ρ and surface tension γ, θ is the contact angle at the ternary interface of the system considered (test liquid, film or plate sample, and air), and g denotes the acceleration due to gravity.

Extrapolating the trend of the total force to the zero depth of immersion (ZDOI), where the buoyancy is zero, and assuming a constant sample weight, we obtain

$$F^1 = F - mg = P\gamma \cos\theta.$$

All DCA data were collected at room temperature, $25 \pm 2°C$. Each value is the mean of 2–5 measurements performed on different samples of the same material.

Unless otherwise stated, all the Wilhelmy runs were also double immersions. The repeated immersions show the real effect of the water on the material surface, frequently modified by this interaction. The problem is fully acknowledged in the Discussion.

The data referring to the pure polymers were obtained on films produced by evaporation of the solvent from a concentrated solution of the polymer in ethyl acetate and then desiccating the material at room temperature in vacuum. The vessel containing the solution was made of Teflon® to reduce absorption and adhesion problems.

2.6. EDXS

EDXS analysis was performed using an SEM microscope (JEOL 4901 model) on gold-sputtered samples through a Be window. The line scan method was used to reduce the time of each analysis, obtaining, however, a map of the elements concentration along a line chosen as perpendicular to the sample boundary.

2.7. Thermal gravimetric analysis (TGA)

TGA measurements were made with a Mettler Tg50 model balance scanning from 30°C to 900°C at a rate of 10°C min^{-1} with sample slices as thin as 1 mm, obtained using a wire saw without any lubricating agent. Untreated and treated samples were compared. The results are reported in Table 1.

Table 1.

Thermo-gravimetric analysis of some stone samples. Percentages of weight variation

	ΔW 30–150°C	ΔW 150–400°C	ΔW 400–900°C	Residue
Noto untreated	0.29	0.71	43.1	55.7
+B72 3%	0.33	1.66	42.4	55.5
+B72 6%	0.35	3.05	41.8	54.8
Candoglia untreated	0.11	0.14	43.4	56.3
+B72 3%	0.11	0.33	42.2	57.4
+B72 6%	0.11	0.35	43.0	56.5

Table 2.

Mercury porosimetry results on Noto samples

Pore diameter (μm)	Volume %
0–0.05	5
0.05–0.1	4
0.1–0.2	1
0.2–0.4	3.3
0.4–0.6	2.8
0.6–0.8	2.5
0.8–1.0	2.1
1–2	9.2
2–4	12.4
4–10	34.5
10–1000	22.8

2.8. Differential scanning calorimetry (DSC)

DSC measurements were made with a Mettler DSC calorimeter at a scan rate of $10.0°C\,min^{-1}$ on about 20 mg of the pure polymers. The temperature range was from -50 to $150°C$.

2.9. Mercury porosimetry

Porosimetry was performed using a 2000 porosimeter by Carlo Erba Instrumentation, Italy. The so-called 'integral open porosity', i.e. the integral of the pore-size diameter from 0 to the maximum value, and the pore-size distribution curve were measured on the untreated samples. The results are reported in Table 2.

2.10. Roughness

Two different kinds of roughness meters were used to characterize the samples. In fact, the two kinds of stones are completely different from this point of view, so that the standard profilometer DEKTAK 3 was able to analyse the Candoglia marble, but not the Noto stone. For Noto stone, we employed a Hommell tester T2000 with a TK100 device (Hommellwerke GmBH, Schenningen). In both cases, a scan length of 1 mm was used, repeating it five times for each face of each sample, with 1000 sampling points.

The raw data were analysed by standard software, obtaining in each case the R_a and W_a parameters [24]. These quantities are respectively called the 'roughness' and the 'waviness' and are defined in an ASTM standard in the cited reference. They correspond to the absolute mean value of the differences among the height of each measured point and the height of a reference line. The waviness is distinguished from the roughness on the basis of a characteristic arbitrary length, *measured along the sampling direction*, and depends on the sampling length, on its spatial resolution, and on the particular surface considered. The characteristic length in these experiments was chosen to be 100 μm, while the spatial resolution was about 1 μm.

3. RESULTS

3.1. Static and dynamic contact angles

Both advancing and receding contact angles of water on untreated and treated samples at different concentrations of the two polymer solutions are shown in Figs 2a–2d.

3.2. Effect of the B72 treatment

As one can see, untreated Candoglia marble is hydrophilic, but apparently less than Noto stone, whose advancing and receding contact angles are zero. It should be noted that to reduce or eliminate the effect of water absorption in the case of untreated material and of the lowest concentration of protecting agent, when absorption is particularly evident, only the first part of the immersion curves with a linear trend has been used to evaluate the contact angles; this interval is indicated in Fig. 3. It appears evident that a systematic error can be present in these two cases. To estimate correctly the actual importance of this phenomenon in other cases, however, we also have to take into account the remark discussed in the Materials and Methods section, remembering that the very first part of immersion is not affected by horizontal suction if the contact angle is greater than 90°.

Moreover, the properties of Candoglia marble also depend on the presence of certain impurities, indicated by the colour of the material; a weak pink, indicating the presence of iron, corresponds to enhancement of the hydrophilic character of the surface.

The advancing contact angles increase very strongly also at the lowest concentration of the protecting polymer and are very near to the static ones. Their values are substantially constant with the increase of the concentration of the protecting polymer solutions.

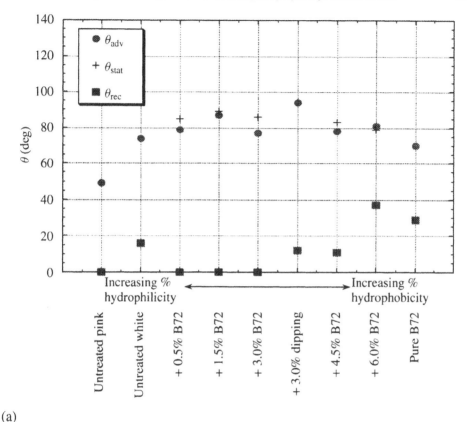

(a)

Figure 2. The advancing and receding contact angles of water on (a) Candoglia marble protected with a solution of B72 of increasing concentration, (b) Noto stone protected with a solution of B72 of increasing concentration, (c) Candoglia marble protected with a solution of B67 of increasing concentration, and (d) Noto stone protected with a solution of B67 of increasing concentration. Immersion speed 105 μm s^{-1} at room temperature. The pure polymer and the untreated material are indicated.

On the contrary, the receding angles are zero until the 3% protecting solution is used. After this concentration, they increase, reaching a value near that of the pure polymer at a 6% concentration.

Some control experiments at only one concentration of the protecting polymer were performed by protecting the stone samples not by capillarity suction, as described in the previous section, but by a dipping procedure. These samples were immersed by fast dipping in the polymer solution, at the same rate as in the subsequent water tests (105 μm s^{-1}); in this case, both the advancing and the receding angles are higher.

A similar trend was also observed on Noto stone; it was fully wetted by water, but even the least concentrated solution of protecting agent induced a strong enhancement of the advancing contact angle. In this case, however (see Fig. 3), the advancing angles obtained at the lowest concentration of protecting agent are

C. *Della Volpe* et al.

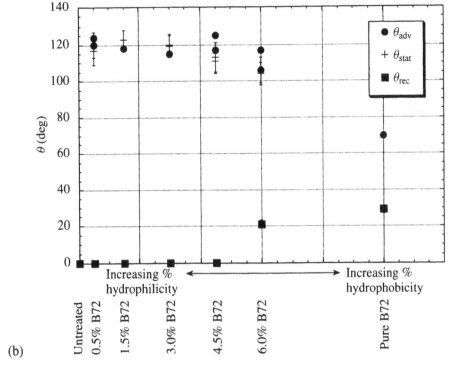

(b)

Figure 2. (Continued).

lower than their real value, due to the effect of water absorption, as previously described.

The general trend of static contact angles corresponds to a slight *decrease* with increasing concentration of the protecting agent. Once again, both the values and the trend of static contact angles are in good agreement with the dynamic advancing contact angles. Similarly to the case of Candoglia marble, the receding contact angles, up to a certain value of the protecting solution concentration (about 4.5%) are constantly zero, while after this value they increase significantly, reaching a value similar to that obtained on pure polymer films.

3.3. Effect of the B67 treatment

The case of B67 on Candoglia marble is illustrated in Fig. 2c. Even the lowest concentration of the polymer is able to induce a very high advancing contact angle, while the receding ones regularly increase with the concentration from the beginning. In all cases, the values of the advancing contact angles are higher than in the case of B72.

The advancing contact angles decrease slightly with increasing concentration of the protecting polymer. This apparently bizarre phenomenon is very evident in the case of Noto stone, for which the receding values are zero, but at the highest concentration of the polymer.

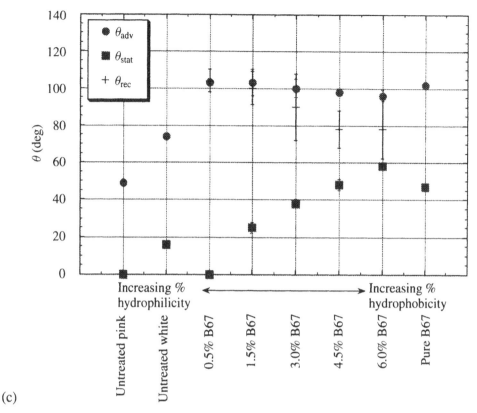

(c)

Figure 2. (Continued).

Some differences between the advancing contact angles and the static ones are detectable for Candoglia marble. Note, however, that exactly for those values of the static contact angles which exhibit the greatest difference with the advancing ones, the standard deviation is particularly large, practically twice the value of the remaining contact angles.

3.4. Roughness and porosimetry

Roughness evaluations are reported in Figs 4a and 4b and confirm that Candoglia marble is much less rough than Noto stone. Note that there is no particular trend in the roughness value with the protection treatment, also considering the significant standard deviations of R_a and W_a. For Candoglia marble, the values of R_a are between 0.23 and 0.32 μm, while for Noto stone R_a varies from 8 to 12 μm, about 40–50 times higher.

In both cases, the waviness parameters are higher than R_a, but again no particular trend is manifested.

Porosimetry of the untreated samples indicates that untreated Candoglia marble has an open porosity of 0.7–0.8% with a mean porous size of 0.1–0.2 μm [5]. In

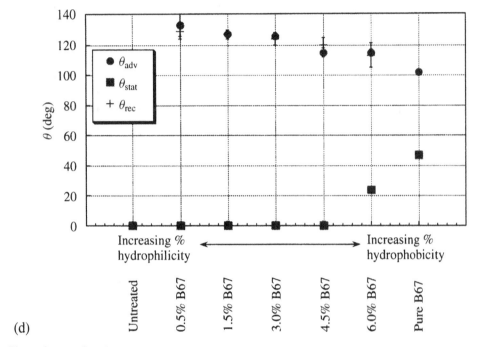

(d)

Figure 2. (Continued).

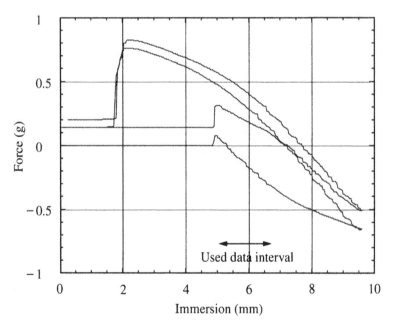

Figure 3. A Wilhelmy run on Noto stone in water. In this case, Noto stone was protected by B72 at the lowest concentration. The effect of water adsorption is the greatest and is visible from the difference in the zero line before and after each of two immersions. In this particular case, only the very first part of the linear portion of the immersion curve was used to calculate the contact angle, as indicated.

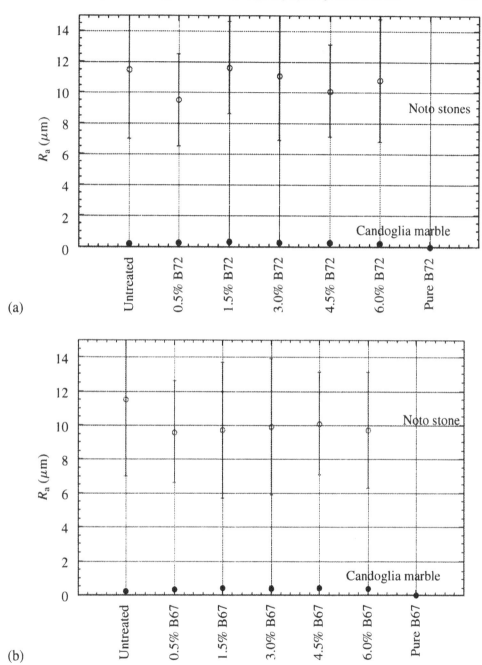

Figure 4. The roughness of calcareous stones protected by (a) B72 at different concentrations and (b) B67 at different concentrations.

the case of Noto stone, the open porosity is between 30 and 35% and the mean porous size is about 4–10 μm. There is, however, a great variability due to the natural origin of the materials. Some data are shown in Table 2.

C. *Della Volpe* et al.

(a)

Figure 5. (a) The line along which the scan of EDXS analysis was made on a Candoglia sample treated with a 6% solution of B72 is perpendicular to the face of the sample. (b) The results of EDXS analysis for carbon, and oxygen, and calcium. The relative abundance of each element is reported versus the sampling length. One can see that only the first 2–3 μm contain an excess quantity of carbon, indicating the presence of absorbed polymer.

3.5. EDXS–SEM

SEM images reveal the micro-texture of the materials; no clear difference, however, is present before and after the protection treatment in simple SEM images. But the situation changes if EDXS is used to investigate the first few micrometres of the sample from the surface.

The case of Candoglia marble is described in Figs 5a and 5b. A line scan, reporting the variation of three different elements, C, O, and Ca, along the first 8 μm

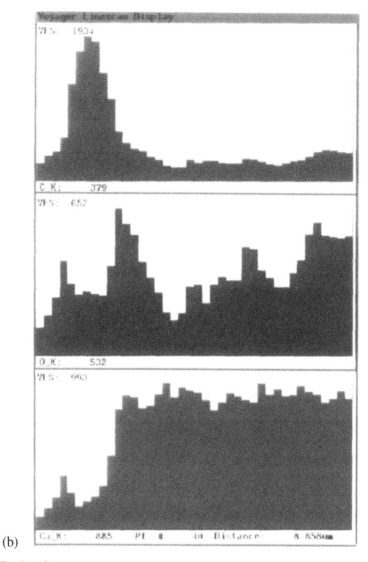

(b)

Figure 5. (Continued).

of a sample, is shown. As one can see, the percentage of C is higher in the first 2 μm of the line scan; after this limit, the element concentrations, are constant. This should correspond to the fact that the absorbed polymer is able to penetrate only in the first few micrometres of the sample, leaving the rest of the material practically untouched.

It should be noted that such a result is different from that obtained from the direct weight of the samples before and after the absorption of the polymer solution. This fact is analysed in the Discussion.

3.6. Thermo-gravimetric analysis (TGA)

This technique, applied to thin slices of materials with a constant thickness of about 1 mm, allows us to detect degradation of the different phases of the material in the untreated and treated stone; the necessity to have a low thickness depends on the very low polymer penetration. As we will see, this problem is probably the origin of the low resolution obtained for Candoglia marble.

The description of the different phases of the material degradation follows: at low temperature, between 30 and 150°C, the transition of the polymers and the loss of water; at higher temperature, between 150 and 400°C, degradation of the polymers and, finally, at about 800°C, carbon oxidation with the production of carbon dioxide. The final result of the transformation is consistent with a low quantity of water and the stoichiometric percentage of carbon dioxide, which transforms the material in calcium oxide.

From Table 1 it is evident that for Candoglia marble the process attributable to the polymer degradation does not depend on the concentration of the protecting solutions, probably because the penetration of the polymer is low and its percentage on a thickness of 1 mm of material is too low with respect to the resolution of the experimental technique. On the contrary, in the case of Noto stone, where greater absorption is measured and also greater penetration expected, the measured value is related to the concentration of the protecting solution.

3.7. Water absorption by capillarity (WAC)

These tests were done on the untreated materials and on the samples protected by B72 polymer. One can easily see from Figs 6a and 6b the enormous differences between the vertical scale of the experiment for Candoglia marble and for Noto stone. A difference of about two orders of magnitude is observed. However, some general considerations can be made:

- the polymer protection reduces the quantity of absorbed water for a certain time, which is longer with an increase of the protecting polymer concentration;
- after this 'lag' time, the absorption of water increases, but never reaches the values of untreated materials; this effect is stronger with increasing protecting polymer concentration;
- in the case of the untreated material, there is a two-slope curve, while for the treated materials a sigmoid or three-slope curve is obtained. Moreover, in the case of Noto stone there is a very sharp transition between the different parts of the curves and the global effect of the protection is very clear; in the case of Candoglia marble, the situation is less clear;
- in all cases, however, even if the advancing or static contact angle is higher than 90° a *positive capillary penetration* occurs, in apparent contradiction with the Jurin law [8]. (In the field of monument protection, the Jurin law is simply the name of the well-known equation which expresses the height of a liquid in a vertical capillary tube.)

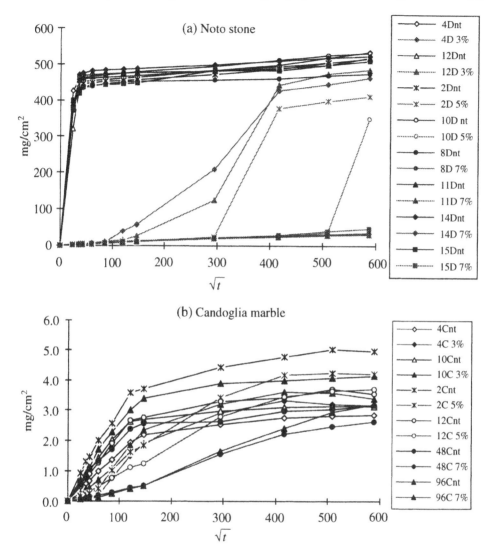

Figure 6. Water capillary absorption on different samples of (a) Noto stone untreated or protected by different films of B72 deposited at increasing concentration of polymer, and (b) Candoglia marble untreated or protected by different films of B72 deposited at increasing concentration of polymer. The samples are indicated by numbers and letters; the symbol nt means untreated and the percentage symbol indicates the concentration of the B72 solution used. The abscissa represents the square root of the absorption time in s.

3.8. Differential scanning calorimetry (DSC)

DSC showed that the T_g of B72 was about 40°C and that of B67 about 58°C, in good agreement with the literature. These data were collected during a second scan, because the first one demonstrated an endothermic peak, due to the so-called 'physical aging' of the material. For B72 the enthalpy of this transition is about 2.7 J g^{-1} at 70°C, while for B67, it is about 6.0 J g^{-1} at 50°C.

4. DISCUSSION

The problem of the effect of roughness and heterogeneity on the values of the contact angles has been widely analysed. In particular, Johnson and Dettre (JD) published some milestone papers [11, 12, 15, 16] about 40 years ago.

The reader coming from the field of stone protection, and not sufficiently expert in contact angles, should remember here that the Young angle is valid only on perfectly plane and homogeneous surfaces. In all other real cases, where heterogeneity and roughness are present (also at a microscopic level), infinite metastable states exist, so that it is not possible to measure a single equilibrium contact angle, but a wide set, comprised between two extremes. The highest one is the advancing contact angle and the lowest the receding contact angle; their difference is the hysteresis. JD extracted some general properties of the hysteresis by analysing some model surfaces with a well-defined roughness or heterogeneity.

In the case of heterogeneity, the advancing contact angle is mostly representative of the low-energy portion of the surface, while the receding one is indicative of the higher-energy portion. For a homogeneous, rough surface only one microscopic angle exists, but due to the geometrical structure of the surface, also in this case a wide set of apparent contact angles is possible. Again, the greatest value of the contact angle corresponds to the advancing and the lowest one to the receding contact angle, the first being larger and the second smaller than the Young angle. Moreover, the increase in roughness generally enhances this phenomenon and thus the hysteresis. These are widely accepted concepts [11].

However, even in recent papers [25] incredible and unacceptable approaches have been adopted to calculate the components of surface free energy directly from contact angles of liquids on rough surfaces, without any consideration of the effect of the roughness on the values of the contact angles, as though roughness modified directly the chemistry of the surface! Criticism has been published [26].

It is noticeable that the JD approach has not been applied to describe contact angle hysteresis on surfaces which are both heterogeneous and rough, probably due to the relatively large number of parameters which would be involved, as well as to the many possible surface morphologies. In any case, the non-linearity of the JD model entails that no simple correlation can be expected among the hysteresis phenomena in the presence of pure roughness, pure heterogeneity, or both.

In the present paper, we analyse the effect that a combination of both roughness and heterogeneity induces in the values of the contact angles of water on porous surfaces made by protecting a stone with a polymer. This analysis will hopefully be able to give practical guidelines for the protection of stone and its testing.

As previously mentioned, many static or advancing contact angle data of water on these kinds of materials are available in the literature [1–10]; these angles are *always* considered as equilibrium or Young contact angles.

It is useful to remember that static angles are not necessarily 'equilibrium' values; they generally correspond to metastable or stationary states. When one places a liquid drop on a surface, one effectively realizes a process of advancing of the liquid meniscus on the solid surface. This is widely proven by the literature data; also the present data are in very good agreement with this point of view. In the case of Candoglia-B67 (Fig. 2c), the greatest differences between the static and the advancing data correspond exactly to the static data with the largest standard deviations (about $\pm 12-13°$ vs. a common value of $\pm 3-4°$).

To measure or to report only static or advancing contact angles is equivalent to looking at an object with one eye only: one misses the correct perspective. Moreover, this was exactly the original motivation of this paper. Some of us (R.P., L.T., and C.C.) found contradictory results: stone surfaces protected with increasing concentrations of polymer absorbed a decreasing quantity of water, *but* the static water contact angles on them were constant or decreasing. This result seemed strange to us, but, however, experimentally well supported; therefore we ask for confirmation from other people working in the field of contact angles (C.D.V. and S.S.).

The static method can be, playfully, defined as the 'lazy' sessile; one can, in fact, obtain advancing and receding values of contact angles also from the sessile method, but with some experimental modification. Some authors define them more correctly as the 'recently advancing' or 'recently receding' contact angles [27].

In our opinion, one of the most useful methods is the Wilhelmy technique at a low speed, by which both the advancing and the receding contact angles can be obtained directly. Moreover, the advancing and receding contact angles are evaluated *instantaneously* during the continuous immersion in water, and thus always on a 'fresh' surface, different from the case of the sessile method. This fact can explain some differences found in our case; in fact, the absorption of the drop in the calcareous stone may produce a *decrease* in the static contact angle.

The advancing and receding contact angle values obtained by the Wilhelmy technique confirmed the static values as fully equivalent to the advancing ones; moreover, they show that the receding values generally *increase* with increasing concentration of polymer in the protecting solution.

This can be easily explained in terms of the JD approach. In fact, we know that the advancing contact angle is more sensitive to the low-energy portion of the surface, while the receding one is more sensitive to the high-energy portion. So, one can expect that the introduction of also a limited portion of low-energy material in a high-energy surface increases the advancing contact angle, the value of which remains, then, practically constant with the percent increase of this portion. On the contrary, the receding contact angles remain very low, until even a small portion of the surface keeps the character of the high-energy surface; only when this portion is covered by a hydrophobic film do the receding angles increase. This is the case when hydrophobic polymers such as B72 (or B67) are introduced on a more hydrophilic surface such as calcium carbonate. As a conclusion, the data of B72 or B67 on Candoglia can be reasonably explained by this application of the JD approach.

It is important to note that in these cases the values of the contact angles obtained on the plane film of pure polymers are very similar to those on protected stones;

the maximum differences between the protected surfaces and the pure film are 15°
for B72 on Candoglia and practically zero for B67 on Candoglia (consider that
the standard deviations of the data are about 2–3° and that only one data point
has a positive deviation of 15°; the other ones have a difference less than 5–10°).
Roughness appears to play a minor role with respect to the other materials.

The static contact angle cannot be totally sensitive to the *increasing* protection
effect of the polymer solutions, but only to the presence of even a low portion of
protected material in a non-protected stone.

To stress, however, the limits of these data, it is noteworthy that they are not able
to fulfil a simple test; in a recent paper [16], Johnson and Dettre proposed a simple
equation to calculate the percentage of the modified surface when the hysteresis
was due to heterogeneity only. Also, in the case of the treated Candoglia marble,
the application of this equation remains difficult, so the conclusion is that the effect
of roughness is low but significant.

It is slightly more complicated to explain the presence of the slight but significant
reduction in advancing contact angles with increasing concentration of protecting
polymer, especially in the case of the system B67-Noto stone, where the phenomenon is more important.

Note that (as also in the case of B72 on Noto stone) the measured advancing
contact angles are now noticeably *higher* than those obtained on the pure polymer
film; the differences can be as large as 30° and 50°, respectively, for B67 and B72.
How is this possible? Following the JD approach, this is the effect of the roughness,
which may enhance the value of the advancing contact angle (and reduce that of
the receding one). In this case, no measured angle is real: all values are 'apparent
contact angles'.

In general, the situation is similar for receding values, but in this case the presence
of a fully wettable material such as calcium carbonate could suggest that those
portions of the surface not well covered by the protecting agent remain at a zero, or
very low, contact angle.

Consider that the roughness of the untreated and treated material in the same set
appears to be very similar and there is no apparent trend in the roughness (R_a) or
waviness (W_a) of the treated material vs. the concentration of polymer solution.
This would correspond to the fact that the thickness of the polymer layer is very low
and does not alter the roughness properties of the material. So, it is not possible to
attribute the decrease in the advancing contact angles (and of the hysteresis) to the
decrease in roughness in the same set.

In the JD approach, a similar effect, the simultaneous slight decrease of the
advancing contact angle and the significant increase of the receding one, is shown
in Fig. 14 of ref. [11]. In that case, the parameter reported on the abscissa is the
ratio between the true surface, A', and the geometric surface A: $r = A'/A$; the two
contact angles seem to become similar to the Cassie-Baxter angle with increasing r.
This phenomenon is explained in terms of the so-called non-composite/composite
transition.

For sufficiently large values of r, a liquid with a high contact angle cannot penetrate the pores and some air remains trapped in them. The surface can also be seen as a heterogeneous one, partially made of 'air-surface' with a very high contact angle (about 180°). These composite surfaces are responsible for the protection of the body of aquatic birds and animals from the effect of prolonged immersion in water: the porous feathers or hairs and the grease repeatedly put on them by the animals are 'composite' surfaces, the most efficient and natural method to be protected by water wetting!

The decrease of hysteresis at the non-composite/composite transition can be interpreted as a consequence of incomplete penetration of the liquid in the surface. When the Wenzel parameter r increases, an increasing portion of the surface is actually inaccessible and only the most external surface remains available to interactions with the liquid. One can expect that such an 'effective' surface has a less rough shape, that energy barriers separating local minima of the surface free energy are smaller, and that a decrease in hysteresis accordingly occurs.

In the present discussion, however, such reasoning is not applicable in this exact formulation, because the roughness of the samples is quite constant vs. the concentration of the polymer protecting solutions. Moreover, there is no simple relation between r and R_a, although it is possible to suppose that they are correlated (at least for many surface textures. A specific model of the surface texture seems to be necessary, anyway).

The existence of a composite surface must satisfy the condition expressed by the JD theory as:

$$\theta° = 180 - |\alpha|,$$

where $\theta°$ is the Cassie-Baxter or Young angle and $|\alpha|$ is the slope due to the roughness at a microscopic level.

Starting from this consideration we can try to solve the problem. We can consider that the increase of the polymer amount on the stone surface is related to a higher Cassie-Baxter or Young angle, $\theta°$, even if we are not able to measure it directly. But it will be intermediate between the advancing and receding angles, so it will increase with the increase of the receding angle. This increase will correspond to the global effect of having a more hydrophobic surface and thus to the possibility of satisfying the previous mathematical condition.

Note that from a purely mathematical point of view the transition from non-composite to composite surfaces becomes easier with the increase of either $\theta°$ or $|\alpha|$, the other quantities being constant.

In the case of Candoglia marble, the roughness is lower, so the considered transition is more difficult, the advancing angles are relatively constant, and their trend can be mainly attributed to the combined effect of the surface heterogeneity and the increasing hydrophobic surface portion due to the B72 or B67 layer. However, some differences due to the roughness are detectable.

In the case of Noto stone, the greater roughness of the surface makes it easier to attain the transition from a non-composite to a composite surface and therefore the advancing angle decreases in a measurable way, simultaneously with the dramatic increase of the receding angle.

The final interpretation of the data can be that if a non-composite to composite surface transition is possible, then this situation corresponds to a greater protection effect; otherwise, the protection phenomenon can be interpreted as the increase of the hydrophobic portion of the surface. Both are, however, revealed by the receding angles.

In all cases, the increase or decrease of the advancing contact angle cannot be considered a direct indication of the degree of stone protection. Its value cannot be higher (at a microscopic level) than that of the pure polymer film. The differences are produced by the roughness. Moreover, when the receding contact angle is low or zero, this is a strong indication that the surface for chemical or for geometric reasons is *wetted*; the absorption of water also remains high if the advancing contact angle is greater than 90°.

The condition of Jurin [8] is written in terms of the Young contact angle; if it is greater than 90°, water cannot penetrate in the porous system unless an external pressure is applied. *But the commonly measured angles are not equilibrium or Young angles*, so that it is not possible simply to input them in the Jurin condition. This can produce the apparent contradiction that also for liquids with contact angles higher than 90° on porous materials there is capillary penetration.

The actual situation is that we cannot measure the equilibrium or Young contact angle, so that we have no possibility of obtaining the strictly necessary condition for the Jurin equation. We can, however, have a kind of 'sufficient' condition; i.e. *if the receding contact angle is greater than 90°, then water cannot penetrate* at all; but this condition is very strict and rarely fulfilled. In particular, it does not hold in our case.

Moreover, it has been shown [10] that when the penetrating liquid is a finite reservoir with the shape of a small drop, liquid penetration in a porous capillary is also possible if the contact angle is greater than 90°, until values of 115°, because the energy 'invested during the formation of the drop can be utilized to enable its penetration into a capillary even for obtuse apparent contact angles'.

Therefore, the results obtained in WAC experiments are not strange. They are illustrated in Figs 6a and 6b for the cases of B72 on Candoglia marble and Noto stone. The curves are reported against the square root of time, following the Darcy law. As shown in the literature [2], each curve can be divided into two parts, the first one with a high slope and the second one with a plateau. In some cases, for increasing protection effects, the global shape of the curve resembles a sigmoidal function.

Consider the first part of each curve. One can see that also in the case of very high advancing contact angles water penetrates in the porous system of the stone. For low values of the concentration of the protecting polymer, the absorption values are similar to or *greater* than those for untreated samples. This is in agreement with the fact that for a low concentration of the protecting agent in the treatment solution *the resulting receding contact angle is low*. Only above a certain critical concentration is the absorption decisively lower than for the untreated sample; this situation is reflected by the receding contact angles, which increase towards strictly positive values only above a certain threshold of the concentration of the protecting polymer.

The situation is very complex in the case of Candoglia marble, for which some samples treated with low concentration solution absorb *more water* than the untreated material. On the basis of the previous analysis, particularly on the fact that the receding contact angle is always zero in these cases, this is not strange. The stone remains able to absorb water and the singular structure of each sample (with a wide variation of properties, due to their natural origin) can explain this phenomenon. Alternatively, one can take into account the formation of a new set of pores, with lower diameters (due to the action of the protecting agent), but not necessarily with a higher contact angle (e.g. consider a pore divided into two parts by a thin wall of polymer); this would correspond to a *higher* capillary penetration.

Only when the protecting action is able to increase also the receding contact angle is the capillary vertical suction effectively reduced.

From the weight data it is easy to see that more concentrated solutions correspond to a greater quantity of polymer absorbed from the material; the penetration depth of the polymer in the stone is probably not so high. The results that we have obtained are different and depend on the technique employed. From the EDXS results we estimate a value of $2-3$ μm in the case of Candoglia marble; from the weight of samples before and after the treatment and considering the porosity of the samples independently of the depth, we easily evaluate by elementary geometric considerations about $200-300$ μm. These different values can be explained either in terms of experimental errors or by the fact that the porosity of the sample is different at the surface and in the bulk.

Considering that Noto stone absorbs by weight a quantity of polymer about $30-50$ times higher than Candoglia marble, and that its open porosity (empty space) is about $30-40$ times higher, the two things reciprocally elide and the final penetration depth cannot be very different in the two materials.

The effective penetration depth of the polymer in the materials is, however, very low, no more than *some hundreds of micrometres* in the best case. Consider, then, what happens when water penetrates for a length greater than the penetration depth of the polymer: in this deeper part of the pores nothing is changed, so that after a sufficiently long time the penetration of water can easily reach values not far from those obtained on untreated materials.

Note that in some cases the values of the contact angles, in advancing or in receding mode, do not show any variation, while different values of absorption at short times are present in WAC experiments, as at the lower concentrations of

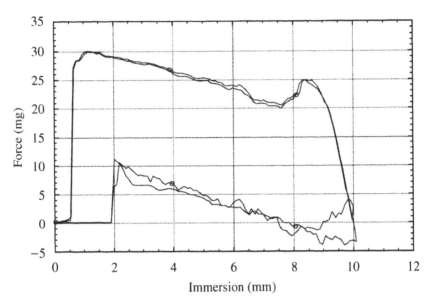

Figure 7. Repeated Wilhelmy run on a sample of B72 film. No significant difference is detectable between the two immersions.

protecting agent on Noto stone. From this point of view, the contact angles appear more correlated with long time protection.

Another issue which is not currently considered in the field of stone protection is the surface dynamics of polymer molecules. Since both B67 and B72 have a T_g slightly higher than room temperature, their mobility at the surface could, in principle, be significant. This is confirmed by the presence of physical aging; this phenomenon works, however, on a time scale that is generally longer than that of surface dynamics.

To test this question the film of pure polymer was repeatedly immersed (Fig. 7), but the effect was not detected by DCA; the difference that one can see between repeated immersion of *treated samples* (which always indicated a more hydrophilic surface) is not simple to explain. Probably it is partially due to the absorption of water by capillarity, as previously described. As noted, this induces a correction in the calculation of contact angles, but it is generally very low, of the same order as or lower than the standard deviation in the data obtained.

Note, however, that if our interpretation is correct, a certain portion of the surface remains free of polymer, and so it is made of calcium carbonate; this portion can be wetted by water and becomes more hydrophilic in repeated immersions. Unfortunately, the percentage of polymer material present in the stone is not sufficient to use DSC as an analytical tool.

The existence of dispersive and polar (or better basic) portions in the analysed polymer molecules can also be invoked.

One can see that in the case of Candoglia marble the receding contact angle on the pure polymer film is *lower* than the receding angle of the polymer present on the stone. This sounds strange because the stone should have, however, a greater portion of high-energy surface than the pure polymer. This could simply be an effect of the roughness, also difficult to prove by the JD calculations. Alternatively, one can *hypothesize* that the polymer is not simply deposited on the stone, but that it weakly but specifically interacts with the calcium carbonate. The most polar (or, better, the most basic) portion of the molecules (the $-COO-$ group) can interact with the positive Ca^{2+} ions. These interactions can work as 'hooks', which allow the dispersive portions of the molecules to be 'oriented' towards the environment in a percentage higher than in the polymer itself; as a final effect, the total character of the obtained surface is more dispersive than the polymer itself. A similar preferential interaction has been invoked in the case of silicate-poly(vinyl alcohol) interaction [28].

As a final remark, we reported in Fig. 2a the advancing and receding contact angles on the B72-protected Candoglia marble obtained by dipping instead of capillary suction, using the Wilhelmy microbalance at the same speed as that used in the measurement of the contact angles. The results show that both the advancing and the receding contact angles of water on the final surfaces are similar to those on the other samples, obtained by dipping, but a repeated immersion gives rise to a greater absorption of water. More accurate experiments will show whether this method of protection is useful.

5. CONCLUSION

The combined effect of heterogeneity and roughness is a complex subject in the field of contact angles. Measurements performed on calcareous stones protected against water penetration by polymeric films show that:

• static contact angles are not simply correlated to the actual degree of protection induced by solutions of polymers of increasing concentration;

• if the surfaces have a low degree of roughness, the situation can be understood in terms of the hysteresis graphs introduced by Johnson and Dettre. If the surfaces have a higher degree of roughness, the surface structure can be transformed into a 'composite' surface, particularly protected against water wetting by the presence of air micro-bubbles entrapped in its structure;

• in all cases, the increase of the receding contact angles can work as a good indication of the presence of a really hydrophobic surface;

• in all cases, however, the receding angles are lower than 90° and water can penetrate, *although at a reduced rate and in a reduced amount.* This also depends on the fact that the penetration depth of the polymer in the porous system cannot be greater than a few hundreds of micrometres.

If measurements are done directly on the protected stone, then the roughness should be measured, because it may strongly modify the values of the contact angles. The obtained contact angle values are not entirely acceptable as an indication of the properties of the surface and a very careful analysis is then necessary.

A final consideration is that from the point of view of surface properties, the ideal protecting agent for these stones should have a *very high* contact angle with water. Due to the difficulty in measuring the Young contact angle of water on *plane surfaces* of the polymer, a sufficient condition for good protection is that the *receding contact angle*, indicating the properties of the highest energy and most easily wettable portion of a plane polymer surface, should be greater than 90°, to prohibit the capillary penetration of water from an infinite reservoir.

Acknowledgements

We are grateful to Ms. W. Vaona, to Ms. C. Gavazza, and to Ms. R. Belli of the University of Trento (Italy) for their skilful technical support. We are also grateful to the Laboratory of the Surface Engineering Center (Rovereto, Italy) for the use of the Hommel tester.

REFERENCES

1. G. Alessandrini and M. Laurenzi Tabasso, *Sci. Technol. Cultural Heritage* **2**, 191–199 (1993).
2. B. H. Vos, *Stud. Conserv.* **16**, 129–144 (1971).
3. B. H. Vos, in: *Proceedings of the International Symposium: Deterioration and Protection of Stone Monuments*, UNESCO-RILEM, pp. 1–10. Imprimé en France par epdgi, Paris (1978).
4. G. Biscontin, in: *Atti del Convegno Internazionale su la Pietra, Interventi, Conservazione, restauro*, pp. 111–127. Museo Provinciale di Lecce, Congedo, Lecce, Italy (1981).
5. R. Peruzzi and R. Bugini, *Rend. Soc. Ital. Mineral. Petrol.* **39**, 71–80 (1984).
6. L. Lazzarini and M. Laurenzi Tabasso, in: *Il Restauro della Pietra*, Ch. 4, pp. 62–122. CEDAM, Padua (1986).
7. Commissione NORMAL (Sottogruppo Sperimentazione Protettivi), *L'Edilizia*, 57–71 (Nov. 1993).
8. C. Atzeni, L. Massidda and U. Sanna, in: *Atti del Convegno di Studi le Pietre nell'Architettura: Struttura e Superfici*, pp. 203–213. Lib. PROGETTO, Brixen, Italy (1991).
9. G. Alessandrini, R. Peruzzi, A. Aldi and F. Fantasma, in: *Proceedings of the 8th International Congress on Deterioration and Conservation of Stone*, J. Riederer (Ed.), pp. 1059–1074. Möller Druck und Verlag, Berlin (1996).
10. A. Marmur, in: *Modern Approaches to Wettability*, M. E. Schrader and G. I. Loeb (Eds), Ch. 12, pp. 327–356. Plenum Press, New York (1992).
11. R. E. Johnson, Jr. and R. H. Dettre, in: *Surface and Colloid Science*, E. Matijevic (Ed.), Vol. 2, pp. 85–154. Wiley Interscience, New York (1969).
12. H. M. Princen, in: *Surface and Colloid Science*, E. Matijevic (Ed.), Vol. 2, pp. 1–84. Wiley Interscience, New York (1969).
13. R. H. Dettre and R. E. Johnson, Jr., *Adv. Chem. Ser.* **43**, 136 (1964).
14. A. B. D. Cassie and S. Baxter, *Trans. Faraday Soc.* **40**, 546 (1944).
15. R. E. Johnson, Jr. and R. H. Dettre, *J. Phys. Chem.* **68**, 1744 (1964).
16. R. E. Johnson, Jr. and R. H. Dettre, in: *Wettability*, J. C. Berg (Ed.), pp. 2–73. Marcel Dekker, New York (1993).
17. J. F. Oliver and S. G. Mason, *J. Colloid Interface Sci.* **60**, 480 (1977).

18. Y. Tamai and K. Aratami, *J. Phys. Chem.* **76**, 3267 (1972).
19. J. B. Cain, D. W. Francis, R. D. Venter and A. W. Neumann, *J. Colloid Interface Sci.* **94**, 123 (1983).
20. Raccomandazione NORMAL 11/85, *Assorbimento d'acqua per capillarità–Coefficiente di assorbimento capillare*. CNR–ICR, Roma (1986).
21. Raccomandazione NORMAL 33/89, *Misura dell'angolo di contatto*. CNR–ICR, Roma (1991).
22. C. Della Volpe and S. Siboni, *J. Adhesion Sci. Technol.* **12**, 197–224 (1998).
23. S. Ross and I. D. Morrison, *Colloidal Systems and Interfaces*. John Wiley, New York (1988).
24. American National Standard ANSI/ASME B46.1, *Surface Texture*. American Society of Mechanical Engineering, New York (1985).
25. M. Lampin, R. Warocquier-Cleroud, C. Légris, M. Degrange and M. F. Sigot-Luizard, *J. Biomed. Mater. Res.* **36**, 99–108 (1997).
26. M. Morra and C. Della Volpe, *J. Biomed. Mater. Res.* **42**, 473–474 (1998).
27. K. Grundke, H.-J. Jacobasch, F. Simon and S. Schneider, *J. Adhesion Sci. Technol.* **9**, 327–350 (1995).
28. J. F. Young and M. Berg, *Mater. Res. Soc. Symp. Proc.* **721**, 609–619 (1992).

Apparent and Microscopic Contact Angles, pp. 377–387
J. Drelich, J. S. Laskowski and K. L. Mittal (Eds)
© VSP 2000.

Microsphere tensiometry to measure advancing and receding contact angles on individual particles

STEFAN ECKE, MARKUS PREUSS and HANS-JÜRGEN BUTT *

Institut für Physikalische Chemie, Universität Mainz, 55099 Mainz, Germany

Received in final form 12 February 1999

Abstract—In this paper, a method to measure the advancing and receding contact angles on individual colloidal spheres is described. For this purpose, the microspheres were attached to atomic force microscope cantilevers. Then the distance to which the microsphere jumps into its equilibrium position at the air–liquid interface of a drop or an air bubble was measured. From these distances the contact angles were calculated. To test the method, experiments were done with silanized silica spheres (4.1 μm in diameter). From the experiments with drops, an advancing contact angle of $101 \pm 4°$ was determined. A receding contact angle of $101 \pm 2°$ was calculated from the jump-in distance into a bubble. Both experimental techniques gave the same contact angle. In contrast, on similarly prepared planar silica surfaces, a clear hysteresis was measured with the sessile drop method; contact angles of $104.5 \pm 1°$ and $93.8 \pm 1°$ were determined for the advancing and receding contact angles, respectively.

Key words: Atomic force microscopy; flotation; contact angle hysteresis; line pinning; sphere tensiometry; surface forces.

1. INTRODUCTION

Sphere tensiometry is an established method to measure the surface tension of a liquid [1–4]. In sphere tensiometry, the force required to pull a sphere out of a liquid is measured. From this force, the surface tension of the liquid can be calculated. Also, the contact angle on the sphere can be determined by measuring the whole force vs. pull-out distance curves [5, 6, 7a]. For sphere tensiometry, usually spheres with typical diameters of several 100 μm are used.

In this paper, we present the results of experiments with microspheres. The aim of these experiments was not so much to measure the surface tension of the liquid (this can be done with simpler techniques), but to measure the contact angle on individual microspheres. The wetting behavior of small particles is important

---

*To whom correspondence should be addressed. E-mail: butt@wintermute.chemie.uni-mainz.de

in many industrial applications and natural phenomena. The mineral processing, pigment, paint, cosmetic, and pharmaceutical industries all require knowledge of the particles' wettability. This property is usually characterized in terms of the solid–liquid contact angle.

Several techniques exist to determine the contact angle on small particles. However, most of these techniques give only qualitative results [8]. The most quantitative results are obtained with the capillary rise method [9–16]. In a capillary rise experiment, microspheres in the form of powder are filled into a tube, which is closed by a filter at the bottom. The bottom of the tube is brought into contact with the liquid. The liquid rises in the tube, due to the capillary effect of the voids between the powder microspheres. From the liquid rising speed or the pressure needed to stop the rise of the liquid, the contact angle can be calculated [17–20]. One limitation of the capillary rise method is that it averages over many microspheres and does not give information about a possible size distribution. In addition, it relies on the assumption that a powder can be treated as a bundle of capillaries and it depends on the specific model applied [21]. Contact angles on nanometer-sized particles have been deduced by spreading particles on a liquid surface in a Langmuir trough [22, 23]. From the surface pressure–area relationship, the contact angle could be calculated. This technique again gives an average contact angle on many particles.

Contact angles on individual spheres of diameters in the 20–50 μm range were determined by Mingins and Scheludko [24]. With an optical microscope, they measured the radius of the three-phase contact line of microspheres attached to the surface of a pendant drop. From this radius, they calculated the contact angle.

Microsphere tensiometric experiments were made possible with the invention of the atomic force microscope (AFM) and the development of microfabricated cantilevers [25]. In these experiments, the microsphere is glued to the end of a tipless cantilever (Fig. 1) as described in ref. [26]. Then it is moved towards a water drop. When the particle touches the air–liquid interface, a three-phase

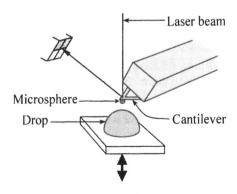

Figure 1. Schematic diagram of the experimental arrangement. The microsphere attached to an AFM cantilever is positioned above the drop. The position of a reflected laser beam is measured to detect the deflection of the cantilever. The drop is moved up and down by a piezoelectric translator.

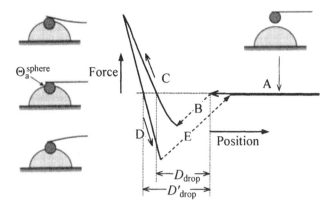

Figure 2. A typical force vs. position curve between a spherical colloidal particle and a liquid drop. The position is the position of the surface on which the drop is placed. It was assumed that no long-range forces were active between the particle and the drop and that the particle surface was not completely wetted by the liquid. At large distances, the cantilever is not deflected (A). When the particle comes into contact with the air–water interface, it jumps into the drop and a three-phase contact (TPC) is formed (B). The reason for the jump-in is the capillary force. Moving the particle further towards the drop shifts the TPC line over the particle surface (C). The relevant factor is the advancing contact angle Θ_a^{sphere}. When retracting the particle again (D), at some point the force is high enough to draw the particle off the air–water interface (E).

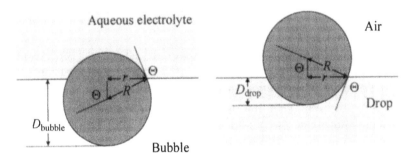

Figure 3. Schematic diagram showing a microsphere at equilibrium in the air–water interface of a bubble (left) or a drop (right). On the micrometer scale, the air–water interface is practically planar. Since the microspheres considered here are small, gravitational forces were neglected. Θ is the equilibrium contact angle; R is the radius of the particle.

contact (TPC) is formed and the capillary force pulls the particle into the drop. The deflection of the cantilever and the position of the drop are measured. Multiplying the cantilever deflection with the spring constant of the cantilever gives the force. In this way, force vs. position curves are measured (Fig. 2).

In general, it is very difficult to analyze force vs. position curves obtained on highly deformable surfaces [26–32]. To determine the contact angle, however, only the equilibrium position of the microsphere in the air–water interface needs to be known. This position is characterized by the penetration depth D_{drop} (Fig. 3). D_{drop} can be directly obtained from force vs. position curves without further analysis. In this way, the advancing contact angle is obtained because the TPC line is advancing over the surface of the microsphere.

Receding contact angles were obtained from experiments with air bubbles as described before [26, 32]. In this case, the microsphere is in the liquid medium and approaches the air bubble. As soon as the particle touches the air–liquid interface and the contact angle is not zero, a TPC is formed. From the penetration depth into the equilibrium position, D_{bubble}, the receding contact angle was calculated.

2. MATERIALS AND METHODS

The experimental set-up was the same as that used for measuring the force between microspheres and bubbles [26, 33]. We describe it briefly for measurements with drops. Drops of typically 1 mm diameter were placed on the bottom of a Teflon cuvette. Then under microscopic control, the microsphere was positioned roughly 10 μm above the drop. To measure force vs. position curves, the cuvette with the drop was moved towards the microsphere with a 15 μm-range piezoelectric translator. During the movement, the deflection of the cantilever was measured by an optical lever technique. The height position of the drop and the deflection of the cantilever were recorded with a digital oscilloscope. After transferring the data to a personal computer, the results were further analyzed. The measurement of one force curve took roughly 4 s. Typically, the drop approached the microsphere with a speed of 7 μm/s.

Silica microspheres of 2 μm radius (Bangs Inc., Carmel, USA) were glued onto tipless cantilevers (Digital Instruments, CA, USA; V-shaped, 200 μm long, 0.6 μm thick) using a small amount of epoxy resin (Epikote 1004, Shell). For details see refs [26, 34]. Then the microspheres were exposed to an atmosphere of dichlorodimethylsilane for 2 min. All spheres were imaged in a scanning electron microscope after the experiment (Fig. 4). In this way, we determined their individual radii. The spring constants of all cantilevers were measured with a reference cantilever as described before [26]. The spring constant of the reference cantilever was calibrated by measuring the shift of the resonance frequency after attaching weights [35]. Typical spring constants were 0.11 N/m. To test the method, experiments were carried out with four microspheres and the corresponding planar surfaces.

The advancing contact angle on a microsphere, Θ_a^{sphere}, was calculated from the distance between the jump-in point and the zero-force point of the force curve (C in Fig. 2). Neglecting the weight of the microsphere and its buoyancy, the contact angle can be calculated with simple geometric arguments from [36, 37]

$$\cos \Theta_a^{sphere} = \frac{D_{drop} - R}{R}, \tag{1}$$

where R is the radius of the microsphere. Receding contact angles, Θ_r^{sphere}, were determined from experiments with bubbles. They could be calculated from D_{bubble}

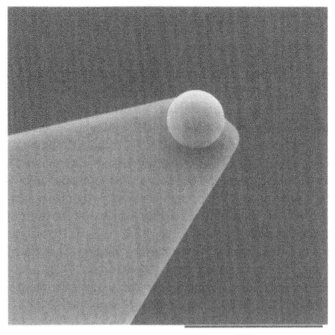

10 μm

Figure 4. Scanning electron micrograph of a silanized silica microsphere glued to the end of a tipless atomic force microscope cantilever.

by

$$\cos \Theta_r^{\text{sphere}} = \frac{R - D_{\text{bubble}}}{R}. \tag{2}$$

Contact angles were also measured on planar surfaces prepared in the same way as the microsphere surface by optical observation of a sessile drop. The profile of the drop shape was imaged with a video camera and from the images the contact angle was obtained. Therefore a commercial goniometer (Krüss, DSA 010, Hamburg, Germany) was equipped with a stepper motor to drive the syringe which controls the drop volume. From repeated experiments with the same sample, the error was found to be roughly 1°.

Water was purified using a commercial Milli-Q system containing ion-exchange and charcoal stages. The deionized water had a conductivity less than 0.1×10^{-6} S/m and was filtered at a pore size of 0.22 μm. All experiments were performed at room temperature without buffer in 5 mM KCl. The pH was around 6.

3. RESULTS AND DISCUSSION

3.1. Measured contact angles

A typical force vs. position curve of a silanized microsphere and a water drop is shown in Fig. 5a. No long-range forces between the water surface and the particle were detected before the jump-in occurred. From the distance between the jump-in position and the zero force point of the TPC part of the force curve during the approach, D_{drop}, the advancing contact angles were obtained (Table 1). The average advancing contact angle was 99°. This average does not take into account sample 2 in Table 1. For an unknown reason but nevertheless reproducibly, we measured a small advancing contact angle of 65° with sample 2. On planar surfaces prepared in the same way as the microspheres, higher contact angles were measured; the average value was 104.5°.

For the interaction of a particle with a bubble, force vs. position curves like the one shown in Fig. 5b were obtained. From the distances D_{bubble}, we calculated an average receding contact angle of 104°. This was higher than the receding contact angle determined on the similarly prepared planar surface of 93.8°.

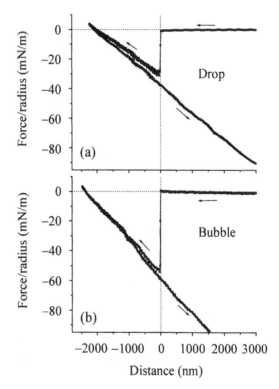

Figure 5. Typical force vs. position curve for a silanized microsphere in air and a drop of aqueous solution with 5 mM KCl (a). For the interaction between the same particle and an air bubble in aqueous medium, a typical force vs. position curve is plotted in b. The normalized force (force divided by the radius of the microsphere) is shown.

Table 1.

The uncorrected results obtained with four microspheres. The radius R was determined from scanning electron micrographs. Using the mean jump-into-equilibrium position distances D_{drop} and D_{bubble}, the advancing and receding contact angles on the microspheres, Θ_a^{sphere} and Θ_r^{sphere}, were calculated with equations (1) and (2), respectively. The values are not corrected for the influence of the Laplace pressure. By taking the Laplace pressure in the bubble or drop into account, Θ_a^{sphere} should increase by $2°$ and Θ_r^{sphere} should decrease by $3°$. The advancing and receding contact angles on planar silica surfaces, Θ_a^{planar} and Θ_r^{planar}, were measured with the sessile drop method on silica plates which were silanized together with the corresponding microspheres

Sample	R (μm)	D_{drop} (μm)	Θ_a^{sphere} (degrees)	Θ_a^{planar} (degrees)	D_{bubble} (μm)	Θ_r^{sphere} (degrees)	Θ_r^{planar} (degrees)
1	2.04	1.67	100.5	106.5	2.51	103.3	94.5
2	2.07	2.94	65.2	103.5	2.58	104.2	93.5
3	2.00	1.85	94.2	103.5	2.48	103.9	93.0
4	2.00	1.57	102.4	104.5	2.46	103.4	94.0

One could imagine determining the receding contact angle Θ_r^{sphere} from the zero-force position during the retracting part of the force curve (D in Fig. 2). When retracting the particle out of the drop again, at a certain position the zero-force line is crossed. We could measure the distance between this point and the jump-in position D'_{drop} (Fig. 2). At this point, the TPC is receding. Hence, one could think of obtaining both advancing and receding contact angles from one force curve only.

This method is, however, not without uncertainty. It relies on the assumption that the TPC is moving over the particle surface and that the TPC line is not pinned. We believe that this is the case for the approaching part of the force curve (C in Fig. 2) but it is not necessarily true for the retracting part of the force curve (D in Fig. 2). The argument is as follows. If the TPC line is pinned, the force curve should be steeper than for a moving TPC line. Usually, the approaching part of the force curve (C) was slightly less steep than the retracting part. Never was the approaching part steeper than the retracting part. Hence, we exclude a line pinning during the approach but not for the retraction. A further hint for a line pinning during the retraction is the observation that often D'_{drop} depended on the maximal load applied.

3.2. Errors and corrections

Before discussing the result further, we consider the error in the contact angle measured. The random error in contact angles determined by microsphere tensiometry is given by the error in the distance D_{drop} between the jump-in point and the point of zero force and by the error in measuring the radius. From repeated measurements of D_{drop} for one microsphere, the random error ΔD was determined to be ≈ 110 nm. The error in the radius, ΔR, is determined by the resolution and the calibration of the scanning electron microscope and we estimate it to be around ≈ 20 nm. The total random error in the contact angle determined is

$$\Delta\Theta^{\text{sphere}} = \left|\frac{\partial\Theta^{\text{sphere}}}{\partial D}\right| \cdot \Delta D + \left|\frac{\partial\Theta^{\text{sphere}}}{\partial R}\right| \cdot \Delta R$$

$$= \frac{1}{\sqrt{2RD - D^2}} \cdot \left[\Delta D + \frac{D}{R} \cdot \Delta R\right]. \tag{3}$$

Inserting typical values of $D = 2\ \mu\text{m}$ and $R = 2\ \mu\text{m}$, a random error in Θ_a^{sphere} of $\approx 4°$ is obtained.

Equation (3) also applies to experiments with the bubble. D_{bubble} could be measured with a higher accuracy than the corresponding D_{drop}. A reason could be that in the bubble the humidity was constant and at 100%. The random error in this case was $\Delta D \approx 60$ nm, which leads to a total random error in Θ_r^{sphere} of $2°$.

In addition, a systematic error due to the van der Waals attraction has to be considered. It is known that attractive forces might cause an instability of the cantilever which leads to a jump onto the sample. This instability occurs when the gradient of the attraction exceeds the spring constant of the cantilever. Deformable samples like drops or bubbles behave similarly to a spring. Therefore, an additional instability could cause the air–water interface to jump towards the microsphere and draw the microsphere into the bubble. This happens when the effective spring constant of the drop or bubble falls below the gradient of the attractive surface forces. Both instabilities lead to a jump-in before the microsphere actually touches the original air–water interface. As a consequence, the parameters D_{drop} and D_{bubble} could appear larger than they actually are.

The van der Waals force can be estimated from

$$F_{\text{vdW}} = -\frac{A \cdot R}{6D^2}, \tag{4}$$

where A is the Hamaker constant. For silica/air/water, A is approximately 10^{-20} J. Hence, the gradient of the van der Waals attraction is $dF_{\text{vdW}}/dD = A \cdot R/(3D^3)$. Equating this to the spring constant of the cantilever of 0.11 N/m leads to an extra jump-in distance of 4 nm, which is negligible.

In the case of experiments with bubbles, the van der Waals force is repulsive and could not affect the results. However, the hydrophobic attraction might have a similar effect. As discussed before, we do not expect an extra jump-in distance larger than 10 nm, which can again be neglected [26, 32].

To derive equations (1) and (2), we assumed that only the capillary force acted on the particle and that the shape of the air–water interface was not affected by the presence of a particle in its equilibrium position. A general calculation of the force acting on a particle in the air–liquid interface is relatively complex, since the shape of the meniscus has to be calculated with the Laplace equation and several forces such as the weight of the particle, the buoyancy of its immersed part, the hydrostatic pressure of the liquid, the capillary force, and the Laplace pressure have to be considered [1–3, 6, 7b, 36, 38, 39].

Fortunately, for small particles the detachment force is dominated by the capillary force and all other contributions are small [40]. The gravitational force due to the weight of the particle is given by

$$F_{\text{weight}} = \frac{4}{3}\pi R^3 \cdot \rho_{\text{sphere}} \cdot g, \tag{5}$$

where ρ_{sphere} is the density of the microsphere and $g = 9.81$ m/s^2 is the acceleration due to gravity. With $R = 2$ μm, a force of the order of 1 pN is obtained. This is much smaller than the capillary forces measured. The buoyancy must be even smaller than the weight since the density of the displaced water is smaller than the density of the particle material.

The Laplace pressure inside the drop (or bubble), $2\gamma/R_{\text{drop}}$, tends to push the particle out of the drop (or bubble). γ is the surface tension of the liquid and R_{drop} is the radius of the drop (or bubble). The force due to the Laplace pressure is obtained by multiplying the Laplace pressure with the cross-sectional area of the particle. At the equilibrium position, this cross-sectional area is given by $\pi r^2 = \pi R^2 \sin^2 \Theta^{\text{sphere}}$ (see Fig. 3). Hence, the repulsive force due to the Laplace pressure is

$$F_{\text{Laplace}} = \pi R^2 \cdot \sin^2 \Theta^{\text{sphere}} \cdot \frac{2\gamma}{R_{\text{drop}}}. \tag{6}$$

With $\gamma = 72.0$ mN/m for water, $R_{\text{drop}} = 0.5$ mm, and $\Theta^{\text{sphere}} = 90°$, one calculates a maximal contribution of 4 nN or a normalized force of 2 mN/m. Usually, the bubbles used had a smaller radius, typically $R_{\text{bubble}} = 0.3$ mm, leading to a Laplace repulsion of 6 nN or 3 mN/m. This is by far more significant than the other terms, but it is still much smaller than the capillary force.

The Laplace pressure pushes the particle slightly out of the drop (or bubble). This might lead to an underestimation of D_{drop} (or D_{bubble}). To calculate the distance to which the microsphere is pushed out, we divide F_{Laplace} by the slope of the force curve at the equilibrium position, $dF/dD|_{\text{eq}}$. Typical values for the slope were 50 nN/μm. For experiments with drops and $F_{\text{Laplace}} = 4$ nN, the underestimation of D_{drop} is 80 nm. For a contact angle near 90°, this would result in an increase of $\Theta_{\text{a}}^{\text{sphere}}$ by 2°. For experiments with bubbles and $F_{\text{Laplace}} = 6$ nN, the maximal underestimation of D_{bubble} is 120 nm. For a contact angle near 90°, this would result in a decrease of $\Theta_{\text{r}}^{\text{sphere}}$ by 3°.

Summarizing, the average value for the advancing contact angle obtained from the jump-into-equilibrium distances into drops is $99° + 2° = 101°(\pm 4°)$. The receding contact angle obtained from experiments with bubbles is $104° - 3° = 101°(\pm 2°)$.

3.3. Interpretation of contact angles

A tentative explanation for the low hysteresis or possibly no hysteresis may be the fact that the zero-force position is used to calculate the contact angle. In this position, the TPC line exerts no net force on the particle. In contrast, on planar surfaces the TPC might exert a force on the surface. In a typical sessile drop experiment, for instance, first the volume of the drop is increased and the contact line advances. During this process, the advancing contact angle is measured. Then the volume of the liquid is decreased. At first, the TPC line is pinned and does not move. Consequently, the contact angle is reduced. When the receding contact angle is reached, the force exerted by the surface tension of the liquid on the TPC is strong enough to move the TPC line again and to overcome line pinning. In microsphere tensiometry, such a gradual increase of the force would immediately be detected. Hence, contact angles measured from the zero-force position by microsphere tensiometry might be close to the equilibrium contact angle.

One could ask why is it not possible to measure the equilibrium contact angle with classical sphere tensiometry? There is also a position of zero-force and from this, one could calculate the contact angle. The advantage of microsphere tensiometry is that the capillary force, which is proportional to R, dominates. In classical sphere tensiometry, the weight of the particle, its buoyancy (which scales with R^3), and the force due to the Laplace pressure (which scales with R^2) all contribute significantly. Hence, at the zero-force position only the sum of these forces is zero, but the capillary force might still be significant.

4. CONCLUSION

Using the technique of microsphere tensiometry, the contact angles on individual spherical particles can be measured. In experiments with liquid drops, an advancing contact angle Θ_a^{sphere} is obtained. Experiments with bubbles gave the receding contact angle Θ_r^{sphere}. The hysteresis is smaller than on similarly prepared planar surfaces. However, the interpretation of Θ_a^{sphere} and Θ_r^{sphere} and their relation to advancing, receding, and equilibrium contact angles on planar surfaces are not yet clear.

Acknowledgements

We thank M. Ruppel, Universität Frankfurt, for scanning electron images. We further acknowledge financial support from the Deutsche Forschungsgemeinschaft (M.P., S.E.).

REFERENCES

1. G. D. Yarnold, *Proc. Phys. Soc. London* **58**, 120–124 (1946).
2. A. D. Scheludko and A. D. Nikolov, *Colloid Polym. Sci.* **253**, 396–403 (1975).
3. C. Huh and S. G. Mason, *Can. J. Chem.* **54**, 969–978 (1976).
4. E. Bayramli, C. Huh and S. G. Mason, *Can. J. Chem.* **56**, 818–823 (1978).
5. E. Bayramli and S. G. Mason, *Colloid Polym. Sci.* **260**, 452–453 (1982).

6. R. Gunde, S. Hartland and R. Mäder, *J. Colloid Interface Sci.* **176**, 17–30 (1995).
7. (a) L. Zhang, L. Ren and S. Hartland, *J. Colloid Interface Sci.* **180**, 493–503 (1996);
   (b) L. Zhang, L. Ren and S. Hartland, *J. Colloid Interface Sci.* **192**, 306–318 (1997).
8. D. Li and A. W. Neumann, in: *Applied Surface Thermodynamics*, A. W. Neumann and J. K. Spelt (Eds), Surfactant Science Series No. 63, p. 509. Marcel Dekker, New York (1996).
9. F. E. Bartell and H. J. Osterhof, *Z. Phys. Chem.* **130**, 715–723 (1927).
10. F. E. Bartell and H. Y. Jennings, *J. Phys. Chem.* **38**, 495 (1934).
11. H. G. Bruil and J. Van Aartsen, *Colloid Polym. Sci.* **252**, 32–38 (1974).
12. D. T. Hansford, D. J. W. Grant and J. M. Newton, *Powder Technol.* **26**, 119–126 (1980).
13. A. Siebold, A. Walliser, M. Nardin, M. Opplinger and J. Schultz, *J. Colloid Interface Sci.* **186**, 60–70 (1997).
14. G. Buckton, *J. Adhesion Sci. Technol.* **7**, 205–219 (1993).
15. R. Varadaraj, J. Bock, N. Brons and S. Zushma, *J. Colloid Interface Sci.* **167**, 207 (1994).
16. D. Diggins, L. G. J. Fokkink and J. Ralston, *Colloids Surfaces* **44**, 299–313 (1990).
17. R. Lucas, *Kolloid Z.* **23**, 15–22 (1918).
18. E. W. Washburn, *Phys. Rev.* **17**, 273–283 (1921).
19. J. Szekely, A. W. Neumann and Y. K. Chuang, *J. Colloid Interface Sci.* **35**, 273–278 (1971).
20. S. Levine, J. Lowndes, E. J. Watson and G. Neale, *J. Colloid Interface Sci.* **73**, 136–151 (1980).
21. J. Van Brakel, *Powder Technol.* **11**, 205–236 (1975).
22. J. H. Clint and S. E. Taylor, *Colloids Surfaces A* **65**, 61–67 (1992).
23. J. H. Clint and N. Quirke, *Colloids Surfaces A* **78**, 277–278 (1993).
24. J. Mingins and A. Scheludko, *J. Chem. Soc., Faraday Trans. 1* **75**, 1–6 (1979).
25. G. Binnig, C. F. Quate and C. Gerber, *Phys. Rev. Lett.* **56**, 930–932 (1986).
26. M. Preuss and H.-J. Butt, *Langmuir* **14**, 3164–3174 (1998).
27. H.-J. Butt, *J. Colloid Interface Sci.* **166**, 109–117 (1994).
28. W. A. Ducker, Z. Xu and J. N. Israelachvili, *Langmuir* **10**, 3279–3289 (1994).
29. M. L. Fielden, R. A. Hayes and J. Ralston, *Langmuir* **12**, 3721–3727 (1996).
30. P. Mulvaney, J. M. Perera, S. Biggs, F. Grieser and G. W. Stevens, *J. Colloid Interface Sci.* **183**, 614–616 (1996).
31. B. A. Snyder, D. E. Aston and J. C. Berg, *Langmuir* **13**, 590–593 (1997).
32. M. Preuss and H.-J. Butt, *J. Colloid Interface Sci.* **208**, 468–477 (1998).
33. M. Preuss and H.-J. Butt, *Int. J. Miner. Process.* **56**, 99–115 (1999).
34. R. Raiteri, M. Preuss, M. Grattarola and H.-J. Butt, *Colloids Surfaces A* **136**, 191–197 (1998).
35. J. P. Cleveland, S. Manne, D. Bocek and P. K. Hansma, *Rev. Sci. Instrum.* **64**, 403–405 (1993).
36. A. Scheludko, B. V. Toshev and D. T. Bojadjiev, *J. Chem. Soc., Faraday Trans. 1* **72**, 2815–2827 (1976).
37. J. Ralston and G. Newcombe, in: *Colloid Chemistry in Mineral Processing*, J. S. Laskowski and J. Ralston (Eds), pp. 173–201. Elsevier, Amsterdam, The Netherlands (1990).
38. H. M. Princen, in: *Surface and Colloid Science*, E. Matijevic (Ed.), Vol. 2, pp. 1–84. Wiley-Interscience, New York (1969).
39. A. V. Rapacchietta and A. W. Neumann, *J. Colloid Interface Sci.* **59**, 555–567 (1977).
40. A. Scheludko, S. Caljovska, A. Fabrikant, B. Radoev and H. J. Schulze, *Freiberg. Forschungsh. A* **484**, 85–97 (1971).

Apparent and Microscopic Contact Angles, pp. 389–404
J. Drelich, J. S. Laskowski and K. L. Mittal (Eds)
© VSP 2000.

Wettability of fine solids extracted from bitumen froth

F. CHEN [1], J. A. FINCH [1], Z. XU [2,*] and J. CZARNECKI [3]

[1] *Department of Mining and Metallurgical Engineering, McGill University, Montreal, Quebec, Canada, H3A 2B2*
[2] *Department of Chemical and Materials Engineering, University of Alberta, Edmonton, Alberta, Canada, T6G 2G6*
[3] *Syncrude Canada Ltd., Research Center, 9421-17 Ave, Edmonton, Alberta, Canada, T6N 1H4*

Received in final form 22 April 1999

Abstract—Production of synthetic crude oil from oil sands deposits in northern Alberta involves open pit mining, mixing the mined ore with water, extraction of aerated bitumen from the slurry, removal of water and solids from the froth formed, and upgrading heavy bitumen to liquid hydrocarbons. The success of the froth treatment operation, aimed at removal of fine solids and water from the bituminous froth, depends on the control of wettability of fine solids by the aqueous phase. Fine solids were extracted from bitumen froth by heptane. The partition of the extracted solids in aqueous, organic, and interphases was measured, and the wettability of the solids by water in various diluents was evaluated from contact angle measurements. The effect of diluent composition, sample drying, and surface washing on the wettability and fine particle partition was examined. The partition of fine particles correlated well with their wettabilities, and the results were found to be useful for interpreting the observations from froth treatment practice.

Key words: Wettability; bitumen; clays; contact angle; interfacial tension; FTIR.

1. INTRODUCTION

Oil sands deposits in northern Alberta contain more hydrocarbons than all OPEC reserves combined. The amount recoverable using the existing technology is slightly larger than the total oil reserves of Saudi Arabia. Today, the total production of oil from oil sands amounts to over 20% of Canadian oil consumption, three-quarters of which is from open pit mining; the remainder comes from *in situ* recovery. Clark's hot water extraction process and its modifications have been used to separate bitumen from the oil sand ore. In this process, the mined oil sand is mixed with hot water and the digested slurry is fed into large gravity separation vessels, where bitumen is recovered as a froth product in a process

---

*To whom correspondence should be addressed.

similar to flotation. The froth produced as such typically contains *ca.* 60% bitumen, 30% water, and 10% solids. The froth is cleaned by adding a diluent (an organic liquid mixture, such as naphtha) to provide a density difference between the water and hydrocarbon phases and to reduce the viscosity of the froth. The diluted bitumen is then fed through a two-stage centrifuge (at *ca.* 250× and 2500 × g, respectively) to remove coarse particles in the first stage by scroll machines and the remaining fine solids and finely dispersed water droplets in the second stage by disc centrifuges. (Inclined plate settlers and/or third-stage centrifuges are also used in commercial operations.) Collectively called froth treatment, this process produces a product still containing $\sim 2\%$ water and 0.5% solids.

It is well documented that the water remaining in the froth treatment product contains dissolved salts, mainly sodium chloride, which are then transferred to the downstream refinery plant. The chlorine carried by finely dispersed water droplets causes a serious corrosion risk, which may lead to catastrophic damages in upgrading operations, if not removed. Fine solids, also suspended in water, tend to accumulate in hydrotreatment reactors, lowering the activity of catalysts and obstructing the flow through reactors. As a result, efforts have been made to reduce the amount of water in the froth treatment product before it is sent to the upgrading operations. It has been found that by changing the diluent characteristics from aromatic to paraffinic, the derived bitumen product is drier (contains less than 0.5% moisture) and practically solid-free, but this is achieved at the cost of lower bitumen recovery. It is thus necessary to understand the mechanism of improved performance of the froth treatment by varying the diluent composition, to optimize the process.

The improved bitumen quality has been speculated to be associated with changes in the wettability of fine solids due to the change in polarity of the diluent and its power to remove (dissolve) naturally occurring surfactants from fine solid surfaces. The wettability of fine solids is anticipated to determine the partition of the solids between organic and aqueous phases and, in turn, to determine the amount of water in the bitumen product [1]. The fine solids, often contaminated by natural surfactants present in bitumen to produce 'asphaltene characteristics' [2], have a strong tendency to collect at the oil–water interface and hence stabilize fine water droplets in the organic phase. This additional factor in water-in-oil emulsion stability may contribute to the increased water content in the froth treatment product [3, 4]. It is therefore necessary to understand the changes in the wettability of fine solids responding to the changes in the dissolving power as well as the polarity of the diluent.

The wettability of fine solids is affected by many factors, including the nature of fine particles (e.g. surface chemical composition), the type and amount of adsorbed surfactant, the composition of the diluent used in the froth treatment, etc. Water contact angles are often used as a benchmark to measure the wettability of solids [1]. Although the contact angle on fine powders can be determined with the capillary rise or pressure compensation method in combination with the Washburn equation, for wet fine particles from oil sands, the sessile drop method remains a convenient technique. In this technique, a sessile drop of water is placed on a powder tablet either in air or in a diluent, the latter providing a unique opportunity to study the bi-wettability of fine solids. The contact angle is often determined directly by a protractor eye-piece on a goniometer or by curve fitting of a digitized image of the sessile drop [5]. Although the wettability can be characterized by contact angles, to understand the nature and variations of the surface wettability of fine solids with different treatments requires a knowledge of surface chemical speciation. This information is often obtained with spectroscopic techniques, such as infrared (IR) spectroscopy [6, 7]. In this paper, the effect of various treatments on the wettability of fine solids extracted from bitumen froth is studied. The measured wettability is interpreted using IR spectroscopic data and correlated to fine solids partition between organic and aqueous phases, which has direct implications in froth treatment practice.

2. MATERIALS AND METHODS

2.1. Sample preparation

A sample of bitumen froth taken from commercial primary separation vessels and stored in CANMET's storage tank in Devon (Edmonton) was transferred to a 1-liter glass bottle and kept in a refrigerator until needed. During the froth treatment, the sample was heated to 85°C to thaw the bitumen, followed by the addition of heptane to a *ca.* 2 : 1 heptane-to-froth weight ratio while under mechanical stirring. The mixture was then transferred to a large separation funnel in which phase separation occurred and three distinct phases formed within *ca.* 2 h. On the top was a nonaqueous-rich phase and at the bottom was an aqueous phase hosting most of the sands entrained in the froth. The middle was most likely a bi-continuous phase (or rag layer) containing fine solids, water, asphaltenes, and bitumen. After separating the coarse solids and water from the bottom, the rag layer was collected into a glass bottle for further washing with heptane. For each wash, the sample was centrifuged at 3000 rpm for 20 min to separate the washing liquid from fine solids. The heptane was used in these stages because it does not dissolve asphaltenes inherently present on the solid surface and therefore it is expected to cause minimal change in the fine solids' surface properties. The treated fine solids were dried in a vacuum oven at 40°C for 4 h in an attempt to remove excess water and washing solvent. The sample preparation up to this point was performed at Syncrude Canada Ltd. (Research Center, Edmonton). The solids prepared as such were finer than 1 μm and contained varying amounts of liquid (not measured). They were stored in several sealed glass bottles for future use. Further washing of the sample with heptane or toluene was conducted as stated.

Table 1.
Comparison of the measured surface and interfacial tensions with those reported
in the literature

Liquid	Surface tension (mN/m)		Interfacial tension (mN/m)	
	Measured	Reported	Measured	Reported
Heptane	20.0	20.1	50.9	50.2
Toluene	28.2	28.5	36.0	36.1

2.2. Surface tension and interfacial tension measurements

The surface tension of a liquid plays a critical role in determining the wettability
of solids. The surface tension of a heptane–toluene mixture at different ratios
(to simulate a diluent of different chemical compositions) and the interfacial
tension of the mixture against water were determined by a platinum plate using
a K12 automatic tensiometer (Krüss, USA). The measurement was conducted at
$24 \pm 0.5°C$. As shown in Table 1, the measured surface tension of spectral grade
liquids (heptane and toluene) and the interfacial tension of these liquids against
water were in excellent agreement with those reported in the literature [8, 9]. A
surface tension of 72.3 mN/m was obtained with the Millipore water used in this
project.

2.3. Contact angle measurements

Samples sufficiently wet (paste-like) were transferred to fill a metal ring hosted in
an aluminum cylinder. A piece of Teflon film on a rigid, molecularly smooth mica
surface was attached to the bottom of this in-house built aluminum cylinder and steel
ring through a movable bottom cap. A centrifugal force of $3000 \times g$ was applied to
the samples in the cylinder. After dismounting the movable bottom cap, a compact
tablet formed in the ring was pushed out from the bottom by a piston mounted on
the top of the cylinder. The smooth bottom surface of the tablet in the ring was
used to measure the contact angle. For samples that were not sufficiently wet, a
small amount of toluene was added to make a paste. Procedures similar to those
described above were applied to the resultant paste to obtain a smooth surface for
contact angle measurements. It is important to note that toluene evaporates rapidly
even at ambient temperatures, so that the effect of residual toluene on the measured
contact angles can be considered negligible.

For contact angle measurements, a water drop was placed on the tablet by a
micro-syringe in either air, heptane, toluene, or a mixture of the two. The image
(photograph) of the water drop was taken by a Sony Video 8 professional video
recorder with a close-up function for macro shooting, or by a Canon AE-1 camera
with macro focus lens. The angle was measured through the water phase, i.e. a
large contact angle represents a more hydrophobic or oil-wettable surface. It should
be noted that the centrifuge method developed in this study to prepare tablets has

proven convenient and suitable, with an average variation of $\pm 3°$ in the contact angle measurements obtained.

2.4. Fine particle partition

To correlate the measured wettability of samples made into tablet form with the performance of the froth treatment, a simple fine particle partition test was conducted with the powder samples. A 50 ml tube was used to hold 24 ml test liquids at a volumetric diluent–water ratio of 1 : 1. A hole was made in the wall of the tube slightly below the interface and blocked by a stopper. A 0.3 g wet or 0.1 g dried sample of fine solids extracted from the rag layer was placed in the liquid and hand-shaken vigorously to mix the three phases thoroughly. The mixture was then left to stand, and phase separation occurred immediately. The solids partitioned to the organic phase, aqueous phase, or organic–aqueous interphase, depending on the wettability of the solids. The organic phase was then removed carefully with a pipette and the stopper was unplugged to remove the solids remaining in the interphase region. The solids that sank in the aqueous phase and those retained in the organic and interphases were collected separately and weighed after drying in an oven at 60°C for 72 h. The weight fraction of the solids partitioned in each phase was calculated.

2.5. Infrared spectra

To understand the nature of the wetting phenomenon requires a knowledge of the chemical speciation and its variation with various treatments. For this purpose, IR spectra of fine solids were recorded using a Bruker IFS 66 Fourier transform IR spectrometer in the mid-IR region ($5000–500$ cm^{-1}). The surface-sensitive diffuse-reflectance sampling technique was used with dried KBr as background. The fine solid samples were mixed with KBr to maximize the optical sensitivity. To measure the species washed off by the solvent, the transmission IR spectra were obtained with a 3M IR card, onto which a drop of the washed supernatant was placed and dried under an ambient environment. In both cases, the spectra were obtained using a mercury cadmium telluride detector with 100 scans and presented with background corrections. For quantitative analysis, the characteristic IR band intensity was determined using the built-in fitting procedures in a commercial OPUS system supplied with the IFS 66 spectrometer.

3. RESULTS AND DISCUSSION

3.1. Surface and interfacial tensions

Figure 1 shows that the surface tension of the heptane–toluene mixture (diluent/air) decreases, whereas the interfacial tension of the diluent against water (γ_{wd}) increases with increasing heptane concentration. A nonlinear relationship between the

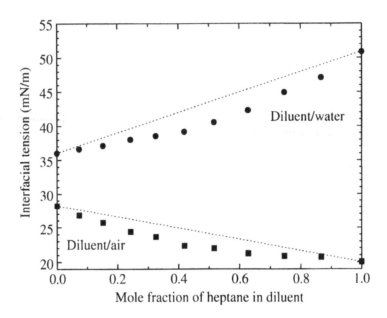

Figure 1. Surface and interfacial (against water) tensions as a function of the heptane concentration in the diluent.

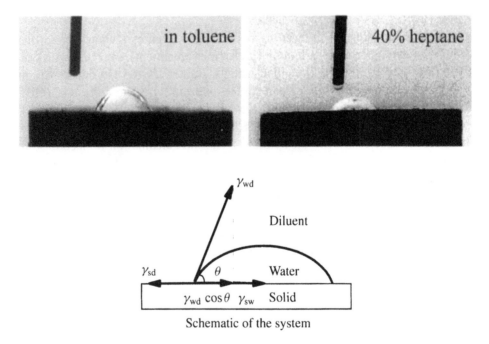

Figure 2. Definition (bottom) and variation (top) of the contact angle of water on fine solids in diluent of varying composition: fine solids washed with toluene six times and dried in an oven at 60°C for 72 h.

surface (interfacial) tension and heptane molar fraction of the diluent was observed, suggesting a system of non-ideal mixing with intermolecular interactions among the various species. The negative deviation of the measured surface (interfacial) tension from that expected for an ideal mixing system [dotted line given by $\gamma_{mix} = \gamma_A f_A + (1 - f_A)\gamma_B$, where f_A is the mole fraction of component A, and γ_{mix}, γ_A, and γ_B are the surface tensions of the mixture and components A and B, respectively] indicates preferred partition of the low surface energy component at the surface (interface). In this case, heptane tends to accumulate at the air–diluent interface, while toluene at the water–diluent interface. It is clear that the decreased polarity of the diluent with increasing heptane concentration weakened the interactions between the diluent and the interfacial water molecules, thus increasing the interfacial tension. This change in diluent polarity plays an important role in the froth treatment, which is illustrated later. From Young's equation applied to a bitumen froth treatment system (see Fig. 2: bottom) [10], i.e. $\cos\theta = (\gamma_{sd} - \gamma_{sw})/\gamma_{wd}$, an increase in the diluent–water interfacial tension (γ_{wd}) would mean a decreased water wettability, i.e. an increase in the contact angle of water on fine solids. This implies that solids should tend to remain in organic phases, which is inconsistent with the observed reduction of fine solids and water in the organic phase with increasing non-polar (paraffinic) nature of the diluent. Apparently, to understand the effect of diluent–water interfacial tension on the froth treatment, it is necessary to consider also the interactions of solid–diluent and solid–water, i.e. the term ($\gamma_{sd} - \gamma_{sw}$) in Young's equation.

3.2. Water contact angle

Typical variation of the water contact angle on tablets with the diluent composition is illustrated in Fig. 2 (top). In this example, tablets were prepared from particles further washed by toluene six times and dried. A significant decrease in the contact angle with the addition of heptane in the diluent (40% by volume) is seen.

The contact angle values measured with 'wet' tablets are shown in Fig. 3 (note: the word 'wet' here is rather qualitative). It was noted during the measurement that moisture on a solid surface had a significant impact on solid wettability and made accurate measurement of the contact angles difficult. Nevertheless, it is evident that the contact angle decreases with increasing mole fraction of heptane in the diluent. In air, water wets the solids with a near zero contact angle. These observations suggest that with increasing concentration of heptane in the diluent, the solids prefer to partition into an aqueous environment, which is consistent with the observed decrease of solids and hence the water content in the organic phase when the froth is treated with an alkane (paraffinic) diluent.

To examine the effect of washing on the wettability of fine solids, the extracted solids were further washed six times by either heptane or toluene, and the contact angle of water was measured on the wet tablets. A similar wetting characteristic was observed with and without this further heptane washing (Fig. 3), suggesting that heptane washing does not alter the surface wetting property of the extracted

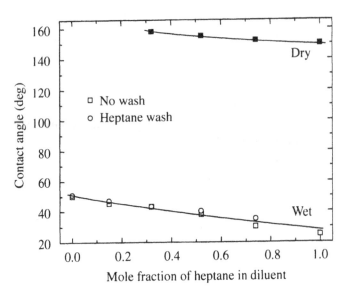

Figure 3. Effect of heptane washing and sample drying on the contact angles of water on fine solids in the diluent.

fine solids substantially, as expected by its poor dissolving power for surfactant and asphaltenes.

When the extracted sample was washed with toluene four times, on the other hand, water spread readily on the 'wet' solids, independently of the diluent composition (not shown). It is clear that the increased dissolving power of a strong solvent (toluene) removes the natural surfactant from the solid surfaces, resulting in an increased wettability by water. The practical implication of this finding is that a good diluent should have a strong dissolving power for surfactant and asphaltenes, to make fine solids hydrophilic and remain in the aqueous phase. This is probably one of the reasons why the diluent used in practice is usually a mixture of poor (paraffinic) and good (aromatic) solvent constituents. One of the functions of the good solvent component is intended to dissolve (remove) natural surfactant from the solid surface (which otherwise makes the surface hydrophobic). The paraffinic component, on the other hand, is to cause a low water contact angle in the diluent and hence a high wettability by water, as illustrated in Figs 2 and 3.

Also evident in Fig. 3 is that the contact angle of water on relatively dry tablets in the diluent is as high as *ca.* 150°, which is significantly greater than that on wet samples (*ca.* 45°). In this case, the water contact angle on solids decreased only marginally with increasing heptane concentration in the diluent. The significant increase in the water contact angle upon drying solids suggests that caution has to be exercised in sample preparation when studying the wettability of fine solids from oil sands in relation to understanding froth treatment performance, i.e. using wet samples is essential in this context. However, the observed change accompanying the sample drying process does shed light on the wetting mechanism. Whether

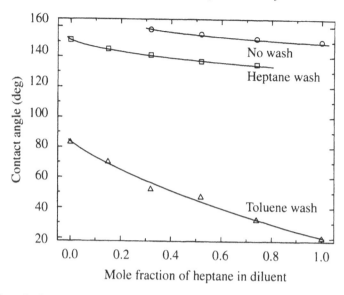

Figure 4. Effect of toluene washing on the contact angles of water on fine solids in the diluent (dried samples).

the observed change in contact angle upon sample drying is due to the removal of water from the solid surface or originates from the alternation of surface chemistry remains to be studied. It appears that the evaporation of moisture and solvent during drying caused a preferential orientation of the low-energy hydrocarbon chain towards the air, forming a closely packed assembly of the adsorbed surfactant molecules. As a result, an increased water contact angle is anticipated, which accounts for the observation in our experiments. This explanation is further supported by the irreversible nature of the drying process, as the dried tablet remained dry even after being dipped in water and the contact angle remained high and unchanged.

For toluene-washed particles, the drying process changed a solid from water- to oil-wettable with a contact angle value as high as 80°, decreasing with increasing heptane concentration in the diluent (Figs 2 and 4). Compared with the contact angles measured using particles after heptane washing, a much lower contact angle at a given diluent composition was measured in this case, confirming the removal of surface hydrophobic species by toluene washing. Also found was a greater diluent composition dependence of the contact angles (changed from 80° in toluene to 20° in heptane), in contrast to the particles washed with heptane. It appears that low surface coverage of the natural surfactant after toluene washing is responsible for the reduced contact angle with increasing heptane concentration arising from an increased penetration of the diluent into surfactant assemblies. The fact that the contact angle of water is smaller on toluene-washed than on heptane-washed samples may also suggest that after removal of hydrophobic species, part of the particle surface responds to the aromatic molecules in the diluent, showing a diluent polarity dependence of wettability. It is clear that the study using dried fine solid samples does provide insight into the role of diluent composition in the froth treatment, although it is less relevant to practice.

Another important observation derived from the above discussion is that the change in surface chemistry rather than the removal of moisture upon drying appears to determine the wettability of fine solids. This is seen as the moisture content in the solids would be at most kept the same, if not decreased, by toluene washing while a significant increase in the wettability of fine solids by water resulted. Of course, the effect of moisture on the wettability of fine solids should not be overlooked. A significant increase in the contact angle (from zero to ca. 80°) was obtained with toluene-washed fine solids after they were dried in air for 48 h (see Fig. 4), although the effect may be secondary or indirect. It is important to note that the effect of diluent composition on the froth treatment is probably more complex than what is presented here, although our measurements have accounted for what is generally observed in froth treatment practice. For example, the important effect of diluent composition on the formation of finely dispersed water droplets in diluted bitumen needs to be considered, in addition to the stabilization of the finely dispersed water droplets by bi-wettable fine solids.

3.3. Fine particle partition

A general observation from the partition experiments was that the water phase was always clear. As to the organic phase, heptane was clear with a light yellow color, in contrast to toluene which was turbid and dark brown in color, caused, probably, by dissolved bitumen. A stable interphase between the organic and aqueous phases, consisting of organics, water, and fine clays in a bi-continuous phase, was observed in all the partition tests.

The results from partition tests with 'wet' particle samples are shown in Fig. 5. It is clear that the fine solids were partitioned mainly between the organic phase and the interphase with a negligible amount in the aqueous phase. With a diluent of low heptane concentration (less than 40%), a majority of fine clays (ca. 80%) resided in the organic phase, suggesting that most particles are strongly hydrophobic and hence oil-wettable, with the rest being of moderate hydrophobicity and bi-wettable. In a more paraffinic diluent (heptane concentration higher than 80%), the partition between the organic phase and interphase reversed. The observed increase of fine particle partition to the interphase is consistent with a decreased contact angle of water on the particles with increasing heptane concentration in the diluent (Fig. 3). A wide range of surface hydrophobicity was seen in these tests.

For the dried particles, a similar variation of partition between the organic phase and the interphase was seen, but to a lesser extent (Fig. 6). This observation corresponded well to a marginal decrease in the contact angle of water with increasing heptane concentration. Compared with the wet samples, the partition to the organic phase increased only marginally unless a pure paraffinic diluent was used, even though a much greater contact angle was measured with dried powders. This finding may suggest the existence of a critical contact angle value (ca. 45°)

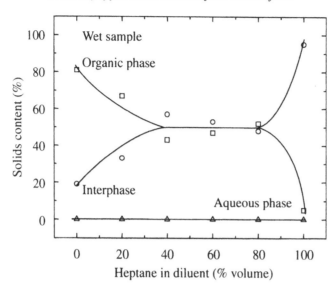

Figure 5. Effect of diluent composition on fine particle partition among the various phases ('wet' samples).

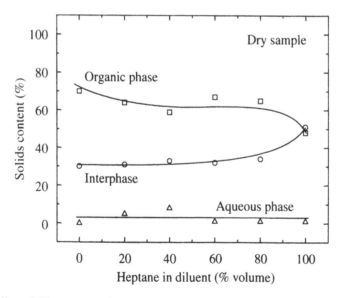

Figure 6. Effect of diluent composition on fine particle partition among the various phases ('dry' samples).

above which the particles respond similarly to partitioning. A similar phenomenon has been reported in the Pickering emulsion [11], although the critical value of the contact angle varied. In the Pickering emulsion, a step increase in the stablization of emulsions by fine solids was observed when the solids had a contact angle value greater than 65° [12]. This discrepancy between the two systems may be attributed to the difference in the physical phenomena concerned. In the Pickering emulsion,

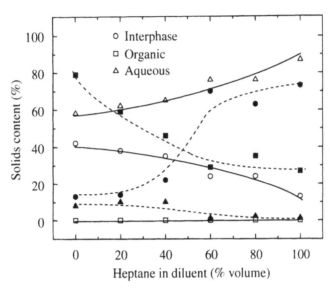

Figure 7. Effect of sample washing on fine particle partition among the various phases ('wet' samples): filled symbols — heptane wash; open symbols — toluene wash.

the accumulation of fine particles at the interface plays a critical role in stabilizing the emulsion. However, in the test of fine particle extraction as encountered here, the emphasis is placed on the migration of particles into the phase, which has little effect on the stabilization of emulsions. The driving force, measured merely by wettability, for particles to migrate from the organic to the aqueous phase may be significantly different from that in a reversed system, which appears to account for the observed differences. Nevertheless, the existence of the critical contact angle emphasizes the role of surface wettability in fine particle partition between the various phases and, therefore, in oil sands extraction practices.

The partition tests conducted with fine solids washed by heptane and toluene, in parallel to contact angle measurements, confirmed the effect of solvent washing on the wettability of fine solids. For heptane-washed particles, a similar partition curve to that without washing was obtained as anticipated from the wettability measurement, although the absolute value changed (Fig. 7). However, with toluene washing, a majority of the fine solids became hydrophilic (i.e. water-wettable), with a small amount being bi-wettable. This observation is consistent with the observed near-zero contact angle on particles in the diluent, independent of the diluent composition in this context. It is the removal of surface hydrophobic species by dissolution in toluene that is responsible for the inversion of partition from the organic to the aqueous phases. Clearly, hydrophobicity is the key to controlling the partition of fine solids to aqueous phases. The accumulation of bi-wettable fine solids at the water–diluent interface would stabilize the phase boundary, and hence the stability of finely dispersed water droplets, which is considered undesirable for bitumen froth treatment. The experimental findings suggest that the key to improving the removal of fine solids and water from the organic phases during bitumen froth treatment is to develop a method capable of converting a solid from being hydrophobic to hydrophilic by, for example, a diluent of controlled dissolving power and polarity.

3.4. IR spectra

The IR spectra of wet solids extracted from bitumen froth using heptane, with and without further washing by heptane or toluene, are shown in Fig. 8, and those corresponding to the dried solids in Fig. 9.

The IR spectra in both figures show similar general spectral features regardless of the treatments (drying or washing), except that the broad bands at 3300 cm^{-1} corresponding to hydrogen-bonded water diminished upon drying. Considering the extremely high sensitivity of the IR technique to water, the amount of water removed during drying is considered marginal, yet a significant impact of drying on the wettability of fine solids was observed. The strong, sharp peaks above 3600 cm^{-1} are indicative of the presence of inherent inner hydroxyls in fine solids [13, 14]. Therefore, the band at 3696 cm^{-1} was used as an internal standard in quantitative analysis of the spectra. Unambiguous assignments of the bands over 800–1600 cm^{-1} are difficult, if not impossible, due to the complex nature of bitumen. Nevertheless, the presence of carboxylic/carboxylate groups is supported by the bands at 1600 and 1450 cm^{-1} [15–18]. The bands at 1005 and 1030 cm^{-1} suggest the presence of sulfur-oxy species [18–20] and that at *ca.* 1110 cm^{-1} may be caused by Si—O vibrations [6]. Strong bands around 2900 cm^{-1} arise from anti-symmetric and symmetric stretching of CH$_3$ and CH$_2$ groups. The presence of these

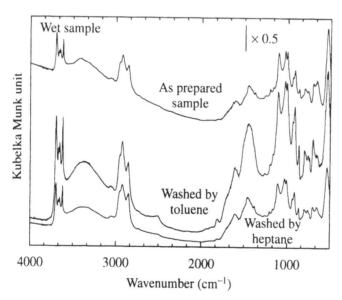

Figure 8. IR spectra of fine solids extracted from bitumen froth before and after washing by heptane and toluene ('wet' sample).

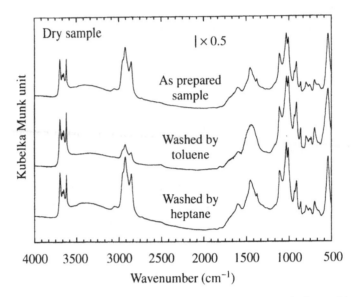

Figure 9. IR spectra of fine solids extracted from bitumen froth before and after washing by heptane and toluene ('dry' sample).

Table 2.
Relative intensity (ratio) of characteristic IR bands with respect to the band of internal OH group at 3696 cm^{-1}

| | Band position (cm^{-1}) | | | |
| | 2925 | | 3368 | |
Sample treatment:	Wet	Dry	Wet	Dry
As-prepared	0.72	1.14	0.16	0.06
Heptane-washed	0.77	1.11	0.15	0.05
Toluene-washed	0.19	0.28	0.17	0.03

bands along with those assigned to carboxylic and sulfur-oxy species confirms the contamination of the fine solids by natural surfactants.

The relative intensity of the bands assigned to CH$_2$ (at 2925 cm^{-1}) and adsorbed water (around 3368 cm^{-1}) with respect to the band of internal OH group (at 3696 cm^{-1}) was calculated for each spectrum and the results are given in Table 2. The removal of adsorbed water during sample drying is evident from the reduction in the relative band intensity at 3368 cm^{-1} from *ca.* 0.16 to 0.05. At this wavenumber the changes in the relative band intensity among the various treatments can be considered negligible. This observation is anticipated as both solvents used for sample washing are immiscible with water and should not have any effect on the removal of water. Also observed in Table 2 is that heptane washing did not change substantially the relative band intensity for hydrocarbons, while a significant reduction from 0.75 to 0.19 was observed with toluene washing. A similar trend was observed for the dried samples, although the absolute ratio changed. Clearly, the toluene wash removed more than 75% of adsorbed hydrocarbon species, mostly as surfactant and/or asphaltenes, from the fine solid surfaces. The removal of hydrocarbons results in a drastic increase in the wettability as described earlier in this paper.

The removal of organic species by washing with various solvents was confirmed by applying IR spectroscopy to the supernatant. In solid-free heptane supernatants, no additional species were detected. However, in the supernatant from toluene washing, the bands attributed to carboxylic and sulfoxy groups associated with surfactant molecules were observed, confirming the removal of organic species as surfactants from the surface. The fact that various treatments did not change the spectral features in spite of the removal of organic species from the surface suggests the contamination of fine solids by natural surfactants in bitumen. These contaminated fine solids are strongly hydrophobic and bi-wettable. They tend to accumulate at the aqueous–diluent interface, stabilizing finely dispersed water droplets in the oil phase and contributing to an increased water content and fine solids in the treated froth. Our study therefore suggests that a practical solution to minimize the water and fine solids contents in the treated bitumen froth is to decontaminate the fine solids, converting them from bi-wettable to water-wettable only. One of the approaches to this is to consider the dissolving power when formulating a diluent.

4. CONCLUSIONS

From this study, the following conclusions can be made:

(1) The wettability of water on fine solids in bitumen froth increased with increasing alkane (paraffinic) components in the organic phase, resulting in an increased partition of the solids into the interphase region. Weaker interactions of apolar molecules with solids compared with that of polar molecules seems to be responsible.

(2) The wettability of water on fine solids remained unchanged by washing the solids with heptane, but increased significantly by toluene washing. An inversion in partition from the organic to the aqueous phase was observed in the latter case. The strong dissolving power of toluene for organic matter (surfactant) made fine solids more water-wettable by removing the adsorbed natural surfactant from fine solids.

(3) A significant decrease in wettability with corresponding changes in the partition characteristics was measured with dried particles. The drying process caused an irreversible change in surface wettability, suggesting a preferential orientation of low-energy hydrocarbon chains towards the air with an enhanced assembly of the surfactant on the solid.

(4) A general correlation exists between the wettability of fine solids and their partition among the various phases. Accumulation of bi-wettable fine solids contributes to the stabilization of finely dispersed water droplets in the oil phase, which are responsible for the entrainment of water and fine solids in the froth treatment product. An ideal diluent should possess a dissolving power for surfactant while keeping an alkane nature. A strong solvent component is to decontaminate the solids from natural surfactant, while an alkane (paraffinic) component is to enhance the interactions of water with the solids, which minimizes both the solids and the water content in the organic phase during bitumen froth treatment.

Acknowledgements

Financial support for this work by an NSERC-Industrial Postgraduate Student Scholarship with Syncrude Canada Ltd. (F. Chen) and an NSERC-Research Grant (Z. Xu) is gratefully acknowledged.

REFERENCES

1. L. S. Kotlyar, B. D. Sparks, J. R. Woods, S. Raymond, Y. Le Page and W. Shelfantook, *Pet. Sci. Technol.* **16**, 1 (1998).
2. L. S. Kotlyar, B. D. Sparks, K. Deslandes and R. Schutte, *Fuel Sci. Technol. Int.* **12**, 923 (1994).
3. S. Levine and E. Sanford, *Can. J. Chem. Eng.* **63**, 258 (1985).
4. N. Yan and J. H. Masliyah, *Ind. Eng. Chem. Res.* **36**, 1122 (1997).
5. J. K. Spelt and E. I. Vargha-Bulter, in: *Applied Surface Thermodynamics*, A. W. Neumann and J. K. Spelt (Eds), p. 379. Marcel Dekker, New York (1996).
6. *The Aldrich Library of FT-IR Spectra*, 1st edn, C. P. Pouchet (Ed.). Aldrich Chemical Company, Milwaukee, Wisconsin (1985).
7. K. Suga and J. F. Rusling, *Langmuir* **9**, 3649 (1993).
8. C. J. van Oss, *Interfacial Forces in Aqueous Media*, p. 171. Marcel Dekker, New York (1994).
9. J. Donhue and F. E. Bertell, *J. Phys. Chem.* **56**, 480 (1952).
10. R. E. Johnson, Jr. and R. H. Dettre, in: *Wettability*, J. C. Berg (Ed.), p. 57. Marcel Dekker, New York (1993).
11. S. U. Pickering, *J. Chem. Soc.* **91**, 2001 (1907).
12. N. Yan and J. H. Masliyah, *Colloids Surfaces A* **96**, 229 (1995).
13. R. L. Frost and A. M. Vassallo, *Clays Clay Miner.* **44**, 635 (1996).
14. C. T. Johnston, S. F. Agnew and D. L. Bish, *Clays Clay Miner.* **38**, 573 (1990).
15. S. H. Pine, J. B. Hendrickson, D. J. Cram and G. S. Hammond, *Organic Chemistry*, p. 163. McGraw-Hill, New York (1980).
16. D. H. Williams and I. Fleming, *Spectroscopic Methods in Organic Chemistry*, 5th edn, p. 49. McGraw-Hill, London (1995).
17. K. H. Rao, K. S. E. Forssberg and W. Forsling, *Colloids Surfaces A* **133**, 107 (1998).
18. T. M. Ignasiak, Q. Zhang, B. Kratochvil, C. Maitra, D. S. Montgomery and O. P. Strausz, *AOSTRA J. Res.* **2**, 21 (1985).
19. L. G. Hepler and R. G. Smith, in: *AOSTRA Technical Publication Series #14*, p. 129. Alberta Oil Sands Technology and Research Authority, Edmonton, Canada (1994).
20. L. H. Ali, *Fuel* **57**, 357 (1978).

Apparent and Microscopic Contact Angles, pp. 405–417
J. Drelich, J. S. Laskowski and K. L. Mittal (Eds)
© VSP 2000.

Modification of calcium carbonate surface properties: macroscopic and microscopic investigations

M. CHAMEROIS *, M. FRANÇOIS, F. VILLIÉRAS and J. YVON

Laboratoire Environnement et Minéralurgie, INPL ENSG and CNRS UMR 7569, BP 40, 54501 Vandoeuvre-lès-Nancy Cedex, France

Received in final form 21 April 1999

Abstract—Understanding the wettability of mineral powders and its modification by the addition of various surface active agents is crucial for many industrial applications. In many cases, wettability is investigated by macroscopic characterization techniques. In this framework, we decided to study non-porous calcium carbonate powders coated by known amounts of water-repellent molecules. A detailed characterization of the interface, focusing on the analysis of surface heterogeneity, was carried out using water vapor, nitrogen, and argon adsorption. It clearly reveals specific adsorption sites for water-repellent molecules. Wettability and immersion enthalpy measurements show that saturation of the carbonate surface by water-repellent molecules is not necessary for obtaining maximum hydrophobicity. It is reached for approximately one-third of surface saturation; at that point, some high-energy surface sites are still available for water and nitrogen adsorption. This suggests that wettability is not only linked to the availability of surface sites for water molecules.

Key words: Calcium carbonate; wettability; water-repellent molecules; gas adsorption.

1. INTRODUCTION

Surface modification of mineral substances with adsorbed organic layers is widely used in various industrial processes, ranging from the polymer industry, where it facilitates filler incorporation in polymer composites [1], to ore flotation, where it permits selectivity between the ore and the matrix [2]. New promising applications, such as thin coating for sensors or optical devices, have also been reviewed recently by Swalen *et al.* [3]. The surface properties induced by adsorbed organic layers are usually investigated on flat surfaces with contact angle or surface force measurements. However, in the case of powders, these methods are inappropriate. Wettability of finely divided solids is then assessed by contact angle measurements on pellets, or by the capillary rise technique [4, 5]. In this paper, we develop

---

*To whom correspondence should be addressed. E-mail: chameroi@ensg.u-nancy.fr

an alternative approach based on the exploitation of high-resolution adsorption isotherms. This technique, primarily applicable to high surface area powders, is illustrated by using a calcium carbonate sample coated with increasing amounts of water-repellent molecules. By combining various experimental techniques, results obtained at the solid–liquid interface, wetting index and immersion enthalpy, are related to those derived from an analysis of nitrogen, argon, and water vapor adsorption at the solid–gas interface.

2. EXPERIMENTAL

2.1. Materials

The calcium carbonate used in this study was a calcite powder obtained by precipitation and commercialized by Solvay under the trade name Socal 31. Chemical analysis, Fourier Transform infrared spectroscopy, and X-ray diffraction investigations revealed that this material was pure. Scanning electron microscope (SEM) observation showed that elementary particles were globular with an average particle size of 0.3 μm.

The nitrogen adsorption at 77 K yielded a BET specific surface area of 22.5 m^2 g^{-1} and showed that the calcite powder was not porous.

The water-repellent molecule (WRM) was an organosiloxane molecule, supplied by Rhône-Poulenc. Its structure is similar to the superwetter surfactants [6]: a trisiloxane hydrophobe head with a hydrophilic group grafted on the central silicon atom. Whereas the superwetter hydrophilic group is a poly(oxyethylene) group of various lengths, the WRM hydrophilic group is a short hydrocarbon tail (three carbons) ending in a succinic anhydride function (Fig. 1). Cyclohexane (Aldrich) and isopropanol (Fisher Scientific) used in this study were of high purity grade (> 99%).

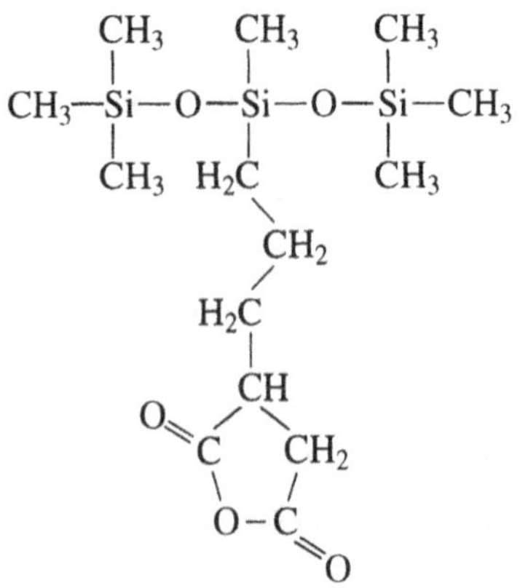

Figure 1. Molecular structure of the WRM.

2.2. Adsorption at the solid–liquid interface

The adsorption isotherms of water-repellent molecules from cyclohexane onto calcite were determined using a batch equilibration technique: 4 g of calcium carbonate was contacted with 120 ml of cyclohexane containing increasing quantities of adsorbate (from 0 to 4 g l^{-1}). Prior to the experiments, solid samples were equilibrated with 0.75 water vapor relative pressure using NaCl-saturated solution at 298 K [7]. The solid/liquid mixtures were stirred (rotary shaker — 20 rpm) for 24 h at 303 K. Equilibrium concentrations were determined after centrifugation (30 min at 15 200 g) from the height of the 755 cm^{-1} band relative to the methyl rocking vibration of the Si$-$(CH$_3$)$_3$ group [8] by transmission infrared spectroscopy (IFS 55 Brucker Fourier Transform infrared spectrometer). The adsorbed quantities (μmol m^{-2}) were then calculated by the following equation:

$$Q_{WRM} = V(C_i - C_e)/m\,S,$$

where V is the volume of solution (in ml); C_i and C_e are the initial and equilibrium concentrations (in mmol l^{-1}), respectively; m is the mass of powder (in g); and S is the specific surface area (22.5 m^2 g^{-1}). WRM-coated samples dried at room temperature were collected for further experiments.

A blank sample, referred to as 'bare' hereafter, was also obtained following the same operations (stirring, centrifugation, drying) in pure cyclohexane solution.

2.3. Wettability

The determination of a wetting index is based on the observation of powder behavior at a liquid–air interface. Coated powders are sprinkled onto the surfaces of liquids with different surface tensions. The liquids used were obtained by mixing isopropanol and water; the surface tension (γ) of the mixture is given by the relation $\gamma = 68.97 - 11.34 \ln(v)$, where v is the % volume of isopropanol [9]. The wetting index is then located in the interval between the surface tension of the liquid in which the particles sink and that of the liquid on which the particles float. The value obtained can be considered the critical wetting surface tension of the particles [10]. This method can only be considered as a guideline for powder wettability estimation as it suffers from a number of limitations, especially preferential adsorption of one component of the liquid mixture, as generaly admitted.

2.4. Immersion microcalorimetry

The heat released from immersion experiments was measured with a Calvet differential calorimeter (Setaram MS70). About 250 mg of powder was placed into a glass bulb closed with a brittle end. After outgassing at room temperature under a residual pressure of 0.1 Pa, the glass bulb was sealed and placed into a calorimetric cell filled with deionized water. When thermal equilibrium at 303 K was reached, the brittle end was broken and the water entered into the bulb and wet the solid.

The resulting heat was recorded as a function of time. The integration of this signal is proportional to the total heat exchanged. After experimental corrections (broken end heat, liquid vaporization), the immersion enthalpy, ΔH (mJ g^{-1}) was obtained. Taking into account the specific surface area of the bare sample, the results were then normalized to mJ m^{-2}.

2.5. Adsorption at the solid–gas interface

2.5.1. Nitrogen and argon adsorption.   Low-pressure isotherms of argon and nitrogen at 77 K were recorded on a laboratory-built automatic quasi-equilibrium volumetric set-up [11]. About 1 g of powder was outgassed under a residual pressure of 0.001 Pa at 343 K. A slow, constant, and continuous flow of the adsorbate was then introduced into the adsorption cell through a microleak. By recording the quasi-equilibrium pressure vs. time, high-resolution adsorption isotherms were obtained with more than 2000 data points for the filling of the first layer. A derivative curve of the adsorption isotherm can then be obtained as a function of the logarithm of relative pressure. As the logarithm of relative pressure is an energy scale, graduated in kT units, the derivative curve can be considered a footprint of the surface energetic heterogeneity for a given gas and can then be used to study surface energetic heterogeneity modifications due to WRM adsorption.

2.5.2. Water vapor adsorption.   The adsorption isotherms of water vapor were recorded at 303 K using an automatic gravimetric apparatus based on a quasi-equilibrium adsorption procedure. This homemade apparatus was built around a Setaram MTB 10-8 Symmetrical microbalance and a Texas fused silica Bourdon tube automatic gauge [12]. About 700 mg of powder was outgassed for 18 h at 303 K under a residual pressure of 0.1 Pa. Water was supplied from a source kept at 314 K through a Granville-Phillips leak valve to ensure quasi-equilibrium at all times. The adsorption isotherms (i.e. mass adsorbed at 303 K vs. quasi-equilibrium pressure) were directly recorded on a simple $X–Y$ recorder.

3. RESULTS

3.1. Adsorption of water-repellent molecules

Figure 2 presents the adsorption isotherm (adsorbed quantity vs. equilibrium concentration) of the water-repellent molecules on calcium carbonate. The adsorbed quantity increases sharply at low equilibrium concentration, up to 3 μmol m^{-2}, suggesting a very high affinity for the non-porous adsorbent. A plateau is then reached at 4.3 μmol m^{-2}, which seems to correspond to the saturation of the surface.

Complementary infrared measurements on equilibrium solutions showed that part of the WRM was hydrolysed into diacid (unpublished results). This hydrolysis probably results from reaction between anhydride end and water molecules of the

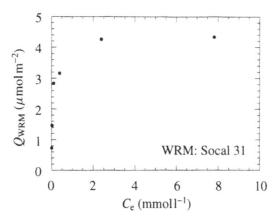

Figure 2. Adsorption isotherm of WRMs from cyclohexane on calcium carbonate at 303 K.

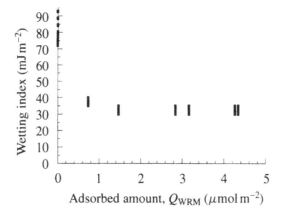

Figure 3. Wetting index of calcium carbonate samples vs. the adsorbed amount of WRMs.

calcite surface. It can thus be assumed that in the first step of adsorption, hydrolysed WRMs presenting a double carboxylic end react strongly with surface calcium ions, which explains the strong affinity of WRMs for calcite suggested by the vertical step at low concentration.

3.2. Wetting index

Figure 3 shows the evolution of the wetting index as a function of the adsorbed amount of WRMs. The wetting index for bare mineral ($Q_{WRM} = 0$) is greater than 72 mJ m^{-2}, which implies that the solid sinks in pure water. A small increase in the adsorbed amount yields a sharp decrease in the index. Above 1.5 μmol m^{-2} of adsorbed molecules, the wetting index ranges between 30 and 35 mJ m^{-2}. Maximum hydrophobicity, as determined by this experiment, is therefore reached before saturation of the surface. In other words, above 1.5 μmol m^{-2} of the adsorbed WRMs, additional WRMs do not induce any further change in the wetting index.

3.3. Immersion microcalorimetry

The evolution of the heat of immersion as a function of the adsorbed amount is shown in Fig. 4. The heat of immersion of the bare mineral is about 350 mJ m^{-2}. This relatively high value is in agreement with the insoluble inorganic salt nature of calcite [13]. The adsorbed WRMs induce a decrease in the immersion heat which agrees with the wetting index determinations. An apparent minimum is then reached for 1.5 μmol m^{-2} adsorbed amount. It is worth noting that this minimum is at about 240 mJ m^{-2}, which is surprisingly high for a hydrophobic powder floating on the surface of a 35 mJ m^{-2} surface tension liquid. Above 1.5 μmol m^{-2}, a slight increase of the immersion heat can be observed which reaches a value of 280 mJ m^{-2} at maximum adsorbed amount of WRMs.

The immersion heat of a solid by water can be divided into two components: (i) adsorption of water vapor molecules and (ii) wetting [14]. In the case of perfect wetting (cos Θ = 1), the heat released during the last step, related to the specific surface area of the powder, is equal to the water surface enthalpy. At 303 K, the corresponding value is derived from water surface tension data given in ref. [7]:

$$\gamma_w - T\frac{d\gamma_w}{dT} = 119 \text{ mJ m}^{-2}.$$

This component is decreased in the case of partial wetting. The fact that the immersion enthalpy, even at its minimum, is twice as high as 119 mJ m^{-2} shows that adsorption of water vapor still occurs. In order to estimate the relative contribution of the two components, the procedure described below can be followed. After being outgassed, the powder is equilibrated with water vapor at high relative pressure (> 0.8) before being placed into the calorimeter. Thus, the heat released is only related to the wetting component [15, 16]. In the case of the 4.3 μmol m^{-2} coated samples, this yields a heat of immersion close to zero (less than 10 mJ m^{-2}). This confirms that wetting of these coated powders is very low; it also indicates that

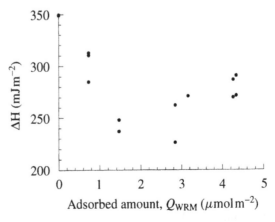

Figure 4. Immersion enthalpy of calcium carbonate samples vs. the adsorbed amount of WRMs.

the high heat of immersion values are mainly due to the water vapor adsorption phenomenon even when WRMs are adsorbed on the surface.

3.4. Argon and nitrogen adsorption

The derivative curves of argon and nitrogen adsorption are shown in Fig. 5. For the bare mineral sample, significant differences between the two gases can be observed: argon reveals two main adsorption peaks, one at high energy ($\ln P/P_0 = -7.5$) and the other at medium energy ($\ln P/P_0 = -4$). Nitrogen adsorption begins at higher energy (peak at $\ln P/P_0 = -13$) with no well-defined peak at lower energy.

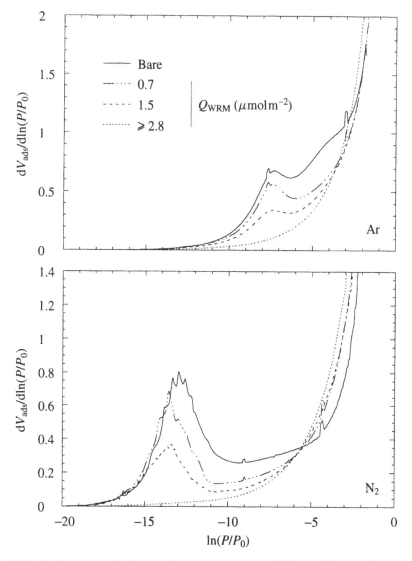

Figure 5. Derivative argon and nitrogen adsorption isotherms at 77 K on calcium carbonate samples coated with different WRM amounts.

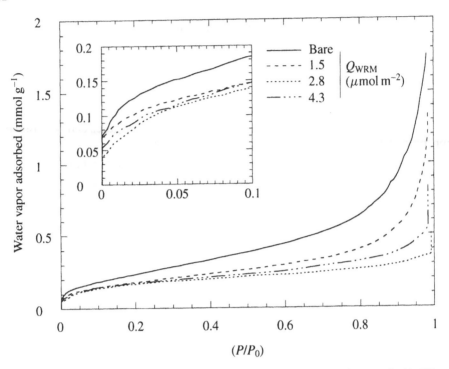

Figure 6. Adsorption isotherms of H_2O at 303 K on calcium carbonate samples coated with different WRM amounts.

Such differences between the nitrogen and argon adsorption isotherms are often observed and are assigned to stronger interactions between some specific surface sites of adsorbents and the nitrogen molecules [17, 18].

The adsorption of WRMs obviously blocks some of the adsorption sites. For argon adsorption, increasing amounts of WRMs affect first medium-energy sites and then high-energy sites. For nitrogen adsorption, the lower energetic part of the high-energy domain first disappears. Finally, for both gases, adsorbed amounts higher than 2.8 μmol m^{-2} lead to similar exponential shapes of the derivative curves which are typical of low-energy surfaces. This suggests that monolayer surface coverage by the WRMs is reached between 1.5 and 2.8 μmol m^{-2}. This result agrees with the packing area of the superwetter trisiloxane group obtained by Ananthapadmanabhan et al. [6], i.e. 70 Å^2, since the theoretical monolayer capacity of 70 Å^2 molecules is 2.4 μmol m^{-2}.

3.5. Water vapor adsorption

The adsorption of water vapor on the bare and precovered surfaces is shown in Fig. 6. All curves exhibit the same shape until P/P_0 around 0.6: at low pressure, a vertical step, followed by a knee-bend, is first observed. The adsorbed amounts then increase approximately linearly with the water vapor pressure until a relative pressure of about 0.6. For higher water vapor pressures, the bare sample

Table 1.
BET parameters and slope values from H_2O adsorption isotherms on water-repellent molecule (WRM)-coated samples

$Q_{WRM}{}^{a}$ (μmol m^{-2})	C^{b}	$V_m{}^{c}$ (cm^3 g^{-1})	Slope value^{d} (mmol g^{-1})
0	37	4.61	0.49
1.5	51	3.48	0.27
2.8	47	3.31	0.15
4.3	74	3.28	0.19

^{a} Adsorbed amount of WRMs.
^{b} BET energetic constant.
^{c} BET monolayer capacity.
^{d} Slope of the H_2O isotherm in the 0.3–0.5 relative pressure range.

shows a quasi-infinite adsorption, corresponding to capillary condensation between particles. Capillary condensation is reduced for the sample coated at 1.5 μmol m^{-2}, and is absent for higher WRM adsorbed amounts as the adsorption curves feature vertical steps near saturation. Thus, with the increase of the WRM amount, the shape of the water vapor adsorption isotherms shifts from type II to type VI, according to the Brunauer classification, extended by Adamson [19].

From a quantitative point of view, the differences between the adsorption isotherms are the following:

- at the end of the first vertical step, the adsorbed water quantities are about 0.07 mmol g^{-1} for both bare and 1.5 μmol m^{-2} samples and decrease to about 0.04 and 0.055 mmol g^{-1} for the 2.8 and 4.3 μmol m^{-2} samples, respectively;

- the apparent monolayer volume (V_m) calculated from the BET procedure decreases from 4.61 cm^3 g^{-1} (bare sample) to 3.28 cm^3 g^{-1} in the case of maximum coverage of water-repellent molecules (Table 1);

- the slopes of the linear portions calculated in the relative pressure range 0.3–0.5 decrease sharply with WRM adsorption (Table 1). For instance, this slope is 2.5 times lower for the 2.8 μmol m^{-2} coated sample than for the bare mineral.

The main effect from WRM adsorption then appears to be the reduction of multilayer formation. Indeed, the first statistical layer is completed in the same relative pressure range (0.1–0.15) for all samples, whereas the second layer is filled at very different relative pressures: 0.55 for bare mineral, 0.64 for 1.5 μmol m^{-2}, 0.86 for 2.8 μmol m^{-2}, and 0.72 for 4.3 μmol m^{-2} coated samples.

4. DISCUSSION

The changes in the wetting properties due to the adsorption of any water-repellent molecule are generally used for determining the optimal surface treatement conditions. In the case of powders, the classical techniques used for the determination of contact angles cannot be applied. The analysis of wetting properties by the powder sinking method as used to assess the wetting index or immersion microcalorimetry shows that a macroscopic optimum is not necessarily linked to complete coverage of the surface by the treating molecules. The powder sinking method gives access only to partial information on the surface wetting properties, due to the fact that it emphasizes the wetting phenomenon corresponding to the advancing contact angle. It is thus very sensitive to the presence of hydrophobic groups on the surface. Hydrophilic sites are most frequently assessed by the use of the receding contact angle technique. Unfortunately, such a technique does not exist for powders. However, the microcalorimetry technique reveals high water-surface interaction energy sites which are necessarily hydrophilic. Therefore, there is no need for a receding contact angle technique. Furthermore, gas adsorption techniques reveal that the optimal hydrophobicity deduced from wetting indices and heats of immersion corresponds only to submonolayer surface coverage. In mineral processing, such a partial surface coverage, leading to maximum hydrophobicity, is sufficient for ore flotation. In other cases, such as surface protection (i.e. water-repellent treatment), complete surface coverage is required to ensure stability and efficiency of the protection film. It is then necessary to use techniques adapted to the determination of the precise conditions leading to complete monolayer adsorption.

4.1. Interactions between coated samples and gases at low pressure

The nitrogen and argon surface energy distributions of samples coated with different amounts of WRMs reveal a stepwise disappearance of carbonate surface sites. The energetic domains first affected by WRM adsorption are not the most energetic ones for either argon or nitrogen. This means that WRMs do not probe the same energy distribution as argon and nitrogen. At high WRM coverage ($\geqslant 2.8\ \mu\text{mol m}^{-2}$), the derivative curves exhibit a shape typical for low-energy surfaces which reveals complete hindrance of the solid surface sites. It can then be assumed that both nitrogen and argon adsorb on WRMs with low-energy interactions.

In the case of water, low pressure adsorption isotherms do not exhibit a type V shape classically assigned to low-energy surfaces. Furthermore, the BET water monolayer capacity is only slightly diminished. Such behavior, different from what would be expected for a hydrophobic powder, has already been observed by Zettlemoyer and Sing [20] in the case of fully hydroxylated silica treated with hexamethyldisilazane. In the system calcium carbonate/WRM, high-energy adsorption sites are still observed for water molecules which correlates with the high immersion enthalpy values measured. The peculiar behavior of water, compared with argon and nitrogen, is probably due to the high temperature in water vapor adsorption experiments. Indeed, thermal agitation of the WRMs at 303 K may allow water molecules to diffuse through the adsorbed layer. On the contrary, at 77 K, WRMs remain stationary on the surface and thus prevent access to argon and nitrogen molecules. The fact that water molecules can diffuse through the adsorbed layer suggests that attractive interactions can take place on the surface.

The adsorption sites for these water molecules can be located on the non-covered mineral surface and/or on some special sites of the adsorbed WRMs. In the first stage of WRM adsorption, up to monolayer completion, the adsorption of increasing amounts of WRMs on the surface induces a continuous decrease in water monolayer capacity, which suggests that water molecules interact less and less with the calcium carbonate surface. However, the high water monolayer capacity indicates that adsorption sites are still present on the calcite surface even in the presence of adsorbed WRMs. This is consistent with the results published by Mielczarski *et al.* [21]. In the case of oleate adsorption on the surface calcium of apatite, these authors showed that water molecules were present on the surface, in the vicinity of carboxylic hydrophilic heads.

Once the WRM monolayer is reached, i.e. the adsorbed amount is higher than 2.8 μmol m^{-2}, the water monolayer capacity remains roughly constant (Table 1), whereas the C BET energetic constant increases (Table 1) together with the immersion enthalpy values (Fig. 4). This indicates that new high-energy sites appear for water that must be assigned to water–WRM interactions. The exact structure of the corresponding adsorbed film remains unknown at present, but these additional adsorption sites for water, above the WRM monolayer, could be assigned to the presence of free carboxylic heads on the surface.

4.2. Effect of WRM adsorption on wetting properties

The wetting indices and immersion enthalpies reveal a maximum macroscopic hydrophobicity for the 1.5 μmol m^{-2} coated sample, i.e. before completion of a WRM monolayer as demonstrated by argon and nitrogen experiments. The water vapor adsorption isotherms show that the wetting properties still change after this maximum. For instance, the decrease in adsorbed water amounts at high water vapor relative pressure induced by a higher adsorbed amount of WRM coating indicates that the reduction in multilayer adsorption and capillary condensation of water is not maximal when 1.5 μmol m^{-2} of WRMs is adsorbed.

The decrease in the slope values (Table 1) can be related to the cluster theory proposed by Zettlemoyer [22]: water molecules first adsorb on some hydrophilic sites and further adsorption occurs around them, through hydrogen bonding, to form clusters. This phenomenon has already been observed in the case of water adsorption on the basal faces of talc [23] and has been proposed to explain water adsorption on activated carbons [24, 25]. The slope in the multilayer adsorption region then depends on the number of hydrophilic sites present on the surface. The adsorption of WRMs on the carbonate surface decreases the number of such sites and induces the decrease in the slope after water monolayer coverage.

In addition, capillary condensation between particles should be related to the direct interactions with the surface of WRMs. This means that droplets of bulk water will have to be formed on this surface. In the case of a hydrophilic surface, capillary condensation occurs before water vapor saturation pressure, depending on the pore size, as described by the Kelvin law. In the case of very hydrophobic surfaces, condensation in pores occurs at a higher pressure than for hydrophilic ones, due to the increase in the solid–liquid–vapor contact angle. This is particularly the case for the 2.8 and 4.3 μmol m^{-2} coated samples, for which the contact angle should be high. Condensation in mesopores occurs as a vertical step close to the saturation pressure, meaning that a finite water film is in equilibrium with macroscopic water droplets on the surface [26].

As discussed in the previous section, free carboxylic groups, which can be considered hydrophilic sites, appear for the 4.3 μmol m^{-2} coated sample. As a consequence, both the slope value of the linear portion of the adsorption isotherm and the condensation at high water vapor relative pressure increase, and the 4.3 μmol m^{-2} coated sample appears less hydrophobic than the 2.8 μmol m^{-2} coated one.

5. CONCLUSION

The study of a carbonate modified by WRM adsorption shows that the degree of hydrophobicity depends on the method used to investigate the surface energetic properties. The WRM coverage needed to obtain the maximum hydrophobicity is lower for macroscopic methods, such as wetting indices or immersion enthalpies, than for methods based on molecular probe adsorption at the solid–gas interface. Monolayer WRM coverage can be assessed by nitrogen and argon high-resolution adsorption and its effect on local wetting properties can be derived from water vapor adsorption isotherms. This last technique shows that the main effect of adsorbed WRMs is the reduction in the formation of water multilayers, while the adsorption of the first layer of water appears only slightly affected.

Acknowledgement

We are grateful to Rhône-Poulenc for financial support. During his PhD, M.C. has been financed by ANRT/CIFRE convention No. 033/96. We are indebted to Dr. L. J. Michot, Dr. B. S. Lartiges, and Professor J. L. Bersillon for helpful discussions.

REFERENCES

1. E. P. Plueddemann, *Silane Coupling Agents*. Plenum Press, New York (1982).
2. A. M. Gaudin, *Flotation*. McGraw-Hill, New York (1957).
3. J. D. Swalen, D. L. Allara, J. D. Andrade, E. A. Chandross, S. Garoff, J. Isrealachvili, T. J. McCarthy, R. Murray, R. F. Pease, J. F. Rabolt, K. J. Wynne and H. Yu, *Langmuir* **3**, 932–950 (1986).
4. A. W. Neuman and R. J. Good, in: *Colloid and Surface Science*, R. J. Good and R. R. Stromberg (Eds), Vol. 11, pp. 31–91. Plenum Press, New York (1979).
5. G. Buckton, in: *Contact Angle, Wettability and Adhesion*, K. L. Mittal (Ed.), pp. 437–451. VSP, Utrecht, The Netherlands (1993).

6. K. P. Ananthapadmanabhan, E. D. Goddard and P. Chandar, *Colloids Surfaces* **44**, 281–297 (1990).
7. D. R. Lide (Ed.), *CRC Handbook of Chemistry and Physics*, 74th edn. CRC Press, Boca Raton, FL (1993).
8. C. N. R. Rao, *Chemical Applications of Infrared Spectroscopy*. Academic Press, New York (1963).
9. P. Stevens, L. Gypen and R. Jennen-Bartholomeussen, *Farm. Tijdschr. Belg.* **51**, 2 (1974).
10. J. Diao and D. W. Fuerstenau, *Colloids Surfaces* **60**, 145–160 (1991).
11. F. Villiéras, J. M. Cases, M. François, L. J. Michot and F. Thomas, *Langmuir* **8**, 1789–1795 (1992).
12. J. E. Poirier, M. François, L. J. Michot and J. M. Cases, in: *Fundamentals of Adsorption*, A. T. Liapas (Ed.), pp. 473–482. AIChE, New York (1987).
13. J. J. Chessick and A. C. Zettlemoyer, *Adv. Catal.* **11**, 263–299 (1959).
14. W. D. Harkins and G. Jura, *J. Am. Chem. Soc.* **66**, 919–927 (1944).
15. S. Partyka, F. Rouquerol and J. Rouquerol, *J. Colloid Interface Sci.* **68**, 21–31 (1979).
16. J. M. Cases and M. François, *Agronomie* **2**, 931–938 (1982).
17. F. Bardot, F. Villiéras, L. J. Michot, M. François and J. M. Cases, *J. Dispersion Sci. Technol.* **19**, 739–759 (1998).
18. F. Villiéras, L. J. Michot, E. Bernardy, M. Chamerois, C. Legens, G. Gérard and J. M. Cases, *Colloids Surfaces A* **146**, 163–174 (1999).
19. A. W. Adamson, *Physical Chemistry of Surfaces*, 5th edn. Wiley, New York (1990).
20. A. C. Zettlemoyer and H. H. Sing, in: *Colloid and Interface Science*, M. Kerker, A. C. Zettlemoyer and R. L. Rowell (Eds), Vol. I, pp. 279–290. Academic Press, New York (1977).
21. J. A. Mielczarski, J. M. Cases, P. Tekely and D. Canet, *Langmuir* **9**, 3357–3370 (1993).
22. A. C. Zettlemoyer, *J. Colloid Interface Sci.* **28**, 343–369 (1968).
23. L. J. Michot, F. Villiéras, M. François, J. Yvon, R. Le Dred and J. M. Cases, *Langmuir* **10**, 3765–3773 (1994).
24. H. F. Stoeckli, F. Kraehenbuehl and D. Morel, *Carbon* **21**, 589–591 (1983).
25. M. M. Dubinin and V. V. Serpinsky, *Carbon* **19**, 402–403 (1981).
26. N. V. Churaev and Z. M. Zorin, *Adv. Colloid Interface Sci.* **40**, 109–146 (1992).

Apparent and Microscopic Contact Angles, pp. 419–430
J. Drelich, J. S. Laskowski and K. L. Mittal (Eds)
© VSP 2000.

Factors influencing contact angle measurements on wood particles by column wicking

MAGNUS E. P. WÅLINDER [1],* and DOUGLAS J. GARDNER [2]

[1] *Wood Technology and Processing, KTH — Royal Institute of Technology, SE-100 44, Stockholm, Sweden*
[2] *University of Maine, Advanced Engineered Wood Composites Center, Orono, ME 04469, USA*

Received in final form 16 February 1999

Abstract—The present work focuses on a capillary rise technique, referred to here as column wicking, for determining contact angles on wood particles. The liquid front rise versus time for different probe liquids has been measured for extracted and non-extracted spruce wood particles packed into glass columns. Wood is a porous, heterogeneous, and hygroscopic material. The sorption process of certain polar liquids in the wood substance, i.e. bulk sorption, is exothermic and causes swelling. This bulk sorption process and the resulting release of heat are observed as a distinct temperature increase within the columns during the wicking of water, formamide, and methanol. No temperature increase is observed for ethylene glycol, diiodomethane, and hexane. In some cases, the increase in temperature is observed in advance of the moving visible liquid front line. This may indicate that vapor is moving in advance of the liquid front, resulting in bulk sorption and the corresponding release of heat. An apparent non-linearity is observed when the square of the capillary rise is plotted versus time, mainly for water, formamide, and methanol. This non-linearity is strongly dependent on the probe liquid used and the variation in wood particle size. For the wicking of water, the bulk sorption, and hence the swelling of the wood particles, seems to appear instantaneously at the wetting front line, but for formamide and methanol a time delay is observed. The bulk sorption and resulting swelling of the wood particles strongly influence the determination of the effective interstitial pore radius between the particles, and thus the determination of contact angles by use of the Washburn equation.

Key words: Wood particles; wicking; capillary rise; contact angle measurements; wettability.

1. INTRODUCTION

The wetting characteristics of different liquids in contact with wood are important information for the processing, in-service behavior, and durability of various wood and wood-based products. Moreover, the wetting characteristics may be determined by contact angle measurements using different well-defined probe liquids.

---

*To whom correspondence should be addressed. E-mail: magnusw@woodtech.kth.se

M. E. P. Wålinder and D. J. Gardner

A liquid may penetrate spontaneously into a porous medium by capillary forces. This process, referred to as wicking, also occurs when the liquid contacts a powdered material or a fiber network. Wicking, i.e. the distance traveled by the liquid front (or the capillary rise of the liquid), in such materials is described by the Washburn equation [1]:

$$h^2 = \frac{R_e \gamma_L \cos \theta}{2\eta} t = Kt, \tag{1}$$

where h is the distance traveled by the liquid in time t, R_e is the effective interstitial pore radius between the packed particles or fibers, θ is the liquid–solid–vapor contact angle, and γ_L and η are the surface free energy (surface tension) and viscosity of the liquid, respectively. K is a constant, i.e. a grouping of all constants in equation (1), and is referred to as the Washburn slope. It is important to note that θ in this situation represents an advancing contact angle, and also that no spontaneous penetration will occur for θ values $\geqslant 90°$.

Equation (1) may be used to determine the contact angles and wetting characteristics of various powdered materials such as clay, polymer and barbiturate particles [2–5], and cellulose fiber networks [6–9]. Early attempts to relate the capillary rise to the wettability of wood particles were made by Freeman [10], but the method was rejected because of inconsistencies, which he attributed to structural variations of the wood flour. Bodig [11], however, overcame such problems by using a corrected water-absorption height method for studying the wettability of wood flour. Recently, Gardner et al. [12] have applied a column wicking method to study the wetting characteristics of various wood particles.

For wood powders and cellulose fiber networks, the pore structure, i.e. R_e, is very irregular. In this case, a liquid will penetrate more rapidly in large pores than in small ones. Hodgson and Berg [6], however, state that the overall matching of the pores in the fiber structure ensures a natural averaging process, and therefore R_e may be considered the actual average interstitial pore size in the matrix. Heterogeneity of the surface composition and differences in surface roughness may also cause local variations of $\cos \theta$, which may influence the movement of a liquid front. Moreover, in contact angle measurements it is important to have pure probe liquids. However, the presence of contaminants in the powder or fiber network may change the surface free energy of the penetrating liquid. It is also understood that probe liquids such as formamide, dimethyl sulfoxide, ethylene glycol, and glycerol are highly hygroscopic, which means that they easily gain water from either air or from the materials that they may contact. In addition, particle or fiber swelling may also occur during wicking, and therefore the pore radius R_e will decrease during testing with a resulting decrease in the Washburn slope with time [8, 9]. Hodgson and Berg [6] further state that despite some of the complications mentioned above, many investigations of paper-like materials have shown that the form of equation (1) is adequate, i.e. the square of the wicking distance is linear over time.

Wood is porous, heterogeneous, and hygroscopic. The last means a high affinity for the sorption of especially polar liquids in the wood substance. The sorption process of polar liquids in the wood substance, i.e. bulk sorption, is exothermic and causes swelling. The wicking of certain polar liquids in wood particles, therefore, causes swelling, which may influence not only the pore radius R_e, but also the entire wetting process. Furthermore, in almost any situation wood will contain a certain amount of water. Collett [13] pointed out that water had come to be recognized as the universal surface contaminant. Water will adsorb onto almost any surface, and Zisman [14] showed that water had greater affinity for both polar and non-polar surfaces than other liquids by a factor of almost 3 to 1. Further, it is clear that the adsorption of water onto solids will strongly influence their surface free energy and adhesion properties.

On the molecular level, the wood surface can be considered to be composed of cellulose, hemicellulose, lignin, and extractives (see, for example, ref. [15]). Furthermore, the extractive constituents in wood are generally hydrophobic in nature and play a central role when analyzing its surface properties. In wicking analyses, they may also act as contaminants of the probe liquids.

A machined wood surface generally consists of layers of crushed and damaged wood cells, i.e. a severely irregular and porous surface. The high surface roughness and the heterogeneous and porous nature of such surfaces make direct optical measurements of contact angles, such as the sessile drop method, inherently difficult. However, machined wood surfaces may also be represented by milled wood particles, which may be studied by other surface characterization methods such as IGC (inverse gas chromatography) and column wicking, which do not involve direct measurements of contact angles.

The objective of the present study was to examine the column wicking method for determining contact angles on extracted and non-extracted wood particles. This method is based on the capillary rise of some commonly used probe liquids in a glass column filled with wood particles. An effort was made to recognize important factors that influenced these measurements. By the development of methods for reliable estimations of contact angles on wood and the resulting characterization of its surface properties, we should be able to add valuable information to the understanding of surface and adhesion properties of not only wood composite products, but also glued, coated, and impregnated solid wood products.

2. MATERIALS AND METHODS

2.1. Sample preparation and probe liquids

Samples of clear heartwood and sapwood of spruce (*Picea abies* Karst.) were prepared. To minimize the effects of variation and heterogeneity within and between the samples, each sample was chosen from approximately the same annual rings and the same part of the tree stem. Half of the solid wood samples were extracted

Table 1.
Measured material data for samples of spruce wood particles. Values for − 60 mesh, − 60/+ 80 mesh particles, and solid wood density are based on 25, 12, and 10 replicates, respectively. Standard deviation are given in parentheses

Sample	Packing density (kg/m³)		Solid wood density (kg/m³)
	− 60 mesh	− 60/+ 80 mesh	
Sapwood	382 (9)	348 (13)	539 (18)
Extracted sapwood	428 (8)	328 (10)	
Heartwood	374 (8)	334 (11)	491 (19)
Extracted heartwood	433 (8)	302 (11)	

by the Soxhlet method. The solvents used were (in order) ethanol, acetone, petroleum ether, once again ethanol, and deionized water. All samples were initially conditioned to a moisture content of about 7–8% (based on oven-dry weight). The solid wood samples were then cut into thin veneers, ground in a Wiley mill, and passed through a 60-mesh screen. Two groups of particle size fractions were prepared. The first group combined all particles passing through the 60-mesh screen, referred to here as − 60 mesh particles. The second group combined the fraction of particles passing through a 60-mesh screen but not through a 80-mesh screen, thus comprising a more uniform fraction of particle sizes between 60 and 80 mesh, referred to here as − 60/+ 80 mesh particles. After the grinding process, to remove volatiles, the powder was conditioned at 70°C for 24 h in a convection oven. By this conditioning process, the moisture content dropped to about 4–5%. The wood particles were packed, using an electric vibrator, into marked glass columns (pipettes) with an inner diameter of about 3 mm. The packing density was determined by weighing the columns before and after the packing procedure. In summary, this resulted in four different spruce samples for analysis: (1) sapwood; (2) extracted sapwood; (3) heartwood; and (4) extracted heartwood. The solid wood density and column packing density for each sample are shown in Table 1. The probe liquids used in the wicking analyses were methanol and hexane as 'wetting out liquids', and water, formamide, ethylene glycol, and diiodomethane for the determination of contact angles. The surface tension and viscosity values at 20°C for these liquids were taken from *CRC Handbook of Chemistry and Physics*, 76th edition. All liquids were of HPLC grade (suitable for chromatography analyses) and purchased from Sigma-Aldrich.

2.2. Wicking measurements

According to van Oss *et al.* [5, 16], various initial hydrodynamic disturbances could affect the initial height rise versus time measurements and cause difficulties in determining the 'real' time zero. To avoid this, according to their suggestion, the wicking experiments were performed by immersing the columns vertically to a depth of about 5 mm in the probe liquid. Thereafter, the vertical movement of each probe liquid was determined by measuring (using a stopwatch) the time t for the visible liquid front line to reach various heights h at intervals of about 3–5 mm, each starting from 5–10 mm above the initial front line.

There are two unknowns in equation (1), the effective interstitial pore radius R_e and $\cos\theta$. By using a liquid with low surface tension, such as methanol or hexane, this liquid is expected to spread over the solid surface of the sample particles during the wicking measurements, so that $\cos\theta = 1$, and thus equation (1) can be solved for R_e. This R_e value is thereafter used for the same sample particles to determine $\cos\theta$ for each non-spreading probe liquid.

In addition to this, the temperature was measured at a low (h_1) and a high (h_2) point within the column during a series of wicking experiments. In this case, columns of polystyrene were used, to enable drilling of holes for positioning thermocouples (type J combined with a DT2801 data acquisition board) at the measuring points. The temperature was measured at intervals of 3 s and recorded in a personal computer using a Quick Basic program.

3. RESULTS AND DISCUSSION

Figure 1 shows an example plot of the wicking measurements for extracted spruce sapwood with a particle size of − 60 mesh. In this figure, the square of height h for the liquid front line is plotted against time t, referred to here as Washburn plots, for wicking measurements using the probe liquids methanol, water, and formamide. As can be seen, the wicking (or the capillary rise) of methanol is faster than that of water, which, in turn, is considerably faster than that of formamide.

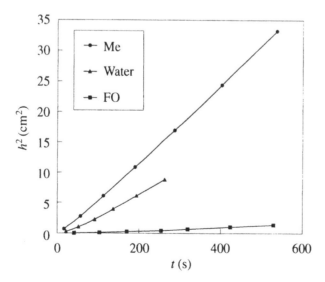

Figure 1. Example of a Washburn plot from column wicking measurements. Sample: extracted spruce sapwood, − 60 mesh particles. Probe liquids: methanol (ME), water, and formamide (FO).

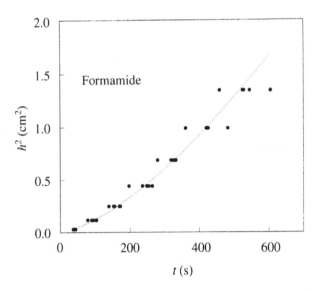

Figure 2. Washburn plot showing a tendency of increasing slope. Sample: extracted spruce sapwood, − 60 mesh particles. Probe liquid: formamide.

Figure 2 shows a more detailed Washburn plot of the same sample for formamide. In this case, it is obvious that the Washburn slope increases with time. This behavior is more pronounced for methanol, water, and formamide than for the other probe liquids. In other words, the constant in equation (1) is changing (increasing) systematically over time. It is possible that formamide and water may be contaminated when they are brought into contact with wood. The contamination originates from the dissolution of various wood extractives into the probe liquids, which in some cases results in a distinct decrease in their surface tension. Such a decrease in surface tension would also easily occur during wicking in wood particles, resulting in a change of the capillary rise behavior, i.e. increase the slope of the Washburn plots. However, no contamination should occur for wicking in extracted samples. In other words, the increasing slope for the extracted sample shown in Fig. 2 cannot be explained by this contamination hypothesis. For the extracted samples, the increasing slope may be due to pre-wetting effects of the wood particles, as discussed later, caused by the vapors moving in advance of the liquid front line which systematically could decrease the contact angle on the wood particles in the higher parts of the columns. The change from − 60 mesh particle size to the more uniform − 60/+ 80 mesh particle size will increase the average pore radius and thus increase the overall penetration rate. Thus, the pre-wetting behavior should be less pronounced, which would explain why the increasing behavior disappears, for most cases, when the − 60/+ 80 particle size is used. It is also clear that the wicking behavior and swelling effects depend significantly on the geometric form of the particles. For swelling particles in the form of powder rather than fibers (or splinters), significant pore blocking occurs during wicking [9]. Hence, during wicking in the columns filled with − 60 mesh particles, very fine

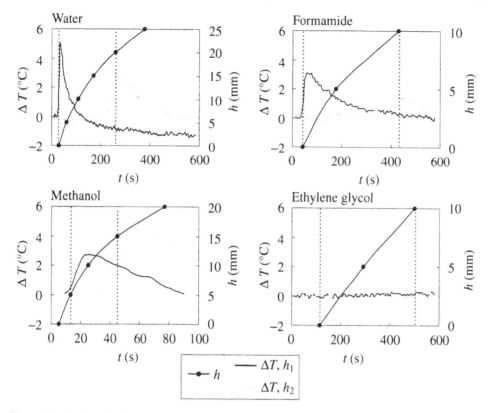

Figure 3. Graphs showing the temperature difference ΔT during wicking measured at two heights within the column versus time t, combined with height h of the liquid front versus time t. The vertical dashed lines represent the location of the two temperature measurement points. The h_1 and h_2 temperature measurement points for formamide and ethylene glycol are located at $h_1 = 0$ mm and $h_2 = 10$ mm, respectively. The corresponding values for water are $h_1 = 0$ mm and $h_2 = 20$ mm, and for methanol $h_1 = 5$ mm and $h_2 = 15$ mm. Sample: extracted spruce sapwood, -60 mesh particles. Probe liquids: water, formamide, methanol, and ethylene glycol.

wood powder particles may cause pore blocking, resulting in a considerably slower capillary rise compared with wicking in $-60/+80$ mesh particles. Thus, this pore blocking could mean that pre-wetting, sorption, and swelling effects will be more pronounced and thus will strongly influence the wetting process and the wicking analysis.

Moreover, the sorption of polar liquids in the wood material, i.e. bulk sorption, is exothermic. Figure 3 shows the results of temperature measurements at two points within a column for wicking measurements of water, formamide, methanol, and ethylene glycol in spruce sapwood, -60 mesh particles. The two 'peak curves' in the graphs for water, formamide, and methanol correspond to the temperature measurement at a low (h_1) and a high (h_2) point within the column, and the continuous increasing curve represents the movement of the visible liquid front line. As can be seen in Fig. 3, water, formamide, and methanol show distinct temperature increases following the movement of the liquid front, whereas ethylene glycol does not show any notable temperature increase. The latter behavior is also observed, as expected, for diiodomethane and hexane. These temperature measurements confirm that a bulk sorption process occurs during the wicking analysis for methanol, water, and formamide, but not for ethylene glycol, diiodomethane, and hexane. Since methanol is used as the 'wetting out' liquid for the determination of R_e, it is clear that this value may not be relevant for any of the other probe liquids.

In addition, for the wicking of water and formamide, the temperature increase seems to be instantaneous with the movement of the visible liquid front line at the lower measurement point, whereas at the higher point it is in advance of the moving front (see Fig. 3). This result may be attributed to the fact that vapor from the probe liquid is moving in advance of the liquid front line, thus resulting in the bulk sorption of vapor with a corresponding release of heat. In other words, this indicates that vapor moves in advance of the liquid front, causing 'pre-wetting' of the particles.

Another issue that arises is concerned with whether there might be an influence of equilibrium film pressure, π_e (see, for example, ref. [17]), caused by the adsorbed vapor of the liquid at the solid–gas interface which would affect the surface energy of the solid. In cases where a liquid is forming a finite contact angle against a 'low energy' surface, it has been common to argue that π_e should be neglected [18, 19]. Jacob and Berg [20], on the other hand, state that there is a large and growing body of evidence that this assumption is wrong. Furthermore, in wicking studies using the Washburn equation, when a liquid is spreading over a solid surface the influence of π_e would mean that the assumption $\cos \theta = 1$ is not valid. This was investigated by van Oss *et al.* [5] for thin-layer wicking analyses of clay particles using a series of low-energy spreading liquids. However, they found a linear relationship when $2 \eta h^2 / t$ was plotted against γ_L, which implies that the assumption $\cos \theta = 1$ is valid. Further, they explained this observation by the hypothesis that small amounts of the spreading liquid, prior to spreading, will pre-wet the solid, which means that the liquid will spread over a thin layer of its own material. In the present investigation, the influence of π_e is not considered on the basis of the conclusions of van Oss *et al.* [5] and the lack of convenient techniques for its determination [20].

Figure 4 shows the Washburn plots for the test series of $-60/+80$ mesh spruce particles. The use of more uniform and larger average particle size increases the rate of the capillary rise through the particle matrix. Consequently, the bulk sorption and swelling effects may be reduced. As shown in these plots, methanol and formamide now reveal a decreasing rate of capillary rise over time. This behavior has also been observed by Schuchardt and Berg [8] in wicking studies of composite cellulose–superabsorbent fiber networks and was attributed to delayed swelling effects, which decrease the R_e value over time. For water, on the other hand, no decreasing rate of capillary rise is observed as shown in Fig. 4. Schuchardt and Berg [8] also reported that the swelling in water of cellulosic fibers during wicking was instantaneous, which explains why no decrease in the slope of h^2 versus t is observed. This observation may also be related to the swelling of wood particles

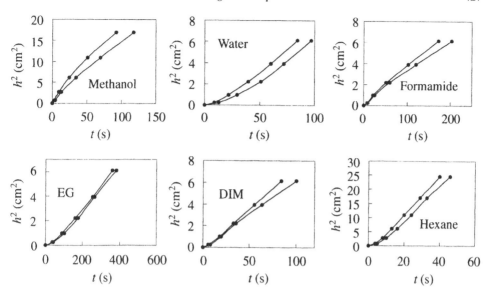

Figure 4. Graphs showing two replicates of the Washburn plots for spruce sapwood, − 60/ + 80 mesh particles, for all the probe liquids used (EG = ethylene glycol; DIM = diiodomethane).

Table 2.
Estimated contact angles θ and effective interstitial pore radius R_e (measured with hexane and methanol) for different spruce samples of − 60/ + 80 mesh particle size. Values are the average of two replicates

Sample	R_e (μm)		θ (degrees)			
	Methanol	Hexane	Water	Formamide	Ethylene glycol	Diiodomethane
Sapwood	0.8	2.2	85	77	50	69
Extracted sapwood	2.1	2.8	83	72	24	58
Heartwood	0.9	2.8	89	79	48	69
Extracted heartwood	1.6	3.0	85	65	24	59

during wicking experiments with water. The tendency of an increasing slope for water, however, may be explained by the pre-wetting phenomenon as discussed above.

It is also evident from Fig. 4 that there exists a pronounced initial time lag for some of the Washburn plots (see especially the initial slopes for ethylene glycol, diiodomethane, and hexane). These plots show a distinct delay of the initial capillary rise, i.e. the slope is considerably lower in the beginning than in the later parts of the Washburn plots. This behavior may possibly be due to the various initial hydrodynamic disturbances in the liquid caused by the immersion of the column as suggested by van Oss *et al.* [5].

Table 2 summarizes the R_e and contact angle values for the samples with − 60/ + 80 mesh particle size, estimated by using equation (1) combined with the measured Washburn slopes as exemplified in Fig. 4. The values are based on the average of two replicates for each sample. R_e values are determined for both methanol and hexane, which are assumed to completely wet out the wood particles. As shown in Table 2, the R_e values determined using methanol are in all cases considerably lower than those based on hexane. The low R_e values for methanol may be attributed to swelling effects, and also because methanol may not completely 'wet out' the particles due to the existence of hydrophobic extractives in the wood substance. The methanol-based R_e values of the extracted samples are also higher than those of the non-extracted samples. An explanation for this may be that if methanol does not completely 'wet out' the non-extracted particles, the removal of extractives will reduce the contact angle, resulting in a faster capillary rise and an apparently higher R_e value. Because of these uncertainties when calculating the R_e values from the methanol measurements, the contact angles for all the probe liquids in Table 2 are instead determined by using the R_e values calculated from the hexane measurements. Because of the decreasing rate of capillary rise with time for formamide, only the first parts of these curves are used to determine the contact angles. The initial time lag, mentioned above, was excluded in the slope determination. The decrease in surface tension and viscosity, caused by the temperature increase during the wicking measurements, may also appear in the $h^2 = f(t)$ relationship at the beginning of the wicking process. This, however, was not considered in these calculations, due to the difficulties in estimating the actual temperature rise at the wetting front line.

The contact angle values for the samples with $-60/+80$ mesh particles, shown in Table 2, are considerably higher than the contact angles measured by other methods on solid wood [21, 22]. Furthermore, Parsons et al. [3] concluded that the capillary rise method for contact angle measurements on particles gave significantly higher values than dynamic and static contact angle measurements for plates covered with the same material. Similarly, Yang et al. [23] concluded that a column wicking method combined with equation (1) for estimating contact angles led to apparent contact angles which were greater than static values. On the contrary, Crawford et al. [24] showed that the advancing water contact angles on quartz particles determined from wicking experiments using the Washburn equation were equal to static contact angles measured on plates of chemically identical material over a wide range of contact angles from 10° to 88°. It is also important to note that in contrast to the ideal contact angle as defined by the Young equation, the measured contact angles are phenomenological, or apparent, angles and do not represent surface energetics exclusively. For example, contact angles measured on rough surfaces are found to be larger than those measured on smooth, and chemically identical, surfaces [25]. It is also important to note that the contact angle estimated from wicking measurements represents an advancing contact angle, and it is commonly accepted [17, 26–28] that this angle on a heterogeneous surface is related to regions with lower-energy components. In other words, the advancing contact angle may be equal to the equilibrium contact angle on a flat homogeneous surface composed of the lower-energy component [27]. Hence, the contact angle estimated using the column wicking method will be strongly related to the hydrophobic components like wood extractives. For example, Carlsson and Ström [29] observed a contact angle of 106° for water droplets on non-extracted high-yield kraft pulp compared with 67° for the corresponding extracted pulp (with a surface composed of about 70% lignin) and 29° for extracted bleached kraft pulp (with a surface composed of about 10% lignin). Hence, a larger amount of both extractives and lignin in the surface will result in a higher contact angle.

4. CONCLUSIONS

The results show that the wicking of water, formamide, and methanol in wood particles gives rise to a considerable temperature increase, caused by bulk sorption, i.e. sorption of the liquid in the cell-wall material. No temperature increase was observed for ethylene glycol, diiodomethane, and hexane. An apparent non-linearity in the Washburn slopes was observed, mainly for water, formamide, and methanol. The apparent non-linearity in the capillary rise depends on the probe liquids used and the variation in particle size. The use of a more uniform wood particle size is desirable for the wicking analyses. The increase in temperature, measured at a high and a low point within the column, is instantaneous with the rise of the visible liquid front at the lower level, but at the higher level it is in advance of the main front. The bulk sorption, and hence the swelling of the wood particles, seems to appear instantaneously at the wetting front line for water, but for formamide and methanol a time delay is observed. The bulk sorption process and the resulting swelling effects strongly influence the determination of the effective interstitial pore radius R_e, and thus the determination of contact angles using the Washburn equation.

Acknowledgements

We thank Professor Ingvar Johansson and Dr. Göran Ström for helpful comments during the preparation of the manuscript.

REFERENCES

1. E. W. Washburn, *Phys. Rev.* **17**, 273–283 (1921).
2. G. Buckton and J. M. Newton, *Powder Technol.* **46**, 201–208 (1986).
3. G. E. Parsons, G. Buckton and S. M. Chatham, *J. Adhesion Sci. Technol.* **7**, 95–104 (1993).
4. Q. Shi, D. J. Gardner and J. Z. Wang, in: *Proceedings of the Fourth International Conference on Woodfiber-Plastic Composites*, 12–14 May. Forest Product Society, Madison, WI (1997).
5. C. J. van Oss, R. F. Giese, Z. Li, K. Murphy, J. Norris, M. K. Chaudhury and R. J. Good, *J. Adhesion Sci. Technol.* **6**, 413–428 (1992).
6. K. T. Hodgson and J. C. Berg, *Wood Fiber Sci.* **20**, 3–17 (1988).
7. K. T. Hodgson and J. C. Berg, *J. Colloid Interface Sci.* **121**, 22–31 (1988).

8. D. R. Schuchardt and J. C. Berg, *Wood Fiber Sci.* **23**, 342–357 (1991).
9. S. Wiryana and J. C. Berg, *Wood Fiber Sci.* **23**, 457–464 (1991).
10. H. A. Freeman, *For. Prod. J.* **9**, 451–458 (1959).
11. J. Bodig, *For. Prod. J.* **12**, 265–270 (1962).
12. D. J. Gardner, W. Tze and Q. Shi, in: *Progress in Analytical Methodologies Applied to Lignocellulosic Materials*, D. S. Argyropoulos (Ed.). Tappi Press (in press).
13. B. M. Collett, *Wood Sci. Technol.* **6**, 1–42 (1972).
14. W. A. Zisman, *Ind. Eng. Chem.* **55**, 19–38 (1963).
15. E. Sjöström, *Wood Chemistry: Fundamentals and Applications.* Academic Press, New York (1981).
16. C. J. van Oss, W. Wu and R. F. Giese, *Particulate Sci. Technol.* **11**, 193–198 (1993).
17. R. J. Good, in: *Contact Angle, Wettability and Adhesion*, K. L. Mittal (Ed.), pp. 3–36. VSP, Utrecht, The Netherlands (1993).
18. R. J. Good, *J. Colloid Interface Sci.* **52**, 308 (1975).
19. F. M. Fowkes, D. C. McCarthy and M. A. Mostafa, *J. Colloid Interface Sci.* **78**, 200 (1980).
20. P. N. Jacob and J. C. Berg, *J. Adhesion* **54**, 115–131 (1995).
21. D. J. Gardner, *Wood Fiber Sci.* **28**, 422–428 (1996).
22. H. J. Zhang, D. J. Gardner, J. Z. Wang and Q. Shi, *For. Prod. J.* **47**, 69–72 (1997).
23. Y.-W. Yang, G. Zografi and E. E. Miller, *J. Colloid Interface Sci.* **122**, 35–46 (1988).
24. R. Crawford, L. K. Koopal and J. Ralston, *Colloids Surfaces* **27**, 57–64 (1987).
25. K. Grundke, T. Bogumil, T. Gietzelt, H.-J. Jacobasch, D. Y. Kwok and A. W. Neumann, *Prog. Colloid Polym. Sci.* **101**, 58–68 (1996).
26. A. Ullman, S. D. Evans, Y. Shnidman, R. Sharma and J. E. Evans, *Adv. Colloid Interface Sci.* **39**, 175 (1992).
27. R. J. Good, in: *Surface and Colloid Science*, R. J. Good and R. R. Stromberg (Eds), Vol. II, pp. 1–29. Plenum Press, New York (1979).
28. J. C. Berg, *Nord. Pulp Paper Res. J.* **8**, 75–85 (1993).
29. G. Carlsson and G. Ström, *Nord. Pulp Paper Res. J.* **10**, 17 (1995).

Apparent and Microscopic Contact Angles, pp. 431–444
J. Drelich, J. S. Laskowski and K. L. Mittal (Eds)
© VSP 2000.

A new model to determine contact angles on swelling polymer particles by the column wicking method

SHELDON Q. SHI* and DOUGLAS J. GARDNER
Advanced Engineered Wood Composites Center/Department of Forestry Management, University of Maine, Orono, ME 04469, USA

Received in final form 9 September 1999

Abstract—The contact angle determination on swelling polymer particles by the Washburn equation using column wicking measurements may be problematic because swelling occurs during the wicking process. The objective of this research was to develop a new model to more accurately determine contact angles for polymer particles that undergo solvent swelling during the column wicking process. Two phenomena were observed related to the swelling effect during the wicking process: (1) a temperature rise was detected during the wicking process when the swelling polymer particles interacted with polar liquids, and (2) a smaller average capillary radius (r) was obtained when using methanol (polar liquid) compared to using hexane (non-polar liquid). The particle swelling will induce both particle geometry changes and energy loss which will influence the capillary rise rate. The model developed in this study considered the average pore radius change and the energy loss due to the polymer swelling effect. Contact angle comparisons were conducted on wood with formamide, ethylene glycol, and water as test liquids, determined by both the new model and the Washburn equation. It was shown that the contact angles determined by the new model were about 4–37° lower than those determined by the Washburn equation for water, formamide, and ethylene glycol. To determine whether the polymer particles are swelling, two low surface tension liquids, one polar (methanol) and the other non-polar (hexane), can be used to determine the average pore radius (r values) using the Washburn equation. If the same r values are obtained for the two liquids, no swelling occurs, and the Washburn equation can be used for the contact angle calculation. Otherwise, the model established in this study should be used for contact angle determination.

Key words: Contact angle; wicking; capillary rise; swelling polymers; Washburn equation.

NOTATION

h	height of the capillary rise (m)
t	time (s)

---

*To whom correspondence should be addressed. Present address: 5793 AEWC Building, University of Maine, Orono, ME 04469, USA. E-mail: sheldon_shi@umenfa.maine.edu

r	average capillary radius (m)
r_s	average capillary radius after polymer swelling (m)
δ_V	volumetric swelling (%)
θ	contact angle (degrees)
γ_L	surface free energy of the liquid (mJ/m^2)
η	viscosity of the liquid (mN·s/m^2)
P_c	capillary pressure (N/m^2)
ρ_m	material density (g/cm^3)
W_c	work done by the capillary pressure (J)
W	work done by the effective driving pressure (J)
Q	energy loss (J)
S	unit energy loss (J/m)
C	energy loss coefficient (J/m)
ΔP_{eff}	total effective driving force (N/m^2)
N	number of capillaries in the column
M_m	mass of swelling material (g)
R	inner radius of the column (m)
G_m	unit column mass of the material (g/m)
dV	change in the liquid volume flow through any cross section of the capillary (m^3)

1. INTRODUCTION

Use of the column wicking method to measure contact angles on particulate materials is primarily based on the Washburn equation [1], which is:

$$h^2 = \frac{r\gamma_L \cos\theta}{2\eta}t,$$ (1)

where h is the distance traveled by the liquid in time t, r is the effective interstitial pore radius [2] or tortuosity constant [3], θ is the contact angle, and γ_L and η are the surface free energy and viscosity of the liquid, respectively. If the packed column is assumed to consist of many small capillaries with different capillary radii, r is the average capillary radius in the column. In equation (1), γ_L and η for different liquids are known parameters; h and t can be measured experimentally. There are two unknowns in equation (1), r and $\cos\theta$. The r value of packed materials can be determined using a low surface tension liquid, hexane or methanol, which can wet out the solid surface ($\cos\theta = 1$).

It has been shown that the contact angles on polymer particles determined by wicking measurements are usually higher than those determined by the Wilhemy plate and sessile drop methods [4, 5]. This is partially because of the material geometry difference between the particles and plate, and the thermal effect on the surface as a result of particle grinding. Also, the sessile drop and the Wilhemy methods measure the outermost polymer surfaces, while the wicking method measures the surfaces of both outer particles and macro-pores within the polymers. For swelling and porous polymers, the contact angle differences between wicking measurements and sessile drop measurements may be even larger. There are two concerns to consider in the wicking measurements on swelling polymers:

(1) the polar liquid not only interacts with the surface of the particles, but also penetrates into the polymers. The polar liquids will be absorbed into the micro-pores of the polymers and the particles will swell. Because of the particle swelling, the average capillary radius in the packed column will be reduced, which will affect the capillary rise rate;

(2) the liquid interaction and penetration into the micro-pores of the polymer may possibly contribute to an energy loss during the liquid rise process. The Washburn equation was derived assuming that the capillary force was the only effective driving force occurring during the capillary rise process. If an energy loss is experienced during liquid movement, the capillary rise rate will also be slowed down.

Based on the above two concerns, the Washburn equation may not be suitable for determining contact angles on swelling polymer particles interacting with polar liquids. It is the goal of this paper to theoretically develop a new equation to determine the contact angles on swelling polymer particles, considering both pore radius change and energy loss.

2. MODEL ESTABLISHMENT

2.1. Consideration of the geometrical changes

The capillary rise process can be described by Poiseuille's viscous flow. For liquid movement in a capillary, if the slip flow is neglected, Poiseuille's law can be expressed as follows:

$$dV = \frac{\pi \Delta P_{\text{eff}}}{8\eta h} r^4 \, dt, \tag{2}$$

where dV is the volume change $(= \pi r^2 \, dh)$, η is the viscosity of the liquid, h is the height of the capillary rise, r is the capillary radius, and ΔP_{eff} is the effective driving pressure.

In the application of Poiseuille's law to column wicking measurements, it was assumed that the packed column consisted of many small capillaries with an average capillary radius of r (Fig. 1: case a). It was also assumed that the liquid movement in each capillary was regarded as viscous flow [6]. The r value used in Poiseuille's law

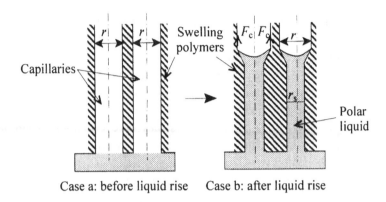

Case a: before liquid rise   Case b: after liquid rise

Note r: capillary radius
       r_s: capillary radius after polymer swelling
       r_c: capillary force

Figure 1. Capillary radius change for polar polymers due to their swelling in polar liquids.

[equation (2)] should be the one behind the liquid front line. During the capillary rise of polar liquids, due to polymer swelling, the geometry of the particles will change and the capillary radius behind the liquid front line will be decreased (Fig. 1: case b). Polymer swelling is a function of time, and maximum swelling will occur after a finite time period. During wicking measurements, because the polymer particles in the column are very small, it can be assumed that a small particle reaches its maximum swelling instantly after it interacts with the polar liquid. Under this assumption, the capillary radius after polymer swelling (r_s) should be used instead of the original capillary radius (r) in Poiseuille's equation. Substituting $dV = \pi r_s^2 \, dh$ in equation (2), Poiseuille's law can be rewritten as

$$dh = \frac{\Delta P_{\text{eff}}}{8\eta h} r_s^2 \, dt. \tag{3}$$

In the derivation of Washburn's equation, the total effective driving force (ΔP_{eff}) of the capillary rise is only the capillary pressure neglecting the hydrostatic pressure. Because the capillaries are very small and only a short height rise is measured, the hydrostatic pressure can be neglected compared to the capillary pressure. In each capillary, the capillary pressure (P_c) can be expressed as

$$P_c = \frac{2\gamma}{r} \cos\theta, \tag{4}$$

where γ is the surface tension of the liquid and θ is the contact angle.

It should be noted that, unlike Poiseuille's equation, the capillary radius used in equation (4) should be the original capillary radius which is ahead of the liquid front line (see Fig. 1: case b). Substituting equation (4) into equation (3), after intergration, we have the following expression:

$$h^2 = \frac{r_s^2 \gamma_L \cos \theta}{2\eta r} t. \tag{5}$$

We call equation (5) the modified Washburn equation considering the effect of particle swelling during wicking measurements.

2.2. Determination of the swelled average capillary radius (r_s)

The total volume of the packed column is the summation of the volume of capillaries and polymer particles in the column, which can be expressed as

$$N \cdot \pi r^2 h + \frac{M_m}{\rho_m} = \pi R^2 h, \tag{6}$$

where r is the average capillary radius, h is the height of the capillary rise, R is the inner column radius, ρ_m is the material density, and M_m is the material mass in the column, which can be obtained by

$$M_m = G_m \cdot h, \tag{7}$$

where G_m is the unit column mass (g/m).

The number of capillaries in the column (N) can be calculated using the following equation:

$$N = \left(\frac{R}{r}\right)^2 - \frac{G_m}{\pi r^2 \rho_m}. \tag{8}$$

After polymer swelling, the number of capillaries in the column (N) should remain the same. If the expansion due to polymer particle swelling in the column height direction is neglected, the following equation also holds:

$$N \cdot \pi r_s^2 h + (1 + \delta_V) \frac{M_m}{\rho_m} = \pi R^2 h, \tag{9}$$

where r_s is the swelled average capillary radius and δ_V is the volumetric swelling of the polymer particles in the liquid.

Combining equations (8) and (9), the relationship between r_s and r can be established as:

$$r_s = \sqrt{\frac{\pi R^2 \rho_m - (1 + \delta_V) G_m}{\pi \rho_m R^2 - G_m}} \cdot r. \tag{10}$$

In equation (10), the radius of the column R, the density of the material ρ_m, and the unit mass in the column G_m are known parameters. The volumetric swelling (δ_V) of the materials can be obtained either from the literature or from measurements.

2.3. Consideration of the energy loss during the wicking process

The energy loss during column wicking is mainly due to the polar liquid penetration into the micro-pores of the polymers, i.e. swelling heat (heat of adsorption) escaping from the wicking system. Considering the possible energy loss during the wicking process, the model should be further modified based on the following derivations.

During the vertical capillary rise, when the liquid rises a certain distance (h), the work done by the capillary pressure is

$$W_c = P_c \cdot \pi r^2 \cdot h = \frac{2\gamma}{r} \cos\theta \cdot \pi r^2 \cdot h. \tag{11}$$

If Q is the amount of energy loss when the liquid rises to a height of h, the total work done by the effective total driving pressure $(W = \Delta P_{eff} \pi r^2 \cdot h)$ can be expressed as

$$W = \Delta P_{eff} \cdot \pi r^2 \cdot h = W_c - Q. \tag{12}$$

Combining equations (11) and (12), we have

$$\frac{2\gamma}{r} \cos\theta \cdot \pi r^2 \cdot h - Q = \Delta P_{eff} \cdot \pi r^2 \cdot h. \tag{13}$$

Therefore, the effective driving pressure during capillary rise is

$$\Delta P_{eff} = \frac{2\gamma}{r} \cos\theta - \frac{Q}{\pi r^2 h}. \tag{14}$$

The energy loss is a function of the height of the rising liquid. In the wicking process, the higher the liquid rises, the more material and liquid are involved in the heat of adsorption, and the more energy loss would occur. To simplify the derivation, it is assumed that the energy loss is proportional to the height of the rising liquid (h), which can be expressed as

$$Q = S \cdot h, \tag{15}$$

where S is the unit energy loss in the packed column.

Substituting equations (14) and (15) into (3) gives

$$dh = \frac{\frac{2\gamma}{r} \cos\theta - \frac{S}{\pi r^2}}{8\eta h} r_s^2 \, dt. \tag{16}$$

Integrating equation (16), the expression for the capillary rise process for swelling polymers considering the energy loss is

$$h^2 = \frac{r_s^2 \gamma \cos\theta}{2\eta r} t - \frac{S}{4\eta\pi} \left(\frac{r_s}{r}\right)^2 t. \tag{17}$$

In equation (17), the first term is equation (5) and the second term is the adjustment term considering the energy loss due to polymer swelling during the wicking process. For non-swelling polymers, $S = 0$ and $r_s = r$, and equation (17) can be converted to the Washburn equation. From equation (17), it is seen that the contact angle determination will depend on the swelled capillary radius (r_s) and the energy loss (S). A smaller r_s value or a larger S value will result in a reduced capillary rise rate.

To theoretically determine the S value is difficult because the energy transfer inside the column during the interaction of the liquid and the swelling polymer particles is a complex process. In the wicking process, the form of energy inside the capillary system will change. For example, some energy will be involved in polymer particle swelling and the heat loss from the wicking system will tend to delay the capillary rise. On the other hand, the swelling heat created in the system decreases the liquid viscosity, which will increase the rate of capillary rise. The S value in equation (17) is related only to the energy lost from the system during the wicking process.

Theoretically, the more the particles swell, the greater the energy loss in the wicking system. Polymer swelling depends on the different polar liquids used in the wicking measurements: therefore, the S value will also depend on the probe liquid used. If the unit energy loss (S) is proportional to the volumetric swelling (δ_V) of the polymer particles in a certain liquid, then

$$S = C\delta_V, \tag{18}$$

where C (in J/m) is called the energy loss coefficient in the wicking system. The energy loss coefficient represents the property of a packed column related to the resistance to the energy loss. Therefore, for a certain polymer and column system, the C value is a constant independent of the liquids used.

Substituting equation (18) into equation (17), we have the following final expression for the new model:

$$h^2 = \frac{r_s^2 \gamma \cos\theta}{2\eta r}t - \frac{C\delta_V}{4\eta\pi}\left(\frac{r_s}{r}\right)^2 t. \tag{19}$$

In equation (19), the volumetric swelling (δ_V) is a known parameter or can be measured. The original average capillary radius (r) can be determined using hexane, for which $\delta_V = 0$, $r_s = r$, and $\cos\theta = 1$. The average capillary radius after particle swelling (r_s) can be calculated using equation (10). The energy loss coefficient (C) can be determined using methanol or other polar liquids with a low surface tension, for which $\cos\theta$ can also be regarded as 1.

3. EXPERIMENTAL

To investigate the effect of swelling on the wicking process and to evaluate the new model, the following experiments were conducted: (1) observation of the swelling phenomenon during the wicking process; (2) the effect of polymer swelling on the average capillary radius and possible energy loss; and (3) comparison of the contact angles determined by the new model and the Washburn equation on swelling polymer particles.

3.1. Materials and methods

Two swelling polymers, Douglas-fir wood (*Pseudotsuga menzisii*) heartwood and nylon 6, were used. A non-swelling material, polyethylene obtained from Polysciences, Inc., Warrington, PA, USA, was also used in the experiment as a non-swelling control. Some properties of these polymers are summarized in Table 1. It is shown in Table 1 that the wood swelling depends on the nature of the probe liquid. The volumetric swelling of wood for the four polar liquids is in the range of 13.8–17.4%. To avoid non-uniform packing due to the wide particle size distribution, all the polymers were ground into particles by a cutting mill and sifted to $-40/+60$ mesh (0.28–0.43 mm). Fisherbrand standard disposable polystyrene serological pipettes were used as columns. The outer diameter of the column was 4.57 mm, while the inner diameter was 2.95 mm. For the wicking measurements, the particles were packed into marked columns using an electric vibrator. The probe liquids used in the experiments and their properties are summarized in Table 2. The unit mass of the particles in the columns, shown in Table 3, was obtained by weighing the column before and after the packing procedure. The packed columns were conditioned at ambient temperature (20°C) before the wicking measurements. In the measurements, the column was vertically immersed in the probe liquid to a depth of about 5 mm. The height reached by each liquid over time was measured. Five to six readings were taken for each column. Six contact angle replicates were averaged for each probe liquid on the corresponding polymer.

3.2. Observation of the swelling phenomenon: temperature rise measurements

To detect the temperature rise during the wicking measurement, thermocouples were placed in the column to measure temperatures at two points along the column by drilling holes at heights of 7.5 and 15 mm. Type J thermocouples were placed in the holes drilled on the wall of the column to collect the temperature changes with a DT2801 data acquisition board at intervals of 1 s. A program written in Quick Basic was used to collect the temperatures in the column. The height of capillary rise at different times was measured, and the temperature changes during the wicking process were also collected using a computer. Measurements of the temperature rise in the columns packed with Douglas-fir wood particles were conducted using water.

3.3. Effect of polymer swelling on the pore radius in the packed column

An experiment was conducted on the determination of the average capillary radius (r) using both hexane and methanol, both of which are low surface tension liquids and can be regarded to wet out the solid surface. Hexane is a non-polar liquid which will not swell the polar polymers, while methanol is a polar liquid which will swell

Table 1.
Some properties of the polymers used in the experiments

Polymer	Density (g/cm^3)	Volumetric swelling, δ_V			
		Water	Formamide	Ethylene glycol	Methanol
Polyethylene	0.95	0	0	0	0
Wood[a]	0.54	0.150[b]	0.170[b]	0.174[b]	0.138[b]
Nylon 6	1.14[c]	N/A	N/A	N/A	N/A

[a] The equilibrium moisture content (EMC) for the Douglas-fir wood was 6.4%.
[b] Volumetric swelling was obtained from 1.5 × tangential swelling [7]. The tangential swelling value was taken from ref. [8].
[c] From ref. [9].

Table 2.
Values of the surface free energy (γ_L), viscosity (η), and density (ρ) of the probe liquids used for the wicking measurements at room temperature

Liquid	γ_L (mJ/m^2)	η (mN·s/m^2)	ρ (g/cm^3)
Hexane	17.9	0.30	0.659
Methanol	24.0	0.54	1.113
Ethylene glycol	48.0	19.90	1.113
Formamide	58.0	4.55	1.134
Water	72.8	1.00	1.000

Sources: refs [2, 12].

Table 3.
Column information and parameters for wood particles

No. of capillaries in one column, N^a	Unit column mass, G_m (g/m)	r value determined by hexane (m) $\times 10^{-6}$	r_s value in methanol calculated using equation (10) (m) $\times 10^{-6}$	r value determined by methanol using equation (5) (m) $\times 10^{-6}$	Energy loss coefficient, C^b (J/m)
96 369	2.085	3.15	2.84	1.84	0.132
(3183)	(0.043)	(0.67)	(0.30)	(0.12)	(0.018)

Standard deviation in parentheses.
[a] Obtained from equation (8).
[b] Obtained from equation (19) using methanol.

the polar polymers. Swelling polymers, wood and nylon 6, and a non-swelling polymer, polyethylene, were used in this experiments. The Washburn equation was used to determine the r values.

3.4. Comparisons of the new model with the Washburn equation

The wicking data for wood particles were used in the comparison of the Washburn equation and the newly developed model. The contact angles of formamide, ethylene glycol, and water for wood particles were calculated using both the Washburn equation and the new model. To investigate the effect of both the particle geometrical changes and the energy loss on the contact angle determination, both equation (5) (considering only the geometrical change of particles) and equation (19) (considering both the geometrical change and the energy loss) were used to calculate the contact angles. The contact angle results were compared and analyzed.

4. RESULTS AND DISCUSSION

The heat of adsorption is a typical phenomenon occurring as the result of polymer swelling. When the liquid penetrates and is adsorbed onto the inner surface of the micro-pores in the polymer (very large surface area), the energy of the polymer and liquid system will tend to decrease. Therefore, some energy in the system will be released in the form of heat, which is called the swelling heat or heat of adsorption. Figure 2 shows an example of the observation of the temperature rise during the wicking measurements on wood particles when the water liquid front line rises to where the thermocouple was placed. It is shown in Fig. 2 that the temperature rises about 3° and decreases to room temperature after approximately 40 s. This temperature decrease occurs presumably because (1) the probe liquid in the column absorbs the heat of adsorption, and (2) some heat of adsorption escapes to the ambient environment. This temperature rise phenomenon indicates that the particle swelling occurs during the wicking process for the wood and water system.

Figure 3 shows the average capillary radius (r) of polyethylene, nylon 6, and wood determined by both hexane and methanol calculated using the Washburn equation.

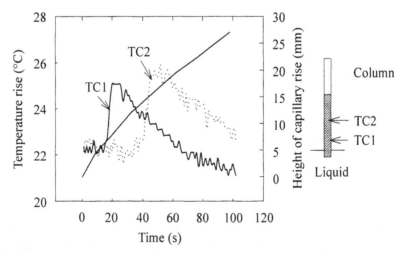

Figure 2. Temperature rise with time at different heights of the capillary rise for water.

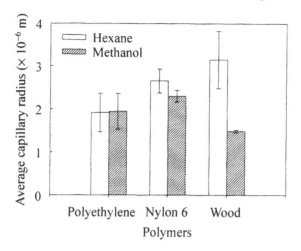

Figure 3. Average capillary radius determined using the Washburn equation.

It is seen from Fig. 3 that the r values determined by both hexane and methanol are similar for polyethylene, while they are different for nylon 6 and wood. As polyethylene is a non-swelling polymer, there is no polar interaction between the particles and the liquid. Therefore, the r value of polyethylene is not affected by using either polar or non-polar liquids. For swelling polymer particles, nylon 6 and wood, methanol interacts with the polar functional groups on polar polymers and causes particle swelling, which causes both geometrical changes (see Fig. 1) and energy loss. It is seen from Fig. 3 that the difference between $r_{methanol}$ and r_{hexane} for nylon 6 is much smaller than that for wood, indicating that the swelling effect on wood is greater than that on nylon 6. Figure 3 also indicates that a comparison of the $r_{methanol}$ and r_{hexane} values is a good way to determine whether the particles in the column are swelling. Therefore, in determining the r value for wicking measurements, it is suggested that both hexane (or other low surface tension non-polar liquids) and methanol (or other low surface tension polar liquids) be used. If the r values determined by the Washburn equation are the same for both hexane and methanol, the swelling effects for the contact angle determination can be neglected, and the Washburn equation can be applied to determine the contact angle for the other liquids. Otherwise, the new model should be employed.

Table 3 shows the measured data on wood columns used in the experiments. The average capillary radius after wood particle swelling in methanol (r_s) calculated by equation (10) and the energy loss coefficient (C) for the wood particle columns determined by methanol are also listed in Table 3. It is seen from Table 3 that the calculated r_s (2.84×10^{-6} m) is still larger than the r value determined directly by equation (5) (1.84×10^{-6} m), which considers only the effect of geometrical changes of the swelled particles. This result indicates that the energy loss played a role in delaying the capillary rise in the wood particle columns. Therefore, both geometrical change and energy loss should be considered in the contact angle determination on swelling polymer particles.

Table 4.
Comparison of the contact angles on wood particles determined by the new model and the Washburn equation

Liquid	S (J/m)	r_s (m) $\times 10^{-6}$	Washburn equation	New model	
				Equation (5)	Equation (19)
Water	0.0198	2.84	86.2	85.3	77.1
		(0.02)	(0.3)	(0.3)	(0.5)
Formamide	0.0224	2.77	42.5	18.5	4.9
		(0.05)	(11.0)	(18.7)	(7.5)
Ethylene glycol	0.0230	2.80	3.8	0.0	0.0
		(0.06)	(8.6)	(0.0)	(0.0)

Standard deviation in parentheses.

The contact angles on wood particles calculated by both the Washburn equation and the newly developed model are compared and summarized in Table 4. It is seen from Table 4 that the contact angles of the three polar liquids determined by the new model are all smaller than those determined by the Washburn equation. For example, the water contact angle was calculated to be 77.1° by the new model, instead of 86.2° by the Washburn equation. From the water contact angle data, consideration of only the geometrical changes due to the particle swelling reduces the contact angle by only 1°. When both the geometrical changes and the energy loss are considered, the contact angle is reduced by about 9°. Therefore, in this case, the energy loss effect has a larger contribution to the water contact angle difference between the Washburn equation and the new model. However, for the formamide contact angle data, the geometrical changes play a larger role for the differences than the energy loss. The geometrical changes reduce the contact angle to about 24°, while the energy loss further reduces the contact angle to about 13° for formamide. The contact angle of ethylene glycol was reduced from 3.8° to 0°. The water contact angles calculated from the new model (77°) are closer to the sessile drop results (60°) from the literature [10, 11]. It should be pointed out that the volumetric swelling value (δ_V) of wood used in the calculation was obtained from ref. [8], in which a larger wood specimen was used. This δ_V value may not be accurate for the contact angle calculation. The amount of particle swelling actually occurring in the wicking measurements needs to be further investigated.

5. CONCLUSIONS AND RECOMMENDATIONS

Swelling polymer particles, such as wood and nylon, undergo swelling when they interact with polar liquids. This phenomenon was confirmed by two observations: temperature rise during the wicking process and the differences in the average capillary radius (r) values determined using a low surface tension polar liquid (methanol) and a low surface tension non-polar liquid (hexane). The capillary rise rate is reduced due to the particle swelling and therefore the contact angles calculated using the Washburn equation are not accurate because of the particle geometrical changes and energy loss that occur during the wicking process. A new model was developed in this study to describe more accurately the wicking process of swelling polymers interacting with polar liquids considering the effects of swelling on the wicking process. Based on the results of the wicking measurements on wood particles, it was shown that the contact angles determined by the new model were about 10° lower than those determined by the Washburn equation for water, and about 37° lower for formamide. The contact angle of ethylene glycol determined by the new model was 0°, compared with 3.8° determined by the Washburn equation. It is shown from this study that the differences between the r values determined by hexane and methanol using the Washburn equation can be used to judge whether swelling occurs during the wicking process. If the same r values are obtained using the two liquids, there is no swelling effect on the contact angle determination and the Washburn equation can be used for the contact angle calculation. Otherwise, the new model developed in the study should be used.

This study is a pilot work for modeling the wicking process of swelling polymer particles. In some aspects, the model can be further refined. For example, it was assumed that the energy loss (Q) was proportional to the height of the rising liquid, while the unit energy loss (S) had a linear relationship with the volumetric swelling (δ_V) of the particles. In reality, these assumptions may not necessarily be true, since the energy loss due to the adsorption during the rise of the liquid may be varied both with time and with height. The relationship between the unit energy loss and polymer swelling may not necessarily be linear. Further studies need to address the relationship between the energy loss and particle swelling. Also, to simplify the model derivation, it was assumed that the small particles in the column swelled immediately when interacting with the polar liquid. This may not necessarily be true because polymer swelling may occur as a function of time. To obtain accurate contact angles on swelling particles, more accurate volumetric swelling data on particles in the column should be obtained.

REFERENCES

1. E. W. Washburn, *Phys. Rev.* **17**, 273 (1921).
2. C. J. van Oss, R. F. Giese, Z. Li, K. Murphy, J. Norris, M. K. Chaudhury and R. J. Good, *J. Adhesion Sci. Technol.* **6**, 413–428 (1992).
3. J. B. He and J. S. Laskowski, *Proceedings of the 1st UBC-McGill Bi-annual International Symposium on Fundamentals of Mineral Processing*. Process Research Associates Ltd., Vancouver, British Columbia, Canada (20–24 Aug. 1995).
4. G. E. Parson, G. Buckton and S. M. Chatham, *J. Adhesion Sci. Technol.* **7**, 95–104 (1993).
5. D. J. Gardner, W. T. Tze and S. Q. Shi, *Progress in Lignocellulosics Characterization*, pp. 263–293. Tappi Press, Atlanta, GA (1999).
6. V. N. Constantinescu, *Laminar Viscous Flow*. Springer-Verlag, New York (1995).
7. J. Siau, *Transport Processes in Wood*. Springer-Verlag, New York (1984).
8. G. I. Mantanis, R. A. Young and R. M. Rowell, *Holzforschung* **48**, 480–490 (1994).
9. D. Braun, *Simple Methods for Identification of Plastics*, 2nd edn. Macmillan, New York (1986).

Part 4

Dynamic Effects in Contact Angle Measurement

Apparent and Microscopic Contact Angles, pp. 447–455
J. Drelich, J. S. Laskowski and K. L. Mittal (Eds)
© VSP 2000.

Determination of the peripheral contact angle of sessile drops on solids from the rate of evaporation

H. YILDIRIM ERBIL*

Kocaeli University, Department of Chemistry, Izmit 41300, Kocaeli, Turkey and TUBITAK, Marmara Research Center, Department of Chemistry, P. O. Box 21, Gebze 41470, Kocaeli, Turkey

Received in final form 10 March 1999

Abstract—A method was developed to determine the initial peripheral contact angle of sessile drops on solid surfaces from the rate of drop evaporation for the case where $\theta_i < 90°$. The constant drop contact radius, the initial weight, and the weight decrease with time should be measured at the ambient temperature for this purpose. When water drops are considered, the relative humidity should also be known. The peripheral contact angle so obtained is regarded as the average of all the various contact angles existing along the circumference of the drop. Thus, each determination yields an average result not unduly influenced by irregularities at a given point on the surface. In addition, the error in personal judgment involved in drawing the tangent to the curved drop profile at the point of contact can be eliminated. The application of this method requires the use of the product of the vapor diffusion coefficient with the vapor pressure at the drop surface temperature. This product can be found experimentally by following the evaporation of fully spherical liquid drops.

Key words: Contact angle; evaporation; diffusion; vapor pressure; surface tension.

1. INTRODUCTION

The surface tension of solids can be estimated from the contact angles measured from the profiles of the sessile drops formed on them. In principle, a given pure liquid drop on an ideal (flat, homogeneous, smooth, rigid, and isotropic) solid should give a unique value for the equilibrium contact angle, θ_e, as determined by Young's equation [1–3]. However, in practice, a whole range of angles between advancing and receding values exist depending on the previous history of the triple line. The difference between the advancing and the receding contact angles is the contact angle hysteresis, which is a measure of the deviation of the solid surface from the ideal state. It is a known fact that instead of measuring the contact angle directly, it can be calculated from the profile of the drop [4–7]. With this method,

*E-mail: yerbil@kou.edu.tr

the error in personal judgment involved in drawing the tangent to the curved drop surface at the point of contact is eliminated. Mack and Lee [5] showed that the contact angle depended not only on the size of the drop given by the drop height and drop contact radius, but also on its shape given by the radius of curvature at the apex. They calculated the radius of curvature from the capillary constant of the drop liquid and found the average peripheral contact angle from the dimensions of the drop. They used the contact radius, r_b, and the height, h, of the drop in their calculations and the angle so obtained was regarded as the integral of the sum of all the various contact angles existing along the circumference of the drop [5]. Since a sessile drop may show varying horizontal profiles having different r_b and h values according to the direction of photographing or when one looks for a revolving drop along its axis, the validity of the Mack and Lee conclusion from a single profile is questionable. Recently, Neumann and co-workers [8] developed a new method (ADSA-CD) to determine the horizontal profile of the drop simultaneously with the top view that was specifically used to calculate the average contact diameter. Then they calculated the mean contact angle by using predetermined drop volumes; this method resulted in contact angles approximately 10° greater than the angles calculated by applying spherical cap geometry, especially for low contact angles [8]. The precise determination of both the advancing and the receding contact angles is very important but the apparent simplicity of the contact angle measurement is, however, very misleading and the acquisition of thermodynamically significant contact angles and hysteresis values requires painstaking effort [4].

On the other hand, the occurrence of liquid evaporation is inevitable unless the atmosphere in the immediate vicinity of the drop is completely saturated with the vapor of the liquid. This fact decreases the reliability of the measured contact angles. An initial advancing contact angle will diminish towards a receding contact angle when the liquid constituting the meniscus starts to evaporate. A more complete understanding of how evaporation influences the contact angle of a drop on polymer surfaces in still air or in controlled atmospheric conditions is very important in the surface characterization processes and is the subject of many recent publications [9–17]. The evaporation behavior of small droplets of volatile fluids from solid surfaces depends on whether the initial contact angle is less or more than 90°. In the former case, the contact radius remains constant and the contact angle decreases for much of the evaporation time. Picknett and Bexon [9] reported two modes of evaporation of drops on surfaces: that at constant contact angle and that at constant contact area. They also developed a very successful theoretical analysis of each mode and compared the theoretical predictions with the experimental measurements of methyl acetoacetate drops on a poly(tetrafluoroethylene), PTFE, surface [9]. Birdi et al. [10, 11] reported the change in the mass and contact diameter of liquid drops placed on solids with time. They observed that the initial rate of evaporation was dependent on the radius of the liquid–solid interface, r_b, by assuming a spherical cap geometry [10]. A model based on the diffusion of vapor across the boundary of a drop was considered to explain their data. Shanahan and Bourges [12] pointed out that the liquid evaporation effect on the contact angle measurements seemed to have been largely neglected. They have shown the existence of three stages in the drop evaporation process in open air conditions. In the first stage, r_b remains constant while θ and the drop height, h, decrease. In the second stage, h and r_b diminish concomitantly, thus maintaining θ more or less constant for smooth surfaces. This stage does not exist on rough surfaces [13]. In the final stage, the drop disappears in an irregular fashion with h, r_b, and θ all tending to zero. They also showed that when the surrounding atmosphere was completely saturated by the vapor of the given liquid, the contact angle remained constant and they proposed a theory to calculate the diffusion coefficient of the liquid vapor in air. They used a drop model in which the vapor concentration varied between saturation on the drop surface to zero over a stagnation layer thickness covering the drop meniscus [13].

Recently, Rowan *et al.* [14] examined the change in mass and the geometry of small water droplets on poly(methyl methacrylate), PMMA, due to evaporation in open air. They examined only the first stage of drop evaporation where r_b was constant, with θ and h decreasing with time. The measurements of θ, r_b, and h with time in the regime of constant contact radius, valid for $\theta < 90°$, were reported. It was shown that the rate of mass loss was proportional to both h and r_b, but not to the spherical radius, R. The results were explained by a vapor diffusion model based on a two-parameter spherical cap geometry and the observed constant value of the contact radius [14]. Later, they applied this model to cases where $\theta > 90°$ for water drops on PTFE surfaces [15].

In all the above evaporation studies, the spherical cap geometry was used. It was assumed that when a drop of fluid was sufficiently small, the influence of gravity became negligible and the drop assumed a spherical cap shape. However, although gravity effects are negligible, shape distortion of liquid drops often occurs, due to internal flow in the drop during evaporation [7], and the ellipsoidal cap geometry was used in order to compare the differences between the surface area, drop volume, and evaporation rates between the spherical and ellipsoidal geometry approaches [16]. Rowan *et al.*'s [14] precise θ, r_b, and h data were used for the calculations. A vapor diffusion model for an ellipsoidal cap geometry was also developed similar to Rowan *et al.*'s and some success over the classical spherical cap geometry model was obtained [16]. In addition, it was realized that when the three different equation sets of the classical two-parameter spherical cap geometry were used to calculate the volume and the surface area of the drops, three different results were obtained, a point which has always been missed before in the existing literature [17]. In order to overcome this deficiency, a new model was proposed using the three-parameter pseudo-spherical cap geometry [17].

More recently, an attempt was made [18] to determine the mean peripheral contact angle from the evaporation rate of water sessile drops on a PMMA surface in still air using the data of Rowan *et al.* [14]. In this approach, the data of the weight (or volume) decreases with time, and the values of the constant drop contact radius, ambient temperature, and relative humidity were required. The product of the vapor diffusion coefficient of the evaporating liquid with the vapor pressure at the drop surface was found directly by following the evaporation of fully spherical liquid drops [19]. The other possible method to estimate the product $D \cdot \Delta p_v$ is the measurement of drop surface temperature to determine Δp_v and to multiply it with the constant D. The volumes of the drops reported in ref. [14] were calculated by applying the three-parameter spherical geometry model to the three given values of h, r_b, and θ [17], and it was found that [18] the arithmetic average of the calculated peripheral contact angle values deviated by only 0.4% from the average of the reported contact angles in ref. [14]. The obvious advantage of the peripheral contact angle determination is that it is not derived from a single profile of the drop on the surface, but is found from the volume to area relationship of the entire drop surface or, in other terms, it is the average of the contact angles of all the existing drop profiles.

In this paper, the determination of the average peripheral contact angle from the evaporation rate of a methyl acetoacetate sessile drop on a PTFE surface in still air is studied using the data of Picknett and Bexon [9]. They reported the weight decrease of a sessile drop having an initial contact angle of 63° and with a constant initial contact area with time until 100 min and also the weight decrease of a spherical drop with time until 45 min at an average temperature of 22.5°C [9]. The approach of determining the mean peripheral contact angle from the evaporation rate [18] is applied to these data in order to compare the resultant peripheral contact angle with the reported experimental tangential value of 63°.

2. THEORY

Based on geometry, the initial volume of a spherical cap drop is given as

$$V_i = \frac{\pi r_b^3 (2 - 3 \cos \theta_i + \cos^3 \theta_i)}{3 \sin^3 \theta_i}, \tag{1}$$

where r_b is the contact radius at the solid–liquid interface and θ_i is the initial contact angle at time $t = 0$ of the evaporation period. The initial surface area, A_i, of the liquid–air interface for this sessile drop is given as

$$A_i = \frac{2\pi r_b^2}{1 + \cos \theta_i}. \tag{2}$$

Then the initial volume-to-area ratio becomes

$$\frac{V_i}{A_i} = \frac{r_b (2 + \cos \theta_i)(\sin \theta_i)}{6(1 + \cos \theta_i)}. \tag{3}$$

When the vapor concentration of a substance is considered, the initial evaporation rate is given as [14–17]

$$-\frac{dm}{dt} = -\rho_L\left(\frac{dV}{dt}\right) = D\int \nabla c\,dA = D\int \frac{\partial c}{\partial n}\,dA, \tag{4}$$

where m is the weight of the drop (in g), D is the diffusion coefficient (in $cm^2\ s^{-1}$), ρ_L is the density of the liquid (in $g\ cm^{-3}$), c is its vapor concentration (in $g\ cm^{-3}$), ∂n denotes the element of the outward normal (in cm), and the integral of the concentration gradient is taken over the entire surface of the spherical cap drop. In order to perform the integration in equation (4), the concentration gradient is assumed to be radially outward, and using the boundary conditions $c = c_\infty$ as $r \to \infty$ and $c = c_0$ as $r = r_{sphere}$, it is assumed that

$$\frac{\partial c}{\partial n} = \frac{(c_\infty - c_0)}{r}, \tag{5}$$

where c_0 is the vapor concentration at the surface of the drop, c_∞ is the vapor concentration at infinite distance, and r is the radius of the spherical drop. Equation (5) is the main assumption of the vapor diffusion models of Rowan *et al.* [14] and Birdi *et al.* [10]. Since $r_b = r \sin \theta_i$ by geometry and $\Delta c = c_\infty - c_0$, then by combining equations (4) and (5), one obtains

$$-\frac{dm}{dt} = D\int \frac{\Delta c \sin \theta_i\,dA}{r_b}. \tag{6}$$

Langmuir assumed that D was practically independent of the partial density of the vapor of the evaporating substance for low concentrations of the evaporating vapors [20]. Using the ideal gas laws, he expressed Δc as

$$\Delta c = \frac{M\Delta p_v}{RT}, \tag{7}$$

where M is the molecular weight, Δp_v is the vapor pressure difference of the evaporating substance, R is the gas constant, and T is the absolute temperature in K. Surface cooling of the drop by evaporation must also be considered in this process. When water drops on solids are considered,

$$\Delta p_v = p_0 - p_\infty, \tag{8}$$

where p_0 is the saturation water vapor pressure at the drop surface temperature and p_∞ is the water vapor pressure in the ambient still air, which can be calculated by multiplying the relative humidity (RH) and the water vapor pressure at the ambient temperature. When organic liquid drops are considered, Δp_v is the vapor pressure difference between saturation at the drop surface and $p_\infty = 0$ at infinite distance. By combining equations (6) and (7) and integrating, one obtains

$$-\frac{dm}{dt} = \frac{(D\Delta p_v)M}{RTr_b}(A_i \sin \theta_i). \tag{9}$$

By combining equations (3) and (9) and rearranging, one obtains

$$\cos\theta_i = \frac{6V_i M(D\Delta p_v) - 2(-dm/dt)RT r_b^2}{(-dm/dt)RT r_b^2 - 6V_i M(D\Delta p_v)}. \quad (10)$$

Thus, it is possible to calculate the initial peripheral contact angle if the constant contact radius r_b is measured, the decrease in the weight (or volume) of the sessile drop is monitored, and the product of $D \cdot \Delta p_v$ is known. It should be noted that D is dependent on the ambient temperature and Δp_v is dependent on both the drop surface temperature and the ambient temperature. (When the volume decrease is monitored, ρ_L, the liquid density, which is dependent on the bulk drop temperature, is also used in the calculations.) The product of $D \cdot \Delta p_v$ can be determined by following the evaporation of fully spherical drops [19] so that

$$D\Delta p_v^* = \frac{0.0712 m_0^{2/3} \rho^{1/3} RT}{M t_{1/2}}, \quad (11)$$

where m_0 is the initial weight (in g) of the spherical drop at time $= 0$ and $t_{1/2}$ is the half-life of the drop, which is the time taken to lose half the initial mass (in s). In general, $\Delta p_v^* = P \ln\{1/1 - (\Delta p_v/P)\}$ and in practice, $\Delta p_v^* \cong \Delta p_v$ for small Δp_v values (less than 3–4 mmHg pressure). Consequently, when $t_{1/2}$ is determined experimentally, it is possible to calculate the product of the diffusion coefficient and vapor pressure of the drop surface at the surface temperature of the drop.

It should be noted that the above equations can be applied to the case where $\theta_i < 90°$. For a non-wetting liquid on a solid where $\theta_i > 90°$ (e.g. a water drop on PTFE), the evaporation data suggest that the contact angle is a slow variable compared with the contact radius r_b; McHale et al. [15] showed that the time dependence of r_b^2 was the important parameter for this case.

3. RESULTS AND DISCUSSION

The data of Picknett and Bexon [9] were used in the calculations. They formed a methyl acetoacetate drop having an initial weight $m_i = 2.0 \times 10^{-3}$ g on a PTFE surface and this drop gave a constant contact radius $r_b = 0.12$ cm throughout the experiment with an initial contact angle $\theta_i = 63°$ which progressively decreased over a period of 90 min (see Fig. 11 of ref. [9]). The average temperature during the experiments was 22.5°C. The weight decrease of the sessile drop by evaporation was extracted from the above figure and is given in Fig. 1. Since an organic liquid drop was used in this experiment, the information on relative humidity is not required. The evaporation rate was calculated from the slope of this plot as $-(dm/dt) = 2.71 \times 10^{-7}$ g s^{-1} and the regression coefficient (R^2) was found to be 0.9987. Picknett and Bexon determined the weight decrease of a fully spherical methyl acetoacetate drop by using a circular frame containing 0.2 mm in diameter fibers crossed at the center. They formed a drop of liquid at the crossing point and

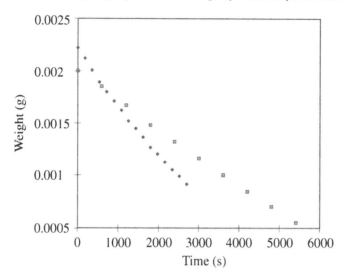

Figure 1. Weight decrease with time for a methyl acetoacetate sessile drop (■) on the PTFE surface and a fully spherical methyl acetoacetate drop (◆) in still air at 22.5°C [9].

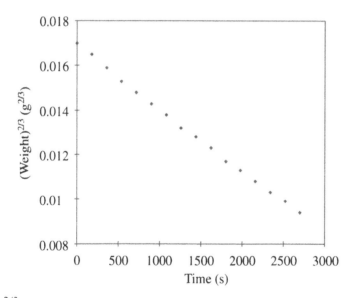

Figure 2. $m^{2/3}$–time plot of the experimental evaporation of a fully spherical methyl acetoacetate drop supported on crossed fibers at 22.5°C [9].

claimed that it had an almost spherical shape for most of its life. The initial weight of this spherical methyl acetoacetate drop was $m_o = 2.217 \times 10^{-3}$ g and the weight decrease of the drop by evaporation was extracted from Fig. 9 of ref. [9] and is given also in Fig. 1. The drop half-life time was determined from this plot as $t_{1/2} = 2120$ s. The linearity of the $m^{2/3}$ change with time of the fully spherical drop can be seen in Fig. 2 with a regression coefficient (R^2) of 0.9989. The molecular weight of methyl acetoacetate was taken as $M = 116.12$ g mol^{-1}, the density at 22.5°C as

$\rho = 1.075$ g cm^{-3}, and the gas constant as $R = 62360$ cm^3 mmHg mol^{-1} K^{-1}. The product of the diffusion coefficient at 22.5°C with the vapor pressure at the drop surface temperature, $D \cdot \Delta p_\mathrm{v}$, calculated using equation (11) was 0.09285 cm^2 s^{-1} mmHg^{-1}. The initial peripheral contact angle was calculated using equation (10) and was found to be $\theta_\mathrm{i} = 60.8°$. The value of the peripheral contact angle was found to deviate by only 3.5% from the reported tangential contact angle [9]. This value is larger than our previously reported 0.4% average deviation [18] from the reported tangential water contact angle of Rowan *et al.*'s [14] data on PMMA, but is well within the $\pm 3°$ experimental error range. In addition, the reported tangential value of 63° may represent the contact angle of a single profile having a comparatively larger θ value than the average. It should be noted that both data were obtained by very precise experimentation and the evaporation rates were slow, of the order of 10^{-7} g s^{-1}. Improper experimentation may give unsatisfactory results because this method is very sensitive to the evaporation rate of sessile drops on solids and the drop half-life of fully spherical drops.

4. CONCLUSION

A method has been proposed to determine the initial peripheral contact angle of sessile drops on solid surfaces from the rate of drop evaporation by using the data of Picknett and Bexon [9]. The values of the drop contact radius, initial weight (or volume), the weight decrease with time, and the ambient temperature were used for this purpose. The method resulted in a peripheral contact angle value which deviated by only 3.5% from the reported tangential contact angle obtained from a single drop profile. The peripheral contact angle so obtained may be regarded as the average of all the various contact angles existing along the circumference of the drop. Thus, each determination yields an average result not unduly influenced by irregularities at a given point on the surface. In addition, the error in personal judgment involved in drawing the tangent to the curved drop surface at the point of contact is eliminated.

REFERENCES

1. H. Y. Erbil, in: *Handbook of Surface and Colloid Chemistry*, K. S. Birdi (Ed.), pp. 259–306. CRC Press, Boca Raton, FL (1997).
2. M. Morra, E. Occhiello and F. Garbassi, *Adv. Colloid Interface Sci.* **32**, 79–116 (1990).
3. R. J. Good, in: *Contact Angle, Wettability and Adhesion*, K. L. Mittal (Ed.), pp. 3–36. VSP, Utrecht, The Netherlands (1993).
4. (a) J. K. Spelt and E. I. Vargha-Butler, in: *Applied Surface Thermodynamics*, A. W. Neumann and J. K. Spelt (Eds), pp. 379–412. Marcel Dekker, New York (1996).
   (b) D. Y. Kwok, A. W. Neumann and D. Li, in: *Applied Surface Thermodynamics*, A. W. Neumann and J. K. Spelt (Eds), pp. 413–440. Marcel Dekker, New York (1996).
   (c) S. Lahooti, O. I. Del Rio, A. W. Neumann and P. Cheng, in: *Applied Surface Thermodynamics*, A. W. Neumann and J. K. Spelt (Eds), pp. 441–506. Marcel Dekker, New York (1996).

5. (a) G. L. Mack and D. A. Lee, *J. Phys. Chem.* **40**, 159–167 (1936).
   (b) G. L. Mack and D. A. Lee, *J. Phys. Chem.* **40**, 169–176 (1936).
6. (a) Y. Rotenberg, L. Boruvka and A. W. Neumann, *J. Colloid Interface Sci.* **93**, 169–183 (1983).
   (b) D. Y. Kwok, W. Hui, R. Lin and A. W. Neumann, *Langmuir* **11**, 2669–2673 (1995).
7. H. Y. Erbil, G. McHale, S. M. Rowan and M. I. Newton, *J. Adhesion Sci. Technol.* **13**, 1375–1391 (1999).
8. (a) F. K. Skinner, Y. Rotenberg and A. W. Neumann, *J. Colloid Interface Sci.* **130**, 25–34 (1989).
   (b) D. Li, P. Cheng and A. W. Neumann, *Adv. Colloid Interface Sci.* **39**, 347–382 (1992).
9. R. G. Picknett and R. Bexon, *J. Colloid Interface Sci.* **61**, 336–350 (1977).
10. K. S. Birdi, D. T. Vu and A. Winter, *J. Phys. Chem.* **93**, 3702–3703 (1989).
11. K. S. Birdi and D. T. Vu, *J. Adhesion Sci. Technol.* **7**, 485–493 (1993).
12. M. E. R. Shanahan and C. Bourges, *Int. J. Adhesion Adhesives* **14**, 201–205 (1994).
13. C. Bourges-Monnier and M. E. R. Shanahan, *Langmuir* **11**, 2820–2829 (1995).
14. S. M. Rowan, M. I. Newton and G. McHale, *J. Phys. Chem.* **99**, 13 268–13 271 (1995).
15. G. McHale, S. M. Rowan, M. I. Newton and M. Banerjee, *J. Phys. Chem. B* **102**, 1964–1967 (1998).
16. H. Y. Erbil and R. A. Meric, *J. Phys. Chem. B* **101**, 6867–6873 (1997).
17. R. A. Meric and H. Y. Erbil, *Langmuir* **14**, 1915–1920 (1998).
18. H. Y. Erbil, *J. Phys. Chem. B* **102**, 9234–9238 (1998).
19. H. Y. Erbil and M. Dogan, *J. Phys. Chem. B* (submitted).
20. I. Langmuir, *Phys. Rev.* **12**, 368–370 (1918).

Apparent and Microscopic Contact Angles, pp. 457–473
J. Drelich, J. S. Laskowski and K. L. Mittal (Eds)
© VSP 2000.

Analysis of evaporating droplets using ellipsoidal cap geometry

H. Y. ERBIL [1], G. McHALE [2,*], S. M. ROWAN [2] and M. I. NEWTON [2]

[1] *Tübitak, Marmara Research Center, Department of Chemistry, P.O. Box 21, Gebze 41470, Kocaeli, Turkey*
[2] *Department of Chemistry and Physics, The Nottingham Trent University, Clifton Lane, Nottingham NG11 8NS, UK*

Received in final form 10 March 1999

Abstract—The evaporation of small droplets of volatile liquids from solid surfaces depends on whether the initial contact angle is larger or less than 90°. In the latter case, for much of the evaporation time the contact radius remains constant and the contact angle decreases. At equilibrium, the smaller the drop, the more it is possible to neglect gravity and the more the profile is expected to conform to a spherical cap shape. Recently published work suggests that a singular flow progressively develops within the drop during evaporation. This flow might create a pressure gradient and so result in more flattening of the profile as the drop size reduces, in contradiction to expectations based on equilibrium ideas. In either case, it is important to develop methods to quantify confidence in a deduction of elliptical deviations from optically recorded droplet profiles. This paper discusses such methods and illustrates the difficulties that can arise when the drop size changes, but the absolute resolution of the system is fixed. In particular, the difference between local variables, such as contact angle, cap height, and contact diameter, which depend on the precise location of the supporting surface, and global variables such as radii of curvature and eccentricity, is emphasized. The applicability of the ideas developed is not limited to evaporation experiments, but is also relevant to experiments on contact angle variation with drop volume.

Key words: Wetting; contact angles; evaporation.

1. INTRODUCTION

The nature of the evaporation of sessile drops of a volatile liquid resting on a solid surface depends on the initial contact angle [1, 2]. For an initial angle less than 90°, for much of the evaporation time the contact radius remains constant as the droplet mass reduces [1–4]. In contrast, for an initial contact angle of greater than 90°, it is the contact angle which remains approximately constant [5]. To explain

---

*To whom correspondence should be addressed. E-mail: glen.mchale@ntu.ac.uk

the observed dependence of the geometrical parameters, Birdi and Vu [2] modelled the evaporation process using a diffusion model and a spherical shape with a single radius of curvature, R. This model was later extended to include the effect of the two radii of curvature (spherical radius R and contact radius r) and was applied to both regimes of initial contact angle [3, 5] (also see ref. [4]). Whilst the model accounts well for the observed data, it has been criticized for its assumption of a spherical cap shape [6, 7]. A generalization of the model based on an elliptical cap model has been given by Erbil and Meric [6]. In this work, it was suggested from a re-analysis of previously published data, for a water-on-poly(methyl methacrylate) (PMMA) system, that such an elliptical shape was more appropriate. It was also concluded that the elliptical shape was more pronounced for the smallest drops. In studying the equilibrium contact angle, a convenient measure of the relative importance of gravity to capillary forces is given by the capillary length $\kappa^{-1} = (\gamma_{LV}/\rho g)^{1/2}$, where γ_{LV} is the liquid–vapor surface tension, ρ is the density of the liquid, and g is the acceleration due to gravity. In the data for water on PMMA, the capillary length was 2.8 mm and the range of contact radii for the droplets studied was 0.3–0.6 mm. It is therefore plausible that gravity would lead to a small flattening of the droplets, but the same considerations would also suggest that as the droplet volume reduced, gravity would become less important and the droplet would become more spherical in shape. Thus, more intriguing than the existence of a small correction to the spherical cap shape is the suggestion that a droplet may deviate *towards* an elliptical shape *as* the droplet evaporates. This would be contrary to initial expectations based on equilibrium ideas and the capillary length.

One possible physical mechanism which may result in droplet flattening is outlined in Fig. 1. When evaporation occurs from a droplet with a pinned contact line, the evaporation is necessarily accompanied by a flow towards the periphery. Such a flow has been quantitatively observed and reported as the mechanism underlying the formation of ring stains from a drying drop [8]. If evaporation

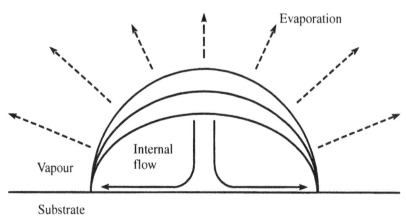

Figure 1. An illustration of a droplet undergoing evaporation with a pinned contact area and possible internal flows. The evaporation may result in a better spherical cap shape as the volume reduces or it may result in greater drop flattening as the internal flows are established.

proceeded with no outward flow towards the contact line of the drop, shrinkage of the contact area would occur. We expect the internal outward flow towards the contact line will necessitate a vertical downward flow in the centre of the drop. Such a flow could lead to a distortion of the spherical cap shape, otherwise dictated by surface tension considerations. The normal expectation would be for the Laplace pressure to act sufficiently rapidly to ensure a minimal surface area that in the absence of gravitational effects would be a spherical cap. However, in the model of ring stains suggested by Deegan *et al.* [8], a divergence occurs in the evaporative flux as the contact angle reduces and this leads to a diverging flow velocity near the perimeter of the drop. If the downward flow does distort the spherical cap shape, then the importance of the distortion could grow as the flow velocity increases. Thus, it is possible for us to envisage at least one dynamic mechanism by which an evaporating droplet could increasingly become distorted in shape as the evaporation proceeds and the actual drop volume reduces.

To determine experimentally whether small evaporating droplets become increasingly elliptical as they evaporate is not simple. The change in a hypothesized shape from a spherical cap to an ellipsoidal cap may increase the agreement between a fit and an experimentally determined profile. However, the change in shape is at the cost of introducing one extra parameter. Since the droplets are assumed to be small, the increased agreement is likely to be marginal and it will be difficult to determine whether such a gain in the quality of a fit is physically meaningful. In this paper, we examine methods by which elliptical distortions can be identified from optical profiles of small droplets and the use of image analysis. In practice, digital image analysis necessarily introduces a finite pixel resolution in an observed drop profile and requires consideration of the quantization error. For systems in which a fixed optical magnification is used to observe droplets with changing volumes, the effective resolution for the different droplet profiles during the course of an experiment changes. Although this paper concentrates on evaporation and ellipsoidal deviations of shape, the discussion applies to some studies of the contact angle dependence on the drop size and hence to the line tension of solid–liquid systems (see refs [9–12]). Initially, the extent of the problem of shape analysis using digitized profiles of droplets is discussed and the sensitivity to small systematic errors is illustrated using direct values of the contact angle, height, and contact radius. Alternative techniques based on complete fitting of profiles are then discussed within the context of determining confidence in an estimate of eccentricity. From these considerations, we then suggest the concept of local and global variables and a method, single image sequencing (SIS), for determining confidence in a deduction of ellipticity from any single image is proposed. The method is illustrated using estimates of the volume, contact radius, and drop height. To enable a clear separation of uncertainties due to pixel resolution from other factors in image processing, such as edge detection algorithms and lighting, the discussion in this paper makes extensive use of simulated data. However, the ideas developed have been tested on experimentally determined profiles. The analysis of two such images, one of a ball-bearing representing a spherical cap and one of a large drop of water resting on a Mylar substrate, are presented to conclude the discussion.

2. ELLIPSIDAL CAP GEOMETRY AND ACCURACY

In contact angle measurements, digital image analysis offers many advantages such as speed of use, ease of processing images, the ability to average images, the ability to analyze dynamically changing shapes, etc. [13–19]. However, digital representations are finite-sized arrays of the light intensity values in an image and each pixel element of the array represents a finite-sized area of the image. Image pixel resolutions, offered by current framegrabber cards, include 320 × 240 and the more usual 'high-resolution' formats of 480 × 512, 640 × 480, and 512 × 512. Advances in technology are now providing even higher resolutions, such as 1536 × 1024, at relatively low cost [15]. The finite resolution of the digital images necessarily introduces a quantization error in any parameter extracted from an image. It also opens the possibility of relatively subtle, but systematic, errors due to calibration and lighting considerations. Effects such as optical distortions and non-square pixels are well known and can be corrected using known calibration grids. Lighting conditions can influence the perceived locations of edges in digital images and the techniques for edge detection need to be carefully considered. Consequently, in this paper we largely present simulated profiles obtained by quantizing exact calculated profiles. To ensure such simulations are representative of true image resolutions, we use resolutions of around 500 × 500; this also simplifies the problem to square pixels. However, the analysis and conclusions that we present using simulated data have also been tested on images of small sessile droplets of water captured using a DataTranslation DT-3152 Scientific framegrabber card. This card provides 256-gray level images of pixel resolution 768 × 576 and with a 1 : 1 aspect ratio. These experimentally recorded images were analysed using UTHSCSA ImageTool software supplemented by in-house routines written both in 'C' and using spreadsheets to implement the algorithms.

An elliptical shape is defined by the equation

$$\frac{(x - x_0)^2}{a^2} + \frac{(y - y_0)^2}{b^2} = R^2, \tag{1}$$

where (x_0, y_0) is the centre of the co-ordinate system and aR and $2bR$ give the maximum height and width of the ellipse. The special case of an ellipse centred on $(0, 0)$ is shown in Fig. 2. Typically, in contact angle measurements, the contact radius r, the height h, and the contact angle θ are measured directly and these depend on the location of the intersecting solid surface (AB in Fig. 2). Alternatively, the profile between points A and B may be recorded and geometrical parameters extracted by analysing the pixel co-ordinates [13–15]. For an ellipsoidal cap, a convenient parameter for defining the extent to which the shape is an ellipse rather than a circle is the ratio of the axes, $\varepsilon_r = b/a$, which we will also refer to as

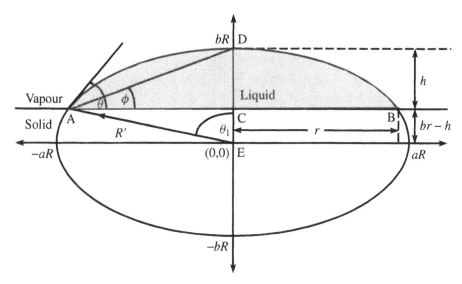

Figure 2. The definition of geometrical parameters for an ellipsoidal cap droplet on a supporting solid surface. The parameters can be classed as either global parameters (such as bR and ε_r) or local parameters (such as r, h, and θ), depending on whether their values depend on the location of the intersecting substrate (AB).

drop ellipticity; values less than unity indicate a flattened ellipsoidal cap shape for a droplet. It should be noted that the drop ellipticity is related to the classical eccentricity by $\varepsilon^2 = 1 - \varepsilon_r^2$ when $\varepsilon_r < 1$, although some texts also use $\varepsilon^2 = 1 + \varepsilon_r^2$. One of the simplest consistency tests for a spherical cap can be found by measuring the contact radius, height, and contact angle independently and then calculating the contact angle from the relation

$$\tan\left(\frac{\theta}{2}\right) = \frac{h}{r}. \tag{2}$$

In any measured data from a spherical cap, exact agreement is unlikely to occur between the directly measured contact angle and that deduced from equation (2). A measurement from a profile will have an intrinsic accuracy and, additionally, a systematic error may exist. For example, identifying the precise height at which the horizontal surface (the baseline) intersects a drop profile may be difficult, particularly for contact angles near 90°. A zero offset in h could then occur and this would produce a systematic error. Since calibration grids often have lines that are at least one pixel wide, one should expect that the normal situation will be to have such a small error. Alternatively, if a system has non-square pixels, a calibration error may exist between the vertical and horizontal directions. The importance of finite precision can be examined by rewriting equation (2) as

$$\tan\theta = \frac{2hr}{r^2 - h^2} \tag{3}$$

and then generalizing it to the following form for an ellipse (see the Appendix for a proof):

$$\tan \theta = \frac{2hr}{r^2 - \left(\dfrac{h}{\varepsilon_r}\right)^2}. \tag{4}$$

Equation (4) allows the ellipticity to be expressed in terms of the parameters most easily measured directly:

$$\varepsilon_r = \sqrt{\frac{h^2 \tan \theta}{r^2 \tan \theta - 2hr}}. \tag{5}$$

To examine the effect of a slight, but systematic, misidentification of the baseline as the drop volume changes, we generated a set of perfect circles, each with a contact angle of exactly 69°. The height and contact radii were then rounded to the nearest integer to obtain pixel estimates. The largest circle was chosen to have a contact radius of 320 pixels and the smallest of 80 pixels. This models the effect of using the maximum optical magnification and maximum resolution of a standard type image capture card for the largest drop. To scale to real units, we used a conversion of 623 pixels to 1 cm and for simplicity assumed a 1 : 1 vertical to horizontal aspect ratio. Figure 3 shows the effect of introducing a systematic 4-pixel overestimate on the drop height of each circle and then using the pixel values of height and contact radius to calculate the contact angle using equation (2). The solid diamond symbols illustrate how $\cos \theta$ apparently changes systematically as the volume increases; the solid line is a straight-line fit and has an R^2 parameter better than 0.99. Figure 3 also demonstrates how the spherical cap shape can be misinterpreted as being an ellipsoidal cap shape if the ellipticity is deduced by measuring the three parameters (r, h, θ) and a small systematic error exists. This is not surprising since the systematic error highlighted in Fig. 3 will have the least percentage influence for the largest drops and an increasing influence as the drop size reduces. The circle symbols show ε_r^{-1} calculated from equation (5), using the overestimated h with the pixel contact radius and the true value of $\theta = 69°$. Again, a very good straight-line fit can be made to these data; a plot of ε_r against $1/r$ also shows a linear trend and can be fitted with a straight-line having an R^2 of better than 0.99. The choice of an overestimate of height gives $\varepsilon_r > 1$ and this would represent the opposite of drop flattening. However, similar results, but with a trend towards drop flattening, arise if we make an underestimate of just a few pixels.

The choice of parameters in Fig. 3 actually represents a simulation of the experimentally observed increase in contact angle due to line tension for dodecane on FC721 [9]. In these very careful line tension experiments, the stated accuracy of the calibration was ± 1 pixel, so that we cannot claim a systematic baseline error as the cause of the observed contact angle increase with volume. However, the range of pixel heights used in Fig. 3 is 55–220 and the baseline error is only 4 pixels, which corresponds to an error of just 2% at the maximum resolution. Figure 3 therefore

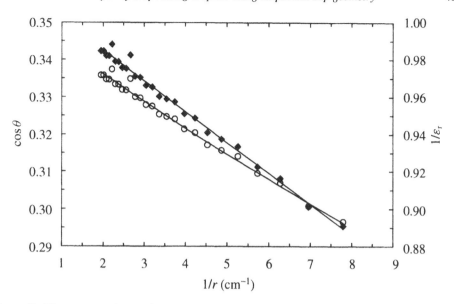

Figure 3. The apparent changes in contact angle (solid diamonds and left-hand scale), shown as a cosine, and the ratio of axes (circles and right-hand scale), shown as inverse ellipticity, as the drop size changes. A 4-pixel overestimate in height was used to generate the data. The solid lines are straight-line fits with R^2 values of better than 0.99.

illustrates how such a small error could lead to difficulties in interpretation of the shape and the precise contact angles in experiments on evaporation [3, 6] or line tension [9], which necessarily involve volume changes. It is worth emphasizing that the simulation assumes that any processing of a real image is capable of identifying the exact profile. The error considered is then only in the extraction of profile parameters from pixel approximations to profile co-ordinates. However, the source of the difficulties which produces an apparent trend as the drop volume changes is not simply whether a systematic error exists, as in the baseline error in Fig. 3. A closer consideration suggests that the quantization of a profile to pixel co-ordinates produces a systematically varying effective resolution as the volume changes. We consider, therefore, the circumstances under which a reliable deduction that a shape is elliptical can be made given a perfectly obtained pixel profile with no errors, random or systematic, in the pixel measurements of height and contact radius. The only influence considered will be the effect of a typical fixed image resolution, although care will be taken to allow for any volume change of the droplet.

3. PROFILE FITTING BY MINIMIZATION

The use of a small number of directly measured parameters to determine shape or contact angle leads to a critical dependence on specific geometric points [14]. One alternative is to digitize the complete profile and use all the points, in equal measure, to determine the best shape. A possible approach is to minimize a defined error

function between a predetermined shape and the pixel co-ordinates. As discussed
by Li *et al.* [14], the choice of error function can be critical when the sessile drop
shape is influenced by gravity forces. Indeed, this motivated them to develop an
alternative method: axisymmetric drop shape analysis (ADSA-P) [14, 20]. In
this technique, the objective error function is based on the sum of the squares
of the normal distances between the measured points and the calculated curve.
The calculated curve itself is not a predetermined ellipsoidal cap shape, but is a
theoretical Laplacian curve, based on numerical solution of

$$\Delta P_0 + \Delta \rho g z = \gamma \left(\frac{1}{R_1} + \frac{1}{R_2} \right), \tag{6}$$

where ΔP_0 is the pressure difference at an arbitrary datum plane, $\Delta \rho$ is the density
difference across the liquid–vapor interface, g is the gravitational acceleration, z is
the vertical difference above the datum plane, and the R_i's are the principal radii of
curvature at any point. In our work, we seek to determine whether a flattening of
drop shape towards that of an ellipsoidal cap shape occurs, as characterized by the
ellipticity parameter ε_r. We have therefore adopted the approach of minimizing an
objective function representing the error between a theoretical ellipsoidal cap and
the measured profile, rather than the ADSA-P type approach.

In determining the fit to a profile for any given image, we wish to find a parameter
set $E = (V, \varepsilon_r, bR, r, h, \theta)$; we have ignored the surface areas in this parameter
set. To do so, we also need to find the origin (x_0, y_0) that maps the measured pixel
co-ordinates onto the theoretical ellipse. Any three parameters in the set E can be
used to define the other parameters in the set. It is interesting to note that the volume
could be directly estimated from a mass measurement rather than be extracted from
an image profile. A profile-fitting routine for an ellipsoidal cap shape needs to
determine five parameters consisting of (x_0, y_0) and three further from the set E.
An example set for the more restricted case of a spherical cap is (x_0, y_0, R, r). The
fourth parameter, r, is not actually required to determine the overall spherical shape,
but rather to give estimates of V, h, and θ. Since for an ellipsoidal cap the shape has
a major and a minor axis, the fit to the profile requires one further parameter. This
extension in the number of fitting parameters as the assumed shape changes from a
spherical cap to an ellipsoidal cap illustrates the difficulty in determining whether
any marginal improvement in fit is physically significant. To examine the effect
of pixel resolution, fitting routines were written in the 'C' language, which used
numerical recipes to minimize objective error functions for ellipsoidal cap shapes.
Since we were only interested in profiles showing both left- and right-hand sides,
we used the average x-pixel co-ordinate to define x_0; this reduces the difficulty of
the minimization algorithm by removing one parameter. Various versions of the
routines with x_0 as a fitting parameter were also tested, but were found to be less
accurate. The fitting routines therefore sought to determine the best values of the
three parameters (y_0, bR, ε_r) and these enabled the shape to be defined. To obtain
a contact angle, θ, from a fitted shape would also require an estimate for either r or

h. The best objective error function tested was based on

$$\sum_{\text{data}} \left[(bR)^2 - \varepsilon_{\text{r}}^2 (x_{\text{data}} - x_0)^2 - (y_{\text{data}} - y_0)^2 \right]^2, \tag{7}$$

which leads to a linear best-fit equation set in the variables $(bR)^2$ and ε_{r}^2 and has the advantage of weighting both x and y co-ordinates equally. The routines used initial guesses with $\varepsilon_{\text{r}} = 1$ and bR as the spherical cap radius estimated using equation (2) and the maximum drop width and height in the measured co-ordinates.

To demonstrate the extent to which a finite pixel resolution can influence the determination of shape, we generated an ellipsoidal cap with $\varepsilon_{\text{r}} = 0.9$ and a maximum height of 250 pixels when the contact angle was 90°. The co-ordinates were then rounded to the nearest integers to represent an image with pixel co-ordinate values. The choice of a maximum height of 250 pixels corresponds to a droplet profile with maximum resolution using a standard framegrabber card. The original pixel co-ordinates were then used with the profile-fitting routine and the data set was reduced by one row at a time, from the base of the shape, to simulate the effect of a reduction in the extent of profile available for fitting. In principle, for a perfect ellipse the ellipticity parameter should not depend on the extent to which a full profile is available for fitting. Figure 4 shows how ε_{r} changes with h. Initially, the value of ε_{r} obtained from fitting was equal to the expected value of 0.9, but as the number of rows decreased, ε_{r} became less consistent. The observed behavior is not a simple systematic increase in underestimate of the drop ellipticity ε_{r}. The contact angle corresponding to the number of rows in the image is also shown in

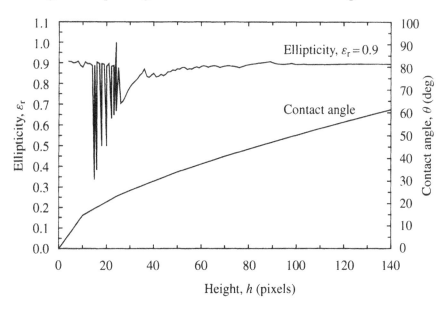

Figure 4. The upper curve shows the ellipticity predicted by fitting ellipsoidal cap shapes to pixel profiles that are systematically reduced one row at a time. The lower curve shows the value of the contact angle as the number of rows of pixels changes.

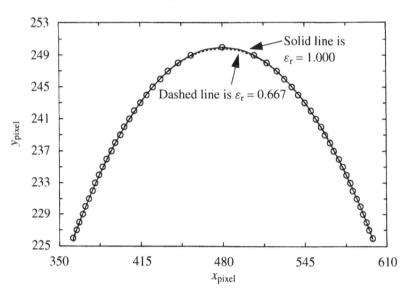

Figure 5. The circles show an example of a profile that can be fitted by ellipsoidal cap shapes with widely differing values of the ratio of axes. The two solid lines show fits with ellipticities of 0.667 (dashed line) and 1.000 (solid line) and these can hardly be distinguished. The contact angles predicted by the two fits are almost identical.

Fig. 4 and this indicates that significant inaccuracies occur once the contact angle falls below 40°. As the contact angle reduces further, it is even possible to find a limited number of cases having more than one apparently good solution for the profile. In this context, a good solution means that for a given tolerance in the error function, a fit can be found which apparently intersects all the data points in the profile and that changing this tolerance marginally does not alter the fit. However, in this limited number of cases, changing the tolerance by a larger margin can give a second apparently stable solution with a quite different ellipticity. An example of two such fits is shown in Fig. 5 and it is difficult to distinguish which of the two solutions is the better description of the profile. Moreover, the fitted value of the ratio of axes closest to 0.9 is not necessarily the one corresponding to the smallest tolerance in the error function. Apart from this limited number of cases, unique fits were found in which changing the tolerance in the error function did not lead to potentially competing fits. When the same fitting routine was used with a data set consisting of the actual co-ordinates, rather than the rounded co-ordinates, fits with the correct ratio of axes were obtained. It is also interesting to note that using the alternative error function

$$\sum_{\text{data}} \left[\varepsilon_r (x_{\text{data}} - x_0) - \sqrt{(bR)^2 - (y_{\text{data}} - y_0)^2} \right]^2 \tag{8}$$

also gave apparently good fits, but with $\varepsilon_r \approx 1.00$. In contrast to equation (7), the best fit based on equation (8) is non-linear and the square root suggests large relative errors when $x_{\text{data}} \approx x_0$. This supports the warning given by Li *et al.* [14] on using the horizontal distance between the measured points and the calculated curve.

4. SINGLE IMAGE SEQUENCING (SIS)

The difficulty in deciding whether a drop becomes more or less elliptical as it evaporates is not in obtaining a fit to an observed profile, but in knowing the reliability of the deduced ellipticity ε_r. Given an image of a drop, parameters for this purpose can be categorized into two types. The first type we define as global parameters, in the sense that their values should not depend in any way on the location of the solid surface (the line AB in Fig. 2) that intersects the shape. For a circle, the spherical radius R is global and for an ellipse, the scaled radius bR and the ratio of axes ε_r are global. The second type of parameter is local in the sense that it depends on the precise location of the solid surface intersecting the shape. Examples of such local parameters are the contact radius r, the height h, and the contact angle θ. The idea that global parameters, extracted from the analysis of an image, should not depend on the intersecting surface leads to a simple scheme for classifying the reliability of a shape determination within the constraints of the image resolution. To analyze an image, we begin with a sub-image consisting of the top few rows of pixels near the apex of the drop and then extract the profile and estimate the global parameters. The sub-image is then increased by a single row of pixels, the profile is extracted, and new estimates are constructed. We term this process single image sequencing (SIS) and it enables the eventual global parameters for the full image to be placed in the context of a sequence of measurements from a single image. For a reliable estimate of a global parameter, the sequence will tend towards a limiting value, although some quantization noise may still be evident. A similar sequence can also be constructed for the local variables, although these will not tend to a limit as the sub-image builds towards the full image. Such a sequence can provide an indication of the importance of the image resolution for the accuracy of the final value of a variable, such as the contact angle.

SIS requires only the profile of a drop and can therefore be performed in a variety of ways. For example, either a minimization routine to fit an elliptical shape to the profile or an axisymmetric drop shape analysis (ADSA-P) could be used. However, we will illustrate the SIS technique by using an axisymmetric drop volume technique (V-SIS) based on an expected ellipsoidal cap shape. In this approach, the geometrical relations for the ellipsoidal shape will be used to express the ratio of axes as a function of the contact radius, height, and volume. The volume of an ellipsoidal cap is given by

$$V_e = 2\pi \int_{y=bR-h}^{bR} \int_{x=0}^{x(y)} x \, dx \, dy, \tag{9}$$

where we have assumed, without loss of generality, that $x_0 = 0$ (Fig. 2). The volume of the ellipsoidal cap shape is therefore

$$V_e = \frac{\pi h^2 (3bR - h)}{3\varepsilon_r^2}. \tag{10}$$

From Fig. 2 we know that $(\pm r, bR - h)$ is a point satisfying the defining equation for an ellipse [equation (1) with $x_0 = 0 = y_0$], so that bR is given by

$$bR = \frac{h^2 + \varepsilon_r^2 r^2}{2h}. \tag{11}$$

Combining equation (11) with equation (10) for the volume and rearranging gives the drop eccentricity as a function of the desired parameters:

$$\varepsilon_r = \left(\frac{\pi h^3}{6V_e - 3\pi h r^2} \right)^{1/2}. \tag{12}$$

In the measured profile of a drop, the axially symmetric pixel volume can be evaluated from

$$V_a = \pi \sum_{p=\text{rows}} r_p^2 \tag{13}$$

and this provides an estimate for the ellipsoidal cap volume V_e. Converting from pixel to real-world units would require a calibration factor, $x_{\text{scale}}^2 y_{\text{scale}}$, in equation (13), which would also account for systems without $1:1$ pixel aspect ratios. Once the ratio of axes, ε_r has been evaluated, the contact angle can also be estimated from equation (4).

The SIS sequence for an image using the volume approach consists of evaluating the axially symmetric volume [equation (13)], and the contact radius and height for each sub-image and using equation (12) to estimate ε_r. The advantages of the volume-SIS approach over, for example, a minimization of the profile are that it is simple and numerically efficient in its implementation. Two examples based on simulations with pixel rounded values calculated from the co-ordinates of a perfect spherical cap ($\varepsilon_r = 1.0$) and a perfect ellipsoidal cap ($\varepsilon_r = 0.9$) are given in Fig. 6. The scaling of the pixel co-ordinates has been chosen to realistically represent a typical image capture card; the contact diameter has been set to 500 pixels when the contact angle is 90° for the spherical cap. If the resolution of the image were infinite, we would expect the calculated ratio of axes to follow a horizontal line in Fig. 6. However, significant underestimates are observed with the values only coming within 10% of the true values, as indicated by the horizontal dotted lines, once more than 75 rows of pixels are encompassed by a sub-image. Sub-images with more than 75 rows of pixels show the ratio of axes saturating at a value close to the true value. Figure 6 also shows the trend in the contact angle as the sequence is constructed and that contact angles of between 40° and 50° are needed before the estimated ratios of axes are within 10% of the true values. However, angles

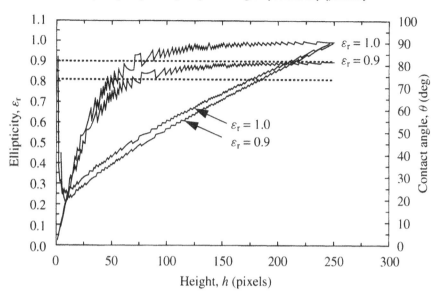

Figure 6. Single image sequences based on axial volumes for simulated data corresponding to ellipsoidal caps with $\varepsilon_r = 0.9$ and $\varepsilon_r = 1.0$. Initially the ellipticities increase rapidly before saturating close to the theoretical values; the horizontal dotted lines show when the ellipticities from the image sequences come within 90% of the true values. The corresponding sequence of contact angles is also shown. The influence of pixel quantization can be seen even after the ratio of axes has saturated.

even greater than this would be necessary before the saturation could be identified, and hence the reliability of the estimates ensured. Once the saturation region has been entered, the saw-tooth oscillations in both ε_r and θ persist and illustrate the effect of the finite resolution. Such quantization errors in the angle can be expected to complicate the analysis of data on line tension if the observed angular increases with volume are only a few degrees.

5. TWO EXAMPLES FROM EXPERIMENTS

In the data captured using experiments, effects due to factors such as lighting, heterogeneity of surfaces, whether a surface is horizontal, etc. are likely to complicate the consideration of the influence of pixel resolution. For this reason, the majority of examples shown in this work have been from simulations of data. However, the conclusions drawn from these simulations have been tested on images captured using a Data Translation DT3152 framegrabber card. Figure 7 shows two such images. The first is a ball-bearing of approximate diameter 1.43 mm, chosen to represent a shape assumed to be close to spherical, and the second is of a drop of water on a Mylar substrate. Both images were taken using the same system with the same optical magnification and the only image processing performed was to use a simple 3 × 3 median filter, followed by a global thresholding operation. The trends in

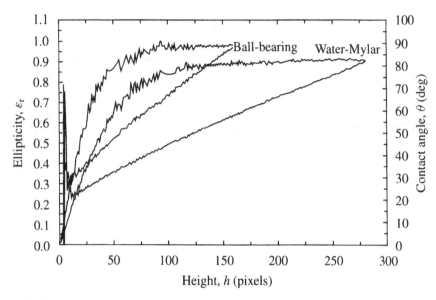

Figure 7. Single image sequences (V-SIS) for experimentally determined profiles of a ball-bearing and of a large drop of water on Mylar. These show the same trends as the simulated data in Fig. 6.

both the eccentricities and the contact angles strongly resemble the simulated data in Fig. 6. Moreover, it is also possible to find example sub-images that can be fitted to ellipsoidal cap shapes with quite different ellipticity parameters (i.e. comparable to Fig. 5). A closer examination of the saturation portion of the single image sequences in Fig. 7 shows that the ratios of axes (ellipticities) are (0.98 ± 0.01) and (0.91 ± 0.02) for the ball-bearing and water droplet, respectively.

To understand, physically, the dramatic reduction in the ratio of axes as $h \to 0$, we note that this limit implies that all rows of pixels are close to the apex of the drop. A single row of pixels can only have the character of a rectangle and even two rows only define a shape which is closer to a rectangle than an ellipse. However, as more rows are added, greater resolution is obtained and a better approximation to an ellipse can be obtained. Nonetheless, the shape will always only be an approximate ellipse and as the saturation of ε_r for the ball-bearing shows, it will tend to give a slight underestimate of the true ratio of axes. The ability of a fitting routine to generate two good fits to an observed profile, but with quite different eccentricity parameters can also be understood from a physical point of view. As the contact angle reduces, the radius of curvature increases and only a fractionally smaller segment of shape will be available to the fitting routine. This means that it becomes easier to approximate the portion of the ellipse by a segment of a circle. The reduction of the ratio of axes as measured by the V-SIS technique is another statement of this effect.

6. CONCLUSION

The problem of determining whether droplets of fluids are spherical or ellipsoidal cap in shape has been considered. The influence of pixel resolution on the deduced parameters, particularly with regard to droplets that change in volume, has been illustrated using simulations of profile data. Difficulties with small systematic errors and fitting routines have been discussed and the concept of local and global parameters has been identified. A method, single image sequencing, has been suggested which enables the confidence in an estimate of drop eccentricity to be quantified. A version of this method, V-SIS, based on the axially symmetric volume has been developed and applied to images obtained experimentally.

Acknowledgements

HYE and GM are grateful to TÜBITAK and the Royal Society for partial funding of this work through a European collaborative exchange. The initial image processing was performed using the UTHSCSA ImageTool program (developed at the University of Texas Health Science Center at San Antonio, TX, and is available from the Internet by anonymous ftp from maxrad6.uthscsa.edu).

REFERENCES

1. K. S. Birdi, D. T. Vu and A. J. Winter, *J. Phys. Chem.* **93**, 3702–3703 (1989).
2. K. S. Birdi and D. T. Vu, *J. Adhesion Sci. Technol.* **7**, 485–493 (1993).
3. S. M. Rowan, M. I. Newton and G. McHale, *J. Phys. Chem.* **99**, 13 268–13 271 (1995).
4. M. E. R. Shanahan and C. Bourgès, *Langmuir* **11**, 2820–2829 (1995).
5. G. McHale, S. M. Rowan, M. I. Newton and M. K. Banerjee, *J. Phys. Chem.* **B102**, 1964–1967 (1998).
6. H. Y. Erbil and R. A. Meric, *J. Phys. Chem.* **B101**, 6867–6873 (1997).
7. R. A. Meric and H. Y. Erbil, *Langmuir* **14**, 1915–1920 (1998).
8. R. D. Deegan, O. Bakajin, T. F. DuPont, G. Huber, S. R. Nagel and T. A. Witten, *Nature* **389**, 827–829 (1997).
9. D. Li, *Colloids Surfaces A* **116**, 1–23 (1996).
10. R. J. Good and M. N. Koo, *J. Colloid Interface Sci.* **71**, 283–292 (1979).
11. J. Drelich, J. D. Miller and R. J. Good, *J. Colloid Interface Sci.* **179**, 37–50 (1996).
12. J. Noordmans and H. J. Busscher, *Colloids Surfaces* **58**, 239–249 (1991).
13. F. K. Skinner, Y. Rotenberg and A. W. Neumann, *J. Colloid Interface Sci.* **130**, 25–34 (1989).
14. D. Li, P. Cheng and A. W. Neumann, *Adv. Colloid Interface Sci.* **39**, 347–382 (1992).
15. B. J. Lowry, *J. Colloid Interface Sci.* **176**, 284–297 (1996).
16. Y.-N. Lee and S.-M. Chiao, *J. Colloid Interface Sci.* **181**, 378–384 (1996).
17. D. J. Ryley and B. H. Khoshaim, *J. Colloid Interface Sci.* **59**, 243–251 (1977).
18. A. Kumar and S. Hartland, *J. Colloid Interface Sci.* **124**, 67–76 (1988).
19. A. Kumar and S. Hartland, *J. Colloid Interface Sci.* **136**, 455–469 (1990).
20. Y. Rotenburg, L. Boruvka and A. W. Neumann, *J. Colloid Interface Sci.* **93**, 169–183 (1983).

APPENDIX. PROOF OF EQUATION (4) FOR $\theta = \theta(r, h, \varepsilon_r)$

The first step in the proof is to review the conversion of the standard half-angle relation, equation (2), into equation (3). A standard trigonometrical formula is

$$\tan 2\phi = \frac{2 \sin \phi \cos \phi}{1 - 2 \sin^2 \phi}, \qquad (A1)$$

and considering the triangle ACD in Fig. 2, we find

$$\sin \phi = \frac{h}{\sqrt{r^2 + h^2}} \qquad (A2)$$

and

$$\cos \phi = \frac{r}{\sqrt{r^2 + h^2}}. \qquad (A3)$$

Substituting equations (A2) and (A3) into equation (A1) and rearranging gives

$$\tan 2\phi = \frac{2hr}{r^2 - h^2}. \qquad (A4)$$

These four equations, (A1)–(A4), are valid whether the cross-sectional shape is a spherical cap or an ellipsoidal cap. To show the close relationship between equations (3) and (2) requires us to prove that $\phi = \theta/2$ when the shape is restricted to a circle. To do so, we consider the triangle AEC in Fig. 2. For a circle, it is well known that $\theta_1 = \theta$ and $R' = R$ (using, without loss of generality, $a = 1 = b$), so that

$$\tan \theta = \frac{r}{R - h}. \qquad (A5)$$

Treating ACE as a Pythagorean triangle with sides of length r, $R - h$, and R leads to

$$R = \frac{h^2 + r^2}{2h}, \qquad (A6)$$

and eliminating R from equation (A5) then gives

$$\tan \theta = \frac{2hr}{r^2 - h^2}. \qquad (A7)$$

Since the right-hand sides of equations (A4) and (A7) are identical, $\tan 2\phi = \tan \theta$ must be true and this allows us to use $\phi = \theta/2$. Hence, for a circle, equation (A4) becomes

$$\tan \theta = \frac{2hr}{r^2 - h^2}, \qquad (A8)$$

which is equation (3). Since we have identified $\phi = \theta/2$ for a circle, the triangle ACD in Fig. 2 immediately gives equation (2), i.e.

$$\tan \left(\frac{\theta}{2}\right) = \frac{h}{r}. \qquad (A9)$$

When the shape is the more general case of an ellipse, the angle θ_1, defined by the triangle AEC, is no longer identical to the contact angle θ. Using the defining relation for an ellipse, equation (1), and assuming without loss of generality that $x_0 = 0 = y_0$, the contact angle is given by

$$\tan \theta = \left(\frac{dy}{dx}\right)_{edge} = -\varepsilon_r^2 \frac{x_A}{y_A}, \tag{A10}$$

where the edge co-ordinates are $(x_A, y_A) = (-r, bR - h)$. Hence

$$\tan \theta = \varepsilon_r^2 \frac{r}{bR - h}. \tag{A11}$$

Considering the triangle ACE, the right-hand side of equation (A11) can be linked to the angle θ_1, i.e.

$$\tan \theta_1 = \frac{r}{bR - h}, \tag{A12}$$

which leads to

$$\tan \theta = \varepsilon_r^2 \tan \theta_1. \tag{A13}$$

We can now use equation (11), obtained by observing that the co-ordinates $(r, bR - h)$ of the point B must satisfy equation (1) with $x_0 = 0 = y_0$, to eliminate bR in equation (A12) and rearrange to obtain

$$\tan \theta_1 = \frac{2hr}{\varepsilon_r^2 r^2 - h^2}, \tag{A14}$$

which with equation (A13) gives the desired relation [equation (4)]:

$$\tan \theta = \frac{2hr}{r^2 - \left(\dfrac{h}{\varepsilon_r}\right)^2}. \tag{A15}$$

Apparent and Microscopic Contact Angles, pp. 475–485
J. Drelich, J. S. Laskowski and K. L. Mittal (Eds)
© VSP 2000.

Surfactant-induced wetting singularities in confined solid–liquid–liquid systems: kinetic and dynamic aspects

H. HAIDARA *, L. VONNA and J. SCHULTZ

Institut de Chimie des Surfaces et Interfaces, ICSI-CNRS, 15 Rue Jean Starcky, B.P. 2488, 68057 Mulhouse Cedex, France

Received in final form 10 March 1999

Abstract—The paper reports some wetting singularities related to surfactant adsorption at solid–liquid 1–liquid 2 interfaces which are specifically driven by the time evolution of the concentration $C(t)$ around the triple line as the surfactant diffuses freely from a localized source. Two model systems have been considered: a hydrophobic methyl-terminated surface and a hydrophilic glass plate, both in contact with a squalane drop and surrounded by external bulk water. On the hydrophobic surface, an intermediate plateau was observed over a certain concentration range in the kinetics of the drop parameters [contact angle $\theta(t)$ and radius $R(t)$]. This singularity is shown to arise from the overlap of the linear domains in the individual kinetics $\gamma[C(t)]$ at the different interfaces. On the hydrophilic plate, an earlier spreading at low concentration is observed before the drop starts to recede as the local concentration around the triple line increases. This inversion in triple line motion, driven by the bilayer adsorption transition at the glass–water interface, combines with the buoyancy in the late stage, leading to the neck rupture of the confined drop.

Key words: Surfactant adsorption kinetics; triple phase contact; wetting singularities; layering transition.

1. INTRODUCTION

The adsorption of surfactants from inhomogeneous solutions onto a solid–liquid 1–liquid 2 triple phase contact [1–3] is often encountered in many fundamental and technological processes. For instance, air bubbles may be trapped in poorly wet domains of a substrate during coating processes. In the same way, fluid particles may adhere to boundary walls, forming a triple phase contact with the surrounding surfactant solution where competitive adsorption can take place. In both of these systems, local fluctuations in the concentration (applied shear stress for coatings, or convection in emulsion) can result in transient wetting events related to either

---

*To whom correspondence should be addressed. E-mail: H.Haidara@univ-mulhouse.fr

conformational rearrangement of the adsorbed layers or local imbalance in surface forces. This paper investigates such transient events through the kinetics of free diffusion of nonionic surfactants from a localized source and their adsorption at the model hydrophobic squalane–water and hydrophilic squalane–water interfaces.

Regarding the experimental results involving the hydrophobic surface, it is worth mentioning that we present here a new interpretation qualitatively different from the former one [1] which was based on the structural rearrangement within the adsorbed layer at the (solid–water) interface.

2. MATERIALS AND METHODS

Two model surfaces, a hydrophobic and a hydrophilic one, were used to study the dependence of the wetting events on the nature of the solid–oil–water (S–O–W) interface. The hydrophobic surface was a silicon wafer with a methyl (CH_3)-terminated self-assembled monolayer of hexadecyltrichlorosilane (HTS), supplied by ABCR Karlsruhe (Germany). The hydrophilic surface was a fresh piranha-treated microscopic glass slide prior to experiments. The dynamic contact angles of doubly distilled and deionized water on these surfaces at an advancing and a receding velocity of 20 μm/s were, respectively, $\theta_a = 115°$ and $\theta_r = 105°$ for the hydrophobic surface and $\theta_a = 30°$ and $\theta_r < 2°$ for the hydrophilic one. The surfactants used were nonionic alkylethylene oxide, C12E7 and C10E8 from Fluka and Aldrich, respectively. Chromatographic grade squalane (C30) from Prolabo and air bubbles were used to study the dependence of the surfactant adsorption and related wetting events on the nature of the hydrophobic fluid drop. Further details and specifications regarding the experiment and materials can be found in ref. [1].

Basically, the experiment relies on the interpretation of the time evolution of the characteristic parameters of the S–O–W system: drop contact angle $\theta(t)$ and radius $R(t)$, as the surfactant diffuses through the aqueous phase (from a localized source) and adsorbs onto the interfaces. The initial system consisting of the substrate and the pure liquids (10 ml of water and a 1.5 μl squalane drop) was first equilibrated for about 0.5 h in the working polystyrene cell. A given amount of a surfactant stock solution ($\sim$ 100 times the critical micelle concentration, cmc) was then introduced in the diffusion sources (compartments) at both sides of the cell (Fig. 1) to achieve the desired diffusion profile and bulk concentration at t_∞, $C(t_\infty)$. Diffusion sources of 0.4, 1.5, and 15 times cmc per compartment (cmc $\approx$ 40 mg/l for C12E7 and 510 mg/l for C10E8) were used, leading to final bulk concentrations $C(t_\infty)$ of 0.25, 1, and 10 times cmc. Each diffusing source was then characterized by a concentration profile around the triple line (TL). This concentration profile $C(TL, t)$ can be varied either by introducing different amounts of a given stock solution or by using identical amounts of stock solutions with different concentrations. We retained for these experiments the former method, which leads to varying final bulk concentrations and thus affects less the dissociation

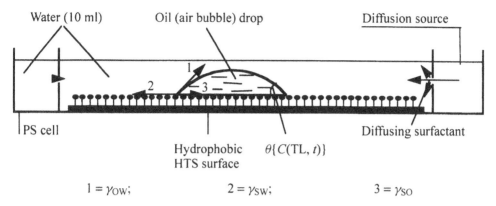

Figure 1. Diagram of the experimental cell unit showing the characteristic parameters chosen for this work. The compartments have an identical size, the central one being enlarged for details.

step which depends on the concentration of the stock solution. An automatic contact angle analyzer (Krüss G2) and a video recorder system were used for the time-dependent variation of the drop profile and parameters.

3. RESULTS AND DISCUSSION

3.1. Kinetics involving HTS–C30 drop (air bubble)–aqueous surfactant solutions

The adsorption kinetics and subsequent reconformation of the triple phase represented by the reduced quantity $[\Delta\theta(t)/\theta_0]$ are given in Figs 2 and 3 for the HTS–C30–W and HTS–air bubble–W systems. Figure 2 represents the time evolution of the HTS–C30–W system for three concentrations of C12E7 in the diffusion source, each leading to a characteristic diffusion profile $C(TL, t)$ and equilibrium bulk concentration (see Section 2). The dependence of these kinetic and competitive processes on the nature of the confined fluid (C30 vs. air bubble) is shown in Fig. 3 for the nonionic C10E8 surfactant at $C(t_\infty) \sim$ cmc (510 mg/l).

For both of these systems, the surfactant transport and adsorption kinetics result in the occurrence of an intermediate plateau which depends on the concentration profile $C(t)$ around the triple phase. The length of this intermediate plateau varies roughly with the nature of the confined fluid drop. It is worth mentioning here that the drop evolves throughout the entire kinetics with a spherical shape, for θ ranging from $\sim 20°$ to $35°$ for the C30 drop and from $\sim 75°$ to $\sim 125°$ for the air bubble. In order to clarify the physics involved in these experiments, one should mention that the capillary, Reynolds, and Marangoni numbers related to the reconformation of the drop (R_a, $C_a \ll 1$, $M \sim 0.8$) are such that no significant contribution is expected from viscous, inertia, or Marangoni effects as regards the main result discussed here (the intermediate plateau). This is especially true for the system involving the air bubble (viscosity $\eta \sim 0$), which as the C30 drops, leads to intermediate plateaus under similar conditions [$C(t_\infty) \sim$ cmc, slow reconformation kinetics].

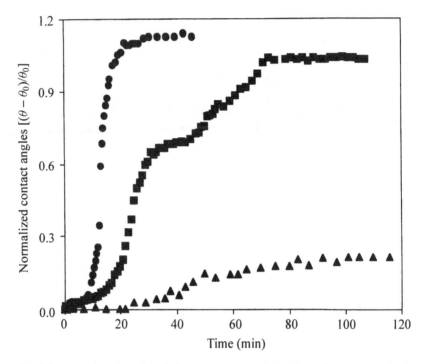

Figure 2. Adsorption kinetics of C12E7 at HTS–C30–W interfaces for three equilibrium bulk concentrations $C(t_\infty)$: (▲) 0.25 cmc; (■) 1 cmc (40 mg/l); (●) 10 cmc. These equilibrium bulk concentrations correspond to increasing diffusion source concentrations of (▲) 0.4, (■) 1.5, and (●) 15 cmc.

To interpret these results, especially the intermediate plateau observed on the hydrophobic surface, the time dependence of the Young equation, $\cos\theta = (\gamma_{SW} - \gamma_{SO})/\gamma_{OW}$, is used to derive the imbalanced driving force of the TL motion and the conditions to its transient stability (intermediate plateau). Upon a few rearrangements, the first derivative of the above Young equation relative to the local concentration $C(TL, t)$ around the TL leads to

$$-\sin\theta(\mathrm{d}\theta) = \gamma_{OW}^{-1}\left[\mathrm{d}(\gamma_{SW} - \gamma_{SO}) - (\mathrm{d}\gamma_{OW})\cos\theta\right].\qquad(1)$$

As shown by the experimental results given in Figs 2 and 3, the contact angle essentially increases ($\mathrm{d}\theta/\mathrm{d}t \geqslant 0$) over the entire extent of the adsorption kinetics, with θ ranging between $\theta_{min} = 20°$ for the pure HTS–squalane drop–water system and $\theta_{max} \approx 125°$ for the HTS–air bubble–aqueous solution at $C(t_\infty) \sim$ cmc. The term on the left-hand side of equation (1) is therefore strictly less than 0, while the denominator γ_{OW} on the right-hand side is greater than 0 for these surfactants [4], even at bulk concentrations higher than the cmc.[1] In addition, based on the strong

---

[1]In our own measurements using the Wilhelmy plate method, we obtained equilibrium interfacial tensions as low as 1 mN/m for the squalane–C10E8 aqueous solution (at the cmc), but never negative values.

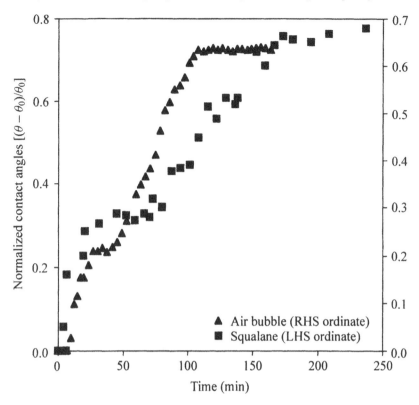

Figure 3. Adsorption kinetics of C10E8 at (■) HTS–C30–W and (▲) HTS–air bubble–W interfaces for an equilibrium bulk concentration $C(t_\infty) = $ cmc (510 mg/l). This corresponds to a concentration of 1.5 cmc in each diffusion source compartment.

hydrophobic character of the CH_3-terminated HTS–C30 and HTS–air bubble interfaces, the insolubility of C12E7 and C10E8 in the drop fluid, and the retracting motion of the TL, we can reasonably assume that these interfaces remain free from adsorption, leading to $d\gamma_{SO} \equiv 0$ in equation (1). This assumption firstly relies on the thermodynamic consideration that the confining free energy of oxyethylene groups at CH_3-terminated HTS–C30 or HTS–air bubble interfaces is by far much larger than their retention in the bulk aqueous phase and adsorption at the O–W or S–W interface [1]. Secondly, the retracting motion of the TL significantly reduces the mean residence time of the surfactant molecules at the moving triple phase, making even more unlikely such a confinement which might be possible in equilibrium conditions, due to some partial solubility of nonionic alkylpoly(ethylene oxide) surfactants in alkanes. Thirdly, the assumption naturally holds at the HTS–air bubble–aqueous solution interface exhibiting the same behavior as HTS–C30–W and where one can hardly imagine under normal conditions (temperature, pressure) the thermodynamic transfer of C12E7 or C10E8 from water into the air bubble or to the HTS–air interface. Lastly, we performed some benchmark experiments involving infra-red (IR) and surface tension measurements on small amounts of squalane which were gently poured onto cmc bulk aqueous C12E7 solutions and allowed to equilibrate for 24 h. These measurements did not reveal any detectable solubility, the IR spectra and surface tension of the equilibrated squalane being identical to those of pure squalane.

Taking into account the above considerations ($-\sin\theta(\mathrm{d}\theta) < 0$, $\mathrm{d}\gamma_{SO} \equiv 0$, and $\gamma_{OW} > 0$), equation (1) is simplified, leading to the following relation for the net Young force f_{TL} applied to the triple phase line:

$$f_{TL} = \mathrm{d}\gamma_{SW} - (\mathrm{d}\gamma_{OW})\cos\theta \leqslant 0. \tag{2a}$$

In terms of the adsorbed film pressure, equation (2a) reads

$$f_{TL} = \mathrm{d}\Pi_{OW}(\cos\theta) - \mathrm{d}\Pi_{SW} \leqslant 0. \tag{2b}$$

Equations (2a) and (2b) correspond to the unbalanced surface force which drives the inward motion of the triple phase line or, equivalently, the expansion of the S–W interface at the expense of the O–W one. What equations (2a) and (2b) show is that the adsorption kinetics at the O–W interface contributes to the drop motion uniquely through its in-plane component. Therefore, for the systems considered in these investigations, which already have quite low initial contact angles, an outward motion of the TL (spreading) should involve surfactants exhibiting much lower adsorption kinetics at the S–W interface compared with those at the O–W interface.

As mentioned earlier at the end of the Introduction, the initial interpretation that we gave to these intermediate plateaus was based on the substitution of the finite quantities $(\gamma_{SW}-\gamma_{SW}^0)-(\gamma_{OW}-\gamma_{OW}^0)\cos\theta$ for $\mathrm{d}\gamma_{SW}-(\mathrm{d}\gamma_{OW})\cos\theta$ in equations (2a) and (2b). This shorthand substitution is correct only where the independent O–W and S–W adsorption isotherms (*kinetics*) are both linear. This, of course, is seldom the case even for adsorption kinetics involving simple fluid–fluid interfaces in quite uniform surfactant solutions (adsorption kinetics at a fresh pendant drop interface vs. a uniform solution) [4, 5]. This substitution has therefore resulted in quite different conditions on the surface pressures for $\cos\theta$ to be constant on the intermediate plateau ($f_{TL} = 0$) and mechanisms [1], as discussed below.

As for any adsorption process, the unique plateau related to a stable thermo-dynamic equilibrium within the system (zero chemical potential gradient) is that at t_∞, where f_{TL}, $\mathrm{d}\Pi_{SW}$, and $\mathrm{d}\Pi_{OW}$ are all zero. At this final plateau at t_∞, one recovers the Young equilibrium equation as a boundary condition, $\cos\theta_\infty = (\gamma_{SW}^\infty-\gamma_{SO}^\infty)/\gamma_{OW}^\infty$, where $\gamma_{SO}^\infty \equiv \gamma_{SO}^0$. The intermediate plateau between $t = 0$ and t_∞ then does not correspond to any thermodynamic equilibrium and should therefore involve some transient event where $\cos\theta = (\mathrm{d}\gamma_{SW}/\mathrm{d}\gamma_{OW}) = $ constant, as $f_{TL} = 0$ [equations (2a) and (2b)]. This is obtained from the condition that the interfacial tension increment is constant (linear slope) over the intermediate plateau for both S–W and O–W interfaces, $\mathrm{d}\gamma_{SW} = \alpha \neq 0$ and $\mathrm{d}\gamma_{OW} = \beta \neq 0$, leading to $\cos\theta$ (and therefrom θ) $= (\alpha/\beta) = $ constant. Physically, this requirement is met when the adsorption profile $\gamma_{ij} = \gamma_{ij}[C(t)]$ at each of the coupled interfaces simultaneously exhibits a linear behavior ($[\mathrm{d}\gamma/\mathrm{d}C(t)] = $ constant) as illustrated in Fig. 4. This

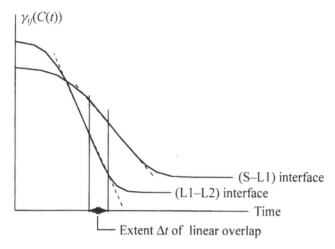

Figure 4. Virtual adsorption kinetics at the independent S–L1 and L1–L2 interfaces showing overlap of the respective linear domains where $(d\gamma_{SW}/d\gamma_{OW}) = (\alpha/\beta) = $ constant. The concentration profile $C(t)$ is the same at both interfaces.

interpretation sems to be well supported by the observed dependence of the time length involved in the intermediate plateau on the nature of the drop (Fig. 3, C10E8 vs. C30 drop and air bubble). In fact, the extent of the overlap between the linear domains (Fig. 4) is expected to vary with the individual isotherms of the adsorption kinetics at the W–air and W–C30 interfaces. These individual kinetics may be shifted with respect to that of the reference S–W interface, leading to a smaller overlap for the air bubble (Fig. 4) and hence a shorter time length for the corresponding intermediate plateau (Fig. 3). If confirmed, this result constitutes evidence that the evolution of a triple phase interface under competitive adsorption may simply derive from the superimposition of their individual adsorption behaviors. However, additional experimental support involving, for instance, independent measurements of $\gamma_{ij} = \gamma_{ij}[C(t)]$ at O–W, air–W, and S–W interfaces (under similar experimental conditions of free diffusion from a localized source) is necessary to confirm these results. If no specific co-operative process is involved, such measurements should allow us to identify (*a priori*) a certain range of 'intermediate' concentration profiles where the linear domains of the isotherms $\gamma_{OW}[C(t)]$ and $\gamma_{SW}[C(t)]$ will overlap, resulting in an intermediate plateau with a characteristic time length. Finally, the existence of this intermediate plateau depends strongly on the concentration profile (see Fig. 2) around the triple phase line $C(TL, t)$ which determines, through the time dependence of the adsorbed amounts Γ_{ij} at the interfaces [2, 4, 5], both the kinetic profiles $\gamma_{ij}[C(t)]$ and the extent of the domain of linear overlap [5]. Furthermore, for a given experimental set-up (cell size, volume of water, etc.) these concentration profiles are controlled by two independent parameters: the surfactant concentration in the diffusion source and the distance between the diffusion source and the TL (Fig. 1).

3.2. Kinetics involving the hydrophilic glass–C30–aqueous surfactant solutions

If we now substitute a hydrophilic surface for the above hydrophobic one, the free diffusion and adsorption of surfactant at the triple phase result in this case in the neck rupture of the drop through the different steps depicted in Figs 5a–5g. Because the water-covered hydrophilic plate is not wetted by the oil phase, the drop is initially maintained against the surface under slight pressure. A small amount of surfactant is then introduced (0.25 cmc of the nonionic C12E7), which produces a small (but finite) amount of wetting upon adsorption, allowing the free standing of the drop and withdrawal of the needle. The adsorption state and related spreading corresponding to these initial steps at low concentration (Figs 5a–5d) are described in Figs 6a and 6b. An additional amount of surfactant is then introduced from the stabilized state (Fig. 5c) to achieve a bulk concentration at t_∞ equal to cmc. The evolution of the system in this second step and the corresponding adsorption states at the interfaces are depicted in Figs 5e–5g and Fig. 6c, respectively. The most relevant feature of these results is the inversion in the motion of the TL, driven by the formation of a second adsorption layer at the S–W interface as the local concentration around the TL approaches or increases above the cmc, $\{C(\text{TL}, t) \geqslant \text{cmc}\}$. As this second surfactant layer adsorbs on top of the initial one with its polar heads within the aqueous phase (Fig. 6c), the S–W interface gradually recovers its original hydrophilic character. The overall picture is that γ_{SW} is significantly lowered, providing the driving force for the inward motion of the TL. On the other hand, the thin outer strip along the drop contact area corresponding to the first adsorbed layer (Fig. 6b) has a hydrophobic character $[-(\text{CH}_2)_n-\text{CH}_3]$ resulting in a low interfacial tension γ_{S^*O} which resists the retraction of the TL. On the molecular scale, the mechanism which causes the macroscopic distortion in the drop profile and related neck rupture is related to this change in surface chemistry (due to adsorption) on both sides of the TL (the subscripts S* and S stand for the hydrophobic and hydrophilic surfaces, respectively). Near (and above) the cmc, γ_{OW} is typically small [4] (< 2 mN/m for C10E8 at the C30–W interface), leading to negligible $\gamma_{OW} \cos\theta$ for $90° \leqslant \theta \leqslant 120°$ as shown in Figs 5e and 5f, which represent the retraction step. The surface force balance which drives the late stage retraction of the TL (at high surfactant concentration) then mainly arises from the competition between γ_{SW} and γ_{S^*O} as given by $F \approx (\gamma_{SW} - \gamma_{S^*O})$. The drop may already be pinned, or will recede very slowly over the thin outer hydrophobic strip, driven by this small (but finite) force. In the latter case, the TL will move until it meets the boundary between this outer hydrophobic strip and the inner spot ($2R_0$, Fig. 6) where the initial (oil–hydrophilic glass) contact was formed and which remained free from adsorption. Around this wetting–nonwetting transition, the retraction of the triple line involves a higher energy cost of the order of $\sim 2\pi R_0(\gamma_{OW} - \gamma_{O/(\text{CH}_2-\text{CH}_3)})$, assuming $\gamma_{O/\text{glass}} \approx \gamma_{OW} \approx 50$ mN/m. As a result of this high energy cost for the retracting motion, the TL is strongly pinned around this transition. The bulb-like profile and subsequent neck rupture of the drop actually arise from the combined effect of the pinning (or strong reduction

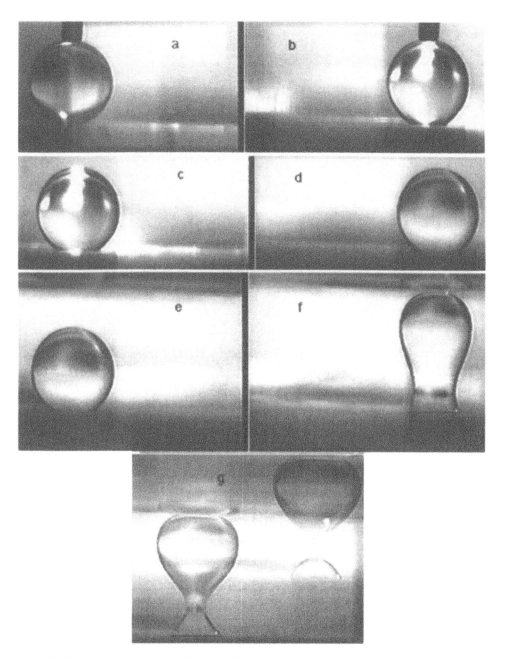

Figure 5. Photographs of the drop shape reconformation during C12E7 adsorption at the hydrophilic glass–C30–W interfaces. (a) Initial system, $t = 0$; (b) about 5 min after introduction of the first surfactant amount to achieve $C(t_\infty) = 0.25$ cmc; (c) about 1 min after step b and withdrawal of the needle; (d) stabilized system under 0.25 cmc; (e) just after adding the complementary surfactant amount to achieve 1 cmc; (f) retraction and formation of the bulb profile; (g) neck rupture due to the combined effects of buoyancy and capillary forces.

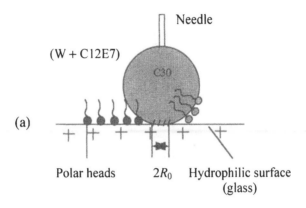

(a)

Needle

(W + C12E7)

Polar heads $2R_0$ Hydrophilic surface
(glass)

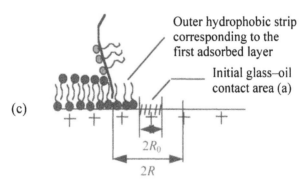

(b)

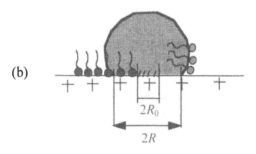

(c)

Outer hydrophobic strip
corresponding to the
first adsorbed layer

Initial glass–oil
contact area (a)

Figure 6. Possible mechanisms involved in the drop shape reconformation of Fig. 5. (a) Initial adsorption at low concentration (0.25 cmc); (b) consecutive spreading of the drop; (c) retraction and vertical protrusion of the drop at higher concentration and related adsorption states around the triple phase line.

in retracting velocity) which maintains a constant contact area and the buoyancy $(\rho_O - \rho_W)gh_d$ applied to the drop which drives the upward liquid flow (at constant drop volume). This experiment illustrates the way that very local and finite changes in the molecular structure around the TL can strongly determine macroscopic events.

4. CONCLUSION

We have investigated the kinetics of competitive adsorption of a surfactant at a solid–water/water–oil drop/oil drop–solid triple phase. In the specific case where the surfactant diffuses freely through the bulk water phase, from a localized source towards the triple phase interfaces, it appears that the concentration profile $C(TL, t)$ around the triple line (TL) is the determining parameter regarding the adsorption kinetics, transient events, and overall time evolution of the system. These results also reveal the richness of the macroscopic behavior which can arise from subtle changes in the local molecular structures around the TL, depending on the nature of the substrate (hydrophobic, hydrophilic). In this respect, these studies confirm how powerful the correct wetting measurement and interpretation can be as far as interfacial phenomena are concerned (layering adsorption, competitive adsorption at a triple phase in both static and dynamic conditions).

REFERENCES

1. H. Haidara, L. Vonna and J. Schultz, *Langmuir* **12**, 3351–3355 (1996).
2. H. Haidara, L. Vonna and J. Schultz, *J. Chem. Phys.* **109**, 2355–2360 (1998).
3. R. J. Good and C. J. Sun, *J. Colloid Interface Sci.* **91**, 341–348 (1983).
4. T. Svitova, H. Hoffmann and R. M. Hill, *Langmuir* **12**, 1712–1721 (1996).
5. S. Y. Lin, R. Y. Tsay, L. W. Lin and S. I. Chen, *Langmuir* **12**, 6530–6536 (1996).

Apparent and Microscopic Contact Angles, pp. 487–496
J. Drelich, J. S. Laskowski and K. L. Mittal (Eds)
© VSP 2000.

An acoustic technique for the monitoring of dynamic wetting behavior

G. McHALE*, M. I. NEWTON, M. K. BANERJEE and S. M. ROWAN

Department of Chemistry and Physics, The Nottingham Trent University, Clifton Lane, Nottingham NG11 8NS, UK

Received in final form 20 March 1999

Abstract—A small stripe of a viscous fluid deposited on a high-energy surface spreads with an increasing contact width that follows a characteristic 1/7th power law with time. At any instant, the fluid has a spherical cap cross-sectional profile with well-defined values of the spherical radius and dynamic contact angle. The evolution of such stripes has been followed using optical interferometry. Simultaneously, the changing contact area has been monitored using an alternative method based on high-frequency ($\sim$ 170 MHz) surface acoustic waves. Such waves have their energy confined to within one wavelength of the surface and are potentially an in-plane technique for monitoring dynamic wetting. Acoustic signals that originate from reflections from the advancing oil, and from transit along the solid–liquid interface, are reported. The changes in these acoustic signals are compared with the optically measured parameters and are interpreted using a viscoelastic model of the fluid.

Key words: Wetting; contact angles; surface acoustic waves (SAWs).

1. INTRODUCTION

The characterization of solid–liquid interactions is important for understanding surface properties, such as adhesional strength. For equilibrium studies, the contact angle produced by various fluids on a solid has been a traditional method of quantifying the type of surface. More recently, measurement of the dynamic contact angle has been important for understanding the approach to equilibrium and film formation properties. The most direct methods of contact angle measurement involve direct visual observation of a liquid droplet using either a profile or a plan view [1]. The side profile allows a simple cross-section to be obtained and is particularly convenient for digital image analysis. However, the spatial resolution can be limited and a cross-section may not be typical of the whole droplet if the surface is heterogeneous. The plan view can involve interferometry and provides

---

*To whom correspondence should be addressed. E-mail: glen.mchale@ntu.ac.uk

a view of the whole periphery of a droplet whilst retaining the potential for digital image analysis. These direct methods can also be combined with video analysis to enable dynamic information to be extracted [2]. If we want to look at microscopic films, rather than macroscopic profiles, then laser interferometry [3], ellipsometry [4], and X-ray reflectivity [5] are all possible methods.

In contrast, indirect methods are based, at least partially, on mass measurement. An example of such a proposed method is the quartz crystal microbalance (QCM). In this technique, a quartz crystal is made into an oscillator circuit whose resonant frequency is sensitive to the mass of liquid deposited on the electrode surface [6, 7]. This is an acoustic technique using a thickness shear mode oscillation of a quartz crystal. The crystal oscillation induces a damped shear wave in the liquid close to the solid–liquid interface; the wave is completely attenuated within a small distance into the liquid. Hence, the resonator response of the oscillator circuit depends only on the mass of liquid within one shear wave penetration depth of the electrode surface, rather than on the mass of the bulk liquid. This method was proposed as a high-speed and highly mass-sensitive technique for measuring surfactant superspreading [8–10]. The technique has also been proposed for the measurement of static contact angles [11]. The perceived potential of acoustic techniques lies not in any ability to measure contact angles better, but in that they may provide information about solid–liquid interfaces both *in situ* and with high sensitivity. A review of the types of acoustic devices used in liquid sensing has been given by White [12]. Potentially, acoustic devices can sense changes in elasticity, in addition to interfacial mass. Rayleigh surface acoustic waves (SAWs) involve an elliptical motion of particles in the surface of a substrate. SAWs propagate along the solid–vapor interface and have their energy confined to within one acoustic wavelength of the interface [13]. SAWs are more sensitive to mass loading than QCMs and this has motivated their use in gas sensing where sub-monolayer coverage sensitivity is required. In addition, a SAW also involves an electric field displacement and may, therefore, be able to provide information on changes in interfacial charge.

The interaction of SAWs with localized liquids has only rarely been considered in the literature. Shiokawa *et al.* [14] have proposed a method for contact angle determination involving SAWs, but this is the only case known to us. De Billy [15, 16] considered the influence of the contact angle on the reflection coefficient of SAWs and demonstrated that it did not change monotonically. Since SAWs provide a natural in-plane technique with high mass sensitivity that can be used *in situ* and in real time, there is clear potential for them to be used to study wetting behavior, particularly of liquid films. However, insufficient understanding currently exists on the nature of acoustic–liquid interactions. This paper therefore reports on the use of a combined optical and SAW technique to monitor the dynamic wetting of a surface by a small stripe of a viscous oil.

2. EXPERIMENTAL

For a model spreading system, we used a stripe of a highly viscous oil, poly(dimethylsiloxane) (PDMS), on a lithium niobate substrate. This system has a number of advantages. PDMS on lithium niobate is a high-energy system, so the equilibrium contact angle vanishes to within a few degrees. Moreover, the choice of oil means that the volume of fluid is conserved. The dynamics of the geometrical parameters therefore follows simple power law behaviors. Since the geometry is a small stripe, gravity can be ignored and capillary forces dominate, thus resulting in a spherical cap cross-section for the oil [17]. Due to the essentially two-dimensional form of the stripe, the resulting dynamics follows 1/7th power laws [18], rather than the Tanner type 1/10th laws [19]. The choice of an oil also enables the time-scale of a wetting experiment to be chosen via the viscosity. In this work, we used oils covering two orders of magnitude of viscosity from 10 000 to 100 000 cSt.

In our system, we use a plan view of the stripe to obtain the geometrical parameters (Fig. 1). Sodium light is used to illuminate the stripe from above and reflections occur from both the top of the oil and the oil–substrate interface. The result is an interference pattern, which we record onto videotape using a CCD camera; an example image has been given in a previous report [20]. After a spreading experiment, images from the videotape are captured into a personal computer for analysis. Initial image processing allows us to measure the contact width of the stripe and to obtain a cross-section of the grey levels from one side of the stripe to the other. The line for the cross-section is chosen to be in alignment with the centre of the acoustic path on the substrate surface. Programs written in 'C' are then used first to reconstruct the shape of the profile from the interference fringes and then to fit a spherical cap shape through the profile. From this analysis, we obtain the contact width and the spherical radius of the stripe. This procedure

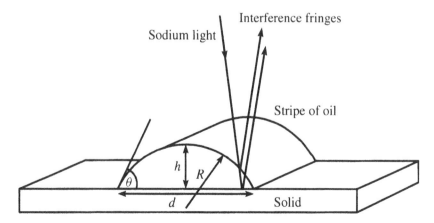

Figure 1. The stripe of oil has a spherical cap cross-section defined by the contact width d, the spherical radius R, the contact angle θ, and the height h. Reflection of sodium light from the air–oil interface and the oil–substrate interface produces interference patterns that enable the profile to be determined.

is typically performed at 30 different instants in time over the course of a single spreading experiment. Hence, it is possible to provide information on the dynamic changes in the geometrical parameters. For the majority of time covered by an experiment, it is not possible to resolve the fringes all the way from the crest of a stripe to the edge. However, it is possible to do so near the end of an experiment and so verify the assumption of a spherical cap shape necessary for a meaningful spherical radius parameter. No assumption of shape is made on the contact width measurement. A knowledge of the contact width and spherical radius enables the stripe height and contact angle to be estimated. It would be possible to measure directly the contact angle by altering the magnification, but then we would lose the direct estimate of stripe contact width. Choosing a magnification suited to the maximal stripe width in an experiment also enables us to monitor the variation in crest height along the stripe.

The acoustics is based on a pulse mode principle and is only designed to measure changes in the magnitude of the acoustic signal rather than the velocity or, equivalently, frequency shifts. The pulse mode allows acoustic reflections from the stripe of oil to be detected, in addition to signals that transit along the solid–liquid interface. It would also be possible to develop a continuous wave system capable of detecting velocity changes as well as losses, although we have not done so. For a given substrate, the frequency of the surface wave determines the acoustic wavelength and so provides a basic length scale in the system. In our case, a 168.8 MHz carrier signal is modulated by a set of pulses to generate a pulsed radiofrequency electrical signal (Fig. 2). A set of interdigital transducers, fabricated

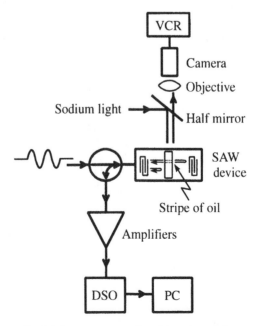

Figure 2. High-frequency Rayleigh waves are produced in pulses and are partially reflected by the oil and partially transmitted along the oil–substrate interface.

on the substrate surface, then converts the electrical energy into a Rayleigh SAW of wavelength ~ 22 μm. A second set of transducers 2 cm from the first set can be used to detect the Rayleigh wave. This second set of transducers also acts as a reflector of the Rayleigh wave and this returned signal can be detected by the original generating transducers. When a stripe of oil is placed across the propagation path, between the two sets of transducers, it also acts as a source of reflections. In addition to providing an acoustic reflection, the stripe causes extra damping of the wave travelling between the two sets of transducers. The detected signal is amplified, displayed on a digital storage oscilloscope (DSO), and the levels of the pulses detected are saved by a computer. The measurement of these acoustic signals is repeated at 7 s intervals throughout the period of spreading. Typically, the spreading of a stripe of oil is followed for 8 h with both the acoustics and the optics recording simultaneously.

3. RESULTS AND DISCUSSION

In cross-section, a stripe of oil has a spherical cap shape which is characterized by the contact width d, the spherical radius R, the contact angle θ, and the height h. Power laws for the time evolution can be obtained by balancing the changes in the surface free energy of the stripe with the losses due to viscous dissipation [17, 18]. Assuming conservation of mass and small angles gives 1/7th laws for the measured quantities:

$$d = k_1(t + c)^{1/7} \quad \text{and} \quad R = k_2(t + c)^{3/7}, \tag{1}$$

where k_1, k_2, and c are constants. The cross-sectional area of the stripe is given by $A = k_1^3 / 12k_2$. From the measured d and R, the contact angle θ and height h can be found:

$$\theta = \frac{k_1}{2k_2(t + c)^{2/7}} \quad \text{and} \quad h = \frac{k_1^2}{8k_2(t + c)^{1/7}}, \tag{2}$$

where θ is in radians. In Fig. 3, we show fits to the optically measured contact width and the contact angle, deduced from the spherical radius, for a 30 000 cSt oil; the contact angle has been converted to degrees in this diagram. The values used in the fits are $c = -23.3$ s, $k_1 = 118.6$ μm s$^{-1/7}$ and $k_2 = 92.38$ μm s$^{-3/7}$; the first two of these parameters were found by fitting to the optically observed contact width, d. The value of k_2 was deduced from the value of k_1 and the average cross-sectional area of $A = 1.505 \times 10^{-8}$ m^2. The power law accuracy is good up to the 3 h point in the spreading process, at which point the contact angles fall below 3°.

The changes in acoustic reflection and transit signals are shown in Figs 4 and 5; in these diagrams, the x-axis has been converted from time to contact width. The reflection from the stripe of oil shows sets of distinct peaks occurring between adjacent points of high transmission loss. The separation of adjacent reflections (successive maxima between A′ and G′) within a set varies systematically from

G. *McHale* et al.

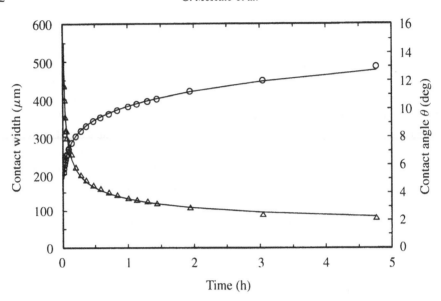

Figure 3. The 1/7th type power laws for the contact width d and the contact angle θ accurately describe the observed dynamics of the 30 000 cSt stripe of oil. The circles indicate the measured contact angle in degrees and the triangles the measured contact width.

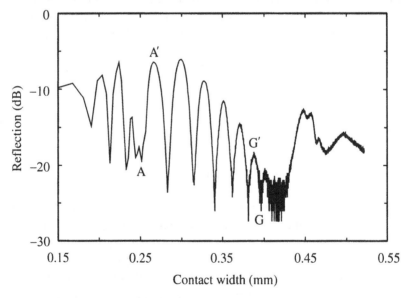

Figure 4. The reflection from the stripe of oil shows a distinct set of maxima and minima with the separation of two adjacent maxima changing from more than λ_{SAW} to less than λ_{SAW}. The reflection maxima are labelled A′–G′ and the minima are labelled A–G.

greater than an acoustic wavelength to less than an acoustic wavelength [21]. In this particular data set for the 30 000 cSt oil, the contact widths at which the reflection maxima A′ to G′ occur are 0.2658, 0.2996, 0.3274, 0.3513, 0.3706, 0.3876, and 0.3980 mm. Hence, this gives a systematic decrease in the separation of the

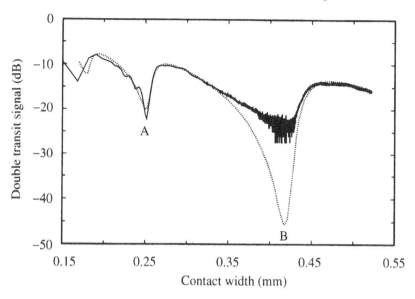

Figure 5. The transit signal, which has passed twice along the oil–substrate interface, shows deep losses at distinct contact widths (A and B). The dotted line is a fit to the data based on the formula for a thin uniform film [equation (3)] with a shear modulus of 1.51 GN m^{-2}. The fit uses a numerical integration to account for the spherical cap cross-sectional shape.

reflection maxima peaks from 34 to 10 μm, which can be compared to the SAW wavelength of 22 μm. The transit signal shows an increasing loss as the stripe widens, but with exceptionally high losses apparent at specific times (indicated by A and B). These points of high loss in the transit signal are separated by contact width changes of much greater than a surface acoustic wavelength, λ_{SAW}. Although only one data set is presented in Figs 4 and 5, these observations have been confirmed in other experiments and for a range of viscosities from 10 000 to 100 000 cSt.

To explain the observed transmission coefficient of the acoustic wave, we have developed a simple model of acoustic losses [22]. This model assumes a damped harmonic oscillator response of the substrate with the damping being due to the shear motion induced in a liquid loading of uniform thickness, t_f. A single relaxation time, τ, is included within the model to allow for viscoelastic effects due to the highly viscous oil. The predicted loss, L, in dB is

$$L = -20 \left(\log_{10} e\right) \frac{d}{\rho_s \xi \lambda_{SAW} v_{SAW}} \mathrm{Re}\left[\frac{\beta}{1 + i\omega\tau}\right], \qquad (3)$$

where ρ_s is the density of the substrate, ξ represents the effective depth of the substrate oscillating, and Re indicates that it is necessary to take the real part of the complex argument. In this expression, v_{SAW} is the speed of the SAW and ω is its angular frequency. Assuming that the shear wave speed in the fluid is much less than the surface wave speed, the function β is given by

$$\beta = \left(\eta\sqrt{2}\mathrm{i}\sqrt{1 + i\omega\tau}/\delta\right)\tanh\left(\sqrt{2}\mathrm{i}\sqrt{1 + i\omega\tau}\,t_f/\delta\right), \qquad (4)$$

where $\delta = (2\eta/\omega\rho)^{1/2}$ is the shear wave penetration depth into the fluid, η is the viscosity, and ρ is the fluid density. For Newtonian-type fluids, $\omega\tau$ is small and the liquid entrained in a shearing motion by the substrate is limited to a film thickness of the order of a penetration depth. Provided that the fluid thickness is greater than δ, the acoustic system acts as an interfacial mass sensor. Indeed, this is the idea used by Ward and co-workers [8–11] in suggesting that the QCM can be used to monitor dynamic wetting and contact angles for small drops of liquids. In their case, the contact area, πr^2, of a small droplet can be related to the frequency shift in a resonant circuit. This frequency shift would be expected to be proportional to the effective mass $\rho\pi r^2\delta$ when the fluid height is much greater than the penetration depth. Such a simple relationship would provide a rapid method of monitoring dynamic wetting. Alternatively, they suggest that the contact angle can be estimated by using a simple spherical cap geometry, valid for small droplets, and calibrating the frequency shift with a fluid of known contact angle. However, for the viscous fluids in our experiments $\omega\tau$ is greater than unity, so the argument in the tanh of equation (4) becomes almost purely imaginary. This converts the tanh term into a tan, which then becomes infinite at specific values of fluid thickness. Since the relaxation time is related to viscosity and shear modulus, μ, by $\tau = \eta/\mu$ and the shear speed in a fluid is $v_s = (\mu/\rho)^{1/2}$, the condition for the infinity is

$$t_f = n\lambda_s/4, \tag{5}$$

where n is an odd number and λ_s is the shear wave wavelength in the fluid.

To describe the transmission data in Fig. 5, it is necessary to allow for the varying thickness of the stripe by integrating β/d, given by equation (3), across the spherical cap shape. A fit to the data is then obtained by computing equation (4), subject to conservation of the cross-sectional area of the stripe. Such a fit, using $\tau^{-1} = 44.1$ MHz and $\xi = 0.87$, is shown by the dotted line in Fig. 5. Data from a range of experiments using the various oil viscosities can all be fitted by relaxation rates obtained from the viscosity and a high frequency shear modulus of $\mu = (1.5 \pm 0.1)$ GN m^{-2}. The model used to describe the data suggests that acoustically we have a creeping amorphous solid rather than a liquid. Hence, with this type of viscous fluid the acoustic method is not acting as a simple mass sensor, although it is probing the elasticity via the shear modulus. The resonant losses observed in our spreading, which are also likely to occur in QCM techniques, suggest that care is needed as the fluids examined may be acoustically solid rather than liquid.

The existence of sets of peaks in the acoustic reflection data appears to correlate with changes in the contact width of the stripe, Δd, of the order of one SAW wavelength, λ_{SAW}. However, close examination indicates that a systematic change occurs with Δd initially greater than λ_{SAW} and then decreasing to less than λ_{SAW} [21]. The variation in Δd may be attributable to the shear wave resonances. The change in the phase speed of the acoustic wave, arising from the model used for the loss equation (3), is

$$\Delta v = -\frac{1}{2\pi\rho_s\xi} \, \mathrm{Im}\left[\frac{\beta}{1 + i\omega\tau}\right], \tag{6}$$

where Im indicates that the imaginary part of the argument must be taken. In the limit $\omega\tau \rightarrow \infty$ of an amorphous solid, equation (6) predicts cyclic changes in v with both positive and negative deviations. Thus, as the fluid spreads and the height of the stripe reduces between two consecutive shear wave resonances in the attenuation such as A and B in Fig. 5, the acoustic wavelength under the fluid-loaded part of the substrate will vary from larger than the unloaded wavelength λ_{SAW} to less than the unloaded wavelength λ_{SAW}.

The experimental results indicate that although the SAW technique does reflect changes in the geometrical parameters of the liquid, the deconvolution of this information is not straightforward. However, these problems may be at their most extreme with experiments that use both highly viscous liquids and macroscopic shapes of fluid. For Newtonian fluids and thin films, the acoustic technique will be highly mass-sensitive and the problem of determining changes in the film thickness or wetted area may simplify. For viscous fluids, the acoustic technique has potential to provide information on the shear modulus and viscosity of small volume samples. It is interesting to note that a viscous fluid, acting as a creeping amorphous solid, provides an elegant way of studying a localized inhomogeneity interacting with a SAW. The production of such solid inhomogeneities with an equivalent range of geometrical shapes by techniques such as thermal evaporation would be extremely difficult. Whilst much work remains to be done on obtaining dynamic wetting information from the acoustic signals, dynamic wetting itself provides potential for studying SAW–solid interactions.

4. CONCLUSION

Changes in the reflection and transmission of Rayleigh SAWs, due to the spreading of small stripes of viscous oils, have been reported. Optical observations confirm that the time evolution of the geometrical parameters defining the stripes follows well-defined 1/7th power laws. Resonances reported in the transmission coefficient of the acoustic wave are interpreted within a Maxwell model of viscoelastic fluid. Acoustically, the stripes of oil behave as creeping solids and resonances occur when the height of fluid matches resonant conditions for the shear velocity.

Acknowledgements

We acknowledge the financial support from the Engineering and Physical Sciences Research Council (Grant GR/L82090). Initial image processing was performed using the UTHSCSA ImageTool program (developed at the University of Texas Health Science Center at San Antonio, TX, and available from the Internet by anonymous ftp from maxrad6.uthscsa.edu).

496   *G. McHale* et al.

REFERENCES

1. D. Li, P. Cheng and A. W. Neumann, *Adv. Colloid Interface Sci.* **39**, 347–382 (1992).
2. Y.-N. Lee and S.-M. Chiao, *J. Colloid Interface Sci.* **181**, 378–384 (1996).
3. J. D. Chen and N. Wada, *Phys. Rev. Lett.* **62**, 3050–3053 (1989).
4. F. Heslot, A. M. Cazabat and P. Levinson, *Phys. Rev. Lett.* **62**, 1286–1289 (1989).
5. J. Daillant, J. J. Benattar and L. Léger, *Phys. Rev. A* **41**, 1963–1977 (1990).
6. S. Bruckenstein and M. Shay, *Electrochim. Acta* **30**, 1295–1300 (1985).
7. K. K. Kanazawa and J. G. Gordon, *Anal. Chem.* **57**, 1770–1771 (1985).
8. Z. X. Lin, R. M. Hill, H. T. Davis and M. D. Ward, *Langmuir* **10**, 4060–4068 (1994).
9. Z. X. Lin, T. Stoebe, R. M. Hill, H. T. Davis and M. D. Ward, *Langmuir* **12**, 345–347 (1996).
10. T. Stoebe, R. M. Hill, M. D. Ward and H. T. Davies, *Langmuir* **13**, 7276–7281 (1997).
11. Z. X. Lin and M. D. Ward, *Anal. Chem.* **68**, 1285–1291 (1996).
12. R. M. White, *Faraday Discuss.* **107**, 1–13 (1997).
13. A. A. Oliner, *Acoustic Surface Waves*. Springer-Verlag, Berlin (1978).
14. S. Shiokawa, T. Yamamoto, S. Yamakita and Y. Matsui, *Proceedings of the World Congress on Ultrasonics*, pp. 90–91 (1997).
15. M. de Billy, *Phys. Lett.* **96A**, 85–87 (1983).
16. M. de Billy and G. Quentin, *J. Appl. Phys.* **54**, 4314–4322 (1983).
17. G. McHale, M. I. Newton and S. M. Rowan, *J. Phys. D: Appl. Phys.* **27**, 2619–2623 (1994).
18. G. McHale, M. I. Newton, S. M. Rowan and M. Banerjee, *J. Phys. D: Appl. Phys.* **28**, 1925–1929 (1995).
19. L. Tanner, *J. Phys. D: Appl. Phys.* **12**, 1473–1485 (1979).
20. G. McHale, M. I. Newton, S. M. K. Banerjee and S. M. Rowan, *Faraday Discuss.* **107**, 15–26 (1997).
21. M. I. Newton, G. McHale and M. K. Banerjee, *Appl. Phys. Lett.* **71**, 3785–3786 (1997).
22. G. McHale, M. K. Banerjee, M. I. Newton and V. V. Krylov, *Phys. Rev. B* **59**, 8262–8270 (1999).

Apparent and Microscopic Contact Angles, pp. 497–520
J. Drelich, J. S. Laskowski and K. L. Mittal (Eds)
© VSP 2000.

Dynamic contact angle explanation of flow rate-dependent saturation–pressure relationships during transient liquid flow in unsaturated porous media

SHMULIK P. FRIEDMAN *,†

Institute of Soil, Water and Environmental Sciences, Agricultural Research Organization, The Volcani Center, Bet Dagan 50250, Israel

Received in final form 27 April 1999

Abstract—The common assumption when modeling transient liquid flow in an unsaturated porous medium is that the capillary pressure–saturation degree relationship is independent of the macroscopic liquid flux. This assumption is not always applicable, and one reason for this is the dependence of the solid–liquid–gas contact angle at the moving liquid–gaseous interface on the flow velocities, as found in systems such as long cylindrical capillaries. In the present theoretical study, a conjecture is made that at a prescribed capillary pressure the criterion for the liquid phase to invade an empty pore is defined by the Young–Laplace equation, but with the expected dynamic contact angle used instead of the static one. An iterative procedure, based on a simplified description of the pore system, enables a quantitative estimation of the extent of the liquid flux dependence of the capillary pressure–saturation degree relationship. For a given capillary pressure, the degree of liquid saturation decreases with increasing liquid flow velocity, for wetting processes, and vice versa for drainage. This effect of the liquid flux is more pronounced as the width of the pore-size distribution increases.

Key words: Dynamic contact angle; porous media; capillary pressure–saturation degree relationship; theoretical analysis.

NOTATION

$(L$ — length, T — time, M — mass)

Ca capillary number in a tube/individual pore $(v\mu/\sigma)$
C_d^m capillary number of the medium $(q\mu/\theta\sigma)$
C_{asat}^m capillary number of the medium at saturation $(q\mu/\theta_{sat}\sigma)$
CV coefficient of variation of the log-normal pore radius distribution (s/m)

---

*E-mail: vwsfried@agri.gov.il

†Contribution from the Agricultural Research Organization, The Volcani Center, Bet Dagan, Israel. No. 618/98, 1998 series.

$f_a(r)$ probability density function for the fraction of cross-sectional area comprising pores of radius $r[1/L]$

g gravitational acceleration $[L^2/T]$

K hydraulic conductivity $[L/T]$

K_r relative hydraulic conductivity, $K_r = K/K_{sat}$

K_{sat} hydraulic conductivity of the (water/liquid) saturated porous medium $[L/T]$

k viscous permeability $[L^2]$

m mean pore radius $[L]$

q water (liquid) flux (specific discharge) in the porous medium $[L/T]$

R maximal liquid-filled pore radius at a prescribed capillary head and liquid flux $[L]$

r pore radius $[L]$

S_e effective saturation degree ($S_e \equiv \theta/\theta_{sat}$)

s standard deviation of the log-normal distribution of the pore radii, $f(r)[L]$

v mean liquid velocity in a pore $[L/T]$

θ solid–liquid contact angle

θ effective volumetric liquid content ($\Theta - \Theta_r$)

θ_{sat} effective saturated volumetric liquid content ($\Theta_{sat} - \Theta_r$)

Θ volumetric liquid content

Θ_r residual volumetric liquid content

μ dynamic viscosity of the liquid $[M/LT]$

ρ density of the liquid $[M/L^3]$

σ surface tension of the liquid–gas interface $[M/T^2]$

ψ capillary (matric) head $[L]$

Subscripts

i   subscript identifying individual pores

D   dynamic (versus static) contact angle

S   static contact angle

1, 2   the adjacent pores (capillaries) in Mualem's [26] model

Superscripts

w   during a wetting (imbibition) process of a porous medium

d   during a drying (drainage) process of a porous medium

A   during the advance of the liquid–gas interface into a single capillary

R   during the recession of a liquid–gas interface from a single capillary

1. INTRODUCTION

The common assumption when modeling transient liquid flow in an unsaturated porous medium is that the capillary pressure–saturation degree relationship is independent of the macroscopic liquid flux. Thus, the usual practice in solving transient flow problems is to incorporate the capillary head (ψ)–saturation degree (S_e) relationships, measured under hydrostatic or steady-state flow conditions, into the continuity equation [1]. The measurement of the dynamic $\psi(S_e)$ relationship under transient conditions is a complex task, mainly because of the finite response time of the devices used for capillary pressure determination. Nevertheless, several experimental studies have demonstrated differences between water retention curves determined under transient conditions and those determined under steady-state or hydrostatic conditions [2–4] as well as differences among determinations at different capillary pressure increments, i.e. at different water fluxes [5–7]. Several possible phenomena have been proposed to explain the liquid flux dependence of the $\psi(S_e)$ relationship: (i) a reduction in the amount of entrapped air when smaller pressure increments are applied (smaller water fluxes) in a wetting process [5]; (ii) 'salt sieving', i.e. greater free salt concentration within the sample and less soil water retained when larger pressure gradients are applied in a drainage process [5]; (iii) a complementary effect of contaminants being accumulated near the water–air interface, causing a reduction in the interfacial tension and, therefore, in the amount of water retained [2]; (iv) the formation of isolated (from the continuous liquid phase) pendular rings, whose size and redistribution rates depend on the rate of drainage [2, 7–9]; (v) limited, local, flux-dependent availability of air to replace draining soil water [4]; and (vi) the disruption of the quasi-crystalline structure of the water molecules near the solid surfaces, caused by flow, which displaces them from their minimum-energy positions [10].

The purposes of the present paper are to propose an explanation for the water (liquid) flux dependence of the $\psi(S_e)$ relationship in transient wetting and drainage processes; to discuss the role of different acting forces and system characteristics affecting the phenomenon; and with the help of a simplified description of the pore space, to attempt to quantify the effect of the macroscopic liquid velocities on the liquid retention [$\psi(S_e)$] and relative hydraulic conductivity [$K_r(\psi)$] functions. The expected effects of the different dynamic $\psi(S_e)$ relationships on the flow were studied previously [11] and will not be discussed here.

2. THEORY

2.1. Physical phenomena and the conceptual conjecture

In the present study, it is suggested that one of the main physical phenomena which should be taken into account when modeling the dynamic retention and conductivity of water or NAPLs (nonaqueous phase liquids) in porous media is the dependence of the advancing and receding solid–liquid–gas contact angles on the velocity of propagation/withdrawal of the liquid–gas interface. Soil physicists have been aware of this phenomenon for at least 40 years [3, 5, 12], but they disregarded it as a possible explanation for the effect of liquid flux on liquid retention and did not attempt to evaluate its extent. This phenomenon has been observed in many experiments on solid–liquid–liquid and solid–liquid–gas systems, for wide ranges of capillary dimensions and flow velocities [13–16], and has been explained by different approaches [16–18]. The concern of the present study is the transient flow of liquids, water and NAPLs, in unsaturated porous media. Therefore, we will refer in the following only to liquid–gas interfaces, moving in domains of constrained geometry. In general, the magnitude of the dynamic contact angle, θ_D, is a result of an interplay between two main forces: capillary (surface tension) forces, which tend to reduce θ_D in the case of an advancing wetting fluid, and viscous forces, which tend to increase θ_D under the same circumstances. Taking into consideration the small velocities of liquid flow in porous materials, it is expected that the dependence of the contact angle on the moving interface velocity will be more significant for liquids of higher viscosity and lower liquid–gas surface tension, i.e. more for NAPLs than for water.

There are various approaches to modeling the retention and conductivity of liquid–gas phases in a porous medium, in most of which the pore system is considered as an ensemble of single-pore domains, usually of simplified geometry, which can be occupied by either the liquid or the gaseous phase. It is also common to assume that under a prescribed capillary (matric) head, ψ (regarded in the following as positive for suction), the pore will be occupied by either the wetting liquid or gas, in accordance with the Young–Laplace equation. For example, for cylindrical pores, this means that all pores of radius r smaller than

$$ r = \frac{2\sigma \cos \theta_S}{\rho g \psi} \tag{1} $$

(where σ is the liquid–vapor surface tension, ρ is the liquid density, g is the gravitational acceleration, and θ is the contact angle, which is taken to be the static one θ_S) will be saturated with the liquid. The static (zero velocity) contact angle is known to depend on the history of wetting/drainage processes, usually with a larger contact angle for wetting than for drainage: $\theta_S^w > \theta_S^d$ [19, 20]. The hysteresis of the static contact angle has already been proposed as one of the reasons for the hysteresis of the static wetting and draining water (and NAPLs) retention characteristics [21, 22]. In the present discussion, it is proposed to refer not only to the difference between the static, advancing (wetting) and the receding (draining) contact angles, but also to their dependence on the velocity of the liquid–gas interface propagation/withdrawal. The conceptual conjecture is that during transient flow processes in porous media, i.e. when there is a macroscopic change in the liquid saturation degree, a pore will be occupied by either the liquid or the gas phase, in accordance with the Young–Laplace equation, but with the

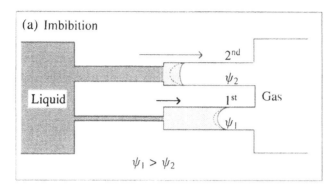

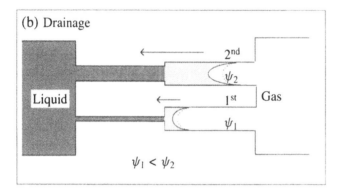

Figure 1. Schematic description of the sequence of the liquid invasion of the pores during an imbibition process (a) and of the liquid withdrawal during a drainage process (b). The dynamic contact angle is larger than the static one (interface marked with a dashed line), being larger for higher flow velocity (long arrow) in an imbibition process (a), and smaller than the static one, being smaller for higher flow velocity in a drainage process (b). (The invasion of the different pores does not take place at the same time, but at different capillary heads, ψ_1 and ψ_2.)

dynamic contact angle, θ_D, rather than with the static one:

$$r = \frac{2\sigma \cos \theta_D(v)}{\rho g \psi}. \tag{2}$$

This is illustrated schematically in Fig. 1 for imbibition (Fig. 1a) and drainage (Fig. 1b) processes. The two cylindrical pores on the right-hand side of Fig. 1a are of the same size, and according to the traditional concept, they should be invaded by the liquid at the same capillary head in a wetting process, starting at large capillary heads (oven-dryness) followed by a gradual decrease of ψ and an increase in the degree of saturation. Nevertheless, the velocity of the liquid invading the upper pore is larger, as that pore is connected to a larger pore in which there is a smaller loss of the hydraulic head. Thus, the contact angle in this pore is expected to be larger, and according to the conjecture offered here, the liquid will invade this upper pore at a second capillary head, ψ_2, smaller than the capillary head, ψ_1, of the earlier invasion of the lower pore where the liquid velocity is expected to be smaller. The situation is analogous, but with an opposite effect of the expected liquid velocity, for the process of draining from full saturation. In this case, the contact angle will be smaller for the pore with the higher liquid velocity (upper right one in Fig. 1b) and the liquid will be drained out of it (the air will invade it) at a higher capillary head, ψ_2, after it has already been drained from the lower pore, in which the flow rates are lower.

The dynamic contact angle is a function of the average liquid velocity within the single pore, $\theta_D(v)$, which, in turn, depends on the macroscopic liquid flux, q, the pore radius, r, and other geometrical factors (e.g. pore-size distribution) determining the relative fraction of the liquid flux passing through pores of that radius. The fraction of liquid-saturated pores determines the relative (to saturation) hydraulic conductivity and thus the macroscopic flux, q, which leads to a physical closure of the problem. This means that the liquid retention (and conductivity, discussed below) of a medium with a given saturated hydraulic (liquid) conductivity depends not only on the capillary head, ψ, but also on another macroscopic state variable, which is the flux, q (specific discharge), or, alternatively, the gradient of the hydraulic head, J. This behavior is considered to describe transient wetting or draining conditions. In steady-state flow, the suggested assumption is that there is no movement of the liquid–gas interfaces, but only flow of the liquid through the liquid-filled pores [23]. Thus, the liquid retention curve should be similar to the static one. The relaxation of the dynamic interfaces to the lower-energy, static/steady-state configuration is thermodynamically favorable and its kinetics is beyond the scope of this article.

2.2. Calculation of the dynamic liquid retention and relative conductivity functions

For the sake of simplicity, the procedure described in this section will refer to the specific case of a liquid, characterized by its viscosity, μ, and surface tension, σ, invading (wetting) an initially dry porous medium, characterized by its porosity, n, pore radius distribution, $f_a(r)$, and saturated hydraulic conductivity, K_{sat}. For most porous materials there is a fraction of the pores which does not contribute to the conductivity. It will be assumed here that this volumetric fraction, Θ_r, does not depend on the saturation degree, and we will refer in the following only to the commonly termed 'effective volumetric water (liquid) content', $\theta \equiv \Theta - \Theta_r$, and effective saturation degree, $S_e \equiv \theta/\theta_{sat}$, θ_{sat} being the effective saturated volumetric liquid content, $\Theta_{sat} - \Theta_r$.

With given macroscopic flux, q, and capillary head, ψ, the following steps outline the calculation of the dynamic saturation degree, S_e, and relative hydraulic conductivity, $K_r \equiv K(\psi)/K_{sat}$.

1. The first stage, according to the conceptual description of the pore system adopted here, is to determine the largest pore which is filled with the liquid. We will assume that the pore is cylindrical and, as an initial guess, calculate its radius, R,

from the Young–Laplace equation with the static contact angle, θ_S [equation (1)]:

$$R = \frac{2\sigma \cos \theta_S}{\rho g \psi}. \tag{3}$$

2. The second stage is to calculate the average velocity $v(R)$ of the liquid flowing in the largest saturated pore, $v(R)$. In order to achieve this task, we have to apply some simplifying assumptions and to refer to a specific geometrical description of the pore system. Possible candidates are a bundle of parallel cylindrical capillaries of varied radii, one of the statistical cut-and-rejoin models [24–26], and more sophisticated three-dimensional pore network models [27–29]. In the present study, Mualem's [26] semi-empirical model was chosen. This model relates the relative conductivity, K_r, at a given capillary head, to the properties of the medium, characterized by the pore radius distribution, $f_a(r)$, defined as the fraction of porosity contributed by pores of radius r, and a parameter α, which accounts for the tortuous pore shape and for the correlation between the radii of pores encountered in two adjacent planes. According to this model, the liquid velocity in the largest liquid-filled pore is given by (derived in the Appendix):

$$v(R) = \frac{R^2 q \theta}{\left[\int_{R_{\min}}^{R} r f_a(r) \, dr \right]^2}, \tag{4}$$

where the integration is over all the liquid-filled pores, from the smallest, $R_{\min}$, to the largest, R, which are filled with the liquid at the prescribed capillary head.

3. From the liquid velocity calculated in stage 2 and a given $\theta_D(v)$ relationship, it is possible now to calculate the dynamic contact angle of the moving interface in the largest liquid-filled pore, $\theta_D(R, v)$. Strictly, θ_D also depends on the viscosity of the gas [16], the temperature [30], the capillary diameter [31], gravity effects [32], and on other microscopic properties, including surface roughness [33–36], surface chemical heterogeneity [34], and surfactants present in the liquid [37]. The dependence of the dynamic contact angles of different solid–liquid–gas systems on the flow velocities and on the other factors is still not fully understood, on either the molecular or the mechanical levels [17, 34, 38]. For the purpose of the present analysis, we will refer to all these additional factors as velocity-independent and adopt a general dynamic contact angle–interface velocity relationship.

4. The calculated dynamic contact angle enables us to calculate the maximal radius by means of the dynamic Young–Laplace equation:

$$R = \frac{2\sigma \cos \theta_D(v)}{\rho g \psi}. \tag{5}$$

5. The new calculated value of R is now re-substituted in equation (4) (stage 2) and this iterative procedure continues until a satisfactory convergence of R [θ_D and $v(R)$] is achieved.

6. With the radius R of the largest liquid-filled pore in hand, the saturation degree is calculated by integrating over the range of the liquid-filled pores:

$$S_e = \left(\int_{R_{min}}^{R(\psi,q)} f_a(r)\, dr \right) \Big/ \theta_{sat}. \tag{6}$$

In spite of the moving interface, we assume a Poiseuille-like, fully developed laminar flow in a long capillary (neglecting the edge effects), as applied previously, in the analysis of the transient spontaneous imbibition of a capillary from infinite [39] and finite [40, 41] reservoirs. Thus, the relative hydraulic conductivity can be calculated from a model based on this single-pore flow description, e.g. Mualem's [26] model, chosen here:

$$K_r = S_e^\alpha \frac{\left[\int_{R_{min}}^{R(\psi,q)} r f_a(r)\, dr \right]^2}{\left[\int_{R_{min}}^{R_{max}} r f_a(r)\, dr \right]^2}. \tag{7}$$

The procedure for calculating $S_e(\psi, q)$ and $K_r(\psi, q)$ for a transient process of drainage from an intially saturated porous medium is fully analogous to that for a wetting process, with a single difference, that the calculated dynamic contact angle is smaller than the static one. The transient reversal of imbibition/drainage processes can be analyzed with the present methodology, but is left out, as the extent of this phenomenon is less pronounced for the internal scanning curves than for the two extreme imbibition (from full dryness) and drainage (from full saturation) processes.

It is most likely that the whole $S_e(\psi)$ and $K_r(\psi)$ functions for a given flux, q, are required, in which case there is no need to repeat the whole iterative procedure (stages 2–5) for each ψ value in the range of zero to infinity. Instead, it is easier to start at saturation and proceed as follows:

1. Calculate R_1 for an infinitesimally small capillary head, ψ_1, according to the static equation (3).

2. Then calculate the flow velocity through the couple $R_1–R_1$, $v(R_1, R_1)$, according to equation (4).

3. Correct R_1 according to the dynamic equation (5).

4. Go back to stage 2 and repeat the iterative procedure only for this single ψ_1 until the convergence of R_1 is achieved.

5. Continue increasing ψ by infinitesimally small increments, without any further iterations. In each step (ψ_i) the radius of the largest liquid-filled capillary, R_i, is calculated from the dynamic contact angle calculated in the previous step as

$$R(\psi_i, q) = \frac{2\sigma \cos \theta_D(v_{i-1}, q)}{\psi_i \rho g}, \tag{8}$$

which enables the calculation of $S_e(\psi_i, q)$ and $K_r(\psi_i, q)$ and the velocity $v_i(\psi_i, q)$ [according to (4)], needed for the next step (ψ_{i+1}).

The calculated radius $R_1(\psi_1, q)$ (stage 4) is that of the largest possible liquid-filled pore in an unsaturated porous medium, according to the proposed model. It should be noted that this radius can be substantially smaller than R_{max}, i.e. there is a discontinuity in the $S_e(\psi_i, q)$ and $K_r(\psi_i, q)$ curves near saturation. This indicates, to some extent, other phenomena which occur near saturation, e.g. the gaseous phase is no longer present as a continuous phase, but as separate bubbles.

2.3. Pore-scale dimensional analysis and the contact angle–velocity relationship

The analysis of the problem at the pore scale can be facilitated by dimensional analysis [42]. The concern of the present study is the effect of the dynamic contact angle on liquid retention and on conductivity in porous media. The dynamic contact angle (θ_D) is determined mainly by the interplay between viscous and capillary forces, the impact of which is commonly expressed by the nondimensional capillary number, Ca, describing the ratio between these forces:

$$Ca \equiv \frac{\mu v}{\sigma}. \tag{9}$$

The other forces which can play a role in determining θ_D are gravitational and inertial forces, their relation to the capillary (interfacial) forces being expressed by the nondimensional Bond and Weber numbers ($B \equiv \rho g r^2 / \sigma$, $We \equiv v^2 r \rho / \sigma$). We refer here to conditions of intermediate fluid velocities where the inertial forces are negligible, but the viscous forces are still significant, and to flow domains (pores) narrow enough for the gravitational effects on the liquid–gas interface to be negligible. Another factor generally affecting θ_D in liquid–liquid systems is the viscosity ratio of the displacing and displaced fluids, μ_1/μ_2. Since we refer here to liquid–gas systems and since the viscosity of water (and of most NAPLs) is much higher than that of air (by two to three orders of magnitude), viscous coupling [43–47] and other viscosity ratio effects can be neglected. Thus, we are left with two dimensionless parameters, the capillary number, Ca, and the static contact angle, θ_S, which incorporates most of the effects of the microscopic solid–liquid interactions and surface roughness, not discussed here. Indeed, Hoffman [15, 18] found the dynamic contact angle, θ_D, of a moving liquid–gas interface to be dependent on Ca and θ_S, in a universal (common to different fluids) relationship similar to the one described in Fig. 2. The receding dynamic contact angle, θ_D^R (not shown in Fig. 2), should vanish for large capillary numbers, and the advancing dynamic contact angle, θ_D^R, should, in principle, approach 180° at high velocities (Fig. 2).

In the demonstrative calculations described below, the empirical $\theta_D(Ca + f(\theta_S))$ relationship experimentally found by Hoffman[15, 18] is applied, using the following function, best-fitted to his data (solid line in Fig. 2):

$$\theta_D = 22.16 + \frac{158.06}{1 + \left(\frac{Ca + 2.405 \times 10^{-3}}{0.0585}\right)^{-0.9255}}. \tag{10}$$

The static contact angle (for $Ca = 0$) was taken to be $\theta_S = 30°$, a value similar to the value $\theta_S = 33°$ found by Bradford and Leij [22] for water on quartz sand,

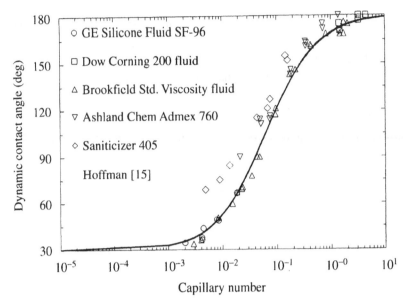

Figure 2. The functional relationship between the advancing dynamic contact angle and the single-pore capillary number, $\theta_D(Ca)$, used in the demonstrative calculations of the dynamic liquid retention and conductivity curves [solid line, equation (10)] and data measured by Hoffman [15] for various liquids displacing air in glass tubes.

to what Fisher and Lark [48] found as a velocity-independent advancing contact angle of water flowing slowly in freshly prepared fused Pyrex glass capillaries, and slightly higher than the value measured by Schramm and Mannhardt [37] for static conditions, also in a system of glass–water–air. A static contact angle of 30° corresponds to a value of $f(\theta_S) = 2.405 \times 10^{-3}$ [equation (10)] for the θ_S-dependent shift factor in Hoffman's [15, 18] universal relationship.

2.4. Porous medium-scale descriptors and dimensional analysis

The heterogeneity of the pore space of the porous medium is represented here by a probability density function of the pore radii: $f_a(r) = dF_a(r)/dr$, where $f_a(r)$ denotes the normalized probability density function for the fraction of cross-sectional area comprising pores of radius r (summing to the value of θ_{sat} when integrated over the whole range of pore radii) and $F_a(r)$ is the cumulative pore radius distribution. A positively skewed log-normal distribution function, often used to describe the pore structure of soils [49, 50] was chosen to represent $f_a(r)$:

$$f_a(r) = \frac{1}{\sqrt{2\pi}\beta r} \exp\left\{-\frac{\left[\ln(r) - \gamma\right]^2}{2\beta^2}\right\}\theta_{sat}, \tag{11}$$

where β and γ are parameters determining the mean, m, and the standard deviation, s, of the pore radius distribution given by

$$m = \exp(\mu + 0.5\beta^2) \tag{12}$$

and

$$s = \left[\exp(2\gamma + 2\beta^2) - m^2 \right]^{1/2}. \tag{13}$$

By analogy to the single-pore scale, the capillary number of the flow in the porous medium can be defined as

$$C_a^m \equiv \frac{\mu q}{\sigma \theta}, \tag{14}$$

where the average liquid velocity is obtained by dividing the flux, q, by the volumetric liquid content, θ.

The prescribed capillary heads can be normalized according to the capillary head, ψ_m, at which the pore of mean radius is invaded, if we assume a zero capillary number, i.e. that the contact angle is the static one, θ_S:

$$\psi_m = \frac{2\sigma \cos \theta_S}{\rho g m}. \tag{15}$$

Thus, the reduced capillary head, ψ/ψ_m, will be used in the presentation of the calculated liquid retention $S_e(\psi/\psi_m)$ and relative hydraulic conductivity $K_r(\psi/\psi_m)$ curves. We do not attempt to characterize the absolute hydraulic conductivity, K, but only the relative one, scaled by the hydraulic conductivity at saturation, $K_r = K/K_{sat}$. For a given capillary number of the flow, the single-pore dynamic contact angles and the resulting macroscopic $S_e(\psi/\psi_m)$ and $K_r(\psi/\psi_m)$ functions do not depend directly on the absolute values of the mean pore size, m, and its standard deviation, s, but only on their ratio, the coefficient of variation, $CV = s/m$, which reflects the broadness of the pore-size distribution.

3. RESULTS AND DISCUSSION

3.1. Demonstrative calculations of the $\psi(S_e)$ and $K_r(\psi)$ relationships

The dynamic liquid retention $S_e(\psi/\psi_m)$ and relative conductivity $K_r(\psi/\psi_m)$ functions were calculated for spontaneous imbibition processes according to the procedure described above, by applying the $\theta_D(Ca)$ relationship of equation (10) (Fig. 2). The calculations were planned to cover wide ranges of capillary numbers of the porous medium, C_a^m (Fig. 3), and widths of the pore-size distribution, expressed as the coefficient of variation of the log-normal pore radius distribution, $CV = s/m$ (Fig. 5). These two factors (C_a^m, CV) are the only ones affecting the retention and conductivity relationships, in their nondimensional forms. In each of the curves presented in Figs 3 and 5, the liquid flux was maintained constant for the whole range of saturation degrees. This means that C_a^m is minimal at saturation (C_{asat}^m, the values marked on Fig. 3) and increases for decreasing liquid contents.

According to the $\theta_D(Ca)$ relationship used in the calculations, the dynamic contact angle at a capillary number of $O(10^{-5})$ is practically the same as the static one (Fig. 2). Therefore, the liquid retention curve calculated for this low C_a^m is the

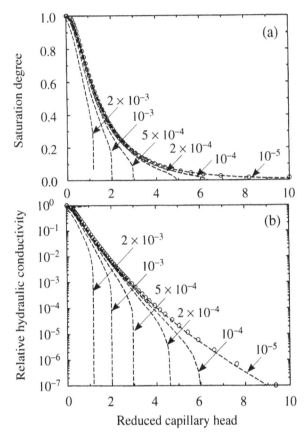

Figure 3. The reduced liquid retention curves, $S_e(\psi/\psi_m)$ (a), and relative hydraulic conductivity functions, $K_r(\psi/\psi_m)$ (b) for hydrostatic conditions, assuming a static contact angle of 30° (circles), and the dynamic relationships for capillary numbers of the medium at saturation, C^m_{asat}, in the range of 10^{-5} to 2×10^{-3} (dashed lines). The coefficient of variation (CV) of the log-normal pore radius distribution is 1.

same as the static one (circles of Fig. 3a). As the capillary number of the medium increases, the calculated dynamic $S_e(\psi/\psi_m)$ functions depart from the static liquid retention curve, giving lower liquid contents for a given prescribed capillary head. The departure is significantly enhanced as C^m_{asat} increases, especially for low degrees of saturation, where C^m_a (the liquid velocity) is higher.

The pattern of the calculated relative conductivity functions (Fig. 3b) resembles those of $S_e(\psi/\psi_m)$, because of the logarithmic presentation of the $K_r(\psi/\psi_m)$ results and the power law dependence of the relative hydraulic conductivity on the saturation degree [equation (7)], but the actual effect of the capillary head on K_r is, of course, more pronounced than its effect on S_e.

The resulting pattern of the $S_e(\psi/\psi_m)$ and $K_r(\psi/\psi_m)$ curves is that of approaching zero saturation degree and hydraulic conductivities at a finite capillary head. The physical interpretation of this feature is that for any prescribed capillary head there is a finite maximum flux that can be conducted through the unsaturated medium.

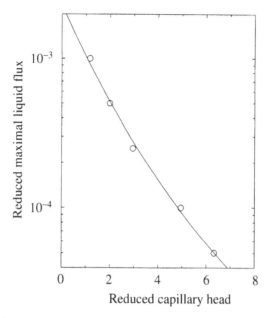

Figure 4. The reduced maximal possible liquid flux, $q_{max}\mu/\sigma$, as a function of the reduced capillary head, ψ/ψ_m, in spontaneous imbibition flow processes. The coefficient of variation (CV) of the log-normal pore radius distribution is 1.

The calculated maximum possible liquid flux, q_{max}, as a function of the prescribed capillary head in the $CV = 1$ medium of Fig. 3 is presented in the nondimensional form, $q_{max}\mu/\sigma(\psi/\psi_m)$, in Fig. 4. One should always bear in mind that the transient flow process analyzed here represents spontaneous imbibition, i.e. capillary forces are the only driving forces, so that the maximal possible contact angle which allows the advancement of the liquid–gas interface is 90°. When we consider other driving forces, such as gravity and liquid pressure gradients, which enable the liquid to invade a pore also at contact angles larger than 90°, the limitation to the liquid flux discussed above should not exist.

The effects of the broadness of the pore-size distribution on the dynamic $S_e(\psi/\psi_m)$ and $K_r(\psi/\psi_m)$ relationships are demonstrated in Fig. 5, which presents results for media of different coefficients of variation of the pore radius distribution, in the range from $CV = 0.25$ to $CV = 2$, with a common capillary number at saturation, C^m_{asat}, of 10^{-3}. According to the concept used here, a pore is occupied either solely by the liquid or solely by the gaseous phase. Therefore, for a medium made of monosize-pores, the liquid retention characteristic and the relative hydraulic conductivity curve are step functions (the dashed-dotted lines in Figs 5a and 5b). As the width of the pore radius distribution increases, the spread of the capillary heads of the static relationships (marked as solid lines in Figs 5a and 5b) also increases, along with an increasing departure of the dynamic functions (dashed lines) from the static ones, especially for lower saturation degree. This departure increases with increasing broadness of the pore-size distribution, because of the larger ratio be-

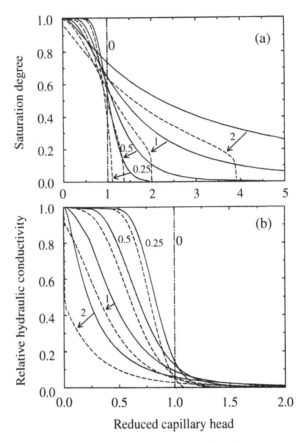

Figure 5. The reduced liquid retention curves, $S_e(\psi/\psi_m)$ (a), and relative conductivity functions, $K_r(\psi/\psi_m)$ (b) for a monosize-pore medium (dashed-dotted lines), and the static (solid lines) and dynamic (dashed lines) relationships for media of various broadnesses of pore radius distribution in the CV range of 0.25–2. The capillary number of the medium at saturation, C_{asat}^m, is 10^{-3}.

tween the maximum liquid velocity [in the larger pores, according to equation (4)] and the average velocity. These higher maximum velocities result in larger contact angles, leading to a smaller maximum radius, R_{max}, of pores able to be invaded by the liquid, which means lower degrees of saturation and lower hydraulic conductivities. Distributions of the liquid velocities in individual pores are described in the nondimensional form of the fraction of the liquid flux flowing through pores of a given single-pore capillary number (Fig. 6). The plotted distributions are for two representative media of narrow ($CV = 0.25$, squares) and wide ($CV = 2$, circles) pore radius distributions at saturation (solid symbols) and at the same liquid fluxes for a saturation degree of 0.1 (open symbols). It is clearly seen that the spread of liquid velocities is wider for the wider pore-size distribution ($CV = 2$), for the entire range of saturation degrees. Two different features characterize the liquid velocity distributions of the lower, $S_e = 0.1$, cases: first, the mean velocity (capillary number) is higher by a factor of 10, and second, since R_{max} is smaller, the velocities are

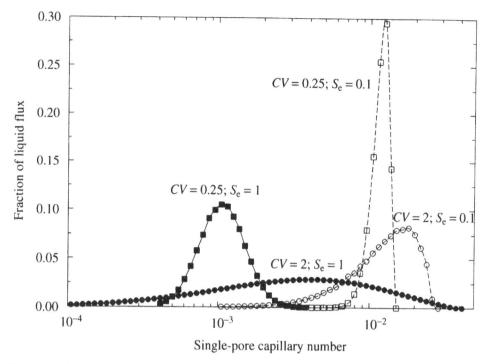

Figure 6. Distribution of the single-pore capillary numbers for two porous media of narrow ($CV =$ 0.25, squares) and wide ($CV = 2$, circles) pore radius distributions, at saturation (solid symbols) and at the same liquid fluxes for a saturation degree of 0.1 (open symbols). The capillary number of the medium at saturation, C_{asat}^{m}, is 10^{-3}.

more narrowly distributed. The first effect enhances the departure of the dynamic $S_e(\psi/\psi_m)$ and $K_r(\psi/\psi_m)$ relationships from their static counterparts more strongly than the second effect reduces this departure, resulting in increasing departure as the volumetric liquid content decreases.

3.2. Other factors affecting the dynamic contact angle and the $\psi(\theta)$ relationships

Strictly, θ_D depends on additional factors, not taken into account here, for example, the macroscopic mode of the liquid invasion, whether by forced wetting or free imbibition [19] and some other microscopic factors mentioned above (gravity effects, surface roughness, chemical heterogeneity). Therefore, it is more appropriate to refer to the contact angle as an *apparent dynamic contact angle* [19], to distinguish it from the real, intrinsic (geometrical/microscopic) contact angle, which varies from one location to another along the pore and depends on the length scale of its determination (resolution of the measurement).

In the present study, the pore structure of the unsaturated porous medium is represented by (cylindrical) pores which can be either fully saturated with the liquid or fully drained. Thus, the saturation degree is determined only by the fraction of pores filled with the liquid, at the given macroscopic capillary head and liquid flux. In reality, a single pore can be occupied by both the liquid and the gaseous phases. Then the solid–liquid–gas contact angle would also affect the volume of liquid at the single-pore scale and, through this, the macroscopic saturation degree. However, there is reason to believe that this effect is of limited significance for the range of contact angles discussed here. The limited effect of the contact angle on the saturation degree can be demonstrated by, for example, geometrical calculation of the saturation degree corresponding to the volume of inter-granular pendular rings at various contact angles [51, 52].

The shape of the pores, other than the cylindrical long capillaries referred to here, affects the liquid invasion processes on both the microscopic and the macroscopic levels. Therefore, the quantitative relationships derived here should differ for porous materials of differing pore shapes and for other pore shapes used in other theoretical studies, e.g. rectangular ducts and crevices [53, 54], four-cusp ducts [44], cylindrical throat–spherical pore bodies [55, 56], cylindrical pores and cracks [57], and bi-conical–spherical pore bodies [52].

Another assumption applied in the present study is that the liquid and gaseous phases are both continuous and are characterized by a given prescribed pressure for each phase; thus, this study neglects air bubble and liquid ganglion dynamics [19]. In flow regimes which encourage the formation of the latter, and when the pressure of the entrapped air is higher than atmospheric [53], these phenomena should, of course, be taken into account.

In principle, depending on the liquid fluxes (Ca numbers) and width of the pore-size distribution, there could be a situation in which, according to the dynamic pore-filling criterion suggested here, the advancing liquid could invade larger pores while smaller ones are still empty. According to the conjugate-pores description in Mualem's [26] model, this is suspected to happen in a small pore conjugate to a larger one within an intermediate range of Ca in which there is a steep increase of θ_D with increasing Ca. This phenomenon could be addressed within the present methodology, but was disregarded, because the same phenomenon could result from other network-connectivity reasons, and a more realistic pore network description would be desirable for future investigation of the subject. Instead, the constraint that the sequence of pore filling is from the smallest to the largest (and vice versa for a drainage process) was applied.

It should also be noted that contact angle dynamics (velocity dependence) and hysteresis (advancing–receding difference), as discussed here, form only one of the factors affecting the liquid flux dependence of the $\psi(\theta)$ [and $K(\psi)$] relationships. Other possible effects of the liquid flow on its retention energies were mentioned in the Introduction. Mokady and Low [10] proposed that the disruption of the quasi-crystalline structure of the water molecules near the solid surfaces, caused by flow, displaces them from their minimum energy positions, thus reducing their interaction energy, a phenomenon described by the general term of the capillary (matric) head. The result of this effect is that for both transient wetting and draining processes and also for steady-state flow, the capillary heads for a given water content should be smaller in flow than in static conditions. In their experiments with a mixture of sand (75%), silica flour (12.5%), and kaolinite (12.5%), Mokady and Low [10] did not observe any significant differences between the static and dynamic $\psi(\theta)$ relationships, but it is reasonable to believe that this effect is meaningful, especially under conditions of thin liquid films, namely for fine-textured soils at lower water contents.

In addition to the dynamic phenomena discussed above, one should also bear in mind the hysteresis found in static $\psi(\theta)$ relationships that stems from 'structural' pore-scale ('ink-bottle') and pore-network-scale factors [52, 56, 58]. These other factors must be taken into account, and this can be done within a methodology similar to the one presented here.

3.3. Correspondence with measured dynamic and static water retention curves in soils

As the $\psi(\theta)$ relationships calculated here were meant to assist in analyzing the effect of the velocity-dependent dynamic contact angle on the resulting dynamic liquid retention characteristic, and not to serve as a comprehensive model for the dynamic $\psi(\theta)$ function, it is meaningless to try to compare them with measured dynamic $\psi(\theta)$ relationships, and especially meaningless to try to calibrate and use them for predicting such relationships. The information gathered in experimental determinations of dynamic $\psi(\theta)$ relationships can be used, at most, for a qualitative verification of the hypothesis and of the results of the present theoretical analysis. The expected features of the dynamic $\psi(\theta)$ relationships are (i) similar $\psi(\theta)$ relationships under static and steady-state flow conditions; (ii) larger liquid contents, for a given capillary head, than under static (and steady-state) conditions in a transient drainage process; (iii) increasing liquid contents, for a given capillary head, with increasing liquid flux in a drainage process; (iv) smaller liquid contents, for a given capillary head, than under static (and steady-state) conditions in a transient wetting process; and (v) decreasing liquid contents, for a given capillary head, with increasing liquid flux in a wetting process.

The experimental evidence corresponding to the above five features comprises measured $\psi(\theta)$ relationships for sands and coarse-textured soils. (i) Similarity of the hydrostatic and steady-state measured $\psi(\theta)$ relationships of coarse materials has been demonstrated by several research groups, e.g. by Elzeftawy and Mansell [4] for Lakeland fine sand and by Topp *et al.* [2] (solid squares and circles in Fig. 7) for an artificial coarse sand mixture. (ii) All the measured water retention curves of Topp *et al.* [2] (Fig. 7) are for a drainage process. The solid symbols denote static and steady-state flow conditions and the open symbols denote dynamic conditions. It is clear that the water contents were always higher in the dynamic experiments. In Fig. 8a, which presents water retention curves of Wray dune sand, measured by Wana-Etyem [7], the dynamic $\psi(\theta)$ curve for initial drainage (ID) is substantially higher than the static one. However, the main drainage (MD) curve is similar to the static one. (iii) The time values in the legend of Fig. 7 denote the total time required

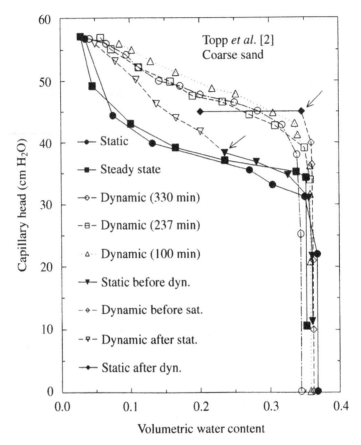

Figure 7. Water retention curves for drainage of an artificial coarse sand mixture, measured by Topp *et al.* [2] under various static (solid circles-solid line), steady-state (solid squares-solid line) and dynamic (open circles, squares, and triangles) flow conditions. Also shown are measured, mixed $\psi(\theta)$ relationships, starting at static (solid triangles-solid lines) followed by dynamic (open inverted triangles-dashed line) conditions, and starting at dynamic (open diamond-dashed line) followed by static (solid diamond) conditions. The times cited in the figure are the total times required for the drainage process to be completed.

for the drainage process to be completed. Thus, a time of 100 min (open triangles) corresponds to the highest water fluxes and, indeed, this curve is higher than those representing two slower dynamic $\psi(\theta)$ relationships, with completion times of 237 and 330 min, which are not distinguishable. Higher water contents for higher drainage fluxes can also be seen for the initial drainage (ID) phase of Fig. 8b, which compares slower and faster flow conditions; the curves for the main drainage (MD) are again nondistinguishable. (iv) Wana-Etyem [7] was the only previous worker who had measured the dynamic wetting $\psi(\theta)$ relationships properly, and his results for the main wetting phase (MW, Fig. 8a) confirm the prediction of lower water contents for dynamic than for static conditions. (v) Likewise, faster imbibition (dashed line, MW, Fig. 8b) resulted in a somewhat lower $\psi(\theta)$ curve than that for the slower wetting conditions (solid line, MW, Fig. 8b).

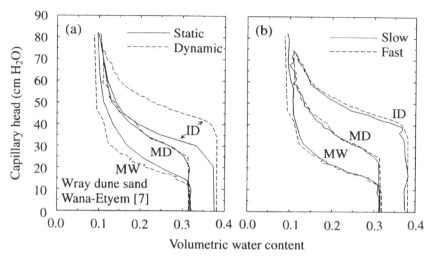

Figure 8. Water retention curves of Wray dune sand, measured by Wana-Etyem [7]: (a) comparison of the $\psi(\theta)$ relationships for initial drainage (ID), main wetting (MW), and main drainage (MD) under static (solid lines) and dynamic (dashed lines) conditions; (b) comparison of the dynamic $\psi(\theta)$ relationships for the same three modes at lower (solid lines) and higher (dashed lines) water fluxes.

Wana-Etyem's [7] results for three other media — Julesburg loamy sand, a fine sand, and Fort Collins clay loam — exhibited trends (features number ii and iv) similar to those of the Wray dune sand, but of a less pronounced dynamic effect. Based on the present calculations (Fig. 5), it is expected that the dynamic effect for a given liquid flux would be more pronounced for media of a broader pore-size distribution (PSD). However, media of broader PSD are also mostly characterized by lower hydraulic conductivities, resulting in lower water fluxes in the experimental set-ups of the studies referred to here. Thus, the dynamic effect was generally more significant for media of narrow PSD. When comparing the patterns of the computed $\psi(\theta)$ functions of Figs 3 and 5 with the measured $\psi(\theta)$ curves (Figs 7 and 8), one should also bear in mind that the computed curves are for a constant water flux, while the measured curves are for conditions of different water fluxes at different water contents.

The expected and measured behaviors described above have also been demonstrated by others. Smiles *et al.* [6] in their experiments with horizontal sand columns, for example, obtained higher water contents for dynamic draining than for static conditions, and also higher water contents in the main draining $\psi(\theta)$ functions, determined with larger capillary pressure increments (larger water fluxes) than those determined with smaller pressure increments.

3.4. Order of magnitude analysis of the acting forces

It is worthwhile to evaluate the magnitude of the various forces for a representative case of, for example, a water–air interface moving at a velocity of 0.1 cm/s through a pore of radius 10 μm. The nondimensional numbers characterizing

S. P. Friedman

Table 1.
The values of the nondimensional numbers reflecting the roles of the various forces acting when water is displacing air from a representative pore of radius 10 μm at a velocity of 0.1 cm/s (using the rounded values of $g = 10^3$ cm/s^2, $\rho = 1$ g/cm^3, $\sigma = 10^2$ dyne/cm, and $\mu = 10^{-2}$ g/cm s)

	Viscous μv	Gravitational $\rho g r^2$	Inertial $\rho r v^2$
Interfacial σ	$Ca = \dfrac{\mu v}{\sigma} = 10^{-5}$	$B = \dfrac{\rho g r^2}{\sigma} = 10^{-5}$	$We = \dfrac{\rho r v^2}{\sigma} = 10^{-7}$
Viscous μv		$Je = \dfrac{\rho g r^2}{\mu v} = 1$	$Re = \dfrac{\rho r v}{\mu} = 10^{-2}$
Gravitational $\rho g r^2$			$Fr = \dfrac{v^2}{rg} = 10^{-2}$

the ratios between the magnitudes of the acting forces are given in Table 1. The lower value of the Weber number, indicating the ratio of inertial to surface tension forces, $We \equiv v^2 r \rho / \gamma = 10^{-7}$, probably justifies the neglect of the inertial effects on the dynamic liquid retention characteristics. The value of the pore-scale capillary number is also relatively small, $Ca = 10^{-5}$, and apparently, according to the dynamic contact angle–capillary number relationship [$\theta_D(Ca)$] of Fig. 2 [equation (10)], the dynamic contact angle practically equals the static one at these lower Ca numbers, and we should not expect any dynamic effect on the macroscopic $\psi(\theta)$ function. However, this is not true, firstly because the flow at the scale of the macroscopic porous medium is characterized by a distribution of liquid flow velocities. According to the theoretical analysis presented here, the extent of the dynamic effect is determined by the maximal liquid flow rates which occur in the largest pores, and the maximal single-pore capillary numbers can be greater than the mean ones by more than an order of magnitude (Fig. 6). Secondly, the basic $\theta_D(Ca)$ relationship used here was based on measurements in straight cylindrical tubes. The intuition of the author suggests that in the case of flow in porous media, the dynamic contact angle should be sensitive to much lower capillary numbers [by analogy, to some extent, to the Reynolds number which distinguishes between laminar and turbulent flow, which is typically of order $O(10^3)$ in large pipes and of order $O(10^0)$ in porous media]. Because of the flow velocity distribution, pore shapes other than cylindrical, and other phenomena mentioned above, the apparent dynamic contact angle, measured in flow experiments, is different from the intrinsic dynamic contact angle. Therefore, it is not possible to use methods based on macroscopic imbibition experiments [59, 60] to determine the value of the latter. The thoughts expressed here about a possible dynamic contact angle effect at capillary numbers of the medium as low as $O(10^{-5})$ contradict those of Yang *et al.* [60], who although not conducting flow experiments at different velocities, claim that at capillary numbers as small as $O(10^{-5})$ the dynamic contact angle of the advancing liquid–gas interface should not differ from the static one.

The value of the representative case Bond number, $B \equiv \rho g r^2 / \gamma = 10^{-5}$, is the same as that of Ca, which means that the viscous and gravitational forces are

55. C. D. Tsakiroglou and A. C. Payatakes, *J. Colloid Interface Sci.* **137**, 315–339 (1990).
56. M. I. Lowry and C. T. Miller, *Water Resour. Res.* **31**, 455–473 (1995).
57. Y. Bernabé, *J. Geophys. Res. B* **100**, 4231–4241 (1995).
58. J. D. Chen and J. Koplik, *J. Colloid Interface Sci.* **108**, 304–329 (1985).
59. M. L. Studebaker and C. W. Snow, *J. Phys. Chem.* **59**, 973–976 (1955).
60. Y. W. Yang, W. G. Zografi and E. E. Miller, *J. Colloid Interface Sci.* **122**, 35–46 (1988).
61. Y. Mualem and S. P. Friedman, *Water Resour. Res.* **27**, 2771–2777 (1991).
62. S. P. Friedman, Ph.D. dissertation, Hebrew University of Jerusalem (1993).

APPENDIX: CALCULATION OF THE FLOW VELOCITY IN A SINGLE PORE ACCORDING TO MUALEM'S [26] MODEL

According to Mualem's [26] model, the pore space of the medium is described as an assemblage of couples of cylindrical capillaries, with the sizes of the individual pores which form the couples determined by the pore-size distribution, $f_a(r)$, defined according to

$$\theta_{sat} = \int_{R_{min}}^{R_{max}} f_a(r)\, dr. \tag{A1}$$

The liquid flow is considered to take place through these couples, each replaced by an equivalent capillary, in parallel. This semi-empirical model was successfully used by Mualem [26] for predicting the hydraulic conductivity and served as a basis also for predicting the electrical conductivity [61] and diffusion coefficients of solutes and gas molecules [62] of unsaturated soils.

At a given saturation degree, the liquid flow takes place through those couples, of which the largest is liquid-filled at the prescribed capillary head, and the fraction of macroscopic flux, dq/q, passing through this couple is given by

$$\frac{dq(r_1, r_2)}{q} = \frac{r_1 r_2 f_a(r_1) f_a(r_2)\, dr_1\, dr_2}{\iint_{R_{min}}^{R(\psi)} r_1 r_2 f_a(r_1) f_a(r_2)\, dr_1\, dr_2}. \tag{A2}$$

The average flow velocity in each capillary of the r_1–r_2 couple, $v(r_1, r_2)$, is equal to the part of the flux flowing through the couple, $dq(r_1, r_2)$, divided by the volumetric liquid content, contained in pores of radius r_1 coupled to pores of radius r_2, $d\theta(r_1, r_2)$:

$$v(r_1, r_2) = \frac{dq(r_1, r_2)}{d\theta(r_1, r_2)}. \tag{A3}$$

This volumetric water content (assuming the correlation between the radii of the two pores is independent of r_1 and r_2, and depends only on macroscopic variables, e.g. θ) is given by

$$d\theta(r_1, r_2) = f_a(r_1)\, dr_1 \frac{f_a(r_2)\, dr_2}{\theta}, \tag{A4}$$

making the velocity $v(r_1, r_2)$:

$$v(r_1, r_2) = \frac{r_1 r_2 q\theta}{\left[\int_{R_{\min}}^{R(\psi)} r f_a(r)\, dr\right]^2},$$
(A5)

with the velocity in the conjugate pore, $v(r_2, r_1)$, being

$$v(r_2, r_1) = v(r_1, r_2)\frac{r_1^2}{r_2^2}$$
(A6)

to maintain the continuity of the liquid flow.

For the analysis presented in the main text, it is required to calculate the flow velocity in the largest liquid-saturated capillary of radius R, belonging to the R–R couple:

$$v(R, R) = \frac{R^2 q\theta}{\left[\int_{R_{\min}}^{R(\psi)} r f_a(r)\, dr\right]^2}.$$
(A7)

Apparent and Microscopic Contact Angles, pp. 521–522
J. Drelich, J. S. Laskowski and K. L. Mittal (Eds)
© VSP 2000.

Subject index

acid–base 149
acid–base interactions 333
acid–base surface free energy 171
adhesion 301
adsorption 47
aging time and temperature 129
air–water contact angle 285
atomic force microscopy (AFM) 245, 377
axisymmetric drop shape analysis 210

bilayer adsorption vs. vesicle adhesion 95
bitumen 389

calcium carbonate 229, 405
capillary pressure–saturation degree relationship 497
capillary rise 419, 431
captive bubble technique 285
clays 389
contact angle 47, 111, 129, 149, 171, 301, 319, 333, 349, 389, 431, 447, 457, 487
contact angle hysteresis 377
contact angle measurements 419
contact angle relaxation 27
contact line tension 3

deformable solid 27
dewetting 261
diffusion 447
dynamic contact angle 95, 497

evaporation 447, 457

fibers 301, 319
flotation 377
FTIR 389

gas adsorption 405
Good–van Oss–Chaudhury theory 171

hemocompatibility 285
heterogeneity 349
heterogeneous surface 27
hydrophobic effect 47
hydrophobic forces 245
hysteresis 129

interactions 95
interfacial film pressure 210
interfacial tension 210, 389
isoelectric point 149

layering transition 475
Lewis acid–base 210
line pinning 27, 377
liquid crystalline polymers 333
liquid droplets 301
liquid films 301
liquids 13

modified Young equation 3
molecular structure 111
monument protection 349

nanocomposites 301
nanotubes 301

Apparent and Microscopic Contact Angles, pp. 523–524
J. Drelich, J. S. Laskowski and K. L. Mittal (Eds)
© VSP 2000.

Author index

Printed and bound by CPI Group (UK) Ltd, Croydon, CR0 4YY

23/10/2024

01778248-0003